The Problem of Animal Generation in Early Modern Philosophy

This book examines the early modern science of generation, which included the study of animal conception, heredity, and fetal development. Analyzing how it influenced the contemporary treatment of traditional philosophical questions, it also demonstrates how philosophical presuppositions about mechanism, substance, and cause informed the interpretations offered by those conducting empirical research on animal reproduction. Composed of cutting-edge essays written by an international team of leading scholars, the book offers a fresh perspective on some of the basic problems in early modern philosophy. It also considers how these basic problems manifested themselves within an area of scientific inquiry that has not previously received much consideration by historians of philosophy.

Justin E. H. Smith is Assistant Professor of Philosophy at Concordia University in Montreal. A scholar of early modern philosophy, he has contributed to the *Leibniz Review, History of Philosophy Quarterly*, and *British Journal for the History of Philosophy.*

CAMBRIDGE STUDIES IN PHILOSOPHY AND BIOLOGY

General Editor
Michael Ruse   *Florida State University*

Advisory Board
Michael Donoghue   *Yale University*
Jean Gayon   *University of Paris*
Jonathan Hodge   *University of Leeds*
Jane Maienschein   *Arizona State University*
Jesús Mosterín   *Instituto de Filosofía (Spanish Research Council)*
Elliott Sober   *University of Wisconsin*
Nils-Christian Stenseth   *University of Oslo*
Daniel W. McShea   *Duke University*

Alfred I. Tauber   *The Immune Self: Theory or Metaphor?*
Elliott Sober   *From a Biological Point of View*
Robert Brandon   *Concepts and Methods in Evolutionary Biology*
Peter Godfrey-Smith   *Complexity and the Function of Mind in Nature*
William A. Rottschaefer   *The Biology and Psychology of Moral Agency*
Sahotra Sarkar   *Genetics and Reductionism*
Jean Gayon   *Darwinism's Struggle for Survival*
Jane Maienschein and Michael Ruse (eds.)   *Biology and the Foundation of Ethics*
Jack Wilson   *Biological Individuality*
Richard Creath and Jane Maienschein (eds.)   *Biology and Epistemology*
Alexander Rosenberg   *Darwinism in Philosophy, Social Science, and Policy*
Peter Beurton, Raphael Falk, and Hans-Jörg Rheinberger (eds.)   *The Concept of the Gene in Development and Evolution*
David Hull   *Science and Selection*
James G. Lennox   *Aristotle's Philosophy of Biology*
Marc Ereshefsky   *The Poverty of the Linnaean Hierarchy*
Kim Sterelny   *The Evolution of Agency and Other Essays*
William S. Cooper   *The Evolution of Reason*
Peter McLaughlin   *What Functions Explain*
Steven Hecht Orzack and Elliott Sober (eds.)   *Adaptationism and Optimality*
Bryan G. Norton   *Searching for Sustainability*
Sandra D. Mitchell   *Biological Complexity and Integrative Pluralism*
Gregory J. Cooper   *The Science of the Struggle for Existence*
Joseph LaPorte   *Natural Kinds and Conceptual Change*
William F. Harms   *Information and Meaning in Evolutionary Processes*
Markku Oksanen and Juhani Pietarinen   *Philosophy and Biodiversity*
Jason Scott Robert   *Embryology, Epigenesis, and Evolution*

The Problem of
Animal Generation in
Early Modern Philosophy

Edited by

JUSTIN E. H. SMITH

Concordia University

CAMBRIDGE
UNIVERSITY PRESS

32 Avenue of the Americas, New York NY 10013-2473, USA

Cambridge University Press is part of the University of Cambridge.

It furthers the University's mission by disseminating knowledge in the pursuit of education, learning and research at the highest international levels of excellence.

www.cambridge.org
Information on this title: www.cambridge.org/9780521840774

First published 2006

A catalogue record for this publication is available from the British Library

Library of Congress Cataloguing in Publication data

The problem of animal generation in early modern philosophy / edited by Justin E. H. Smith.
    p.   cm. – (Cambridge studies in philosophy and biology)
Includes bibliographical references and index.
ISBN-13: 978-0-521-84077-4 (hardback)
ISBN-10: 0-521-84077-5 (hardback)
1. Reproduction.   2. Domestic animals – Reproduction.   3. Animals (Philosophy)   4. Embryology.   I. Smith, Justin E. H.   II. Series.
QP251.P76   2006
612.6 – dc22   2005021617

ISBN 978-0-521-84077-4 Hardback
ISBN 978-1-107-40728-2 Paperback

Contents

List of Contributors   *page* xi

Introduction   1
Justin E. H. Smith

I. THE DAWNING OF A NEW ERA

1. The Comparative Study of Animal Development:
   William Harvey's Aristotelianism   21
   James G. Lennox

2. Monsters, Nature, and Generation from the Renaissance to the
   Early Modern Period: The Emergence of Medical Thought   47
   Annie Bitbol-Hespériès

II. THE CARTESIAN PROGRAM

3. Descartes's Experimental Method and the Generation
   of Animals   65
   Vincent Aucante

4. Imagination and the Problem of Heredity in
   Mechanist Embryology   80
   Justin E. H. Smith

III. THE GASSENDIAN ALTERNATIVE

5. The Soul as Vehicle for Genetic Information: Gassendi's
   Account of Inheritance   103
   Saul Fisher

Contents

6. Atoms and Minds in Walter Charleton's Theory
   of Animal Generation   124
   Andreas Blank

IV. SECOND-WAVE MECHANISM AND THE RETURN
OF ANIMAL SOULS, 1650–1700

7. Animal Generation and Substance in Sennert and Leibniz   147
   Richard T. W. Arthur

8. Spontaneous and Sexual Generation in Conway's *Principles*   175
   Deborah Boyle

9. Malebranche on Animal Generation: Preexistence and
   the Microscope   194
   Andrew Pyle

10. *Animal* as Category: Bayle's "Rorarius"   215
    Dennis Des Chene

V. BETWEEN EPIGENESIS AND PREEXISTENCE: THE DEBATE
INTENSIFIES, 1700–1770

11. Explanation and Demonstration in the Haller-Wolff Debate   235
    Karen Detlefsen

12. Soul Power: Georg Ernst Stahl and the Debate
    on Generation   262
    Francesco Paolo de Ceglia

13. Charles Bonnet's Neo-Leibnizian Theory of Organic Bodies   285
    François Duchesneau

VI. KANT AND HIS CONTEMPORARIES ON DEVELOPMENT AND
THE PROBLEM OF ORGANIZED MATTER

14. Kant's Early Views on Epigenesis: The Role of Maupertuis   317
    John Zammito

15. Blumenbach and Kant on Mechanism and Teleology in
    Nature: The Case of the Formative Drive   355
    Brandon C. Look

VII. KANT AND THE BEGINNINGS OF EVOLUTION

16. Kant and the Speculative Sciences of Origins   375
    Catherine Wilson

Contents

17. Kant and Evolution   402
     Michael Ruse

Bibliography   417
Index   447

Contributors

Richard T. W. Arthur is Professor and Chair of Philosophy at McMaster University in Hamilton, Ontario. Among his recent publications are *Leibniz: The Labyrinth of the Continuum* (Yale University Press, 2001) and "The Enigma of Leibniz's Atomism," *Oxford Studies in Early Modern Philosophy*, vol. 1, 2003.

Vincent Aucante is Director of the French Cultural Center of Rome. He teaches philosophy at the Gregorian University. He has published a critical edition of the *Écrits physiologiques et médicaux de Descartes* (Presses Universitaires de France, 2000).

Annie Bitbol-Hespériès is the author of numerous publications in the history of philosophy and in the history of medicine, notably *Le Principe de vie chez Descartes* (Vrin, 1990), and is the editor with J.-P. Verdet of a critical edition of Descartes' *Le Monde, L'Homme* (Le Seuil, 1996).

Andreas Blank is Lecturer in the Department of Philosophy at the Humboldt University of Berlin and currently Visiting Fellow at the Cohn Institute for the History and Philosophy of Science and Ideas at Tel Aviv University. He is the author of *Der logische Aufbau von Leibniz' Metaphysik* (2001) and *Leibniz: Metaphilosophy and Metaphysics, 1666–1686* (forthcoming).

Deborah Boyle is Assistant Professor of Philosophy at the College of Charleston in Charleston, South Carolina. She has published articles on Descartes and Hume and has a forthcoming article on Margaret Cavendish's natural philosophy.

Francesco Paolo de Ceglia is Assistant Professor of History of Science at the University of Bari (Italy). He is the author of *Introduzione alla fisiologia di Georg Ernst Stahl* (Pensa, 2000).

Dennis Des Chene is Professor of Philosophy at Washington University in Saint Louis. Recent publications include papers on machines in the science of living things and Descartes' theory of mind. His current projects concern seventeenth-century theories of the passions and the history and philosophy of twentieth-century mathematics.

Karen Detlefsen is Assistant Professor of Philosophy at the University of Pennsylvania. Her article "Supernaturalism, Occasionalism, and Preformation in Malebranche" appeared in *Perspectives on Science*, and she is completing a book on generation, individuation, and final causes in the seventeenth century.

François Duchesneau is Professor of History and Philosophy of Science at the Université de Montréal. He is the author of *L' Empirisme de Locke* (M. Nijhoff, 1973), *La physiologie des Lumières: empirisme, modèles et théories* (M. Nijhoff, 1982), *Genèse de la théorie cellulaire* (Bellarmin; Vrin, 1987), *Leibniz et la méthode de la science* (Presses Universitaires de France, 1993), *La dynamique de Leibniz* (Vrin, 1994), *Philosophie de la biologie* (PUF, 1997), and *Les Modèles du vivant de Descartes à Leibniz* (Vrin, 1998).

Saul Fisher is Director of Fellowship Programs for the American Council of Learned Societies. He is the author of *Pierre Gassendi's Philosophy and Science* (Brill, forthcoming).

James G. Lennox is Professor of History and Philosophy of Science at the University of Pittsburgh. He is the author of *Aristotle's Philosophy of Biology: Studies in the Origins of Life Science* in the Cambridge Studies in Philosophy and Biology series (Cambridge University Press, 2001) and is the translator, with commentary, of Aristotle's *On the Parts of Animals I–IV* in the Clarendon Aristotle Series (Oxford University Press, 2001).

Brandon C. Look is Associate Professor in the Department of Philosophy of the University of Kentucky. He is the author of *Leibniz and the Vinculum Substantiale* and editor (with Donald Rutherford) of the *Leibniz–Des Bosses Correspondence*, which will appear as part of the Yale Leibniz series.

Andrew Pyle is Senior Lecturer in Philosophy at the University of Bristol. He is the author of *Atomism and Its Critics: Democritus to Newton* (Thoemmes Press, 1995) and of *Malebranche* (2003), in Routledge's Arguments of the Philosophers series.

Michael Ruse is Lucile T. Werkmeister Professor of Philosophy and Director of the Program in the History and Philosophy of Science at Florida State

University. He is the author of numerous books in the philosophy of biology, including *Darwin and Design: Does Evolution Have a Purpose?* (Harvard University Press, 2003).

Justin E. H. Smith is Assistant Professor of Philosophy at Concordia University in Montréal. A scholar of early modern philosophy, he has contributed to the *Leibniz Review, History of Philosophy Quarterly*, and the *British Journal for the History of Philosophy*. He is currently at work on a book on the problem of species membership in seventeenth-century philosophy and biology.

Catherine Wilson is currently Professor of Philosophy at the Graduate Center, City University of New York. She is the author of *The Invisible World: Early Modern Philosophy and the Invention of the Microscope* (Princeton University Press, 1995) and of a number of articles on seventeenth- and eighteenth-century history and philosophy of science.

John Zammito is John Antony Weir Professor of History and Chair of the Department of German and Slavic Studies at Rice University. He has published two works on Immanuel Kant which take up issues of biology among other matters: *The Genesis of Kant's Critique of Judgment* (University of Chicago Press, 1992) and *Kant, Herder, and the Birth of Anthropology* (University of Chicago Press, 2001).

Introduction

JUSTIN E. H. SMITH

πῶς δ' ἄν κε γένοιτο; εἰ γὰρ ἔγεντ', οὐκ ἔστ', οὐδ' εἴ ποτε μέλλει ἔσεσθαι.
τὼς γένεσις μὲν ἀπέσβεσται καὶ ἄπυστος ὄλεθρος.
(How could [what is] come to be? For if it came into being, it is not:
nor is it if it is ever going to be in the future. Thus coming to be is
extinguished and perishing unheard of.)

Parmenides, from Simplicius, In Phys. 78, 5; 145, 5, 19–21.

οὐσία δε ἐστιν ἡ κυριώτατά τε καὶ πρώτως καὶ μάλιστα λεγομένη, ἣ
μήτε καθ' ὑποκειμένου τινὸς λέγεται μήτ' ἐν ὑποκειμένῳ τινί ἐστιν, οἷον
ὁ τὶς ἄνθρωπος ἢ ὁ τὶς ἵππος.
(Substance in the truest and strictest, the primary sense of that term, is
that which is neither asserted of nor can be found in a subject. We take
as examples of this a particular man or a horse.)

Aristotle, Categories 2a 11–14.

At first glance, the unifying theme of the essays collected here may easily appear to the historian of philosophy to reside in one of the narrower alleyways of this history and certainly not along one of its grand avenues. By the nineteenth century, to be sure, embryology had come into its own as an area of scientific investigation, one whose questions were to be answered by experiment and whose answers were not seen as granting insight into any deep and timeless philosophical mysteries.

The perception could not have been more different in classical Greek thought. Parmenides, for example, offered perhaps the most hard-line formulation of a widespread pre-Aristotelian conviction that whatever *comes to be* can never really be said to be at all. Plato shared in this view, to the extent that for him particular creatures, like you and I, are approximations of what truly is, and what truly is is, among other things, eternal and unchanging. A significant aspect of Aristotle's revolutionary stance vis-à-vis his forebears consisted in his rejection of the absolute dichotomy between being and becoming, between that which fully is and that which can never quite get there. For Aristotle, through matter as the vehicle of change, what at one point existed merely potentially can at a later point exist actually – and thus generable entities, like horses and men, are thought by Aristotle to participate fully in being, to be *ousiai* or substances in the truest sense.

But if the door be opened to coming-into-being, and if entities such as horses that earlier did not exist and now do are to be granted the status of full-fledged beings, then it is not hard to see why attention would soon turn to the details of biological reproduction – for here something of tremendous metaphysical significance takes place, and it behooves any philosopher, such as Aristotle, who bites the bullet and admits that horses *are*, to determine how it is they got to be that way, that is, to determine what takes place at the moment of conception and in the subsequent stages of fetal development that brings about this ontological event consisting in the addition of one more entity to the list of what there is. Thus, the horses and men cited as examples in the *Categories* are the same horses and men that figure, along with all the eels, fish, testacea, etc., in *On the Generation of Animals*. In the one work, it is simply posited that this is what is meant primarily when one speaks of *ousia*; in the other, Aristotle sets about explaining, by careful observation of minute details – if not by experiment in the sense of modern science – how it is that these creatures attain this status.

What we would like to show in this volume – with each contributor doing so in a very different way – is that even in the early modern period embryological investigations, no matter how much they were motivated in many respects by a fierce rejection of Aristotelianism, remained Aristotelian at least to the extent that their results were seen as bearing on a cluster of distinctly philosophical questions inherited from the Greeks concerning the nature and origins of substances or beings. The name by which the study of animal reproduction and fetal development was called up through the eighteenth century, the "science of generation," reminds us of its philosophical legacy. This volume has as its concern the way in which the early modern science of generation, which included the study of animal conception, heredity, and fetal development, influenced the contemporaneous treatment of traditional philosophical questions, and, conversely, the way in which philosophical presuppositions about, for example, mechanism, substance, and cause, informed the interpretations offered by those conducting empirical research on animal reproduction.

In this brief introduction, we shall seek to sketch out in somewhat more detail how the distinctly philosophical problem of animal generation emerged in classical Greek thought, how it developed over the course of the next several centuries, and, finally, how it was addressed in the early modern period by thinkers who shared some of the same concerns the Greeks had but also came to have several of their own, new questions about the nature and mechanisms of generation.

1. ARISTOTLE AGAINST THE PANGENESISTS

In ancient generation theory, much effort was devoted to discovering the nature of semen or seed (in Greek, as in Latin and French, there is no distinction between the two terms). The seed of a thing was presumed to be its first cause, that beyond which one could not inquire about the thing's coming into being. Two questions dominated the discussion of semen:

1. Does each of the parents contribute semen or only one of them?
2. Does the semen originate in a certain part or system of the body or does it originate from all parts?

Aristotle's great legacy in the history of generation theory is a result of his firm answers to both of these questions: As regards the first, he maintained that only the father contributes semen while the mother contributes only matter for the fetus's body. As regards the second, Aristotle maintained that the semen does not come from all parts of the body, as had the Hippocratics. The two issues are intimately connected for Aristotle. Pangenesists – those who thought that semen must come from all parts of the body – believed that it contained miniature particles or traces of all of the bodily organs and even of the particular features of each parent's face, skin, etc. This was the only way, they presumed, that a creature's resemblance to its parents could be explained. But if this materialist account of like begetting like was correct, then, as Aristotle perceived, the mother would have to contribute particles of her face, skin, bones, etc., as well. Thus, for Aristotle the rejection of a maternal semen stands together with the rejection of pangenesis.

Darwin, in offering his own pangenetic theory of "cell-gemmules," traced the theory back to Hippocrates, and indeed is the one who introduced the term "pangenesis" in the 1860s. 1868. Among ancient writers, the theory enjoys its most complete articulation by Aristotle, who in *On the Generation of Animals* offers an account of it only in order to argue against it. Aristotle himself sees the theory as distinctly atomistic and attributes it, as did other ancient thinkers, to Democritus. Hippocratic pangenesis is rooted in the theory of the four humors. Blood, gall, mucus, and water are extracted from all parts of the bodies of both the mother and father in order to combine as semen. Inherited illnesses, like illnesses in the parents, arise from an imbalance of the humors. We do find some reference in the Hippocratic corpus to the separation off of "hard" (πυκνόν) and "soft" (ἀραιόν) particles from the parents' body parts that enter into the constitution of the semen; but it was the atomistic version of pangenesis offered by Democritus that rooted the theory in distinct physical units, the true forerunners of Darwin's gemmules.

For Democritus, writing in the fifth century BCE, atoms are infinite in number and of the smallest conceivable mass, and each has its own unique figure. As Aristotle understands Democritean atomism, atoms and the void together constitute the material cause of all things. All modification of things in the world can be traced back to three basic ways in which atoms in the void can differ from one another: shape, arrangement, and position (*Met.* A 4 985b4). Aristotle, as reported by Simplicius, worries that on this pared-down view of the variety of nature, atoms might be able to become entangled and cling closely to one another but could certainly never "form one substance . . . in reality of any kind whatever; for it is very simple-minded to suppose that two or more could ever become one" (*De caelo* 295, 11; KRS 425 f.).

In sum, pangenesis theory is well able to account for the fundamental problem in ancient thinking on generation concerning the resemblance of offspring to parent, for in a very literal, physical sense, the developing fetus begins from a number of chips off the old parental blocks. There is no lingering mystery of how, to speak anachronistically, "information" about the traits of the parents is conveyed to the fetus; the traits of the parents are conveyed directly rather than being somehow encoded and then recreated out of a material that, while from one or both of the parents, does not in any way resemble the parts of the parents that the offspring will later, evidently, share by way of replication. What pangenesis could not do, though, is ensure the sort of substantiality or true being that Aristotle believed needed so urgently to be secured for all those creatures subject to generation and corruption. It is not surprising, then, that pangenesis enjoys its greatest revival in the mechanism of early modern thinkers such as Gassendi and that it retreats from view only after the rise of Mendelian genetics and the development of an account of the inheritance of traits in terms of genetic information.

But if pangenesis is rejected, as in Aristotle, and if it is determined that the semen comes from the male alone, there is still the question as to the particular system of the male body from which the semen is extracted or concocted. The three most common options were the encephalomyelogenic seed doctrine, the hematogenic seed doctrine, and the pangenetic seed doctrine.[1] The pangenetic theory, as we have seen, has it that the differentiated organism can only develop from seed if the seed actually contains traces of each of the bodily organs. According to another view, the male semen is a product of the brain and bone marrow, which were seen in ancient medicine as part of one and the same bodily system. This view is attributed to the Pythagoreans; a succinct aphorism preserved by Diogenes Laertius has it that "the semen is a drop of the brain [τὸ δε σπέρμα εἶναι σταγόνα ἐγκέφαλου]."[2] The hematogenic theory, in turn, identifies the blood as the source of sperm.

Aristotle was a hematogenesist who believed semen to be a concoction out of the blood produced in the male body, but not the female, due to the greater heat of the former. Significantly, in this connection, the semen is not for Aristotle of a totally different kind than the female's menstrual blood; it is simply more refined by the male heat and thus a suitable vehicle for the *pneuma* that will ultimately impart the form to the mother's menstrual blood and set fetal development in motion. Aristotle himself understood hematogenesis first and foremost as constituting a part of his extended argument against pangenesis. For Aristotle, again, pangenesis had to be refuted lest the essence of the individual human substance amount to a mere coming together of material parts rather than being identified with the form.

For Aristotle, the bodily material is contributed exclusively by the female's menstrual blood. The male contributes, through the material vehicle of the semen, a source of change, the *pneuma*, which prompts the blood to form into a being of a certain kind. There is no material within the semen that combines with the menses to bring about a new material being; rather, the *pneuma* possesses an inherent warmth that stimulates growth in the way that sun brings about growth in plants. As Aristotle explains, "[T]he body of the semen, in which there also comes the portion of the soul source ... dissolves and evaporates, having a fluid and watery nature."[3] The semen carries the immaterial soul source, delivers it to the menses, and, having done its job, simply disappears.

In this connection, Aristotle has occasion to deny explicitly the actual preexistence of the future numerical individual in the semen, arguing instead that the individual can only potentially preexist: "And has the seed soul or not? The same reasoning applies to it as to the parts. For there can be no soul

---

[1] For a detailed discussion of these different theories, see Erna Lesky, *Die Zeugungs- und Vererbungslehren der Antike und ihr Nachwirken* (Mainz: Steiner, 1951).
[2] Ibid., 1235.
[3] *On the Generation of Animals* II 3 737a 8–12.

in anything except in that of which it is in fact the soul. . . . Clearly therefore it does have soul and exists potentially."[4] Aristotle's theory of potentiality enabled him to believe in a sort of preexistence that did not require the actual form of the future individual to be there and localizable in either of the parents. Instead, what preexists for Aristotle is the kind, say, humanity as such, and this kind only yields a particular human once the semen goes to work on the portion of matter contributed by the mother.

The result of the male's contribution is what Lennox calls *formal replication*, whereby the offspring reproduces its progenitor in kind though not in number.[5] Aristotle holds that, at least in sexual, nonspontaneous generation, particular individual substances reproduce their kind – but not themselves – because this is, in a sense, as close as they can get to sustaining themselves in existence. Perpetual generation of a kind by the individual members of the kind is the next best thing to eternity for each and every creature. The capacity for reproduction softens the misfortune of death. In other words, although absolute being is best, since some things "are too far removed from [this principle]" God has instead "filled up the whole in the only way that remained by making generation perpetual. This was the way to connect being together as much as possible, since coming to be continually and generation are the nearest things there are to being."[6] The closest thing to being that the individual primary substance can attain is a temporal, finite, embodied being; and just what it is to be a substance is to be a perishable, embodied soul, a soul no more capable of eternal existence independent of the body than, as Aristotle suggests in one memorable passage, a wax impression is capable of existence independent of the wax.[7]

2. GALEN'S TWO-SEED THEORY AND AVICENNA'S SYNTHESIS

If Aristotle's greatest battle in the debate over generation concerned the origins of semen within the body, setting himself up in opposition to Democritus and others, the third-century physician and philosopher Galen would set himself up, in turn, in opposition to Aristotle on the question whether one or both parents contribute seed. Galen would maintain that both parents are capable of contributing a formal principle to the development of the fetus, in contrast to Aristotle, who, again, thought the mother alone could contribute matter.

---

[4] Ibid., 735a 5–10.
[5] Ibid., 731b 35.
[6] *On Generation and Corruption* 336b 25–35.
[7] *On the Soul* II 1 412b 5–10.

Galen had more thorough anatomical knowledge than did Aristotle, as a result of his extensive observation of the internal parts of dissected animals. For better or worse, Galen was prevented from investigating human corpses by prevailing social taboos, and some of his errors concerning human anatomy result from excessive extrapolation from what he saw in pigs and Barbary apes to what he assumed would appear likewise within the forbidden human body. Most seriously, he mistakenly identified a fluid – the actual nature of which remains in dispute – in the "horn" of a Barbary ape's uterus as the female semen and presumed that such would be found in other similar species as well.[8]

In *On the Usefulness of the Parts of the Body* and much more extensively in *On Semen*, Galen elaborates a theory according to which the maternal semen is a residue of concocted blood, but is nonetheless distinct from the menstrual blood that Aristotle had thought was the sole contribution of the mother to generation. According to Galen, both parents are capable of contributing form – for seed remains for him, as it had been for Aristotle, the formal principle of generation. Galen saw this sort of equal partnership as the only way to account for the obvious fact that offspring are just as likely to resemble their mothers or their maternal ancestors as they are their fathers or their paternal kin.

Nonetheless, it would not do to overemphasize Galen's egalitarian view of animal generation. He believes that sexual difference is reducible to thermal difference: women are naturally colder than men and thus smaller, less hirsute, and have internal rather than external organs of generation (in men the organs are forced out by heat). The female testes (i.e., the ovaries) are thus smaller and their seed is less perfect, indeed so imperfect as to need the efficient-causal influence of the male seed in order for generation to take place and fetal development to commence. Of course, this is necessary for Galen, since if the female seed were not imperfect, since it is already in the womb, female animals would be able to generate on their own and males would become superfluous. Still, the attribution to the mother of any responsibility for the contribution of form to the offspring constitutes a significant departure from Aristotelian generation theory.

Both Aristotle and Galen held tremendous authority in the context of medieval Islamic medicine and natural philosophy. While a vast number of treatises were written in Arabic on the subject of generation, one of the most significant authors in this tradition to take on generation, and certainly the

---

[8] C. G. Kühn, *Galeni Opera Omnia* (Leipzig: C. Cnobloch, 1821–33, rpt. Hildesheim, 1965), 4: 600–1.

one who would turn out to have the most influence in the Latin tradition, was the eleventh-century Persian philosopher Avicenna. Avicenna's great accomplishment was to bring about a synthesis of Aristotle's and Galen's generation theories.[9]

Avicenna was impressed by what he perceived as Galen's greater authority in questions of anatomy, including the question concerning the function of the ovaries, but he was nonetheless committed in his natural philosophy to an Aristotelian picture that fundamentally distinguishes between form and matter and that assimilates the male principle to the former and the female to the latter. Avicenna's resolution of this conflict was to adopt Galen's two-seed theory but to fundamentally transform the notion of "seed," so that ultimately the maternal seed does nothing more than the menstrual blood had in Aristotle's picture.

From the eleventh century up to the sixteenth, first in the Arabic world and ultimately in the Latin West as well, Galen would reign supreme in medicine while Aristotle would be considered the authority on the great majority of questions of natural philosophy. As Jim Lennox discusses in Chapter 1 of the present volume, the Paduan school of medicine in the sixteenth century would constitute an exception to this general division of authority. In the work of Hieronymus Fabricius d'Acquapendente and others, an intense effort was undertaken to bring about an Aristotelian renaissance in the study of the parts, motion, and generation of animals. One of the most prominent students of this school was William Harvey.

3. THE BEGINNINGS OF MODERN GENERATION SCIENCE

Harvey is credited with the motto "Ex ovo omnia," and it is often assumed that the claim that all things come from an egg amounts to an explicit rejection of the theory of spontaneous generation. The real insight motivating the claim, though, was that many more entities in nature qualify as "eggs" than had previously been assumed:

> We, however, maintain (and shall take care to show that it is so), that all animals whatsoever, even the viviparous, and man himself not excepted, are produced from ova; that the first conception, from which the foetus proceeds in all, is an

[9] For a very good and significantly more detailed account of Avicenna's generation theory, see Basim Musallam, "The Human Embryo in Arabic Scientific and Religious Thought," in *The Human Embryo: Aristotle and the Arabic and European Traditions*, ed. G. R. Dunstan (Exeter: University of Exeter Press, 1990), 32–46.

ovum of one description or another, as well as the seeds of all kinds of plants. Empedocles, therefore, spoke not improperly of the *oviparum genus arboreum*, "the egg-bearing race of trees." The history of the egg is therefore of the widest scope, inasmuch as it illustrates generation of every description.[10]

The difference between oviparous and viviparous animals, for Harvey, is just that the former "have their beginning within their parents, and there become ova, . . . [while it is] beyond their parents that they are perfected into the foetal state."[11] Viviparous creatures, on the other hand, "have their completion in the uterus itself." Finally, for Harvey the difference between spontaneous and nonspontaneous generation is not so great. He agrees with Aristotle that, in his own words, nature can allow to "take place by chance or accident [things] which otherwise are brought about by art," so that whether creatures derive their first matter from "putrefaction, filth, excrement, dew, or the parts of plants and animals," there is nothing fundamentally different going on.

> [M]any animals, especially insects, arise and are propagated from elements and seeds so small as to be invisible (like atoms flying in the air), scattered and dispersed here and there by the winds; yet these animals are supposed to have arisen spontaneously, or from decomposition because their ova are nowhere to be found.[12]

In the same chapter, Harvey suggests that conception is a sort of "contagion" and that it is of little import whether contagion spreads through the air or whether it is sexually transmitted.

Still working within the framework of Aristotelian substantialism, thinkers such as Harvey developed theories of what may be called microsubstantiality: the view that there are vastly more true, fully real, particular primary substances or individuals than meet the eye, indeed, than Aristotle had ever dreamed. This discovery had an important influence on the way people thought about the question concerning the ultimate starting point of generation and in fact seems to have motivated many to turn away from the Aristotelian commitment to a starting point for the existences of particular primary substances. Harvey identified the egg as the primordium of life and explained sexual reproduction and spontaneous generation in terms analogous to those of the reproduction of oviparous creatures. But he was unable to answer the question concerning the ultimate origins of the primordia of viviparous creatures,

---

[10] William Harvey, *On the Generation of Animals*, in *The Works of William Harvey*, trans. R. Willis (London: Sydenham Society, 1847), 170.

[11] Ibid., 171.

[12] Ibid., 321.

the discovery of the production of mammalian eggs in the ovaries being made centuries later (by Karl Ernst von Baer in 1827).[13] Though Harvey's early works were written before the microscope was widely used, it was not until 1651 that he published *On the Generation of Animals*, decades after the new technology had begun to make its impact. The neologism "microscope" was coined by Johan Faber in 1625,[14] but Galileo had been looking through his "fly-glass," presumably as a form of recreation in between what he took to be the scientifically more important observations through the telescope, since at least 1610. Catherine Wilson reports that the Jesuit priest and *Universalgelehrter* Athanasius Kircher, who was to have a significant influence on the young Leibniz, "had microscopes in his possession by 1634."[15] Kircher devoted an entire chapter of his 1646 work *Ars magna lucis et umbrae* to the study of nature by means of the microscope. But for the most part microscopy appears to have been an activity that natural philosophers pursued in their spare time, without thinking to publish their findings in this area for the first several decades following the development of the technology that made their observations possible.

The founding document of early modern microscopy as a domain of scientific inquiry in its own right may fairly be said to be Robert Hooke's *Micrographia; or, Some Physiological Descriptions of Minute Bodies made by Magnifying Glasses*, which was not published until 1665. In this work, Hooke attributes to the microscope a significant role in opening up the secrets of nature as it really is:

> [We can add] artificial Organs to the natural, which has been of late years accomplisht with prodigious benefit to all sorts of useful knowledge, by invention of Optical Glasses. By means of Telescopes there is nothing so far distant but may be represented to our view; and by the help of microscopes, there is nothing so small, as to escape our inquiry; hence there is a new visible world discovered to our understanding. . . . By this Earth it self, which lyes so neer us, we now behold almost as great a variety of creatures, as we were able to reckon up in the whole Universe it self.[16]

Ancient medicine had treated biological individuals within or on other such individuals as outsiders. In the *On the Generation of Animals* Aristotle

---

[13] K. E. von Baer, *De ovi mammalium et hominis genesi* within the ovum (Leipzig, 1827). Translated by C. D. O'Malley in *Isis* 48 (1956): 148.

[14] For a nice sketch of Faber's work in this area, see Irene Baldriga, "Il museo anatomico di Giovanni Faber Linceo," in *Scienza e miracoli nell'arte del Seicento* (Milan, 1998), 82–7.

[15] Catherine Wilson, *The Invisible World* (Princeton: Princeton University Press, 1995), 76.

[16] Robert Hooke, *Micrographia; or, some Physiological Descriptions of Minute Bodies made by Magnifying Glasses* (London: Martyn & Allestry, 1665), preface. No page numbers available.

mentions on at least three occasions (548b, 602b, 611b) the problem of parasitism in various animals (sponges, fish, dogs), and nowhere does he mention any sort of nonharmful mutualism or symbiosis between individual organisms. In the seventeenth century, new estimations of the number of organisms out there had to be made as a result of microscopic discoveries. Very generally speaking, this caused a shift in thinking about paradigmatic substances, which were considered to be no longer only macroscopic mammals such as the horses and men of Aristotle's *Categories*, but also "worms" in the general, early modern sense of the term, which covered all small creeping or slithering creatures. This shift did not immediately bring about an identification of the source of future offspring with a spermatozoid worm – the spermatozoon was not discovered until the early 1670s, by Anton van Leeuwenhoek. But it did cause researchers to set out to find microscopic worms or wormlike organisms at the source of biological phenomena, the investigation of which the ancients could only carry as far as their natural faculty of vision would allow. These phenomena included epidemics, which in the seventeenth century finally came to be understood as the result of the presence of microscopic contagion in the environment rather than in terms of the Hippocratic miasma theory. And they also included the phenomena of generation.

Even as Harvey was continuing his work, an ontological austerity program was being promoted by figures such as Descartes and rapidly becoming the order of the day. Mechanism, like ancient atomism, emphasized the explicability of natural phenomena in terms of the mass, figure, and motion of particles. With respect to the question of generation, mechanism meant that the old, Aristotelian account in terms of the combination of a formal and a material principle – the male semen and the menstrual blood, respectively – no longer made any sense: for entities that are generated, such as animals and plants, simply do not have a formal principle at all.

Given the shift toward scientific parsimony, together with the new emphasis on experimentation over reliance upon authority and the development of new technologies that facilitated experimentation and observation, new questions came to dominate in the science of generation. It is on these questions that the essays in this volume focus. Most broadly speaking, they are concerned with the problem of the self-organization of entities in nature and with the question whether this might be accounted for in the austere terms of modern mechanism or whether special considerations must come into play in the domain of vital phenomena.

It is hoped that this volume will help historians of philosophy – even, or perhaps especially, if they did not previously have any particular interest in the problem of animal generation – to gain a fresh perspective on some of the basic problems of early modern philosophy by looking at the way these basic problems manifested themselves in an area of scientific inquiry that has not previously received much consideration. Following is a brief sketch of these problems.

The Limits of Observation and the Limits of Knowledge

The historical period we are surveying is one in which theorists began to place new emphasis on the empirical confirmation of claims about the natural world. Many of the thinkers treated here considered the concrete discovery of such entities as spermatozoa and such phenomena as insect metamorphosis to be highly relevant to their theoretical explanations of generation. Nonetheless, despite the recent enhancement of the human power of observation by the invention of the microscope at the beginning of the 1600s, many of the entities – such as the mammalian ovum – needed for a full account of the mechanism of generation remained elusive. In the absence of empirical confirmation, theorists often relied on a priori, eliminative arguments to defend their particular theories of generation. Malebranche's theory of preexistence provides a good example of this: without empirical knowledge of the existence of ova, Malebranche nonetheless deduces that all creatures must have preexisted in the ova of the first females of their species, since, he thinks, any other account would be incompatible with the mechanism to which he was committed on other, a priori grounds.

Among the so-called rationalists a common strategy seems to have been to adduce a priori arguments for one theory or another of generation and to welcome empirical evidence as it became available. As Leibniz tellingly writes to De Volder in 1703, "No primitive entelechy whatsoever can ever arise or be destroyed naturally, and no entelechy ever lacks an organic body. As far as my consideration of these matters goes, these things could not be otherwise; they are not derived from our ignorance of the formation of fetuses, but from higher principles."[17] Retrospectively, though, one may ask whether it would have dawned on the philosophers at all to support theories of generation on purportedly a priori grounds were it not for the profusion in their time of new empirical data from the work of the experimental generation scientists. After all, there was not a debate raging between spermist and ovist preformationists in, say, the fifteenth century. The relationship between generation science and philosophy provides a nice illustration of the extent to which the rationalists

---

[17] *Die philosophischen Schriften von G. W. Leibniz*, ed. C. I. Gerhardt (Berlin: Weidemann, 1875–90), 2: 251.

were in fact reliant upon, and keenly interested in, empirical confirmation of their philosophical claims, notwithstanding the avowed source of their philosophical commitments in "higher principles."

The Problem of Cause

In a certain sense, Leibniz was no doubt right: it was not just the limitations of microscopic observation that gave generation science its fundamentally speculative character. This character also stemmed from the fact that generation science was dealing not with the description of given natural entities but with the *causes* of their coming-into-being and development. Whether working in the tradition of Aristotelian substance metaphysics, as were Harvey and, arguably, Leibniz, or under the influence of Cartesian mechanism, the speculative nature of generation science was not just an artefact of the inadequacy of technology: as some of the theorists treated in this volume believed, any causal claim necessarily extends beyond the empirical world into speculative metaphysics. As Kant learned from Hume, and as Malebranche seems to have suspected before both of them, causal connections cannot be discovered purely empirically, and so, one might think, any causal claim will remain, in any circumstances, fundamentally speculative.

Generation is, as it were, the mother of all causal events. If nothing comes into being, there is nothing there to undergo subsequent changes. It is thus not surprising to observe that, for many early modern thinkers, the problems of cause and of generation were seen as different manifestations of one and the same problem: generation amounted to a special instance of cause.

If generation science just is the study of the causes of the coming-into-being of creatures, it should not be surprising to find that the study of generation in the era extending from Descartes to Kant – the era in which the problem of cause was, arguably, more hotly debated than in any other period of history – was a domain of scientific inquiry the discoveries of which would be seen by philosophers as highly relevant to their own discipline (which, it is important to remember, had also not yet been segregated from the natural sciences as its own distinct field of inquiry), from Leibniz's and Malebranche's keen interest in the discoveries of the microscopists to Kant's illustration of the possible accounts of experience by appeal to the available theories of animal generation in the *Critique of Pure Reason*.

The Limits of the Mechanistic Program

For many early modern theorists, generation seemed to be the one natural phenomenon that could not be comprehended by mechanistic natural philosophy. As emphasized by the essays in Part IV on "Second-Wave Mechanism and the Return of Animal Souls," though early seventeenth-century thinkers were optimistic about explaining every natural process in mechanistic terms (using form, figure, and motion alone), Descartes himself was able to offer only a very inadequate account of the details of the process of generation, and this failure led many of the theorists working in his wake to seek to exempt generation as the one natural phenomenon that could not be given a mechanistic explanation. And for some, such as Malebranche and Leibniz, it was precisely this intractability in terms of mechanism that served as one of the philosophical motivations for supporting the theory of preformation: if one wishes to remain a mechanist but acknowledges that generation cannot be mechanistically explained, one is forced to deny the possibility of true generation.

The mechanism of later seventeenth-century thinkers would be greatly transformed, in relation to the original intentions of the mechanists, largely as a result of the contemplation of animals, both the problem of their generation as well as their anatomical complexity and capacity for self-motion. Leibniz, in particular, while claiming the label "mechanist" for himself until the end of his life, would greatly alter the meaning of mechanism through his development of a unique concept of organism. On his view, an organism is not simply a very complicated contraption (as in the earlier mechanistic physiology of Borelli) but rather a "natural machine" that differs from artificial machines in that it remains mechanical no matter how far one breaks it down into component machines. An organism is literally an infinitely complex machine. As François Duchesneau shows in his essay, Leibniz's concept of organism set the tone for much of the discussion of animal generation, among, in particular, the Francophone naturalists of the eighteenth century, such as Charles Bonnet, while pre-Leibnizian mechanism had by this time fallen away as an untenable theoretical framework.

In the second half of the seventeenth century, there would be still greater deviations from the original mechanistic program than that represented by Leibniz. The Cambridge Platonists would feel it necessary to invoke plastic natures, or *archaei*, to explain the growth and development of biological entities. By the eighteenth century, some such invocation of an immaterial principle, even among the most narrowly medical thinkers, would be deemed an unavoidable part of embryology. G. E. Stahl, for example, treated in the essay by Francesco Paolo de Ceglia, invokes the power of the soul in fetal development. J. F. Blumenbach (treated in the essays of Brandon Look and John Zammito) hypothesizes a self-propagating formative power in the developing fetus. By the end of the period of history we are surveying, we find Kant writing in the *Critique of Judgment*, "[O]rganized being is not a mere machine, for that has only a motive power, while the organized being possesses in itself a formative power" (see Chapter 5 for more on this passage). Descartes would perhaps have been disappointed to learn that his sharpest philosophical successors found his physical theory so inadequate for describing such a common natural phenomenon as reproduction and fetal development. Descartes himself, as Vincent Aucante explains in his essay, promised throughout his life that he was on the verge of producing a treatise on the mechanism of animal generation, though, not surprisingly, he never managed to do so.

For the 200 years following the emergence of mechanism, it is rare to find anyone who writes on the generation of animals insisting on the possibility of a strictly mechanical account of generation, even among those who claim the label "mechanist" for themselves (e.g., Leibniz and Malebranche). So long as one is focusing on the phenomena studied by physics, the mechanistic dream of explaining all of nature by appeal to form, figure, and motion did not seem so implausible. Once attention is turned to biological phenomena, however, the inadequacy, the naïveté even, of mechanism quickly becomes apparent. It is for this reason, in large part, that in going back to study the relationship of philosophy to the natural sciences, historians of early modern philosophy would do well to devote more attention to the prehistory of the science of biology, even though – from the point of view of early modern thinkers themselves – physics was the science of foundational importance, biology not even having come into existence as an independent discipline.

The Demise of Substance Metaphysics

Within the tradition of substance metaphysics, the fundamental change known as generation could easily seem impossible: as explained earlier according to many in this tradition, whatever truly is, that is, whatever deserves the name "substance," could never *not* have been, since whatever is one and simple, while generation seems to occur by way of composition. Transposed into the seventeenth-century debate about generation, it would seem that those with Platonic leanings would thus also lean toward preexistence. To some extent, this is the case. Leibniz, for example, believes with the Platonists that to be is to be one and simple and, for obvious reasons, that whatever is one and simple cannot be brought into being through any new rearrangement of previously existing ingredients, as epigenesis seems to propose. He thus leans strongly toward preexistence, and is happy to learn from Leeuwenhoek in 1676 of his recent discovery of animalcula in animal semen, believing that these must

15

contain the monadic souls of animals that may subsequently enter into what he calls "the larger theater." Ric Arthur's essay on Leibniz and Sennert gives a nice account of this aspect of Leibniz's thought.

The connection between a Platonic view of substance as simple and ungenerable, on the one hand, and preexistence, on the other, is arguably unique to Leibniz, since, for the most part, the rise of generation science was contemporaneous with, and complementary to, the demise of substance metaphysics. In the eighteenth century, many of the French natural philosophers (discussed in the essays by Zammito, Wilson, and Duchesneau), devoted to what they took to be a Lockean skepticism and inspired by the discovery of parthenogenesis in freshwater polyps by Abraham Trembley in the early 1740s, came to think of nature as a sort of living, plastic mass that separates out temporarily into spatiotemporally discrete bits but never into unified entities that exist in the metaphysically rigorous sense that Leibniz had insisted must hold of anything worthy of the name "substance" or "individual." *In Le rêve de D'Alembert*, Denis Diderot has a character propose of individuals that "[t]here simple are none, no, none at all. There is only one great individual, and that is the all. In this all, as in any machine, or in some animal, there is a part that you will call by such and such name, but when you give the name of individual to this part, this is as false as if you were to give the name 'individual' to the wing, or to a feather of the wing, of a bird."[18] In the same work, we find the suggestion that there is no fundamental difference between the sexual reproduction of human beings and the asexual reproduction of "lower" organisms by budding. Thus in Diderot we find the idea of cloning *avant la lettre*, with the one distinction that, lacking the understanding of the difference between sexual and asexual generation that we have today from the science of genetics, he thought of *all* reproduction as tantamount to cloning.

There are already rumblings of such a vision of nature in the seventeenth century; Deborah Boyle's essay offers a good example of this in her account of Anne Conway's monism. Leibniz also uneasily anticipates the view of nature to come; in answer to a question from Arnauld concerning the fate of the soul of a bisected worm, Leibniz expresses doubt that both halves really go on living but maintains that, if they do, the soul endures only in one of the halves, while the other half is taken over by a previously subordinate monad, much as, for example, Tashkent, once a regional capital, became a

---

[18] Denis Diderot, *Le rêve de D'Alembert*, in *Oeuvres philosophiques de Diderot*, ed. Paul Vernière (Paris: Ediitions Garnier, 1964), 312.

national capital after the breakup of the Soviet Union. It is clear from his response to Arnauld that even the thought of such a phenomenon makes Leibniz uneasy. For the vision of nature as consisting in nested individuals ad infinitum, that is, Leibniz's view, is just one easy step away from a view of nature in which individuals no longer function as the basic ingredient, and the door is opened to the eighteenth-century vision of nature as, to put it bluntly, a blob. It was, indeed, in large part the progress made in the field of generation science, in which Leibniz himself took such an interest, that inevitably led to the picture of the world celebrated by the *philosophes*, a picture so inimical to the philosophical tradition that extends from the Greeks through Leibniz.

Species, Essence, and Individuals in the Debate over Inheritance

Aristotle had maintained that like must beget like. For him, this rule was of central metaphysical importance, for it was only through the perpetual recycling of particular primary substances in reproduction that the secondary substances or forms (which Plato, in contrast, had believed to be the more fundamental of the two) are sustained in existence. Reproduction of like by like is nature's way of providing to perishable substances the closest thing they may have to eternity, that is, to the sort of full, absolute being enjoyed by the Unmoved Mover.

But like clearly does not always beget like, at least not "exactly like." No offspring is ever an exact copy of its parent, and often offspring are, in Catherine Wilson's phrase, rather "bad copies." And it is in light of the background Aristotelian conviction and its perpetual disconfirmation in reality that the history of theories of birth defects, as well as of standard nondefective variety among offspring, is of more relevance to our understanding of the history of philosophy than it may at first appear. In the sixteenth century, monstrosities were viewed as deviations from the species essence that could only be explained by appeal to the evil intervention of the devil. In the seventeenth century, birth defects were gradually medicalized, and by the eighteenth, as Wilson notes, "bad copies" came to be celebrated as an indication of the plasticity of nature. Buffon, as Wilson also notes, came to see certain categories of mammal as modifications of one underlying form.

The next chapter of this story, of course, is Darwinian adaptationism, which lies beyond the scope of the present volume but toward which Michael Ruse's concluding essay points in anticipation. What we see in the pre-Darwinian discussion of deviant offspring, though, is a slow shift from what Stephen Jay

Gould describes as "Platonism" to a world in which variety is the order of the day. Gould writes that

> Darwin's revolution should be epitomized as the substitution of variation for essence as the central category of natural reality. . . . What can be more discombobulating than a full inversion, or "grand flip," in our concept of reality: In Plato's world, variation is accidental, while essences record a higher reality; in Darwin's reversal, we value variation as a defining (and concrete earthly) reality, while averages (our closest operational approach to "essences") become mental abstractions.[19]

As shown by the discussion of the history of theories of species diversity – whether normal, in the case of, say, siblings with diverging traits, or pathological, in the case of birth defects – there was, contrary to Gould's claim, no sudden discombobulating moment with Darwin: the new way of seeing things we now attribute to Darwin was in fact centuries in the making.

The reader will likely discern still other respects in which these essays shine new light on the history of philosophy. We have attempted to highlight what we take to be the most important respects. Overall, we hope what this volume will most help to illuminate is the gradual, and difficult, path by which the modern scientific understanding of the problems of generation emerged out of a long, complicated history of interplay between the scientific advances of the early modern period, on the one hand, and, on the other, the interpretations of these advances within the philosophical community, steeped in tradition as it was, even when it was struggling, in the new spirit of the early modern period, to break free and look at the world with unbiased eyes.

---

[19] Stephen Jay Gould, *Full House: The Spread of Excellence from Plato to Darwin* (New York: Three Rivers Press, 1996), 41.

I

The Dawning of a New Era

1

The Comparative Study of Animal Development

William Harvey's Aristotelianism

JAMES G. LENNOX

> Aristotle is my general, Fabricius my guide.
>
> Wm. Harvey,
> Preface to *Exercitationes*
> *de generatione animalium*

1. INTRODUCTION

Aristotle saw the study of nature as one of three kinds of theoretical investigation, that is, investigation aimed at knowledge for its own sake (the other two being first philosophy and mathematics; cf. Aristotle *Metaphysics* E 1 1026a7–22). The most central ontological distinction among the objects of "natural" study is that between eternal natural objects and those that come to be and pass away. Indeed, Aristotle begins his justly famous encomium for the study of animals with just that division:

> Among the beings constituted by nature, some are ungenerated and imperishable throughout all eternity, while others partake of generation and perishing. Yet it has turned out that our studies of the former, though they are valuable and divine, are fewer (for as regards both those things on the basis of which one could examine them and those things about them which we long to know, the perceptual phenomena are altogether few). We are, however, much better provided in relation to knowledge about the perishable plants and animals, because we live among them. For anyone wishing to labor sufficiently can grasp many things about each kind. (644b22–31)[1]

For Aristotle, the fact that animals and plants are generated and perish raises a number of special questions about how they are to be investigated, and

---

[1] All translations from the *De partibus animalium* are from J. G. Lennox, *Aristotle's Philosophy of Biology: Studies in the Origins of Life Science* (Cambridge: Cambridge University Press, 2001).

21

James G. Lennox

most of the first chapter of *De partibus animalium I*, Aristotle's philosophical
prolegomenon to the study of animals, is devoted to raising and answering
those questions. They, in turn, are part of a larger set of questions whose
answers will provide us with a set of standards for judging whether an inquiry
into nature has been carried out properly or not (Ibid. 639a1–14). Those
standards dictate, among other things, the place of the study of generation in
Aristotle's zoology. In this essay I want to explore the extent to which William
Harvey, through his education in Padua and his careful life-long study of
Aristotle, adopted those standards for his own "philosophical anatomy" and
in particular for his study of animal generation.

2. ARISTOTLE ON THE STUDY OF ANIMAL GENERATION

Aristotle's zoological studies are self-consciously presented as a set of distinct interconnected works. There is a series of histories of animals (*Historia
animalium* [*HA*]) (which were apparently accompanied by "dissections").[2]
These are elaborately organized comparative studies, partly based on systematic dissection, of the differentiae that distinguish the many different kinds of
animals from one another, with no attempt to provide causal demonstrations.
On the Parts of Animals (*PA*) comprises, along with the philosophical first
book already mentioned, three tightly argued books of causal explanations
of many of the differences discussed in the histories (and of some that are
not). *On the Generation of Animals* (*GA*) consists of five books, the first
on the different generative parts found throughout the animals, followed by
four devoted to a careful description and explanation of the differences in the
generative processes themselves.[3]

   Aristotle's reasons for treating generation separately go to the heart of his
philosophy of nature. Before we look at the extent to which Harvey follows
Aristotle's lead here, it will help to understand Aristotle's reasons for doing so.

   For Aristotle, the actual is prior to the potential in every respect – ontologically, causally, and epistemologically.[4] A process of coming-to-be is the
actualization of a potential for a specific, actual being, and consequently to

---

[2] Aristotle often cites the *Histories* and *Dissections* together, as at *On the Generation of Animals*
(GA) 719a11, 740a23–4, 746a15–16 and at *On the Parts of Animals (PA)* 650a31–2, 668b28–30,
680a1–3, 684b4–5, 689a18–20, 696b14–16.
[3] It would be a mistake to include *De anima* in this list. While it is important to be familiar with
it as background, Aristotle is quite clear that a study that focuses on the mind, as a good part
of *De anima* III does, is outside of natural philosophy. His argument for this is to be found in
PA I.1 641a33–641b9.
[4] This is the topic of *Metaphysics* 9. 8.

understand a process of becoming you must understand *what* it is becoming. A fertilized egg, for example, is potentially an actual animal of a specific kind. Moreover, that potential is present due to the actions of the parents – that is, it is not a *bare* potential but a potential for a specific outcome contributed by the *temporally prior* actualization of the parents' generative capacities. Aristotle's teleology rests on the fact that such potentials are "potential *for form*"[5] – as he repeats on a number of occasions (probably quoting Plato's *Philebus*), *generation is for the sake of being*, not the other way around.[6] This means that a thoroughgoing study of actual animals exercising their living capacities has priority over a study of their generation.

Aristotle has equally carefully thought out reasons for using the tools of comparative dissection and multidifference division to organize information about animals before attempting to discover causal explanations for *why* animals differ as they do. This same division of labor applies equally to the study of generation.[7]

Since animal generation is goal-directed, grasping "that-for-the-sake-of-which" (the causal goal) plays a central role in the development of a causal understanding of generation. Aristotle identifies a distinct form of necessity governing such processes, which he refers to as "conditional necessity" (*anagkaion ex' hypotheseos*). In natural generation, the goal determines what "cannot not be" *if* the goal is to be – the nature of the goal specifies the developmental preconditions for its own existence:

> [J]ust as, since the axe must split, it is a necessity that it be hard, and if hard, then made of bronze or iron, so too since the body [of an animal] is an instrument (for each of its parts is for the sake of something, and likewise also the whole body), it is therefore a necessity that it be of such a character and constituted from such things, if it is to be. (*PA* I. 1 642a9–12)

---

[5] I borrow this felicitous phrase from A. Gotthelf, "Aristotle's Conception of Final Causality," *Review of Metaphysics*, 30, 1976/7, revised and reprinted, with an appendix responding to alternative views that appeared in the intervening decade, in A. Gotthelf and James Lennox eds., *Philosophical Issues in Aristotle's Biology* (Cambridge: Cambridge University Press, 1987).

[6] *PA* I. 1 640a17–18; *GA* V. 1 778b5–6; compare *Philebus* 54a8, c4.

[7] The priority of "factual" and "existential" inquiries to those in search of essences and causes is the theme of *Posterior Analytics* 2. 1–2; compare *HA* I. 6 491a6–15; *PA* II. 1 646a8–12; *APr*. I. 30 46a20–28. The centrality of these ideas to the structure of Aristotle's zoological studies is the theme of James G. Lennox, "Divide and Explain: The *Posterior Analytics* in Practice," in Gotthelf and Lennox, *Philosophical Issues in Aristotle's Biology*, and Lennox, "Between Data and Demonstration: The *Analytics* and the *Historia Animalium*," in A. Bowen, ed., *Science and Philosophy in Classical Greece* (New York: Garland Publishing, 1991), both reprinted in Lennox, *Aristotle's Philosophy of Biology: Studies in the Origins of Life Science* (Cambridge: Cambridge University Press, 2001), chaps. 1 and 2.

In the study of life, then, causal demonstrations of developmental processes rest on premises referring to the goals of those processes (cf. *PA* I. 1 640a1–11, 640a32–640b3, 642a32–642b3).

Finally, behind teleological causation governed by conditional necessity lies a fundamental demonstrative first principle, that "nature does nothing in vain." Its importance can be seen by asking the following question: "If teleological necessity is conditional, what reason do we have to think that the materials or processes that are necessary for the goal will in fact be present?"[8] Aristotle's answer is that, in so far as form is in control, nature selects the best from among the possibilities for the being of each animal. Yet another way in which William Harvey carries the Aristotelian banner in the seventeenth century is in his understanding of the importance of this principle.

In sum, an Aristotelian zoological program has a number of distinctive features that derive from Aristotle's unique philosophical perspective. These features are as follows:

1. The demarcation of *historia* – the organization of information about animals using division of multiple differences and searching for universal relations among differences – from the search for causal explanation.
2. The demarcation of the study of animal generation from the investigation of animal being.
3. The articulation of causal investigation into an investigation of material and formal natures, on the one hand, and of sources of change and goals and functions, on the other.
4. The prioritizing of the study of formal natures over the study of materials and the prioritizing of the study of goals and functions over the study of efficient causes guided by an overarching thesis about animal natures: "nature does nothing in vain, but always the best among the possibilities for the being of each kind of animal; wherefore if something is better in a certain way, it is also disposed to be that way by nature" (*IA* 2.704b14–17).

Before turning to Harvey's theory of animal generation, it may be helpful to see how Aristotle puts these general principles to work. *Historia Animalium* is organized around the goal of providing information on four sorts of

---

[8] For a defense of the claim that Aristotle considers this a suppositional first principle in the sense defined in the *Posterior Analytics*, see Lennox, "Nature Does Nothing in Vain...," in H. C. Günther and A. Rengakos, eds. *Beiträge zur antiken Philosophie: Festschrift für Wolfgang Kullman* (Stuttgart: Franz Steiner Verlag, 1997), reprinted in Lennox, *Aristotle's Philosophy of Biology*, chap. 9.

differences across the entire animal kingdom (cf. *HA* 1.1–6 for a discussion of the overall plan of the work): parts (bk. I–IV), activities and ways of life (bks. V–VII, IX), and character traits (bk. VIII).[9] Reproductive *parts* are discussed separately from the other parts (in *HA* 3.1), and reproductive *activities* (copulation, breeding cycles, sexual maturation, and generative processes) are, according to the above plan, discussed in books V–VI and IX. One familiar example from book VI, discussed in some detail by William Harvey, can be used to highlight the methodology of *HA*:

> Generation from the egg occurs in an identical manner in all the birds, while the time to completion differs, as has been said. In the case of the domestic hen, the first signs of the embryo are seen after three days and nights; in the larger birds in a greater time, and in smaller birds in less. During this time the yolk has already migrated toward the pointed end of the egg, wherein lies the source of the egg and whence the egg hatches; and the heart is a mere blood-spot in the white. This point of blood beats and moves as if ensouled; and growing from it two blood-vessels wind their way toward each of the surrounding tunics. At this time a membrane with blood-like fibers coming from the blood vessels already surrounds the white. A little later the body can be distinguished, at first small and entirely white. The head is visible, and its eyes in an extremely developed state; this persists for a long time, though eventually the eyes become small and contract. (*HA* VI. 3. 561a4–21)

This passage illustrates the method of organizing data in *HA*: beginning with a highly abstract generalization about all birds (they are oviparous), then using division to differentiate them (according to a quantitatively varying property, time to hatching) and correlating that differentia with another, the size of the bird. Observations on the initial appearance of the embryo are then reported, but again at a level of abstraction high above particular observations. Though the observational basis of this report is a study of the domestic hen's egg, *except for the timing* Aristotle takes this to be a report of what happens in *all* birds' eggs. It is equally important to note what is *not* here – Aristotle is providing no causal explanation, no analysis of the egg into matter and form, and no claims that certain things happen necessarily. This is a *historia* of the

---

[9] All of the Greek manuscripts place the book dealing with human generation at the end of the treatise. See D. M. Balme, *Aristotle: History of Animals, Books VII–X* (Cambridge, MA: Harvard University Press, 1991), 18–19, and Balme, *Aristotle: Historia Animalium, Volume I: Books I–X: Text*, prepared for publication by Allan Gotthelf (Cambridge: Cambridge University Press, 2002), 1–2. In the fifteenth century, Theodorus Gaza moved the book on human generation to immediately follow the other books on generation, and all editors from Gaza to Balme followed Gaza. Following Balme, I number according to the manuscripts.

development of the bird's egg, not a causal investigation (*HA* I. 6 491a6–15; *PA* II. 1 646a8–12; *APr*. I. 30 46a20–28).

The causal investigation of animal development is reported in *GA*, which opens by looking back to the accomplishments of *PA* and then looks to the task ahead:

> [I]t remains for us to speak about the parts that contribute to generation in animals, which still await clarification, and about the source of the moving cause. To inquire about the latter and about the generation of each animal is in a way the same inquiry; which is why our exposition has brought them together, putting these parts at the end of our account of the parts, and the beginning of our account of generation next after them. (*GA* 1. 1 715a11–18)[10]

Having focused attention in *PA* on the *being* of animals, Aristotle is now going to explain the parts (both nonuniform and uniform) involved in their *coming to be*, search out the *causal source* of animal generation, and provide a causal account of the process of animal generation across the full range of different modes of reproduction.

Aristotle (famously) argues that the male and female contributions differ, with the female contributing material (in the form of menses) possessing the potential to develop into another organism like in kind to the parents, while the male contributes a power (*dunamis*) or motive source (*archê tês geneseos*), transported by semen to the female contribution, that initiates development toward the form of the kind. He considers both semen and menses to be concocted from residual blood.[11]

Books II and III of *GA* are a systematic presentation of Aristotle's theory of the causes of biological generation from fertilization through the development of uniform parts and organs. He develops his general theory while considering viviparous animals in book II, extends it to the various egg-laying kinds in book III, and concludes in chapters 9–11 with a discussion of the development of insects and hard-shelled animals, some of which are generated from creatures which themselves arise "spontaneously" from nonliving matter. Book IV takes up the causes of subkind differences, including those distinguishing male and female; and the same causes are then used to account

[10] Though present in every manuscript and consonant with Aristotle's views about causation, the opening eighteen lines of *GA* are bracketed by David Balme in his translation of that work. He writes, "This stylized preamble, so different from the pointed introduction to P.A. I, may be post-Aristotelian" [D. M. Balme, *Aristotle: De Partibus Animalium I and De Generatione Animalium I (with passages from II 1–3)* (Oxford: Clarendon Press, 1992), 127]. Though I cannot here go into the details, I do not agree with Balme's reasons for doubting its authenticity.

[11] The theory is elaborated gradually in *GA* II 1–4 and then applied to vivipara through the remainder of book II and to ovipara in book III.

for various things that happen "contrary to nature." Book V sets out to explain the coming to be of naturally *variable* features of animals, and since animals seem to be indifferently affected by such variations, Aristotle insists that they will not be explained teleologically but by reference to material and motive causes alone. As he says at one point, "Thus an eye comes to be and is for the sake of something, but it does not comes to be *blue* for the sake of something – unless this character were to be a peculiarity proper to the kind" (778a32–33).

In outline, if not in the details, Aristotle's causal theory of animal generation is clear.[12] The *goal* of biological generation, which Aristotle views as a special exercise of the nutritive capacity, is a kind of self-maintenance. Nutrition maintains organisms for a time as individuals, and though their bodies pass away eventually, sexual generation allows each organism to make another like itself in form and thus to be everlasting "in a way"– not in number but in form (*GA* II. 1 731b24–732a1).[13] The goal of "male" and "female" parts and capacities is to accomplish this formal self-maintenance (*GA* 732a3–11).

The first four chapters of book II provide a question-driven exploration of the causes of animal generation. The female's contribution, which is a specially prepared portion of the menses in vivipara, has developmental potentials specific to the kind, but in order for those potentials to be fully realized, that material must be acted upon by heat born by a form of air (*pneuma*) inherent in the male semen. Within the male parent, the heat within the blood, perhaps contributed by the heart, is responsible for providing nourishment to the various organs; transported by the semen to the female menses, it initiates the process of development by forming the heart, which from that point on is the source of development. But Aristotle is clear that this heat is an instrument that acts in the complex way that it does because it is governed by the form of the male parent:

> Heat and cold make the iron hard and soft, but it is the movement of these instruments, that possesses the *logos* of the art, that makes a sword. For the

---

[12] For details, see Balme, *Aristotle: De Partibus Animalium I and De Generatione Animalium I*; R. Bolton, "Definition and Scientific Method in Aristotle's Posterior Analytics and Generation of Animals," in Gotthelf and Lennox, *Philosophical Issues in Aristotle's Biology*; James G. Lennox, "Are Aristotelian Species Eternal?" in A. Gotthelf, ed., *Aristotle on Nature and Living Things* (Pittsburgh: Mathesis Publications, 1985), reprinted in Lennox, *Aristotle's Philosophy of Biology*.

[13] Compare the following from *De anima* II 4. 415a22–29: "So first we should speak about nutrition and generation; for the nutritive soul . . . is the first and most common power of soul, in virtue of which life is present in all things. Its functions are generation and the use of food; for the most natural of functions for living things, so long as they are complete and not defective or spontaneously generated, is to make others like themselves, animal animal, plant plant, in order that they may partake, so far as they can, in the everlasting and divine."

27

art is the source and form of the generated thing, but in another; while the movement of nature is in the thing itself, derived from another nature having the form in actuality. (*GA* II. 1 735a1–5)

In the case of a sword, the art of sword-making determines when and how much heating and cooling are to be applied; in the case of natural generation, the form of the male parent plays an analogous role but in a way Aristotle leaves somewhat underspecified. What must be underscored, however, is that the making of a sword, however complex, is one process; similarly, the development of an organism is a single, though very complex, process.

Thus a simple and uniform material contributed by the female undergoes a long and complex process of differentiation, with various different uniform parts (e.g., bone, flesh, hair, nail, skin, fat, tendons) coming to be at just the right time and in just the right places to form those internal and external organs appropriate to the kind of organism in question.

The resulting change is *generative emergence*, in a number of respects. At the simplest level, there is the generative emergence of *many* uniform parts from a single one. But due to the temporally and spatially coordinated character of this emergence, parts of a more complex *kind* emerge, the nonuniform instrumental parts. Each of these, moreover, comes to be as and when it does in order to play its role in a coordinated and hierarchically organized set of *living capacities*. That is, not only is there a generative emergence of a complex structure from a simple and uniform material, there is also the generative emergence of a living being, an *empsychon*. Moreover, soul emerges in stages: the earliest stages of the embryo have nutritive soul, and "as they progress they have also perceptive soul in virtue of which they are animals" (*GA* II. 3 736b1–2). To understand biological development in Aristotelian terms is to conceive of it as the goal-directed actualization of a potential, as the gradual, continuous emergence of a complexly structured, functioning living being. In this, as we are about to see, William Harvey is perhaps even more of an Aristotelian than his Paduan teachers.

3. WILLIAM HARVEY AND PADUAN ARISTOTELIANISM

A high percentage of the scholarship on William Harvey focuses exclusively on his research into the movement of the heart and blood. Even works of quite general and philosophical scope all but ignore his studies of animal generation.[14] I do not want to dwell on this fact or the reasons for it – given the topic of

---

[14] Even Roger French's excellent *William Harvey's Natural Philosophy* (Cambridge: Cambridge University Press, 1994), a work that I greatly admire, has few references to *EGA*, though Harvey's

the current volume, at any rate, no further justification need be given for focusing almost exclusively on Harvey's *Exercitationes de generatione animalium.*

In the epilogue to his monumental work, *William Harvey's Biological Ideas*, Walter Pagel cautions against the tendency to discuss Harvey's "Aristotelianism" in terms of whether Harvey adopted this or that doctrine rather than in terms of shared philosophical outlook.[15] I applaud this caution; to it, however, I wish to add the caution implicit in Charles Schmitt's (1983) use of the plural "Renaissance Aristotelianisms."[16] Harvey was exposed to at least three: that of John Case during his undergraduate years in Cambridge; that promoted in Padua in the late Renaissance; and that found in the texts of Aristotle. In this section I briefly review the results of recent research on the form of Aristotelianism to which Harvey was exposed during his years in medical school in Padua. In the following two sections, the focus will be on Harvey's study of animal generation, *Exercitationes de generatione animalium*, where the influences from his careful study of Aristotle's zoological works are apparent. My primary aim is to understand Harvey's theory of animal development on its own terms; but one cannot do that without reflection on Harvey's place in the Aristotelian tradition.

In recent decades an able team of scholars has deepened our understanding of the philosophical and methodological background to William Harvey's philosophical anatomy.[17] By the time Harvey left Cambridge for Padua in 1599, the teaching of anatomy had been decisively reshaped by Hieronymus Fabricius ab Acquapendente into what Andrew Cunningham has dubbed "the Aristotle Project." By that label, Cunningham intends to draw our attention

---

understanding of causality, scientific method, and nature are far more fully elaborated in *EGA* than elsewhere in his published or unpublished writings.

[15] Walter Pagel, *William Harvey's Biological Ideas* (Basel: S. Karger, 1967), 332.

[16] Charles Schmitt, *Aristotle and the Renaissance* (Cambridge, MA: Harvard University Press, 1983), 10–33.

[17] Without being exhaustive, I am most indebted to the work of Jerome Bylebyl, Andrew Cunningham, Robert Frank Jr., Roger French, Walter Pagel, Charles Schmitt, Andrew Wear, and Gwenneth Whitteridge. This work shows a palpable struggle to abandon the debate over whether Harvey is "the last of the ancients or the first of the moderns" in favor of understanding him on his own terms and in his own context. Much of what I say here is an extension of Andrew Cunningham, "Fabricius and the 'Aristotle Project' in Anatomical Teaching and Research at Padua," in A. Wear, R. French, and I. M. Lonie, eds., *The Medical Renaissance of the Sixteenth-Century* (Cambridge: Cambridge University Press, 1985); French, *William Harvey's Natural Philosophy*; and C. Schmitt, *Reappraisals in Renaissance Thought* (London: Warburg Publications, 1989). I am especially indebted to C. Schmitt, "Science in the Italian Universities in the Sixteenth and Early Seventeenth Centuries" (1984), reprinted as chapter 6 of Schmitt, ed., *The Aristotelian Tradtition and Renaissance Universities* (1989). The untimely death of Charles Schmitt in 1986 was an enormous loss to Aristotelian scholarship.

to the Aristotelian character of Fabricius's lifelong project of producing a *Theatrum totius animalis fabricae*. The goal was to produce a systematic study of the parts of animals organized by reference to biological functions. This began with *historiae* based on dissection; proceeded to a study of the actions of the parts; and terminated with an investigation of their uses or utility, that is, their final cause.[18] Moreover, the project should aim first of all at achieving a universal understanding of each part across the full range of animals that have it and only then proceed to a study of the way the part is differentiated in different animals:

> His [Fabricius's] account of organs deals with the norm, with what can be said securely about eyes-in-general, or larynxes-as-a-whole: about their structure, their function and their causes... universally true explanatory accounts of the causes of phenomena.[19]

In approaching anatomy in this way, Fabricius thought of himself as faithfully following Aristotle's idea of a true *historia*, searching out the widest kind to which a part belongs and then tracing its differentiation in various subkinds by means of division. Fabricius also shared Aristotle's vision of an investigation aiming at a causal and especially teleological demonstration of the information so organized.

Fabricius sees such philosophical anatomy as the culmination of natural philosophy and the philosophic foundation for medicine, and in this, as Charles Schmitt has shown, he found support in one of his philosophical colleagues at Padua, Jacopo Zabarella.[20] Like Fabricius, Zabarella saw the study of animal anatomy as that part of natural philosophy directly relevant to the practice of medicine. Among philosophers, however, Schmitt finds Zabarella to be exceptional in this respect. Medical writers of the same period found support for this understanding of the relationship between natural philosophy and medicine in Aristotle's *Parva naturalia*, at *De sensu* 1.436a18–436b1, and in *De juventute* 27. Mainetto Mainetti (1515–72) and Simone Simoni (1532–1602) sum up the lessons of these passages in the phrase "Ubi desinit physicus, ibi medicus incipit" (Where the natural philosopher finishes, there begins the physician). Schmitt also notes this phrase in the work of another of Fabricius's students, Caspar Bartholin (1585–1629).[21]

---

[18] Cunningham, "Fabricius and the 'Aristotle Project,'" 202.

[19] Ibid., 203.

[20] C. Schmitt, "Aristotle among the Physicians," in Wear, French, and Lonie, *The Medical Renaissance of the Sixteenth Century* 1–16.

[21] For possible philosophical influences on Harvey at Padua, see G. Whitteridge, *William Harvey and the Circulation of the Blood* (London: MacDonald Publishers, 1971), 33–55.

Moreover, in addition to Padua's official curriculum of natural philosophy, Fabricius insisted that his students study Aristotle's investigations of animals. This had been possible ever since Gaza's printed Latin edition of the zoological works had been produced in 1476; a series of commentaries began to appear in the sixteenth century.[22] But to include Aristotle's investigations of animals in the core curriculum for medical students was to take a stand on the place of these works in relation to the science of nature, on the one hand, and to medicine, on the other. There is some evidence that not all of Fabricius's students appreciated their good fortune in the way William Harvey apparently did.[23]

4. HARVEY'S ARISTOTLE

As the quotation heading this essay indicates, among previous students of the generation of animals, Harvey had two touchstones: Aristotle and Fabricius.[24] Fabricius was likely working on *De formato foetu* when Harvey was his student, since it was published in folio in 1604, two years after Harvey graduated. Two years after Fabricius's death in 1619, a folio edition of *De fomatione ovi e pulli tractatus accuratissimus* was published.[25] Harvey displays an intimate familiarity with both works. Nevertheless his philosophical approach to animal generation is quite distinctive and, I will argue, based on a distinctive reading of Aristotle. In the preface to *EGA*, Harvey gives us a detailed and

---

[22] For a careful study of three of these, cf. Stefano Perfetti, "Three Different Ways of Interpreting Aristotle's *De Partibus Animalium*: Pietro Pomponazzi, Niccolò Leonico Tomeo and Agostino Nifo," in C. Steel, G. Guldentops, and P. Beullens, eds., *Aristotle's Animals in the Middle Ages and Renaissance* (Leuven: Leuven University Press, 1999), 297–316.

[23] For more on the intellectual context in sixteenth-century Padua, see Howard Adelmann, *The Embryological Treatises of Hieronymus Fabricius of Aquapendente* (Ithaca: Cornell University Press, 1942), 3–35; J. Bylebyl, "The School of Padua: Humanistic Medicine in the Sixteenth Century," in C. Webster, ed., *Health, Medicine and Mortality in the Sixteenth Century* (Cambridge: Cambridge University Press, 1979), 335–70; R. Frank, *Harvard and the Oxford Physiologists: Scientific Ideas and Social Interaction*, (Berkeley: University of California Press, 1980), 16–20; French, *William Harvey's Natural Philosophy*, 18–70; Schmitt, *Reappraisals in Renaissance Thought*; Whitteridge, *William Harvey: Disputations Touching the Generation of Animals* (London: Blackwell Scientific Publications, 1981), xix–lxiii.

[24] There are, of course, references to many other authors ancient and contemporary, but these are infrequent and typically on matters of detail.

[25] Adelmann includes the texts and translations of *De formatione ovi et pulli* and *De formato foetu* by Fabricius and duplications of the original plates. For clear discussions of the historical background and context of philosophical anatomy in Padua when Harvey was studying there, see the introduction in Adelmann, *The Embryological Treatises of Hieronymus Fabricius of Aquapendente*, 3–35.

---

explicit presentation of his own epistemology; his understanding of its basis in Aristotle; and its implications for proper methodology in the study of animal generation. It will pay rich dividends in examining this study to attend carefully to its preface.

Harvey begins by contrasting the views on generation of "all Physicians, following Galen" with those of "Aristotle (Nature's most diligent searcher)." But so as immediately to set his readers straight on his attitude to authority, he warns that it is "unsafe, nay base . . . to be tutored by other men's commentaries without making trial of the things themselves, especially since Nature's book is so open and so legible" (*EGA* Preface 8).[26]

As in *De motu cordis*, Harvey continually urges the readers of *EGA* to perform the necessary observations, dissections, inspections, and experiments for themselves. Though there may be other reasons for Harvey to eschew his teacher Fabricius's heavy reliance on figures, one obvious one is that it may mislead readers into imagining they have observed "the things themselves." Harvey is *not*, however, denying the value of being tutored by the commentaries of others – he is stressing that the value of being so tutored is lost unless one also makes use of one's own eyes. We shall see how precisely he puts these words into action shortly.

A new method (*methodo nova*) for investigating the nature of things is to be presented in *EGA*, one that provides a more difficult but more secure path to *scientia*. It involves finding the nature of things by the things themselves (*ex rebus ipsis*) rather than by reading the opinions of philosophers (*EGA* Preface 8–9). Since the ancients themselves based their conclusions on *experimenta varia*, it is both lazy and unfaithful to their memory to rely on their words rather than to follow the way of experience.

The chief concern of this preface, then, is to recommend a method for proceeding from the experiences of an individual researcher to a universal cognition of the natures of things (*a rerum naturalium cognitione procedit*) – from what is better known to us to *scientia*. Harvey put the point this way:

> [I]n both Art and Knowledge *that which we perceive* in sensible objects differs from *the thing perceived* and is that which is retained in the imagination and memory. The former, the thing perceived, is the exemplar, the idea, the form informing; the latter is the representation, the *eidos*, the abstracted notion. The

---

[26] In most cases I quote from the Whitteridge translation of Harvey's 1651 Latin edition of *Exercitationes de generatione animalium* (Whitteridge, *William Harvey: Disputations Touching the Generation of Animals*). In the references, I use the abbreviation *EGA* followed by the number of the *Exercitatione* and the page number in the Whitteridge translation.

former again is a natural object, a real entity; the latter is a resemblance or likeness, an entity of the mind. The former is concerned with a particular object and is itself a particular and an individual; the latter is a universal and common thing. (*EGA* Preface, 12)

Harvey was, in scholastic parlance, a "nominalist." The "form informing," that is, the form in the object perceived, is particular and individual; only the "abstracted notion" is universal and common. Moreover, these "entities in the mind" are only likenesses and resemblances of the form, the real entity. Unlike Aristotle, Harvey believes the mind takes on not "the form without the matter" but only a likeness or resemblance of the form. Nevertheless, the legitimacy of these likenesses rests in the fact that they "derive from [sensible objects], nor could we attain to these intelligibles at all without the help of the sensibles" (*EGA* Preface 12).

This understanding of the variable relationship between real, singular natural beings and their natures or forms, on the one hand, and the abstracted universal in the mind, on the other, has important methodological consequences. Firsthand observations that are made both frequently and repeatedly on many different animals are the only secure basis for reaching general conclusions:

> For in every discipline, diligent observation is a prerequisite and the senses themselves must frequently be consulted. We must, I say, rely upon our own experiences and not those of other men. Without this, no man is a fit disciple of any part of natural knowledge, nor a competent judge of what I have to say concerning the generation of animals. Without experience and skill in anatomy, he will no better understand me than a man born blind can judge of the nature and difference of colours, or one born deaf, of sounds. Therefore, gentle Reader, take on trust nothing that I say of the generation of animals. I call upon your eyes to be my witnesses and my judges. For since all perfect knowledge is built upon those principles which originate from things perceived by the senses, we must have a special care that those principles are safely grounded by using frequent dissections of animals. (*EGA* Preface 13)

This thoroughly philosophical preface concludes with a discussion *de methodo in cognitione generationis abhibenda*. In good Aristotelian fashion, Harvey tells us that he began with the perfected animals and aimed to trace their generation back to their primordial material cause and efficient causes. He is highly critical of Fabricius, who "relies on probable arguments rather than ocular inspection, and laying aside the judgement of the senses which is grounded upon dissections, he flies to petty reasonings borrowed from mechanics" (*EGA* Preface 18).

We can conveniently outline the method Harvey plans to follow in five steps.

1. Describe the generation of the parts in temporal order as observed in the egg of a hen and then in other animals.
2. Relate what can be observed about the material and efficient causes of the fetus and of the various steps in the generative process.
3. Derive from these observations conclusions about the faculty of the formative and vegetative soul involved in generation.
4. From observations and experiments based on the dissection of animals that are familiar, available, and large enough to be observed, make cautious inductive inferences to all other animals of the same kinds.
5. Finally, contemplate the "hidden nature of the vegetative soul" and the manner, order, and causes of generation in all animals.

In Harvey's explanation of the final step, he shows himself a true Aristotelian.

> [F]or all the other animals agree with those I have enumerated, or at least with some of them, either generically or specifically, and they are procreated after the same manner of generation, or at least in a manner which may be compared with it by analogy. For Nature being divine and perfect is always consistent with herself in the same things. And just as her works either agree or differ, that is in genus or species or some proportion, so her working, namely generation or fabrication, is in all of them either the same or different. (*EGA* Preface 19)

Harvey invokes the consistency of nature to justify inductive inferences from the cases that he will carefully study to those that are like them in the three Aristotelian categories of likeness: in genus, in species, or by analogy (cf. *PA* I. 5 645b1–27). And he also argues that not only are nature's works going to be the same or different in these three ways, but so are nature's ways of generating and fabricating and ultimately the underlying faculties or powers of the soul. It is Harvey's ultimate goal to understand the nature of the generative power, which he suspects will be the same in all animals.

By arguing that constant and repeated comparative dissection is the basis for true inductive knowledge of causes, and by insisting that this is the message to be gleaned from the canonical texts of Aristotle's *Physics* and *Analytics*, Harvey is sending a clear message: natural philosophy requires not a *novum organum* but a new appreciation of Aristotle's *Organon*.[27]

[27] Cf. Pagel, *William Harvey's Biological Ideas*, 40–7; French, *William Harvey's Natural Philosophy*, 326–8. Regretfully Pagel also follows G. Plochman, "William Harvey and His Methods," *Studies in the Renaissance* 10 (1963), in seeing Harvey's teleology as somehow detached from the Aristotelian empiricism endorsed in this preface, as though Harvey's

5. HARVEY AT WORK ON GENERATION

Harvey's *Exercitationes de generatione animalium* opens with a careful *historia* of the generation of the chick and treats it as an exemplar of oviparous generation generally (Exs. 1–23). The *historia* is followed by an extended discussion of the material and efficient causes of generation. The method of reporting the investigation highlights one of Harvey's clearest departures from Aristotle. He is at pains to describe in great detail the development from fertilization to parturition in two species – varieties of domestic fowl and deer. When at Ex. 63 Harvey turns from oviparous to viviparous generation, he explains his intent as follows:

> And so, as I did for the egg, I shall single out one kind of these as an exemplar of all the rest so that, being very well known to us, it may shed light on the rest and be the pattern to which all the others may be related by way of analogy. (*EGA* 63 331)

By contrast, Aristotle presents us with accounts of "widest class universals," discussing particular species only rarely, when they are peculiar in some respect. The contrast reflects, I believe, Harvey's constant concern for bringing his readers actively into the investigation. His descriptions are given in such detail that the readers can use them as guides to carrying out their own investigations, which is what Harvey constantly stresses they should do. Aristotle presents us with the already achieved universals but tells us frustratingly little about the investigations that grounded these achievements. With Harvey we have the opposite problem: we have a great deal of information about how to investigate the development of the chick but very little information about the extent of the generalizations that can be grounded in these "exercises."

And yet at the level of epistemological doctrine, Harvey is following Aristotle perfectly. The discussion of the generation of the chick opens by rejecting Fabricius's justification for beginning with the egg of the hen in *De formatione ovi e pulli*. Harvey gives a completely different justification, which announces to his reader the message of his preface:

> [B]ecause we may borrow therefrom more certain statements which, as being more known to us, may shed light on our investigation of the generation of other animals of different kinds... it is an easy matter to observe from them

views about the purposes and functions of the structures and movements he investigated were "speculations" derived "rationalistically" rather than conclusions resting on experimental dissection.

what are the first things that are clear and obvious in generation, by what steps Nature proceeds in the formation and with what wonderful foresight she steers the whole process. (*EGA* 1 21)

He will begin, that is, in true Aristotelian fashion with what is "better known to us" and use that to shed light on generation in other kinds. But the inductive power of this approach goes far beyond what Aristotle or Fabricius imagined, for it is Harvey's view that

all animals whatsoever, even viviparous creatures, nay man himself, are all engendered from an egg, and that the first conceptions of all living creatures from which the fetus arises are some kind of egg. . . . And so the account of the egg has a wider application because it yields an insight into all manner of generation whatsoever. (*EGA* 1 22)

To understand the role of the egg in generation, then, is to understand the key to *all* generation.

Since our focus will be on Harvey's causal theory of generation, I shall simply outline the rich *historia* of the day-by-day development of the chick within the egg in the first twenty-four "experiences" (*exercitationes*). Exs. 2–8 discuss the female reproductive organs across a wide range of oviparous animals, generalizing as widely as possible whenever possible, though focused on birds, and especially the domestic hen. Exs. 9–10 discuss the generation, growth, and movement of the egg within the reproductive tract and then its exit. Harvey occasionally looks forward to his later discussion of the causes of generation by insisting that the movement, growth, and development of the egg take place "by virtue of the vegetative innate heat and the faculty wherewith it is endowed" and that while "the office and use of the womb was ordained for the procreation of eggs . . . the immediate effective agent is contained in the egg . . . an innate natural principle proper to it" (*EGA* 9 59; 10 60). Harvey feels he must insist on this here in order to distance his view from a popular one that attributes the source of growth to external agents, especially to a power in the womb itself. His view is the truly Aristotelian one, namely, that the source of natural generation is a power inherent in the thing itself, originating in the act of fertilization. But the evidence and argument for this will come later.

Again in truly Aristotelian fashion, Harvey announces in Ex. 11 that "having explained how an egg is generated, it is now time to treat of its parts and their differences" – first the shell (Ex. 11), then the egg's other parts (Ex. 12), and finally differences (Ex. 13), that is, the differences found among hens' eggs, in particular differences in size, shape, and number.

The major transition of Harvey's *historia* of generation occurs between Exs. 13 and 14. Ex. 13 reviews the achievements of the previous discussion in a manner that underscores his inductive caution:

> Thus far then [I have given an account] of the uterus of the hen and its office, of the generation of the hen's egg, its differences and accidents, in which I have spoken of the things I have myself experienced and proved, and from their example you may form your own judgment concerning all other oviparous creatures. (*EGA* 13 83)

He has demonstrated certain things about the hen and her egg – but the inductive leap to all ovipara from this "exemplar" is left in the reader's hands. He then looks forward:

> It remains for me to continue my account of the generation and formation of the fetus out of the egg. For, as I explained before, the whole study of the race of cocks and hens resides in these two things, namely, how the egg is procreated from the cock and hen, and how the cock and hen arise out of the egg, and how by this circle of eternal existence of their race may be preserved by the grace and favour of Nature. (*EGA* 13 83)

He has accomplished the first task; in Exs. 14–24 he moves to the second (explaining how the cock and hen arise from the egg); and from that point on he first collects the results of his *historia*[28] and proceeds to discussions of "What an egg is" (Ex. 26) and then to the efficient and material causes of egg, fetus, and chick. The *historia* regularly compares the results achieved by direct observation and experiment with the views of his "leader" (Aristotle, *Dux*) and his "guide" (Fabricius, *Praemonstrator*).[29] During the causal inquiries, this becomes formalized – typically, separate exercises are devoted to what "Nature" or "Experience" teaches, what Aristotle teaches, and what Fabricius teaches.[30] Harvey finds himself more often in agreement with Aristotle than with his Paduan mentor.

Throughout the discussion of the efficient causes, there are constant references to the ends for the sake of which various parts develop as and when

---

[28] Harvey's occasional dismissive references to Francis Bacon are well-known, so it is perhaps worth quoting what he says at this point in the text: "And so I think it is convenient to explain here what fruit my labour has born and, to use the words of our most learned Verulam, proceed to my 'second vintage'" (*EGA* 25 134).

[29] *EAG* Preface, 20.

[30] I do not mean to imply others are not also discussed. Harvey pays special attention to the views of Aldrovandi, Parisanus, and Coiter and is highly critical of Daniel Sennert; among ancients, he cites Galen, Pliny, and Hippocrates regularly, though not nearly to the extent that he does Aristotle.

they do and to the generally goal-directed nature of animal generation. But in Ex. 59 there is an explicit introduction of the topic "Of the *uses* of the whole egg," followed by an introduction to the uses of the yolk and white (Ex. 60) and the other parts (Ex. 61). (Following Galen, reference to the "use" of a part in the Renaissance is explicated as identifying the final cause.)

The transition to discussing development in vivipara is accomplished by laying out the thesis that is perhaps Harvey's second most familiar "claim to fame" – *ex ovo omni* (everything from an egg). Harvey did not distinguish oviparous, viviparous, and equivocal (spontaneous) productions in the standard way. For him, these differences turn on *how* the egg is produced, not on *whether* an egg is produced. This is a radical position, and Harvey is cautious in introducing and defending it. Nevertheless, he concludes that "just as the chick is fashioned in the egg by *an internal formative agent*, so likewise is the fetus constituted out of the egg of the doe" (*EGA* 69 359).

It is time to explore Harvey's thoughts about this internal formative agent. Is its agency goal-oriented, and if so, in what manner? As was clear from the earlier quotation from the end of Ex. 13, Harvey follows precisely Aristotle's view that "the end of the generative act" and "the aim proposed by nature" is "that all individuals, while they procreate their like for the sake of the race, endure forever" (*EGA* 26 135).[31]

The account of the causation of animal generation in *EGA* is, at one level, profoundly Aristotelian, though, at another, it is a radical departure. Harvey feels he has conclusively established that the egg does not begin to develop until long after coition; that the male's semen is no longer present at all when the egg begins to develop; and therefore that whatever the male has contributed to the fertilization of the egg has been accomplished by an effect on the uterus. Like Aristotle, then, he holds that it is a causal power, rather than material, that is contributed by the male; indeed, it is a kind of innate heat. Unlike Aristotle, he views the female as a co-efficient cause with the male, producing eggs in the ovaries. But analogously to Aristotle's characterization of the female contribution, Harvey's account treats the unfertilized eggs as having only vegetative soul, potentially; the male renders the egg fertile, after which the egg itself takes charge of its own generation (cf. *EGA* 29 153–4; *EGA* 31 158–9). Thus, Harvey's dissections convince him that Aristotle is in error about the actual contribution of the female; and yet the characterization

---

[31] Note that Harvey follows the "individual eternality" interpretation of these passages – through in procreation it is the *individual* that endures forever. (Cf. James G. Lennox, "Are Aristotelian Species Eternal?" reprinted in Lennox, *Aristotle's Philosophy of Biology*, 131–59.)

of that contribution is remarkably like Aristotle's:

> The egg moreover . . . corresponds by analogy to the seeds of plants . . . as being not only that out of which, as out of material, but also by which, as efficient cause, the chick arises. In it however, there is no part of the future offspring actually in being, but all parts are indeed present in it potentially. (*EGA* 26 136)

And he reminds us that this conforms precisely to Aristotle's account of "natural body" in the *Physics*.

> An egg, therefore, is a natural body endowed with animal virtue, that is with the principle of movement, transformation, rest and conservation. And lastly it is such a thing as, all impediments being removed, will turn into the shape of an animal. (*EGA* 26 136)

So far Harvey's theoretical perspective is Aristotelian, though his meticulous study of the production of the ova in the ovaries, and indeed of the reproductive organs of the female in general, leads him to a very different view from Aristotle's about the contributions of the male and female to the developmental process.

What about their accounts of the initial stages of development within the egg? Recall Aristotle's observations of the critical fourth day after the egg has been laid, quoted earlier. Harvey carried out the same experiments, but he interprets what he sees quite differently:

> At that time, the fetus in the egg passes from the life of a plant to that of an animal. Then already the limbus or hem of the colliquament begins to turn purple and is outlined with a tiny line of blood, and almost in its centre there leaps a capering bloody point which is yet so exceedingly small that in its diastole it flashes like the smallest spark of fire, and immediately upon its systole it quite escapes the eye and disappears. (*EGA* 17 96)[32]

In his discussion of the causal explanation of generation, Harvey marks the key disagreement with Aristotle sharply:

> I cannot agree with Aristotle either when he maintains that the heart is this first genital and animate particle. For the truth is, I am persuaded, that this prerogative is due only to the blood, for the blood it is which is first seen in generation. And that not only in an egg, but in every fetus and animal conception whatsoever, as shall soon plainly appear. (*EGA* 51 241)[33]

---

[32] Cf. *De motu cordis*, chap. 17, p. 106.

[33] For what it is worth, Aristotle's view is nearer the truth – at three days the chick embryo already possesses a rudimentary heart with two chambers, and what is being observed is the pumping of blood from the lower to the upper chamber.

In Harvey's view, the living female produces, *in utero*, a very pure fluid that he terms the "primigenial moisture" of the egg. In it "all parts of the chick are present *in potentia* though not *in actu*"; it stands ready "by the law of Nature, to be transformed into all the parts of the body" (*EGA* 72 386). The first part to appear in actuality is blood, which "acts above the powers of the elements," serves as the "instrument of the great Creator," is "the substance whose act is the soul," and is "the cause and author of youth and old age, of sleeping and waking and of breathing also" (*EGA* 71 382) – that is, of all the attributes common to body and soul discussed in the last half of Aristotle's *Parva naturalia*. Again, the conceptual tools are Aristotelian, but the content of the theory is not: in Aristotle's theory the heat transferred to the female material via the semen animates the heart, whereas in Harvey's theory it animates the blood.

Harvey next provides a *hylomorphic* analysis of blood, apparently inspired by Aristotle's remarks about blood at *PA* 2.3. 649b22–33. Blood is the principal part, not only because it is the origin of movement and pulsation in the fetus, but even more because "in it animal heat is first bred, and vital spirit engendered, and the soul itself takes up its abode" (*EGA* 51 243). Harvey is now prepared to go further:

> Materially and in isolation it [blood] is called nutriment, but formally and in so far as it is endowed with heat and spirits (the immediate instruments of the soul), and with soul itself, it is to be accounted the tutelary deity of the body, its preserver, its prinicipal, primigenial and genital part. (*EGA* 52 257)

Not surprisingly, given Harvey's disagreement with Aristotle over whether heart or blood is the first part to exist actually in the egg,[34] the characterization of the blood in Harvey's *EGA* sounds remarkably like the characterization of the heart in Aristotle as well as in Harvey's *De motu cordis*.

As Harvey puts it in the "dessert" (*epidorpidis loco*) that he has prepared for us, called "Of Innate Heat,"

> For the blood alone is the true innate heat, or first-born animal heat, as is excellently well proved from my observations concerning the generation of animals... nor are those spirits which some men distinguish from blood anywhere to be found apart from blood, and blood itself without spirit or heat is no longer to be called blood but gore. (*EGA* 71 374)

---

[34] Though it is worth remembering that both the male and female material contributions to generation are concocted forms of blood.

The characterization of heat as a power that acts "beyond the powers of the elements" and "corresponds to the element of the stars" (cf. *EGA* 71. 378, 379, 381; this doctrine already is mentioned on pages 157, 183, 184, 211) is reminiscent of Aristotle's *De anima* and *GA*.[35] Again, the conceptual framework and the tools of analysis are Aristotelian – blood analyzed into material and formal aspects, innate heat as the instrument of soul – but within that framework the results of Harvey's dissections override Aristotle's conclusions whenever there is disagreement. Harvey is nevertheless fond of quoting Aristotle's occasional remarks on how, when facts and theories differ, the theories must bend to the facts (e.g., *GA* III. 10 760b28–32 on bee reproduction) to show that using observation to overrule Aristotle's theories is itself deeply Aristotelian.[36]

6. HARVEY ON THE GOAL-DIRECTED NATURE OF GENERATION

We can see Harvey zeroing in on the source of the goal-directed nature of animal generation in his discussion of the causation of fertility:

> Now seed and the conception and the egg are all things of the same kind, and that which renders all these fertile is the same in all of them, or something of a like nature and that is some divine thing, an analogue of the heavens, of art, of intelligence and of foresight. This is plainly to be seen from its wondrous operations, its contrivance and wisdom in which there is nothing done to no purpose or rashly or by chance, but all things are established for the sake of some good and to some end. (*EGA* 30 157)

Harvey invokes the strong version of Aristotle's "nature does nothing in vain" principle, and, like Aristotle, he treats it as an obvious result of the study of biological development – "plainly to be seen," as he here puts it. Something "analogous to art, intelligence and foresight" renders the egg fertile. But what is it, and how does it do so? As we have seen, he thinks both male and female are efficient causes, she of the egg in the ovary, he by making the egg fertile; but Harvey is also certain that the egg and its fertilization "are first begun long after coitus." He puts the puzzle poignantly in Ex. 41:

> What is this thing thus handed on, that cannot be found either remaining, or touching, or perceptibly contained, yet performs its office with consummate wisdom and foresight, beyond all the bounds of art, and which even when

---

[35] The key Aristotelian texts are *GA* 2. 3, 736b33–737a1 and *De An.* 2. 4, 416a10–19.
[36] The *GA* 3. 10 passage is in fact quoted in the preface of *EGA* (p. 17).

it has departed and vanished makes the egg fertile ... and not only the egg
that is perfected and completed, but also the imperfected and rudimentary egg
when it is only a pimple, nay, that renders even the hen herself fertile before
she has produced these pimples, and that so in a moment, as if the Almighty
should say "Let there be offspring," and straightway it is so? (*EGA* 41 189;
cf. *EGA* 49 228)

EGA provides no clear solution to this puzzle, which following Harvey we
may call the puzzle of *epigenesis*: through what agency do the parts of the
chick in the egg arise in an established order prepared to perform their proper
functions?[37] The presence of such an agency in the hen's egg is clearly
revealed in that pulsing, capering spot of blood. Because generation proceeds with "contrivance and wisdom" such that everything is "established for
the sake of some good and to some end," the power transmitted in the act of
fertilization must be an *analogue* of art, intelligence, and foresight. Harvey
refuses the dualist solutions proposed by Fernel and Scaliger (cf. *EGA* 71
375). The foresight and design displayed by active blood, "the instrument of
the great Creator," suffices:

> If by the dominion and rule of Art such excellent works are daily produced as
> do surpass the powers of the materials themselves, what then shall we think
> can be done by the precept and rule of Nature, whom Art does merely imitate?
> (*EGA* 71 380)

Broadly speaking, the answer to that question is to be found in Aristotelian
natures, not Platonic demiurges.

Having established that blood at its first appearance is pulsing in a manner
that allows one to distinguish a systolic and diastolic moment, Harvey thinks
he has located the first appearance of soul in the developing thing:

> Moreover, "the heart is erected for this purpose only, that by its ceaseless
> pulsation, and with the ministry of the veins and arteries, it may receive this
> blood and drive it forth again into every part throughout the whole body." (*EGA*
> 51 246; cf. 241, 243)[38]

But long before blood is formed, the goal-directed nature of generation
is seen in changes in the uterus after coitus, changes that take place in the
absence of any obvious causal agent but are nonetheless clearly for the sake of
generation. The egg is first engendered in the ovary, then begins its movement

---

[37] See *De motu cordis*, chap. 17, pp. 95, 106–7, where he looks forward to discussing this question
while reporting his observations *de formato foetu*.

[38] Compare the wording in *De motu cordis*, chap. 14, p. 87.

into the uterus and develops its membranes, white, yoke, and shell "for the sake of the fetus" (*EGA* 60 313; 61 323) – all this shows that a goal-directed agent is present from the start, long before blood is produced. Harvey leaves the nature of this goal-directed agency a mystery in *EGA*. However, he takes one last stab at the solution to the puzzle of its nature in his wonderful little treatise *De conceptione*.

7. TELEOLOGY IN *DE CONCEPTIONE*

De conceptione (*DC*) is openly speculative; Harvey asks for the same freedom he grants to everyone else: to render a probable opinion on "this whole dark business" of the true agency of conception so that others may prove its truth or falsity. He begins by reviewing the evidence from Ex. 51 of *EGA* that nothing corporeal remains after coitus and suggests that the uterus is rendered fertile by an incorporeal power, just as iron touched by a magnet is endowed with the power of the lodestone (*DC* 443). Yet he remains puzzled about what corresponds to the iron and to the power it acquires – what is it *in utero* that acquires the power of generation? He reviews the physical and behavioral changes in the female, especially in the uterus, after coitus, all of which indicate fertilization has occurred even before the first signs of development. In *EGA* there are vague hints of a sort of "contagion" model.[39] Here, however, he has something very different in mind, based on anatomical similarities he sees between the uterus prepared for conceiving and the brain:

> [W]hy may we not justly surmise that the function of each of them [the uterus and the brain] is also alike, and that what imagination and appetite are to the brain, that same thing, or at least something analogous to it, is awakened in the uterus by coitus and from this proceeds the generation or procreation of the egg? For the functions of both are called conceptions [*conceptiones*] and both are immaterial though they be the principles of all the actions of the body, the one of all natural actions, the other of all animal actions, the uterus the cause and first principle of all actions which pertain to the generation of an animal, the brain of all those which tend to the animal's preservation. (*DC* 445)

The cognitive act of the artist is the analogue of the act of generation: just as the external object produces an immaterial concept of its form in the artist's brain, which in turn causes the production of its likeness at the hands of the

---

[39] E.g., *EGA* 49. 227–8. Interestingly, a note on a back flyleaf of a copy of *EGA* in the library of his nephew Eliab develops the contagion analogy in more detail (cf. Whitteridge, *William Harvey: Disputations Touching the Generation of Animals*, 463).

artist, so does the male genitor produce an immaterial "idea" or "appearance" *in utero*, which is then fashioned into a material likeness of the form of the genitor (*DC* 446). Harvey defends his conjecture on anatomical grounds. Why, he asks, should we not think that since nature gave the womb a similar constitution to the brain, she "designed it also to a like function, or at least one that is analogous to it?"

> [F]rom the substance and form of the instruments a man may easily judge of their use and action, no less certainly than Aristotle has taught us to know their natures from the bodies of animals and the shape of their parts. (*DC* 446)

Harvey points out another similarity in the two cases – the form can have its effect in the absence of its original source (*DC* 447). This, it will be recalled, was the initial source of Harvey's puzzles about animal generation. Unlike atomist or dualist conjectures, this theory solves the problem of generation without invoking "incorporeal spirits like so many demiurges" (*DC* 448).

DC closes with a formal model of teleological causation against which any theory of animal generation should be measured. It rests on the Aristotelian assumption that the mover is always together with what it moves. All explanation of unqualified generation must involve a material that is the generated thing potentially, an efficient cause adequate to creating the generated thing, and a final cause, the generated thing. In animal generation, the egg is the chick *in potentia*, the source of fertility is the efficient cause, and the chick is the final cause.

Harvey's two puzzles about generation can now be formulated in terms of this model.

1. Though the agent of fertility must, on the causal assumption we are asked to adopt, be present with the egg if it is to fertilize it, the agent seems long gone before conception begins.
2. The chick, being a final *cause*, must also be present at the act of fertilization if it is to be causally relevant.

Harvey thinks his conjecture at once overcomes both puzzles. First, the incorporeal form of the male parent remains to direct development, informing the uterus and then the egg, using vital heat as its tool. Second, "[i]n every efficient there inheres in some [manner] the concept of the end and by this final cause, the efficient operating with foresight is moved" (*DC* 451).

To help us understand this idea of the end "inhering" in the efficient cause, Harvey cites the argument of *PA* 1.1 639b12–19 for the priority of the final to the efficient cause. Aristotle's key idea is that the final cause is the *logos* (*ratio*) *alike in art and in nature* – with finality being more evident in nature.

Generations in both realms are instituted for the sake of some end; both display foresight in their movement to the proposed end. It is the initial production of the form of the agent without its matter, in either the brain or the uterus, that accounts for the teleological character of generation in both realms: the end "inheres" in that agent.

Harvey's "fable," as he calls it, has a number of problems. When he first introduces his causal model, he denotes the actual outcome of generation (e.g., the chick) as the final cause. Yet he quickly realizes that the actual *outcome* does not exist at the time of conception; thus, in instantiating the model Harvey substitutes the *idea* or *concept* of the end for the actual outcome – it is the concept of the end that inheres in the efficient cause rather than the end itself. In what respect, then, is it a *distinct* cause? For after all, the immaterial form of the kind is the true *efficient cause* created during coitus. Harvey falls prey to a common problem in relying on metaphors of conscious intention to clarify natural teleology: finding a plausible biological analogue for the *intentional* object in cognitively directed action. Harvey typically characterizes the teleology of generation as owing to nature's *foresight* and *wisdom*. In Aristotle this is an analogy – in fact, the *absence* of actual foresight and wisdom in biological generation is one its distinguishing features. Nature's teleology is to be understood in its own terms; the teleology of art is understood by reference to nature's. Harvey's deism and his Aristotelianism appear to be in conflict.

Another problematic feature of this model is that there is no aspect of the Aristotelian philosophy that is more "full of shadows" (Harvey's description of *his* subject) than the view that the mind acquires the form of its objects without the matter. In relying on such a theory of mental "conception" in order to clarify animal conception, Harvey may be accused of removing the shadows by blowing out the candles.

8.   CONCLUSION

Aristotle, on rather basic philosophical grounds, constructs a scientific investigation of animals in which that part of the inquiry aimed at organizing the hierarchy of generalizations about animal differences is distinct from and prior to that part of the inquiry focused on the causes of those differences. Furthermore, it is an investigation in which the inquiry into the parts of fully developed animals and their causes was kept distinct from and prior to the inquiry into the generation of those animals and its causes. The particular style of philosophical anatomy that emerged in the sixteenth century in Padua owes a great deal to Galen and interpreters of Galen; but as more and more of Galen's anatomical claims were challenged, the readily available texts and Latin translations of Aristotle's zoological works became the source for an ever more prominent alternative model. Envisioning anatomy as a part of natural philosophy, its scope as "universal" and comparative (i.e., covering all the animals), and its methods as inductive and experimental was part and parcel of the Aristotelian program in anatomy in the school of medicine and philosophy at Padua. But so too was thinking of the study of the generative parts and generation as an autonomous part of the project.

William Harvey arrived in Padua when Fabricius was at the height of his powers and influence. Not only did Fabricius require his students to treat the *Historia animalium, De partibus animalium,* and *De generatione animalium* as integral to Aristotle's natural philosophy, he trained them to think of the relationship between natural philosophy and medicine in a new way, that taught by his colleagues Zabarella and Cremonini. In reading William Harvey's writings, two things stand out: First, Harvey has the biological works of Aristotle, and indeed Aristotle's corpus as a whole, constantly before his mind; second, he consistently adheres to his oft-repeated principle that true knowledge comes not from reading the words of others but from using comparative vivisection and its manipulative techniques to wrest the truth from nature. It was Harvey's belief that it is only in the light of experience and experiment that anything of value can be learned by studying the words of one's predecessors, a belief he put into practice in his monumental studies of animal generation.

2

Monsters, Nature, and Generation from the Renaissance to the Early Modern Period

The Emergence of Medical Thought

ANNIE BITBOL-HESPÉRIÈS

One of the most striking features of the discussion of monsters in the sixteenth century lies in the variety of texts in which they were discussed. In the second part of the sixteenth century, illustrations of monsters were featured with increasing frequency in the new editions of the *Cosmographia universalis* of Sebastian Münster (first edition 1552), in the volumes of the *Historia animalium* of Conrad Gesner (first volume 1551; fourth volume 1558), in the 1557 edition of the *Prodigiorum ac ostentorum chronicon* of Lycosthenes and in the enlarged editions of the *Histoires prodigieuses* of Pierre Boaistuau (first published in 1560). It would still be a number of decades, following the middle of the sixteenth century, before monsters and their causes were to become the specific subject of study by surgeons and physicians.

Jakob Rüff, a Zurich surgeon, and Ambroise Paré, a French surgeon, were the first to include studies of monsters in medical treatises. These treatises were reprinted and translated many times after their first editions, Rüff's in German in 1553 and Paré's in French in 1573. When describing and drawing monsters, Rüff and, to an even greater extent, Paré borrowed information on the subject from many sources.

Monsters were considered rare and generally were thought not to live long. Since very few doctors or surgeons directly observed the monsters they reported on, most rested content with gleaning as much information as they could from the works of physicians such as Hippocrates, philosophers such as Aristotle, and philosopher-theologians such as Augustine and Albertus Magnus, as well as from the books of historiographers such as Hartman Schedel, Conrad Lycosthenes, and Pliny the Elder; cosmographers such as Sebastian Münster, and even novelists such as Heliodorus and poets such as Ovid.

In his famous *Des monstres et prodiges* (first published in 1573 under the title *Des monstres tant terrestres que marins*), the French surgeon Ambroise

Paré describes two girls, born in 1475 in Verona, who were joined at the rear. He notes that they had been brought to several cities in Italy by their poor parents and exhibited in order to collect money from the people, 'who were very enthusiastic about seeing this new spectacle of nature'.

When quoting this example in his chapter about monsters produced when the quantity of seed is too great, Paré stresses two points, confirmed many times later on in his book as well as in other writings by different authors: the link between monsters and nature, to which we shall return later, and the notion of 'spectacle'. To be sure, one of the etymologies of the word 'monster', deriving from the Latin word *monstrum*, means 'show'. After his treatment of the conjoined twins, Paré discusses another noteworthy teratological case, a 'heteradelphia', in which one of the conjoined twins is not as well developed as the other and is attached to his brother at the thoraco-epigastric region.

This monster, 'observed' in 1530 in Paris, was a man about forty years old who had the body and members of his brother, save for the head, protruding from his belly. This man was reported to carry his parasitic twin in such a 'marvelous' way that people assembled in large groups to 'see' him. In his *Essais*, Montaigne describes with precision a similar teratological case in his chapter about a 'monstrous child' of fourteen months that was shown by his father, uncle and aunt, in return for pay. Monstrous creatures travelling from one place to another caused astonishment, amazement and also fear, as can be seen from the famous example, first quoted by Lycosthenes, of the two-headed girl. According to Paré, echoing Lycosthenes and Boaistuau, these two heads 'had the same desire to drink, eat, sleep; had identical speech, and had the same emotions'. The girl with her two heads was reported to go begging from door to door and get money because of the 'novelty of such a strange and new spectacle'. However, she was driven out by the Duchess of Bavaria because, it was said, she could 'spoil the fruit of the pregnant women' by the 'apprehension and ideas which might remain' in their imaginations.

Since there was generally very little information about them, illustrations, often full of ambiguity, played a crucial part in the public conception of monsters in the sixteenth century. It is also important to observe that in most cases there was a complete lack of information about the conditions of the births of these monsters. This leads us to consider the third main feature of the conception of birth defects in the sixteenth century, namely, that monsters and generation were not always closely linked. True, the noteworthy appearance of a monster was taken as a rupture in the order of generation, which, according to a view stemming from Aristotle, requires the 'reproduction' of 'the same' from one generation to another, a kind of substitute for eternity in our world that is made visible through the resemblance between children and their parents. In the *De generatione animalium*, Aristotle had asserted that 'a monster . . . is unlike its parents'. Because of the preeminence given to the role of the male seed in generation, in the Renaissance doctors continued to emphasize the resemblance between fathers and children. Indeed, before the existence of genetic tests, physical resemblance was of paramount importance in any effort to establish filiation.

In the sixteenth century, explaining the causes of monsters, such as a two-headed girl or conjoined twins, seemed beyond the domain of research of doctors and surgeons, since this was taken to be a matter that involved the consideration of divine designs. The first section of this essay highlights the limitations of efforts to explain the generation of monsters in the sixteenth century. It examines in particular the works of Jakob Rüff and Ambroise Paré in this context. Section 2 sketches the new way of investigating monsters that appeared in some medical writings from the very beginning of the seventeenth century. The goal here will be to chart the split between medicine and theology that developed at the end of the sixteenth century and during the seventeenth century. Section 3 shows the shift of attitudes toward nature and generation that is discernible in the work of Descartes.

1. THE LIMITATIONS OF TERATOLOGICAL EXPLANATION IN THE SIXTEENTH CENTURY

It is important to emphasize the central role of theology in the medicine of the sixteenth century, especially as concerns the generation of monsters. According to Augustine, who was often quoted both by Catholic and by Protestant authors, a monster (*monstrum*) – a notion synonymous with *prodigium* – shows (*monstrat*) God's will.

Since monsters are understood as a divine 'remonstrance', the explanation of their generation, that is, of their conception and development in the womb, not only mingles medicine and theology but also clearly subordinates medicine to theology. This is all the more significant since, in the sixteenth century, the 'generation' (what we call conception and embryonic development) of human beings still belonged, broadly speaking, among the 'secrets of Nature' (*Naturae arcana*). At the beginning of the seventeenth century, André du Laurens (Laurentius), echoing Galen, writes in his *Historia anatomica* that trying to know how the parts form in the womb was 'such a difficult question and so full of obscurity that only God or Nature can comprehend it'. He adds, 'What could be more divine than the first form of man? What could be more admirable? What could be more secret and hidden?'

It is thus important to remember the perception of mystery surrounding the complexity of the phenomena occurring in the generation of a new human being as well as the perception of mystery concerning the role of the organs of generation. 'Generation', a word derived from the Latin *genitum* – supine of the verb *gigno*, whose infinitive, *gignere*, means 'to be born' – refers to a process that requires a male and a female (not necessarily with equally important roles), the use of specific organs, and male seed acting on menstrual blood in the womb. There was one exception to this: the generation of Christ, born of Mary the Virgin, as Jakob Rüff reminds us in his *De conceptu et generatione hominis*.

In the sixteenth century these considerations did not apply to 'less perfect' beings such as lizards, frogs, and worms, which needed neither seed nor menstrual blood, as it was thought that they were formed spontaneously through putrefaction from decaying matter. Among 'monsters created through corruption and putrefaction', Ambroise Paré mentions, for instance, 'a snake... engendered in the putrefaction' of a 'dead body', a toad, as well as other animals 'engendered from some humid substance... which putrefied' to 'produce such creatures'. Spontaneous generation was first contested only with the experiments performed in 1625 by Aromatari and subsequently in the second part of the seventeenth century by Francesco Redi.

To return to the generation of human beings, the role of women was generally perceived to be subordinate to that of men, even if in the sixteenth century it was no longer believed, as Aristotle had asserted, that women only contribute to generation through the provision of the material from which the fetus develops. The role of the ovaries, referred to as the 'female testicles', was much discussed, as well as the function of the male testicles. The 'male seed', the sperm with its 'foamy' (*spumosum*) substance, was of paramount importance in generation, giving a 'form' to the fetus. It was also held to be possessed of an 'innate heat', which, according to Aristotle, was not to be confused with 'ordinary' heat, in so far as it was 'analogous to the element of the stars'.

The conception of twins posed difficult problems, such as whether they come from a single coupling or two different couplings. Many theorists were also concerned with the division of the superabundant seminal material in the womb. This was called 'superfetation' and was invoked in the case of 'distinct and separate' twins, conjoined twins being produced by 'too great a quantity of seed'. The generation of more than two children at the same time was generally considered a variety of monstrosity.

As seen in the example cited above of the two-headed girl driven out by the Duchess of Bavaria, doctors and laypeople alike believed that the mother's imagination could act upon the fetus and produce monsters. As also discussed in the chapter 4 of this volume, the power of maternal imagination, the '*vis imaginativa*', was very often cited in relation to the story of Heliodorus, in which the black queen of Ethiopia, Persina, gave birth to a white child, Chariclea, very unlike herself and her black husband, the king Hidustes. The white child resembled Andromeda, whose portrait Persina had had before her eyes during conception. This example – likely in fact an instance of albinism – was cited by Ambroise Paré at the beginning of his chapter 9, which deals with monsters that are generated as a result of the influence of the imagination. Among many other examples, he also mentions a princess who, while she and her husband had white skin, nonetheless gave birth to a dark-skinned child. Here, Paré (wrongly) invokes the authority of Hippocrates, saving the woman from the accusation of adultery by appeal to 'the portrait of a Moor' that had hung above her bed. At the end of this chapter, Paré also evokes a more recent case, of a young boy with the face of a frog, born in 1517 in France, not far from Fontainebleau. This came about, Paré maintains, because his mother had gone to bed with her husband with a frog in her hand in order to cure her fever.

In this period, of course, the developing fetus was completely inaccessible to direct obeservation. This did not prevent theorists from speculating about the order in which the principal organs appeared in the course of development. This was in fact a very ancient question. According to Aristotle, the heart, considered to be the 'principle of life', was the first organ to be formed, while Galen thought that liver was formed first, before the heart and the brain. With the formation of these principal organs, the fetus began to exist from the seed.

In the Renaissance, the development of the fetus remained strongly linked with the activity of a soul, generally divided into its 'vegetative' component, found in plants as well as animals, and its 'sensitive' component, found exclusively in animals. Finally, human beings alone were held to have a rational or intellective soul. Jakob Rüff appeals to this division in his *De conceptu et generatione homine*, first published in German a year before the Latin edition of 1554 and reissued in 1580 and 1587. Paré also makes use of it in his treatise on the *Generation of Man*, first published in 1573. Paré writes that 'plants live by a vegetative soul, beasts by a sensitive and men by an intellective soul' and asserts that 'the human intellective soul includes the virtues of all the inferior'. It was thus believed that in the course of fetal development, a human being acquires these three souls and their corresponding organs: the liver, seat of the vegetative life; the heart, seat of animal life; and the brain, seat of the intellective soul.

In his *De conceptu et generatione hominis*, Rüff offers many illustrations of fetuses in the womb and also presents various illustrations of monsters. For Rüff, as is obvious from the illustrations, the human 'matrix' (uterus) has one cavity. Rüff thus follows Vesalius, who had been the first to depict a human uterus, in his *De humani corporis fabrica*, written in Padua and first printed in Basel in 1543. In this work, Vesalius wants to reverse the decline of anatomy, in part by depicting 'the fabric of the human body' from human sources alone, not from non-human dissections as in Galenic anatomy.

The title page of the *Fabrica* depicts the public dissection of a woman's body performed by Vesalius himself. From the very beginning of his book, Vesalius, discussing the female genital organs, insists on the importance of internal anatomy as well as the study of generation. In book V of the *Fabrica*, Vesalius explains that the uterus has only one cavity, which was not at all in accord with the prevailing medieval doctrine of the division of that organ into seven 'cells' or cavities. Vesalius criticizes the traditional depiction of the uterus, borrowed from the dissection of animals and not based upon the dissection of human bodies. He notes that 'Galen never inspected a human uterus' but 'only' examined the uterus of animals (e.g., of the cow, the goat, and sheep).

Like Vesalius, Rüff also shows the membranes surrounding the fetus, maintaining that there are three of them. In his capacity as a very expert surgeon and deliverer of infants, he offers illustrations of the various and sometimes extremely perilous positions of single fetuses and twins in the womb. In book V, chapter 3, of his *De conceptu et generatione hominis*, Rüff shows various monsters, such as a two-headed child, conjoined twins in various formations, an elephant-headed boy and, at the very end, a composite monster, the famous monster of Ravenna.

Rüff deals with the 'physical' or 'natural' causes of monsters, such as the 'lack' of seed in the case of a child with only one arm or the 'excess' of seed in the case of a child with two heads or three legs, while ultimately referring all these causes to the will of God. In introducing the famous monster born in 1552 near Oxford (a case of conjoined twins fused in the pelvic region, with four arms and three legs, one ending with ten toes – a genuine case of ischiopagus), Rüff asserts that the 'providence of God almighty' allows the birth of monsters in order to 'punish and admonish human beings'.

Rüff also maintains, significantly, that human beings are able to conceive with animals. He explains this primarily by the 'attractive virtue' of the matrix being the same among human beings and animals. Rüff offers some examples of hybridism, claiming to know of a fruitful union in Switzerland between a mare and a bull and of one in France between a mare and a stag. In this connection, great importance is ascribed to Pliny the Elder. Pliny's vast compilation, including the most fantastic monsters, was much more influential in Rüff's time than, for example, Lucretius's denial of the existence of centaurs and other fantastic creatures. In rejecting the possibility of hybrids between two different species, Lucretius was echoing Aristotle's analysis in his *De generatione animalium*. In the sixteenth century, it was as if this aspect of Aristotle and Lucretius had been completely forgotten.

This remains clear when reading Paré's treatise *On Monsters and Prodigies*. Paré emphasizes the importance of experiments that would permit the confirmation or refutation of the possibility of a woman 'carry[ing] several chidren during one pregnancy'. After having referred to the famous Dorothea, who in Italy was suppposed to have given birth to 'twenty children in two confinements', 'nine at one time' and 'eleven at another', Paré mocked those 'wholly ignorant of anatomy' who wanted to make others believe that in the human uterus there are 'several cells and sinuses, namely seven, three on the right side receiving males, three on the left side for females and the seventh one in the middle being for hermaphrodites'. Paré denounces this as 'untruth' and maintains that it is supported neither by 'reason' nor 'authority' and thus is 'contrary to sense and observation'. Here Paré sides with Vesalius and Rüff. In spite of this correct observation, and in spite of some descriptions of monsters he personally saw and examined (such as the conjoined twins born on the last day of February 1572 between Paris and Chartres), it is for the most part not the case that Paré emphasizes anatomical knowledge and observation when writing about monsters.

In his treatise Paré ascribes the most expansive meaning ever to the word 'monster', including not only seamonsters such as whales, with their 'monstrous' size, and exotic animals such as the chameleon but also 'celestial monsters' such as comets. There is also a great number of monsters that are half-human and half-animal, as well as monsters having the body of an animal and the head of a man. At the beginning of a chapter dealing with the 'mixture of seed', Paré asserts that these latter creatures are 'produced by sodomists and atheists joining together'. One of the most well known of such creatures was the child thought to be conceived in 1493 from a union between a woman and a dog. This child, with a human upper body and a canine lower body, 'was very complete, without Nature's having omitted anything'.

The teleological outlook of Vesalius, common in anatomy since Galen, was enhanced by his more accurate description of the parts of the human body and the beautiful illustrations of perfect human bodies. Paré's aim in writing his treatise about monsters was not to comment with precise description on the differences between perfect human bodies, on the one hand, and deformed bodies, on the other. According to Paré, the illustrations of monsters sufficed.

Paré's aim in writing a book about monsters and prodigies was to 'acknowledge the greatness of nature', which he understood to be 'the chambermaid to this Great God' (faire reconnaître la grandeur de Nature, chambrière de ce grand Dieu). Personified nature remains mysterious, however, especially as concerns the question of monsters. For Paré, in producing monsters, nature is 'playing' in a way that can only give rise to amazement and to awe. Paré would agree with the formula of Pierre Boaistuau, author of the *Histoires extraordinaires*, one of Paré's source books, that, with its monsters, nature is 'a strange spectacle'.

2. THE SPLIT BETWEEN MEDICINE AND THEOLOGY AND THE RISE OF EMBRYOLOGY

From the very end of the sixteenth century and into the seventeenth century – leaving aside Aldrovandi's posthumous treatise on monsters, published by Ambrosini (*Monstrorum historia*, 1642) – some doctors and surgeons began to write about monsters in a more specific way. They progressively rejected two adjectives that Paré had commonly associated with monsters: 'divine' and 'hidden'.

In his *De ortu monstrorum commentarius*, published in 1595, Martin Weinrich promoted a split between medicine and theology. He explained that the theory of monsters needed to become 'physiological', for it belonged to the study of nature by means of human reason. He insisted on the specificity of his task, since being a 'physician' meant, in reference to the Greek etymology of the word, having to deal only with nature (*physis*) and with 'physical' explanations. He contrasted 'physicians' with 'theologians' and explained that their principles as well as their aims were different: theologians 'based their knowledge on the word of God and raised it to the heights' whereas physicians 'relied on the grounds of reason alone' and dealt with the immediate causes of things. In one chapter, Weinrich fought against both Augustine and the Stoics, who had claimed the necessity of less perfect bodies and of monsters in the world, and he asserted that these do not destroy the 'beauty' of the world. According to Weinrich, monsters are 'ugly' but studying them is important because it improves our knowledge of nature, since they are to be traced back to natural causes and to physiological explanations.

In his *Historia anatomica*, published in Latin in 1600 and translated into French in 1610 (by Sizé) and in 1613 (by Gelée), André du Laurens, mentions the variety of the causes of monsters. After having stated that the theologians relate monsters to God's vengeance, and astrologers to the stars, he explains that he himself puts aside theological causes as well as metaphysical causes and instead traces monsters only to 'natural' causes.

Emphasizing the importance of 'natural' causes did not seem to be the main objective of Caspar Bauhin when, in 1614, he published his work on hermaphrodites and monsters, *De hermaphroditorum monstrosorumque partuum natura et theologia, jurisconsultorum, medicorum, philosophorum et rabbinorum sententia libri II*. Here, Bauhin, a teacher of anatomy in Basel, made an inventory of writings about monsters and hermaphrodites and emphasized the minor place then occupied by medical writings about monsters, in comparison with those of theologians, philosophers and lawyers. Bauhin distinguishes between 'inferior' and 'superior' causes of monstrosities. For him, the superior causes always come from God (his wrath, his judgment, his curse); from the influence of stars and planets; or from the influence of the winds. The much more numerous inferior causes are in turn divided into 'internal' and 'external', and the internal into three subordinate ones: the first concerning the 'matter' (of the father and of the mother), the second the 'place' (especially the size of the womb), the third the 'efficient' causes arising from the weakness of the 'formative faculty'. The 'external' causes are divided into those imputable to the parents and those imputable to the mother alone.

Referring to Rüff, Bauhin explains that the generation of monsters is linked with the judgment of God. Theology remains an important topic in his vast historical inquiry into hermaphrodites and monsters, even though Bauhin discusses many medical arguments as well. Contrary to Rüff and in reference to Aristotle, Bauhin deems the birth of a compound monster having the parts of two or three animals 'impossible'. He insists that no animal may be born after a length of time different from its proper gestation. According to Bauhin, infants born with too many limbs are to be classed as monsters. He writes that the expression 'monsters of nature' is improperly used in the case of hermaphrodites, who are a mixture of the masculine and feminine sexes. Bauhin also explains that in monsters the matter 'is conserved' while the 'natural form' is changed, as shown by the examples of monsters having 'four eyes, four arms, and four legs'. Bauhin's treatise is noteworthy, in so far as he questions the view that monsters are *contra naturam* and explains that monsters are in fact *secundum naturam*, precisely because of this conservation of matter and simultaneous change in the form.

The question whether monsters are *contra naturam* was also explicitly raised by Riolan the Younger in his *De monstro nato a Lutetiae anno Domini 1605, Disputatio Philosophica*. According to Riolan, monsters, such as conjoined twins, do not seem to be *contra naturam* in so far as they are produced by nature, which in this case intends to make a perfect product but is unable to do so as a result of some impeding cause, such as a defect in matter.

In 1609, between the two editions of Bauhin's treatise, a very different book about monsters was published, the *Monstrorum historia memorabilis* by Johann Georg Schenck. Schenck shows many human and animal monsters similar to human beings, not to mention the world of the seamonsters that had been so important in the fourth volume of Gesner's *Historia animalium* and in Paré's *Monsters and Prodigies*. Pursuing the work initiated by his father in the two volumes of *Observationum medicarum, rararum, novarum, admirabilium et monstrosorum*, where pathologies and monstrosities had been dealt with in relation to the anatomy of the human body, Schenck also set about classifying the illustrations of monsters in relation to pathologies in the different parts of the human body.

Even if Schenck dealt with many of the same monsters found in Lycosthenes' book, his aim was not at all the same. In showing monsters as examples of various pathologies in the bodies of human beings and animals, whose causes were to be explained only by natural reasons, he sought to introduce a new teratological discourse into medicine. This new discourse contained no references to the traditional task of adducing 'supernatural' or 'supranatural' causes, that is, divine signs, God's wrath or even the wrath of the Devil, as in Lycosthenes, Boaistuau, Rüff, and Paré.

This new discourse also ceased to link the arrival of monsters to meteorological and astronomical phenomena such as parhelia or comets, which had been a common strategy in Lycosthenes. The year 1557, which saw the first edition of the *Chronick of Prodigies* by Lycosthenes, had also witnessed a triple parhelion and the birth of a boy who had neither a forehead nor a neck. In Lycosthenes' book, these two events were not only closely linked but also directly related to God's infinite might as well as to the impotence of human beings. In displaying monsters and prodigies, Lycosthenes wished to admonish Christian readers to pay attention to the salvation of their souls. Thus, the meaning of the word 'monster' was truly linked with its second etymology, derived from the Latin verb *moneo*, whose infinitive, *monere*, means 'to warn'. As is well known, this etymology had been widely illustrated by Cicero in his *De divinatione*, a text often quoted in relation to monsters. In contrast, Schenck concentrated on medical problems. In the illustration of the child born in Switzerland in 1557 and missing both forehead and neck, Schenck passes over the phenomenon of the parhelion altogether.

The tradition of subordinating medicine to theology was also clearly questioned by Fortunio Liceti in his treatise *De monstrorum causis, natura et differentiis*, first published in 1616 in Padua without any illustrations, then printed in 1634 with several, then reprinted in 1665 in Amsterdam and translated into French in 1708. In his treatise, Liceti rejects the conception of monsters as being produced by God's will. He asserts this clearly, though without refuting Augustine openly. He states that the study of monsters is within the proper scope of medicine. Liceti also pays attention to the extension of the word 'monster', a question that mattered for 'those wanting to speak about monsters in a proper way' Liceti distinguishes monsters from prodigies: monsters are animals who are rather close to man and 'whose disposition and arrangement, with respect to their members, is extraordinary; who are different from those from whom they have been begotten', and whose birth can only 'rarely' occur. Monsters give rise to 'suprise' and 'admiration'.

Liceti's definition of a monster becomes all the more significant when compared to Paré's treatise *Monsters and Prodigies*. The difference between their views is shown clearly in the manner in which Liceti classifies the causes of monsters. While explaining that monsters are 'effects' and 'works' of nature, Liceti emphasizes their Aristotelian causes. But Liceti's explanation of the causes of monsters is still underpinned by anthropomorphic statements about nature as the 'mother of all things and thereby of monsters'. In Liceti's opinion, nature, when producing monsters, can 'wander in a marvelous way' (*peut s' égarer merveilleusement*). He also writes that what is distinctive about monsters is that their 'matter [is] organized in a different way' and that the 'error of nature when producing monsters may be seen in the disposition of the organs'.

Liceti and Paré do not differ very much in their belief in the existence of monsters that are half-man and half-animal. According to Liceti, the duration of gestation in women may vary from seven to twelve months, which would allow them to be made pregnant by any of the animals belonging to a species whose gestation occurs within these limits. His treatise offers a number of examples arising from 'the unfortunate coupling of a human and a beast'. Besides the famous child with the canine lower body (already shown by Boaistuau and Paré) and the elephant-headed boy, Liceti shows in his book, among other examples, a pig with a human head, hands and feet and a monster that is half-man and half-pig, already familiar from Paré's treatise.

One of the most outstanding successors of Vesalius in Padua was Fabricius d'Acquapendente, a well-known teacher and the author of the first carefully illustrated treatises in embryology. When he began to publish, at the age of sixty, his aim was to write a *Theatrum totius animalis fabricae*. In 1604, in

Venice, he published the *De formato foetu*, and two years after his death the *De formatione ovi et pulli* was published in Padua in 1621. These treatises 'illustrate on almost every page the conflict between observation and authority'. 'Authority' refers in this case to Hippocrates, Aristotle, and Galen. The carefully made plates illustrate Fabricius's many precise observations.

In the second treatise, Fabricius begins by specifying the three ways in which animals arise: from eggs, from semen and spontaneously from decaying matter. He describes the formation of the chicken in the egg after many patient observations and experiments, stating, against Aristotle, that the chicken embryo is produced from the yolk and nourished by the albumen. He also explains that the matter of the male seed remains at the entrance of the uterus and does not penetrate it. For him, this is not problematic, since the 'spirituous substance' alone produces an effect, through 'radiation'.

In this treatise, published in 1621, two years after Fabricius's death, the Paduan anatomist deals with the formation of monsters such as a chicken with 'four legs and two heads, or four wings' He also explains the production of monsters from an egg having a double yolk. This would subsequently be discussed by William Harvey and by Nathaniel Highmore.

The *Anatomy* of du Laurens, Bauhin's treatise on hermaphrodites and monsters, Riolan's study of the monster born in Paris in 1605, and to a greater extent Liceti's treatise on monsters and Fabricius's books on embryology, as well as his important treatise *De venarum ostiolis*, all involve a conception of nature inspired by Aristotelian texts, especially the *De generatione animalium* and the *De partibus animalium*. This means, among other things, that we find in these books the expression of a very important role for teleology in nature. According to Aristotle, nature 'always does the best she can in the circumstances', does 'nothing superfluous nor in vain'. Monstrosities, 'though not necessary in regard of a final cause and an end', are 'necessary accidentally'. In the medical tradition, the teleological point of view was enhanced by Galen, especially in his treatise *De usu partium*, in which each organ is praised both for itself and for its utility to the whole. And in Renaissance medicine, this tradition was still very much alive, as shown by the treatise of Vesalius first published in 1543. The word *fabrica* itself conveyed a link between nature and the perfectly crafted human body it produces. In the beginning of the seventeenth century, du Laurens' *Historia anatomica* and Fabricius's treatises on embryology are the successors to this tradition. But on one point, as we have just seen, some physicians were beginning to question Aristotle's authority, namely, whether monsters belong to 'the class of things contrary to nature'.

3. THE EMERGENCE OF LAWS OF NATURE

The anthropomorphic conception of nature, very common in medical treatises of the time, and particularly in the question of monsters, was precisely the one René Descartes wanted to eradicate in the first work he wrote in French, which he decided not to publish when he learned of Galileo's condemnation. In *Le monde*, which includes *L'homme*, Descartes writes that by 'nature' he does not 'mean some goddess or any other sort of imaginary power'. Rather for him, nature signifies 'matter itself'. These claims, at the beginning of the important chapter explaining 'the laws of nature', were significant not only for physics but also for medicine. The conception of nature as a 'goddess', or as 'dame Nature', as Paré writes in his treatise on generation, was unambiguously rejected by Descartes.

Descartes also insists that scientific thought, in physics and astronomy as well as in medicine, must lead to the eradication of the admiration of natural phenomena. This is clear from the beginning of the First Discourse of *Les météores*. In this text, whose last discourse deals with the explanation of parhelia, admiration is linked to the ignorance of the causes of phenomena. Most of the anatomical treatises published in the sixteenth and seventeenth centuries insisted on the role of nature, sometimes associated with providence, in the making of the human body, its most remarkable production. Even William Harvey, in his *De motu cordis et sanguinis in animalibus*, writes about the 'skil of Nature'.

But it must also be remembered that in the medical tradition admiration is often taken as the opposite of comprehension. For instance, when Galen marvels at the opening of the uterus during parturition, he writes that it 'surpasses human intelligence' and that 'we can indeed marvel at it, but we cannot understand it'. Du Laurens also writes that understanding the 'admirable' formation of a human fetus in the womb 'surpasses the powers of the human mind'.

On the contrary, Descartes wants to 'explain all the phenomena of nature, that is to say the whole of physics' and to 'explain all the main functions in man'. When describing the human body and explaining its organic functions, Descartes insists on the 'disposition of the organs' and refers to mechanical models instead of praising nature or the 'skil of Nature' and positing occult 'faculties' in the body. These mechanical models are used by Descartes in his writings from *L'homme* onwards in order to explain physiological functions. They are often linked with his use of the expression 'there is no wonder' (*ce n'est pas merveille*), in *L'homme*, for instance, in the *Description du corps humain* (an updated version of *L'homme*, published posthumously with it in 1664), and in his correspondence.

Descartes's medical explanations are grounded in the laws of physics, which for him includes physiology. In his *Primae cogitationes circa generationem animalium*, Descartes reflects on the causes of the generation of monsters and, in an important passage, evokes the laws of nature which he had discussed in *Le monde*. He explicitly subordinates monsters to 'the eternal laws of Nature'. The novelty of this statement is noteworthy, and all the more remarkable when contrasted with the traditional importance of theological explanations.

The *Cogitationes* on generation were first published in 1701, and Descartes's remark found an echo in 1703, at the Académie des Sciences, when Fontenelle stated; 'One commonly regards monsters as games of nature [*jeux de la Nature*], but philosophers are quite persuaded that nature does not play [*que la Nature ne se joue point*], that she always invariably follows the same rules [*les mêmes règles*].' It is also interesting to note that Voltaire, in his article on the subject of the '*nécessaire*' in the *Dictionnaire philosophique*, confirmed that 'the general laws of nature brought some accidents which created [*qui ont fait naître*] monsters'.

Certainly, Descartes was not the first to use the expression 'laws of nature' in medicine. But if such an expression can be found in medical treatises, for instance, in du Laurens' *Historia anatomica*, its meaning is very different from the one Descartes intends. When du Laurens invoked the 'laws of nature', he had in mind regular movements, the causes of which remained completely unknown to human beings. In contrast, in Descartes the laws of mechanics that rule the human body are identical to the laws of nature and are derived from the immutability of God. This is stated explicitly in *Le monde*, alluded to at the beginning of the fifth part of the *Discours de la méthode* and explained at great length in the second part of the *Principia*.

Descartes was the first to link a conception of the principle of life with a mechanical explanation of the generation of living beings. This important task was begun in the *Primae cogitationes circa generationem animalium* and pursued in the *Description du corps humain*, written at the same time as Descartes was preparing his treatise on the *Passions de l'âme*. In his *Primae cogitationes*, Descartes attempts to explain the generation of man, including cases of hermaphrodism, while appealing only to mechanical principles.

Descartes's explanations in terms of matter and of the various motions in matter reject the postulation of various formative faculties that had been so common in medical texts. His mechanical account of the formation of the fetus in the *Primae cogitationes*, and in a much more precise way in the four parts of the *Description du corps humain*, does away with the common themes of, on the one hand, the soul considered as the principle of life and, on the other hand, the 'astral' or 'divine' heat that was supposed to be found in the heart and in the male semen. For Descartes, the heat in the heart, as well as the heat in the semen, is a natural phenomenon. In this connection, it is interesting to compare two parts of the same text, namely, the beginnings of the second and fourth parts of the *Description du corps humain*. In these two passages, Descartes makes use of exactly the same comparisons, appealing to 'liquor' (*liqueur*) and 'yeast' (*levain*) to explain, first, the heat in the heart and, second, the small 'particles of the seminal material' produced by 'the coupling of the two sexes'.

To explain the mixture of the seminal material, Descartes makes his well-known comparison with new wines and with the heat in damp hay. To explain the formation of the fetus, he refers to natural phenomena occurring in the human body, such as a process similar to fermentation, the result of the mixture of the seeds. This fermentation model, important in Descartes's biology both for the explanation of the heat in the heart and the formation of the various organs in the fetus, amounts to a total rejection of the previous medical tradition.

The approach to medicine Descartes prescribes is also linked with the rejection of the search for final causes, so common in medical treatises from Galen through Harvey's 1628 treatise, *De motu cordis et sanguinis in animalibus*. Descartes contrasts physics and physiology, where 'such conjectures are futile', with ethics, 'where we may often legitimately employ conjectures' and where 'it may admittedly be pious on occasion to try to guess what purpose God may have had in mind in his direction of the universe'.

With his particular interest in the most complex questions of anatomy (the structure of the brain, the eye and the pineal gland) and physiology (the movement of the heart and the blood), Descartes played an important role in the renewal of medicine in the seventeenth century. On the complex subject of the generation of animals, we see a vivid example of Descartes's effort to redefine nature, to eradicate the various formative faculties, to demonstrate the impossibility of explaining the vital functions and the movements of the body through the action of the soul and of the faculties, to unify the principle of life and to eliminate teleological assumptions in medicine (*L'homme, Discours de la méthode, Principia* and *Description du corps humain*), offering instead a mechanical account of the phenomena of the generation of monsters and hermaphrodites (*Primae cogitationes*).

With respect to the generation of monsters, it must also be noted that in the first of his *Meditations* Descartes rejects the existence of compound monsters and attributes their representation only to the imagination of the painters 'who simply jumble up the limbs of different animals'.

In the seventeenth century, just after Descartes's death, other important texts were published by physicians dealing with nature and the generation of monsters. Among them, William Harvey's on the generation of animals, *Exercitationes de generatione animalium*, published in Latin in 1651, was indeed an important text, while very different from Descartes's work on the subject. As he had already done in his short and brilliant treatise of 1628 demonstrating the motion of the heart and blood in animals, Harvey here insists on the importance of experiments, in this case for understanding both the embryonic development of the chicken and the conception and embryology of mammals. However, as had been the case in his treatise of 1628, and because he had been very much influenced by his studies in Padua, especially by Fabricius d'Acquapendente's anatomical lectures and demonstrations, he did not question the conceptual framework inherited from Aristotle. As we have seen, Descartes, on the contrary, rejected this framework completely.

4.   CONCLUSION

While precise answers to a number of the particular questions of embryology would take some time yet to become clear, what was clear and irreversible by the late seventeenth century, largely but not only as a result of the work of Descartes, was that generation, both of healthy and defective offspring, is a phenomenon that is properly studied within a scientific and medical context rather than speculated about by religious authorities. Monstrosities now were something to study not because they had the power to reveal the future but for the simple reason that they, like any other product of nature, were produced in accordance with nature's eternal laws.

II

The Cartesian Program

3

Descartes's Experimental Method and the Generation of Animals

VINCENT AUCANTE

From the time of Thomas Aquinas until the late sixteenth century, the bound-aries of human knowledge could easily have seemed fixed, and the sciences grew only very slowly against a rigid metaphysical background. Investigation was limited either to the straightforward observation of nature or to the study of what the ancient authorities had observed. But concerning the question of the generation of animals, it seemed impossible to take the first of these two approaches: the internal body parts in which fetuses grow remained closed to the gaze, and life was considered among the more mysterious and occult subjects. This is why many of the statements made about the generation of animals by the Hippocratic writers, Galen and Aristotle were still discussed during the Middle Ages and the Renaissance, even if they were incompatible with existing evidence and with each other.[1]

As a result of the scientific revolution, in the seventeenth century thinkers adopted a new approach to this old question. With Francis Bacon, Galileo and Descartes, the universe came to be seen as reducible to mechanical laws. The discovery of these laws, it was presumed, would make possible the explanation of the generation of animals. This is particularly the case for Descartes, who assumed that all of nature could be explained by a handful of physical laws. This reduction enabled him to ground the modern science of optics and to introduce new ideas concerning the movement of solids. The first step in his method was based on the clear understanding of principles – for instance, that bodies are moved by material causes – and this understanding was part

---

[1] On this question, see the studies of H. B. Adelmann, *The Embryological Treatises of Hieronymus Fabricius of Aquapendente* (Ithaca: Cornell University Press, 1942), 45–52; M. A. Hewson, *Giles of Rome and the Medieval Theory of Conception* (London: The Athlone Press, 1975); P.-G. Ottosson, *Scholastic Medicine and Philosophy* (Naples: Bibliopolis, 1984). The ancient physicians did not ignore this difficulty (see, e.g., Galen, *De foetuum formatione liber, Ioanne Bernardo Feliciano Interprete*, c. 3, in *Opera* [Venice: Juntas, 1625], vol. 1, f° 322r).

of his metaphysics. In a subsequent step, the first principles of physics – for example, the thesis of the plenitude of matter, which rejects any kind of vacuum – could be deduced. But in the explanation of all the specific facts of nature that could be observed, the method of deduction was confronted with so many alternatives that it was necessary to introduce some specific hypotheses and then to confirm them by experimentation.

Confronted with the complexity of living creatures, the streamlined character of Cartesian physiology seemed to vanish. Its only significant result was the discovery of the formation of an inverse image on the retina and not in the crystalline humor, as most of Descartes's contemporaries believed.[2] Descartes also claimed to have explained the movement of the heart,[3] and his theory of the nervous system based on the physiology of the famous pineal gland occupied a significant position in his works.[4] But for a whole system of physiology, he had to cope with the question of generation, which proved embarrassing, as he himself confessed in 1646:

> The formation of all the parts of the human body ... is something so difficult that I dare not undertake [to explain it] yet.[5]

This doesn't mean that he didn't try to understand 'such a difficult thing', but it seemed at first that this problem was altogether beyond the possibilities of his method. This is confirmed in two important ways: (1) He never published any text on this subject during his life, but from 1638 to 1649 he did announce many times that he was close to achieving a solution.[6] (2) All the unfinished writings that we have on the subject are mutually contradictory and reveal merely tentative efforts at resolving this acute problem.[7] This

---

[2] The discovery of the real location of the image of objects in the eye is the result of the inquiries of both the physician Plempius and Descartes. See my 'La vision chez Descartes et Plempius', in *Il Seicento e Descartes: Dibattiti cartesiani* (Florence: Le Monnier, 2004), 233–54.

[3] Descartes arrived at the hypothesis of the circulation of the blood only by observing the valvules in the arteries and veins, which allowed him to reconstruct the path of the blood's circulation in the whole body. He read Harvey later and retained from him only the discovery of "anastomosis" at the extremities of the members. See my article 'Le rôle du cœur de Fludd à Descartes et Harvey', in *Pour une civilisation du cœur* (Paris, 2000), 51–61.

[4] On the pineal gland, see F. Meschini, *Neurofisiologia cartesiana* (Florence: Leo Olschki, 1998), 121–38; and my appendix 9 to Descartes, *Écrits physiologiques et médicaux* (Paris: Presses Universitaires de France, 2000), 267 f.

[5] René Descartes, *Oeuvres de Descartes*, 11 vols., ed. C. Adam and P. Tannery (Paris: J. Vrin, 1964–76). Cited by 'AT' and then volume and page number. Descartes to Elisabeth, May 1646, AT IV, 407, 18–20.

[6] Cf., e.g., Descartes to Huygens, May 25, 1638, AT I, 507, 16; Descartes to Mersenne, November 23, 1646, AT IV, 566, 24–6.

[7] See my edition of the posthumous Cartesian fragments dedicated to generation in Descartes, *Écrits physiologiques et médicaux*, 28–163.

difficulty has led several commentators either to banish these writings from serious consideration[8] or to view them as juvenilia, as did the author of the first edition of a selection of the fragments, published in 1701 under the title *Primæ cogitationes circa generationem animalium*.[9]

To understand Descartes's original approach, we must first classify all his manuscripts concerning the subject chronologically. As I have shown elsewhere, such classification reveals three main stages of study, beginning in the years 1630, 1637 and 1648 respectively.[10] Each amounts to nothing more than an attempt at reaching a clear understanding of the problem of generation. His ideas on the subject were based on the systematic appeal to physical laws, on the rejection of Aristotelian immaterial principles, on attention to the findings of the physicians, and, above all, on specific experiments.

After considering Descartes's understanding of the nature of seed in section 1 and surveying the two first stages of the elaboration of generation in section 2, I discuss in section 3 how Descartes appealed to experiments on eggs, then in section 4 I examine in detail the third stage of his research, which resulted from these experiments and provides a particularly revealing illustration of the role of observation for Descartes in proving or justifying a theory. Finally, I argue that the very fact that Descartes never succeeded in offering anything like a theory of the generation of animals itself stands as proof of his scientific caution.

1. THE NATURE OF SEED

From the Cartesian point of view, seed consists in nothing but material particles. Thus, all its actions are for Descartes to be explained in terms of the laws of motion accounted for in his physics, without any reference to Aristotelian souls or Galenic faculties.

One of the posthumous fragments of November 1637 refers to the process of formation of the seed:

It can also happen that the parts of the semen do not immediately give rise to something similar to them, but to certain others that then give rise to others from these, and these in turn give rise to others wholly like the semen; which

---

[8] See, e.g., A. Pichot, *Histoire de la notion de vie* (Paris: Gallimard, 1993), 377.

[9] In fact some of these fragments were written in 1648! Cf. for example the Fragments [73] to [76] of the *Cogitationes* (AT XI, 538, 11–18), which are present in the *Excerpta* with the date 1648 (AT XI, 608).

[10] See my introduction in Descartes, *Écrits physiologiques et médicaux*, 9–23.

we see more in animals than in plants. And from this it is easy to understand why the majority of animals and of plants produces a semen different from the rest of the body.[11]

In this fragment, Descartes seems to be rejecting the idea, well established in his time, that the seed comes from the whole body, a theory that explained how it is that the child looks like its parents. This is confirmed by a note of 1630, in *L'homme*, where it is maintained that the seed is produced by the most energetic parts of the animal spirits.[12] On his view, the seed is formed in the testicles, the 'parts serving to the generation',[13] and there only. But Descartes is here confronted by an old question: what is the contribution of the female to generation? Is she only the receptacle of the male seed, is her menstrual blood a sort of depleted seed, or does she have an altogether different seed of her own? Descartes adopts the third solution,[14] like many of his contemporaries, founding his theory on the existence of 'female testicles' commonly 'observed' by the anatomists of his time.[15]

From Descartes's point of view, the seed, in the end, is nothing but the mixing of material particles of two kinds: first, heavy particles which gather and crowd together;[16] second, the fine particles which form the blood and the spirits and which move and separate easily.[17]

---

[11] *Cogitationes*, [69], AT XI, 598, 8–15. 'Potest vero etiam contingere, ut partes seminis non immediate sibi similes producant, sed alias quasdam quæ postea alias, & tandem hæ alias omnino similes iis seminis producant; quod in animalibus videtur potius contingere, quam in plantis. Atque ex his facile intelligitur, cur maxima pars animalium & plantarum semen a reliquo corpore diversum excernant.'

[12] 'Vous pouvez aussi remarquer en passant, qu'après celles qui entrent dans le cerveau, il n'y en a point de plus fortes ny de plus vives, que celles qui se vont rendre aux vaisseaux destinés à la génération. Car par exemple, si celles qui ont la force de parvenir jusques à D, ne peuvent aller plus avant vers C, à cause qu'il n'y a pas assez de place pour toute, elles retournent plutôt vers E, que vers F ny vers G, d'autant que le passage y est plus droit. En suite de quoy je pourrois peut-estre vous faire voir, comment de l'humeur qui s'assemble vers E, il se peut former une autre machine, toute semblable à celle-cy, mais je ne veux pas entrer plus avant en cette matière' (*L'homme*, AT XI, 128, 19–31).

[13] *Cogitationes*, [36], AT XI, 533, 10; *Description*, AT XI, 257, 13–14.

[14] See ibid., [27], AT XI, 524, 14; [31], AT XI, 525, 13, 16; [37], AT XI, 533, 17; [52], AT XI, 518, 6; etc.

[15] See, e.g., V. Coiter, *Externarum et internarum principialium humani corporis partium tabulæ* (Nuremberg: Theodorici Gerlatzeni, 1572), 27; J. Fernel, *Physiologie*, l. VII, *De la génération de l'homme et de la semence*, c. VI, (Paris: Jean Guignard le Jeune, 1655), 712–19; J. Riolan, *Humani fœtus historia*, c. IV, in *Schola anatomica* (Genève: Joannis Celerii, 1625), 287–8; L. Vassé, *Tabulæ anatomicæ* (Venice: Vincentium Vaugris, 1544), f° 30; etc.

[16] *Description*, AT XI, 254, 6–7; 254, 31–255, 1; 258, 2–3.

[17] Ibid., 254, 30–1; 256, 28–257, 24; 258, 1–2.

For Descartes, fecundation is the result of a successful mixing in the uterus of the male and female seeds. The antagonistic properties of the seed – aggregation and separation – were thought to lead to the formation of the parts of the fetus. After fecundation, there was only a 'confused mixing of the two liquors',[18] which were moved as a consequence of heat. Descartes remained uncertain concerning the origin of this heat. At first, he thought in a Galenic manner that it came from the womb of the mother,[19] but he soon became convinced that the mixing of the two seeds was itself the origin of heat, bringing about as it did a kind of fermentation where each seed served as the 'leaven' for the other.[20] Descartes thought that the fermentation of the seed was a chemical phenomenon that could be compared with the famous '*feu sans lumière*' located in the heart.

After fecundation, the material particles constituting the seeds were then moved by the heat of fermentation, and their movements obeyed physical laws.[21] But Descartes also offers a criterion of subtlety: 'the subtler, the faster'.[22] It is according to this criterion that he was able to explain two opposite actions, the union and the separation of the particles of the seed.[23] It is clear that this criterion was the direct consequence of the combination of physical laws, fermentation and the geometrical properties of the seed. The contrariety of the properties of the parts of the seed (some gathering and others separating, as we have seen) was explained by the different sizes and shapes of these parts. The parts which gathered easily were more branchy,[24] and

---

[18] Ibid., AT XI, 253, 15. This idea of a confused mixing of the two seeds is present in a number of physicians. See, e.g., U. Aldrovandus, *Monstrorum historia* (Bononiae: Marcus Antonius Bernia, 1642), 49.

[19] *Cogitationes*, [5], AT XI, 507, 18–21. This Galenic thesis is adopted by most of the physicians. See, e.g., C. Arantius, *De humano fœtu libellus*, c. V (Basel: Laurentius Scholzius, 1579), 27; G. Bouchet, *Serees*, 23 (Lyon: Simon Rigaud, 1616), vol. 2, p. 262; J. Fernel, *Physiologie*, l. VII, c. VIII, P. 732–4; N. Sacco, *De calore naturali*, l. I, c. XI (Papiae: J. Rubeum Baptistam, 1628), 72.

[20] *Description*, AT XI, 253, 13–19; 254, 2–9; 280, 30–281, 8; 281, 27–30; *Cogitationes*, [3], AT XI, 505, 18–20; [4], 507, 10–13; [7], 509, 31–510, 1; [45], 516, 17; [16], 534, 13–15. This thesis is also developed by R. Colombus, *De re anatomica*, l. XII (Venice: Bevilacqua, 1559), 245.

[21] On physical laws in Descartes and their evolution, see D. Garber, *Descartes's Metaphysical Physics* (Chicago: University of Chicago Press, 1992), 197–230.

[22] 'Quod subtilius est, celerius movetur' (*Cogitationes*, AT XI, 507, 1). This criterion may come from Fernel: 'Les parties de la semence les plus subtiles, les plus chaudes & qui abondent d'avantage en esprits, forment un amas. . . . [L'esprit,] qui est le plus subtil, est le premier véhicule de la faculté générative' (*Physiologie*, l. VII, c. IX, pp. 736–8).

[23] See, e.g., *Cogitationes*, [6], AT XI, 509, 17–23; [8] 512, 20–3; etc; *Description* AT XI, 254, 5–13; 258, 1–5; 261, 20–31; 274, 7–9; etc.

[24] *Description*, AT XI, 274, 27; 275, 1; 276, 13, 21.

Vincent Aucante

the parts prone to separation were more fluid.[25] Thus, geometrical difference alone could allow him to understand the different behavior of the parts of the seeds. If Descartes never pretended to fully explain the nature of the seed,[26] he nevertheless was able to conceive a hypothetical structure of the different parts of it.

2. INFLUENCES ON DESCARTES

The first group of fragments dedicated to generation were written after Descartes settled in Amsterdam in 1629 and came under the influence, indirectly, of two famous physicians: Jean Fernel and Fabricius d'Acquapendente. In fact, in dating the fragments by looking at which organ he held to develop first, we can identify a first group of fragments[27] as developing a Hippocratic thesis, namely, that the brain is the first formed organ of the fetus.[28] Here, we find Descartes's main thesis concerning the autonomy of the seed – formed, as we have seen, from material particles moved by their own fermentation after fecundation – presented in a Hippocratic fashion.[29] The circular movement of the particles of seed brings about the formation of the main organs, the arteries and the veins. This is then replaced by the parallel circulation of the animal spirits and of the blood.

We find in these fragments several echoes of the treatise on generation by Jean Fernel. In particular, Descartes appropriates from Fernel the view that the male and female seeds are material particles mixed through fecundation, causing a fermentation, which organizes the movements of particles and so forms the fetus.[30] These material particles form the first organs through their own *virtus*.[31] With his thesis of the autonomy of the seed, a matter acting independently of any soul, Fernel was certainly closer to the seventeenth century than to his contemporaries, and he was particularly close to Descartes.[32] Fabricius d'Acquapendente was the other physician of great importance for Descartes's medical works. He is mentioned in the fragments,[33] and we can

[25] Ibid., AT XI, 274, 13.
[26] 'Je ne détermine rien touchant la figure & l'arrangement des particules de la semence' (*Description*, AT XI, 253, 7–8).
[27] AT XI, 507, 18–516, 15; 535, 5–537, 13.
[28] See Hippocrates *De caro* c. IV, 1–c. VIII, 1.
[29] Hippocrates *De generatio* IV, 3; V, 1.
[30] Fernel, *Physiologie*, l. VII, c. IX, pp. 696, 735.
[31] Ibid., 736–9.
[32] See my study 'La théorie de l'âme de Jean Fernel', *Corpus* 41 (2002), 36–9.
[33] AT XI, 511, 19.

be certain that Descartes first read his treatise *De formatione ovi et pulli* at this time.[34] But evidently this reading did not convince him, and he did not yet seem curious to try his own experiment on eggs.

The second group of fragments, which we may be certain were written in 1637, after the publication of the *Discours de la méthode*,[35] are more developed and constitute another attempt to understand the first steps of generation. Here, the lungs and the liver are thought to be the first organs formed in the fetus.[36] This very view had been the thesis of Arantius, who had himself modified the Galenic view.[37] The legacy of Galenism is further confirmed when we consider three other stages of generation laid out by Descartes: (1) the seeds form the first organs; (2) veins and arteries are formed, allowing the circulation of spirits and blood; (3) the fetus is connected to the umbilical cord for nourishment. We have here the three well-known steps in generation according to Galen: formation (*generatio*), growth (*auctio*), and nutrition (*nutritio*).[38] But this Galenic distinction was grounded in the action of a specific 'faculty' at each step, whereas Descartes retains only the movements of particles and their material effects. Thus, the distinction between Cartesianism and Galenism remains. Why did Descartes change his view? No reason can be given with certainty, but it is clear that he was dissatisfied with his first attempt and resolved to try something else.

These first two stages in Descartes's research reveal the real disappointment the philosopher faced: he did not succeed at finding an explanation strong enough to be offered to his correspondents, but only conjectures. His main difficulty was in finding a good way to choose between competing explanations: the experiments on eggs would provide this to him.

---

[34] 'Pour la formation des poulets dans l'œuf, il y a plus de 15 ans que j'ai lu ce que Fabricius ab Acquapendente en a écrit' (Descartes to Mersenne, November 2, 1646, AT IV, 555, 9–12).

[35] Fragments [58] to [68] (AT XI, 526, 3–531, 21) form a coherent whole, a sort of unfinished treatise. In fact, Fragment [68] (AT XI, 530, 28–531, 21) refers to the *Institutiones anatomiae* of Bauhin, as does a fragment of the *Excerpta anatomica* (AT XI, 587, 19–594, 18) dated 1637. Both were certainly written at the same time. We know that at the same time Descartes performed a number of dissections on animals (*Excerpta anatomica*, AT XI, 583, 5–594, 18; 596, 1–600, 17).

[36] AT XI, 516, 16–520, 1; 526, 6–531, 21; 598, 8–600, 17.

[37] Arantius, *De humano foetus libellus*, c. VIII, P. 49–51; Galen, *De foetuum formatione liber*, c. 3, in *Opera*, vol. 1, f° 323r.

[38] Galen, *De naturalibus facultatibus, Thoma Linacro Anglo interprete*, l. I, c. 5, in *Opera*, vol. 1, f° 291v. These three steps were commonly used by physicians contemporary to Descartes. See, e.g., Fabricius d'Acquapendente, *De formatione ovi et pulli* (Patavii: Aloysii Bencii Bibliopolæ, 1621), 41; J. Riolan, *Anthropographie*, l. VI, *L'histoire du foetus humain*, c. 5 (Paris: Denys Moreau, 1629), 930.

3. DESCARTES CRACKS SOME EGGS

Experimentation on eggs was common in the seventeenth century, and its guiding principle is easy to understand: the chick grows slowly in the egg, and it is easy to determine when it had been laid; this allows the researcher to date its age and to observe its growth by breaking eggs day after day and then observing the developing chick on each successive day of incubation. The Hippocratic writers had already mentioned this experiment, as did Aristotle.[39] During the sixteenth century, Volcher Coiter conducted experiments on eggs and systematically observed the growth of the chick.[40] His observations were sometimes wrong, but he was the first in the modern period to emphasize this kind observation rather than simply offering comments on the ancient authorities. In 1601, following Ulysses Aldrovandus,[41] Fabricius conducted the entire experiment in private and also repeated it in part with some of his students, such as Peiresc and Harvey.[42] He then wrote his famous book *De formatione ovi et pulli*, with several wonderful plates illustrating it (Fig. 3.1).

With Fabricius's book, Descartes had quite a bit of help in figuring out how to conduct the experiment himself. He tried it at least twice. This led to Descartes's observation of the gradual formation of the chick from day to day, chronicled in the manuscripts known as the *Excerpta anatomica*. I list here his main observations, along with data from Coiter and Fabricius:

- Descartes: The lungs, the heart and the blood appear on the second day.[43]
  Coiter: On the third day, it was clear that there was a point or globule of blood pulsating.[44]
  Coiter: On the tenth day the lungs and the ribs were formed.[45]
  Fabricius: No blood enters into the egg.[46]

---

[39] Hippocrates *De natura pueri* c. XXIX, 2–3; Aristotle *Historia animalis* l. VI, c. 3.

[40] V. Coiter, *De ovis et pullis gallinaceis*, in *Externarum et internarum principialium humani corporis partium tabulæ*, 32–9. See also the commentary of H. Adelmann in *Annals of Medical History* (1933), 444–57.

[41] Aldrovandus, *Ornithologiæ* (Bonoiae: Nicolaus Tebaldinus, 1634), vol. 2, 214. The experiment is also mentioned by Ambroise Paré, who certainly did not perform it himself.

[42] P. Gassendi, *Viri illustris Nicolai Claudii Fabricii de Peiresc vita* (The Hague: A. Vlacq, 1651), l. I, 25. The dating of Prevotius's experiment in his dedication to the *De formatione ovi et pulli* to 1604, when the physician was in Florence, could not refer to his first observations. See Adelmann, *Embryological Treatises of Fabricius*, 74–5.

[43] AT XI, 615, 17–19; 620, 9–11.

[44] *De ovies et pullis gallinaceis*, 33.

[45] Ibid., 34.

[46] *De formatione ovi et pulli*, 39.

Fig. 3.1. From Fabricius, *De formatione pulli in ovo.*

- Descartes: The head, the spine and the eyes appear on the third day.[47] Coiter: It is clear that the head of the chick appears on the fifth day and on the sides there appears a black eye.[48] Fabricius: The head and the spine are generated first in the chick.[49] Fabricius: What nature makes evident to us is the cause of admiration for a great number [of people].[50]
- Descartes: The heart beats regularly from the fifth day on. The stomach, the feet and the wings appear at that time as well.[51] Coiter: On the fifth day I saw a globule of blood pulsating; to the side, there were globules moving toward the black ... that all together represented a sort of brain.[52] Fabricius: After three days we see ... the heart beating.... At the same time, the liver and the heart are generated.[53]

---

[47] AT XI, 620, 12–13.
[48] *De ovis et pullis gallinaceis*, 33.
[49] *De formatione ovi et pulli*, 35.
[50] Ibid., 44.
[51] AT XI, 620, 14–18.
[52] *De ovis et pullis gallinaceis*, 33.
[53] *De formatione ovi et pulli*, 44–5.

Fabricius: After four or five days everything already mentioned can be perceived, and the wings and feet, and everything that composes them . . . as muscles . . . is missing.[54]

- Descartes: The beak and the brain appear on the seventh[55] or the eighth day.[56]

  Coiter: On the sixth day . . . there was something under the eyes that resembled a beak.[57]

  Fabricius: If the marrow of the spine is formed before the spine, it must also include the brain.[58]

- Descartes: On the ninth day, there is not yet any muscle or intestine to be seen.[59]

  Coiter: On the seventh day . . . nothing could be discerned of the viscera.[60]

- Descartes: On the tenth day the liver and the gall-bladder appear, and apparently also the intestine; the other organs have begun to move to their fixed places.[61]

  Coiter: On the ninth day . . . the liver, the ventricle, the intestines, the ribs, and all the other viscera could be located.[62]

- Descartes: The spleen appears on the twelfth day.[63]

- Descartes: On the fifteenth day, the bird is formed and a sort of umbilical cord is now visible.[64]

  Coiter: On the eighth day . . . the sufficiently full veins were visible, and the umbilical cord penetrated the chick.[65]

  Fabricius: There was frequently seen a moving humor, like a gout, which in the duck studied was certainly the umbilical cord, since no other part could be found that could be this vessel.[66]

- Descartes: On the sixteenth day there appears a kind of placenta, as well as an external intestine of sorts.[67]

---

[54] *De formatione ovi et pulli*, 45.
[55] AT XI, 619, 19–20.
[56] Ibid., 620, 19–21.
[57] *De ovis et pullis gallinaceis*, 33.
[58] *De formatione ovi et pulli*, 43–4.
[59] AT XI, 619, 25–6.
[60] *De ovis et pullis gallinaceis*, 34.
[61] AT XI, 620, 22–7.
[62] *De ovis et pullis gallinaceis*, 34.
[63] AT XI, 620, 28–9.
[64] Ibid., 615, 6–17.
[65] *De ovis et pullis gallinaceis*, 34.
[66] *De formatione ovi et pulli*, c. I, 24.
[67] AT XI, 620, 30–621, 1; 621, 6–9.

Coiter: On the sixteenth day there hung from the stomach two vessels a bit like those worms that could be said to resemble the intestines.[68]
- Descartes: By the nineteenth day, the yolk has disappeared.[69]
  Coiter: On the nineteenth day . . . the yolk was absorbed.[70]

Given the works of Fabricius and, even more, those of Coiter, Descartes's observations do not yield any discoveries and even miss several important details. Nonetheless, they inspire Descartes to undertake a new attempt to describe the process of generation.

Descartes was not alone in his interest in these experiments on eggs, and Mersenne also claimed in his correspondences to have come up with several noteworthy results in this area. This experimental approach must be contrasted with the other prevailing approach, which would prefer to comment on the opinions of Galen, the Hippocratic writers and Aristotle. There is a typical example of this attitude in the *Dissertatio de ovo et pullo* of Martin Schoock, published in 1643: there are only 12 out of 183 pages on the results of observations, and even these do not lead the author to any decisive argument but only provide another occasion to comment on the ancient authorities.

4. DESCARTES'S FINAL ATTEMPT

In contrast, it is clear that the experiments on eggs led Descartes to his final attempt to tackle the problem of the generation of animals, in the last part of his book *Description du corps humain*. The writing of the text that we have inherited began, we know with certainty, in 1638 and was taken up again in 1648 after a long interruption.

This work was begun after the publication of the *Discours de la méthode,* where we read,

> I've resolved to use the time that remains to me in this life to do nothing else than attempt to gain some knowledge of Nature, such that one could draw from it rules for Medicine that are more certain than those we have had up to the present.[71]

We saw earlier that 1637 had been for Descartes a period of intense study of medicine. This study led him to announce to Huygens that he had nearly

---

[68] *De ovis et pullis gallinaceis,* 35.
[69] AT XI, 621, 1–3.
[70] *De ovis et pullis gallinaceis,* 35.
[71] *Discours,* AT VI, 78, 8–13.

finished an *abrégé de médecine*.[72] It was also the beginning of an intense and complicated discussion with his old friend Plempius.[73] His interest in medicine can be identified easily in the *Description* thanks to certain details. First of all, Descartes mentions his *Dioptrique* and his *Météores*, which implies that he began this study at least at the time of publication of these studies, in 1637, and perhaps earlier.[74] Second, he uses the phrase '*vapeurs de sang*' (vapors of blood) a number of times in describing the state of the blood coming from the heart after having been exposed to the *feu sans lumière* which heats it.[75] This phrase, already present in the *Traité de l'homme*, had been severely criticized by Plempius.[76] When Descartes replied to him in February, 1638, he stated more precisely that the result of the rarefaction of the blood was something distinct from a vapor, and he never used this expression again.[77]

The criticism coming from Plempius and Fromondus, followed by the great opposition to the Cartesian thesis defended by Regius in 1638, as well as the incomplete state of his own work, led Descartes to halt his physiological project for a time. But he didn't stop his medical studies altogether, particularly his studies concerning the question of generation. Thus we have, for example, quite a bit of testimony of his interest in the question of the *marques d'envie* (marks of desire) in his letters and in his medical fragments.[78]

---

[72] Descartes to Huygens, May 25, 1638, AT I, 507, 15–6. The word '*abrégé*' implies a small treatise, which is not the case of the *Description*, where physiology is treated in great detail. The text mentioned to Huygens could be a preliminary summary of Cartesian medicine that was subsequently lost, of which we find a later example in the first part of the *Passions de l'âme*.

[73] Descartes and Plempius met at least once, in 1630, in Amsterdam. As the former worked on his *Dioptrique*, which would be published later with the *Discours*, the latter was finishing his *Ophthalmographia*, published in 1632. Descartes's discovery of the actual location of the images of objects in the eye, on the cornea, is the result of their mutual effort. See my article 'La vision chez Descartes et Plempius', in *Il Seicento e Descartes*, 233–54.

[74] *Description*, AT XI, 254, 15, 17. It is thus not possible to agree with Pierre Ménard, who dated the writing of the three first parts of the *Dioptrique* to 1636 (P. Mesnard, 'L'esprit de la physiologie cartésienne', in *Archives de philosophie* 13 [1937]: 191).

[75] *Description*, AT XI, 232, 23; 282, 20; 286, 6, 9, 12.

[76] Plempius to Descartes, January 1638, AT I, 498, 17–20; id., *Fundamentes physices*, 1638, 265.

[77] Descartes to Plempius, February 15, 1638, AT I, 528, 17–529, 1. La Forge followed this correction in his *Remarques sur le traité de l'homme* (Paris: Michel Bobin, Nicolas Le Gras, 1664), 187–8. But Jean Riolan also held to the same conclusion as Plempius and believed that Descartes was in fact close to Primerose on this subject (J. Riolan, *Opuscula anatomica*, c. IX [London: Milonis Flesher, 1649], 44).

[78] Descartes to Regius, May 24, 1640, AT III, 68, 21–30; id. to Mersenne, April 1, 1640, AT III, 49, 21–7; July 30, 1640, AT III, 120, 13–121, 9; id. to Meyssonnier, February 1, 1640, AT III, 20, 27–21, 2; Fragment [74], *Cogitationes*, AT XI, 538, 3–10. The problem of the sympathy

Thanks to a letter to the Princess Elisabeth, we are able to determine that work on the second edition of the *Description* began in 1648:

> I now have another work in my hands, which I hope may be more agreeable to your Highness: it is a description of the functions of animal and of man. For, in view of my muddled attempt of twelve or thirteen years ago, which was seen by your Highness, and which passed through many people's hands and was poorly transcribed, I felt obliged to clean it up, which is to say to do it over again.[79]

This dating is confirmed by another letter of the same period and by several medical fragments,[80] as well as by a reference to the *Principia*, which were published in 1644.[81]

We find in the Cartesian thesis of 1648 elements of the two first attempts. The nature of the seed is still purely material, and the particles of seed are first moved as a result of their own fermentation. Now, however, the first organ formed is the heart, which in turn forms the blood and moves it through its own 'fire'. Next, the circulation of the blood is mixed with the circulation of the seed and soon replaces it, bringing about the formation of the organs. In this respect, we see an echo of the first attempt of 1630. As in the second attempt of 1637, moreover, the formation of the fetus is broken down into three steps: (1) the action of the seed alone;[82] (2) the action of the blood and of the animal spirits;[83] (3) the fetus being nourished by the umbilical cord.[84]

Descartes finally came to adopt the Aristotelian thesis that the heart is the first formed organ,[85] but this agreement is of course fortuitous. The implicit reference to the first two stages of his research in the last edition of the *Description* proves that the elaboration of Descartes's thesis had in fact been made step by step, with systematic reference to the laws of physics, appeal to

---

between the mother and her child, which could have led to malformations, was common in the seventeenth century. Descartes tried to explain it only by the laws of motion, as we already showed (see Appendice 4, "La sympathie," in Descartes, *Ecrits physiologiques et médicaux*, 227–37).

[79] Descartes to Elizabeth, January 31, 1648, AT V, 112, 12–19.

[80] Descartes (to?), 1648 or 1649, AT V, 260, 29–261, 12; id., Fragments 73–76, *Cogitationes*, AT XI, 537, 10–538, 10.

[81] *Description*, AT XI, 248, 2 and 255, 5 and 16.

[82] Ibid., 252, 18–254, 26.

[83] Ibid., 256, 28–286, 14.

[84] Ibid., 273, 16–27. The third step is only briefly mentioned and was not developed, as the *Description* remained unfinished.

[85] Aristotle, *Historia animalis* VI, 3; *De la génération des animaux* II, 4. Cf. also Pliny *Historia naturalis* l. X, c. LIII. Cf. especially Descartes, *Description du corps humain*, AT XI, resp. 254, 8–9; 259, 18; 261, 5. Descartes sometimes seems to follow Solinus, *De rerum toto orbe*, c. IV (Basel: Michælem Insignum, Henricum Petri, 1538), 10–11.

the ancient authorities when this proved useful (as in the Galenic account of the three steps in the formation of the fetus), and the carrying out of experiments with the aim of coming to decisive conclusions beyond reliance upon ancient authority.

5. CONCLUSION

While for Aristotle the vegetative soul required the activity of the seed,[86] and while Galen had recourse to innate faculties in order to account for the formation of the organs, there was for Descartes only a mechanical process, devoid of soul or any other innate faculty or virtue. His study of generation throughout the three periods we have surveyed was based on a double certitude: that there simply are no occult forces such as Aristotelian souls or Galenic faculties that come into play and that on the contrary everything happens as a result of the properties of the material particles of the seed. Their size, their figure and above all their fermentation were the initial source of force, and this led to the formation of the organs.

In giving an account of the progressive formation of the fetus, we have seen that Descartes first imagined two sequences, based on his readings of Fernel, Fabricius, Arantius and Galen. But he remained unconvinced by his own results, and his realism forbade him to refer only to these authorities: his purpose was to arrive at certitude, not likelihood. This is why the experiments on eggs convinced him that the first organ formed was the heart, a thesis that implied a number of modifications of the first sequences which culminated in the final part of his *Description*. We find here another example of the Cartesian experimental method, which relies on experiments in two respects: for the observation of the things that need to be explained and for the confirmation of the thesis leading to the explanation.[87] This enables us to propose an outline of Descartes's science of medicine. The laws of motion, grounded in metaphysics, must explain the motion of the particles; the explanation of physical questions must refer only to these laws. Observation was to open the

---

[86] Aristotle *De anima* l. II, c. 4 (415 a 24–415 b 7).

[87] Elsewhere I have developed another example of this experimental method, in the integration of the discovery of the 'white veins' by Asellius. Before learning of the results of the works of Asellius, Descartes held to what was basically the Galenic physiology of digestion, which is clear in the *Traité de l'homme* (AT XI, 121, 28–122, 6; cf. La Forge, *Remarques sur le traité de l'homme*, 180–1). After his reading of Asellius (probably in 1641), he himself attempted the experiment that had led Asellius to his discovery, and he then inserted this result in his own explanation, as we see in the *Description* (AT XI, 267, 18–20). See my article 'Les médecins et la médecine', in *La biografia intelletuale di René Descartes attraverso la Correspondance* (Naples: Vivarium, 1999), 608–13.

way, and experiments were to confirm, as necessary, the theses proposed. But this ideal project was never achieved, ultimately, due to the complexity of the physiological questions posed.[88]

If we compare the achievement of the theory of generation with parts of his physics, we must admit that ultimately Descartes never arrived at any definite results. But if we consider the complexity of the questions he had to answer and the relative poverty of the technological aids to which he had access, this weakness is above all a sign of Descartes's realistic caution.

---

[88] The union of mind and body, which is beyond the scope of medicine but at least comes within the scope of pathology and therapeutics, in another way also prevented the success of Cartesian medicine, but it is not possible to discuss this matter here.

4

Imagination and the Problem of Heredity in Mechanist Embryology

JUSTIN E. H. SMITH

1. INTRODUCTION

It may seem improper to propose to speak of theories of heredity in seventeenth-century science. The transmission of traits through material units did not become a central topic of concern until the eighteenth century, in, for example, the work of Charles Bonnet and others impressed with Abraham Trembley's discovery in 1741 of the freshwater polyp's ability to regenerate itself from amputated bits – evidence that each bit of an organism carried in it some sort of plan for the organism as a whole. Indeed, in the early seventeenth century the idea of a program or a blueprint that specifies what a developing creature must become is precisely what the prevailing anti-Aristotelian spirit compelled researchers to reject in their accounts of animal development.

"Heredity," in the narrower sense that this term has today, refers only to those theories of the acquisition of traits by creatures that account for the mechanism by which *information-bearing material* causes a new creature to acquire the traits it does. But by "heredity" in a broader sense we may also understand any number of different kinds of transmission in sexual reproduction, all of which were clearly on the minds of early modern generation scientists. Generation theorists in the seventeenth century were, namely, intensely interested in determining the following:

1. How it is that parents generate offspring with traits similar to their own. Any creature is more than just a perfect blend of equal parts of father and mother. It also bears resemblance to more distant relatives and frequently appears to have traits altogether dissimilar to those of any known ancestors. All copy is to some extent bad copy. Without knowledge of the mechanism of dominant and recessive genes, this fact presented a number of problems.

2. How it is that members of a species regularly generate members of the same species. Not only do we inherit green eyes or attached earlobes, we also "inherit" our humanity, cows "inherit" their bovinity, etc.
3. How the phenomena described in (1) and (2) occasionally fail to occur: how it is that offspring frequently turn out to have traits unshared with either of their parents.
4. How it is that members of the same species are regularly differentiated by sex. The mechanism of sexual differentiation is perhaps the first and most general question early generation theorists faced in dealing with the problem of the variety within individual members of a species.
5. How it is that creatures sometimes appear to have features that are not even shared with other members of their species; that is, how "monsters" are generated.

In this short chapter we will investigate the way in which, in the interim between, at the one end, the banishment of Aristotelian immaterial forms from theories of fetal development and, at the other end, the rise of genetics, embryologists were constrained to understand how it is that offspring regularly resemble parents with respect to species membership, sexual differentiation, and particular variable traits such as hair and eye color. Without any conception of a blueprint by means of which the developing fetus could reproduce traits of the parents, early modern theorists were in effect forced to conceive of the development of particular features as resulting from forces acting upon the fetus in the course of its development rather than as having been set from the start simply by virtue of the character of the parental seed. What we see in early modern embryology is thus the remarkable inclination to account for *all* the traits of offspring, including those which offspring share with parents, in the same way that we account for congenital birth defects caused, for example, by smoking or thalydamide poisoning: that is, traits are conceived as emerging in the course of fetal development rather than being what we would describe as "genetic."

As we will see, one of the consequences of this radical development was that, in spite of the intense effort on the part of the mechanist physiologists to eradicate Aristotelian formative virtues from their account of sexual generation, in seeking to explain heredity in terms of congenital acquisition alone, something very much like the Aristotelian notion of a formative virtue persists under a new guise. In order to see how this could be, we will do best to begin with a brief survey of the relevant features of Aristotle's theory of sexual generation. Since this has already been covered in some detail in the Introduction and in the first chapter, our goal here will be only to draw attention to those aspects of Aristotle's generation theory that would come to bear on the seventeenth-century debate over how like begets like.

2. THE ARISTOTELIAN BACKGROUND

For Aristotle, the semen contributed by the father is the formal cause of generation. The father imposes both a formal and final cause, through the vehicle of the semen, upon the material of the mother's uterus. This brings about what Lennox calls "formal replication," in which the father ensures his own eternity in kind, as Aristotle puts it, if not in number.[1]

While the product of generation is not an exact copy of the father, it is the closest thing possible to this, and the extent to which it falls short of this mark is traceable in large part to the degrading influence of the mother. This view contrasts sharply with the other prominent ancient theory of generation still current in the early modern period, namely, that of Galen, which holds that in intercourse both parents eject their seminal fluids into the cavity of the uterus, where generation is initiated by the mingling of the two parents' seminal contributions.[2] In Aristotle, there can be no such equal partnership. The menstrual blood possesses *in potentia* all of the possible qualities of a human being that may be actualized once the blood is acted upon by the semen; but still the menses are not pure seed, "for it lacks one thing only, the source of the soul."[3] This soul source is at the same time the source of the blueprint or program that determines how the fetus will develop; the semen determines the end toward which the fetus will develop, since, in Aristotle's view, in natural beings such as animals the formal and final causes are one and the same.

For Aristotle, a woman is a degradation of man, "a male deformed." A female animal is generated when the movements introduced by the male's semen prove to be weak and are consequently overcome by the movements of the menstrual blood. Monstrosities occur, finally, when the movements of both of these are very weak and lack even the efficacy to render the menstrual blood into a distinct particular human, leaving it instead at the level of

---

[1] Aristotle, *On the Generation of Animals*, trans. A. L. Peck. (Cambridge, MA: Loeb Classical Library, 1963), 731b 35.

[2] See Galen, *Three Treatises on the Nature of Science*, ed. Michael Frede (Indianapolis: Hackett, 1985).

[3] *On the Generation of Animals* II 3 737a 27.

development common to all animals in general.[4] If no degradation were to occur, the male would presumably reproduce itself perfectly. This can never happen, since particular primary substances are not eternal; perishable substances approximate eternity by reproducing their kind without in so doing reproducing *themselves*:

> [T]hat which comes into being is eternal in the way that is possible for it. Now it is not possible in number (for the being of existing things is in the particular, and if this were such it would be an eternal) but it is possible in form. That is why there is always a kind – of men and of animals and of plants.[5]

The mechanism by which such perfect self-reproduction is hindered is the mutual interaction of the movements of the semen and the menstrual blood at conception. Thus there is no need to explain where defective fetuses come from. They are merely privations of form, which manifest themselves as recapitulations, not of earlier evolutionary stages, as some modern theories would later hold, but of a more general category – animal – to which particular members of any animal species belong.

3. IMAGINATION AS A FORMATIVE FACULTY IN PREMECHANIST EMBRYOLOGY

In the *Politics*,[6] Aristotle recommends to pregnant women that they keep their bodies at rest, since, as he explains, developing fetuses are influenced by their mothers in just the same way that plants are influenced by the soil in which they are growing. Galen and, later, Ibn al-Jazzar (in reference to Galen) both echo Aristotle's observation in the *Politics* with the view that, as the latter of these writes, "the connection between the foetus and the uterus is like that between the fruit and the tree." In the beginning and end of pregnancy, the

---

[4] *On the Generation of Animals* IV 3. For an interesting discussion of inheritance in Aristotle, see D. M. Balme, "Ανθρωπος ἄνθρωπον γὲννα: Human is Generated by Human," in *The Human Embryo: Aristotle and the Arabic and European Traditions*, ed. G. R. Dunstan (Exeter: University of Exeter Press, 1990), 20–31.

[5] *On the Generation of Animals* II 1 731b 33–6. Such an account of the reason for sexual reproduction continues to have some currency in the early modern period. Thus in 1668 Mauriceau writes, "C'est une verité tres-grande, & reconnuë d'un chacun de nous, que tout ce qui est en ce bas monde, est sujet à la corruption, & enfin contraint de souffrir la mort, c'est ce qui a obligé la nature providente & soigneuse de sa conservation, de donner à toutes choses un certain desir de s'éterniser, ce que ne pouvant faire en l'individu, dautant qu'il est mortel, par une necessité indispensable, elle le fait par la propagation des formes & des especes" (*Des maladies des femmes grosses et accouchées* [Paris, 1668], 69).

[6] Aristotle, *Politics* Vii 16, 1335b, 16–19.

connection between the two is weak, but throughout the middle it is very strong. "Therefore, a woman who is in the beginning of her pregnancy should not be confronted with the mentioning of different kinds of food not available at that time, for fear that her soul might desire and crave them."[7] In subsequent medieval Arabic Aristotle commentaries, we see an increasing literalism with respect to the womb-soil analogy and a parallel emphasis on the power of the maternal imagination to influence fetal development. As late as the end of the sixteenth century, the concept of formative faculties was perceived in the Christian West as having a distinctly Islamic pedigree. André du Laurens, for example, in his *Histoire anatomique*, writes,

> [Monsters] can arise in a variety of ways from the agent. The principal agent is the formative virtue, or the imagination. It will suffice to remark here, following the doctrine of the Arabs, that a powerful imagination can produce elms no more and no less than the superior intelligences produces the forms of metals, plants, and animals.[8]

It is telling here that du Laurens identifies what might appear to be the very distinct concepts of formative faculty on the one hand and imagination on the other. In both the Arabic and Latin traditions stemming from Aristotle's natural philosophy, there is a common presumption that the formative power at work in nature in general is but a different manifestation of the formative faculty traditionally held to govern biological growth and development and that each involves a sort of imposition of an image or form upon malleable matter. This is confirmed in the standard Scholastic account of the origins of fossils. The Dominican Antoine Goudin, in his *Philosophy, Following the Principles of Saint Thomas*, argues that there are both efficient and final causes at work in the earth's production of rocks that resemble animals or parts of animals. Their efficient cause is a sort of cooking brought about by exhaltations from the depth of the earth that makes the strata where fossils are found into a furnace of sorts. Their final cause, in turn, is

> a certain force earth itself possesses variously, following the different places in which the mixed body is formed. This force, similar to the maternal bosom from which animals arise, assuredly plays a great role in the formation of these bodies; this is why, according to Aristotle and Saint Thomas, earth and water furnish to everything arising from the bowels of the earth their matter and

---

[7] Ibn al-Jazzar, *Ibn al-Jazzar on Sexual Diseases and Their Treatment: A Critical Edition of* Zad al-musafir wa-qut al hadir, ed. and trans. Gerrit Bos (London: Kegan Paul International, 1997), 285 ff.

[8] Ambroise Paré, *On Monsters and Marvels*, trans. Janis L. Pallister (Chicago: University of Chicago Press, 1982), 38.

bosom, as would a mother, while heaven and the stars fulfill the office of the father, who imparts the form.[9]

A "male" formative principle exercises its influence over the "maternal" matter of the earth and thereby gives rise to forms in earth that resemble living beings. For Goudin, there is real symmetry between the way in which the celestial influences the terrestrial on the one hand and the way in which the formal principle in generation influences the material principle on the other. This parallelism is also a commonplace in the Platonic tradition, extending from at least as early as Marsilio Ficino in the fifteenth century to figures such as Henry More in the seventeenth century. Ficino asks rhetorically, "From the beginning of any thing that is to be generated, do not celestial influences bestow wonderful gifts in the concoction of the matter and its final coming together?" Perhaps the limit case of the formative power of the stars is spontaneous generation, in which the rays of the heavens concoct suitably disposed matter into complex organisms – presumably entities higher on the scale of being than the zooid but ultimately merely mimetic fossils that interested Goudin. Thus Ficino goes on to ask, "Do not innumerable frogs and similar animals often, when the face of the heavens favors it, leap forth out of the sand in a moment? Such is the power of the heavens in well-disposed material." Ficino adduces a number of other earthly phenomena involving the influence or power of vision and imagination in order to make the case that, *a fortiori*, celestial rays have the power to influence the form of earthly things.

> I pass over fascinations achieved by a sudden glance and very passionate loves instantly kindled by rays from the eyes. . . . Nor will I mention how quickly an inflamed eye afflicts whoever looks at it and how a menstruous woman affects a mirror by looking in it. . . . What can a mad dog accomplish even without an apparent bite? . . . In the light of all this, are you going to deny that the celestials with the rays of their eyes with which they both look at us and touch us, achieve wonders in an instant? But now a pregnant woman instantly by touch stamps a bodily part of the person who is about to be born with a mark of something she desires. Are you then going to doubt nevertheless that rays touching in this way or that accomplish diverse things?[10]

---

[9] A. Goudin, *Philosophie suivant les principes de Saint Thomas*, trans. T. Bourard (Paris, 1864 [original ed. Paris, 1668]), 301: *Des corps mixtes inanimés, dit fossils.* Thanks to Roger Ariew for bringing this work to my attention. In his article "Leibniz on the Unicorn and Various Other Curiosities," *Early Science and Medicine* 4 (1998): 267–88, Ariew cites this same passage with the aim of comparing Goudin's view with Leibniz's theory of fossilization.

[10] Marisilio Ficino, *Three Books on Life*, edited, translated, and with an introduction and notes by Carol V. Kaske and John R. Clark (Binghamton: Medieval and Renaissance Texts and Studies, 1989), 323–5.

In humoral medicine too, vision and the formative power of the stars are seen as different manifestations of the same force: both involve the emission of rays. Ficino maintains that "the immense size, power, and motion of celestial things brings it about that all the rays of all the stars penetrate in a moment the mass of the earth." These rays penetrate to the center of the earth. "By the rays' intensity, the material of the earth there – being dry and far from any moisture – is immediately kindled and once kindled, is vaporized and dispersed through channels in all directions and blows out both flame and sulfure." He describes this fire as "very dark" and "like a flame without light," using the same vivid phrase that Descartes would later echo in describing the "fire" that burns in the heart at the moment the fetus is quickened. In short, in medieval and Renaissance embryology, the organization of the matter of the uterus in embryogenesis was seen as a whole. Each process is governed by natural "intelligence"; in the case of embryogenesis, this natural intelligence could be dirempted through the interference of the mother's passions, which are frequently triggered by the sense of sight.

In premechanist embryology, then, there is an independent formative power with which the mother's imagination might interfere. In the sixteenth-century Coimbran Aristotle commentaries, for example, the imagination is held to "occasionally [make] the formative faculty wander from its target, and imprints upon the fetus absurd or alien figures."[11] In mechanist embryology, as we shall see, without any notion of an immaterial force working upon the matter contained in the uterus, or of a teleology toward which this form may conspire with the matter to move, the only formative power left to appeal to would be the imagination. In other words, up to the sixteenth century, the imagination theory was convenient, primarily in the explanation of aberrations; in the seventeenth century, deprived of formal and final causes in the account of organic growth, it was necessary in the explanation of the regular course of sexual reproduction.

4. DESCARTES: FETAL DEVELOPMENT AS THE ACHILLES' HEEL OF MECHANISM

Reports of the influence of the imagination in embryogenesis abound in the *Philosophical Transactions* of the Royal Society. In a letter of March 1687, the Lord Bishop of Cloyne writes of "a girl called Elizabeth Dooly . . . whose mother being with Child of her, was frighted by a Cow as she milked it, and

---

[11] Coimbra, *In phys.* Cited in Dennis Des Chene, *Physiologia: Natural Philosophy in Late Aristotelian and Cartesian Thought* (Ithaca: Cornell University Press, 1996), 206 ff.

hit with the Teat on the left Temple ... in which very Place the Girl has a piece of Flesh growing exactly like the Cows Teat in Bigness, Shape, &c."[12] Most such reports are alike in their reliance on the eyewitness of presumably trustworthy contributors generally writing in at some distance from the journal's place of publication. Monsters always show themselves conveniently elsewhere.

Most accounts tell of the presence on the newborn child of some mark resembling a part of some nonhuman living being, while in some cases the whole child resembles a whole animal. In some extreme cases, there is no child to be spoken of as resembling something else: in the case of changelings – which Locke famously adduced as a central argument against the reality of species and in favor of his nominalism – what is born is an actual member of some other species, usually an animal, though perhaps also a plant. Each such case provides an occasion for reflection on philosophical issues concerning the relation of mind and body and the problem of species membership; on theological issues, such as whether a changeling born of a human is suitable for baptism; and on legal issues, such as whether a monster can inherit property. As an example of the first of these, in the *Philosophical Transactions* of 1686, a woman in Paris is featured who gave birth to an ape after seeing one at a carnival, as a result of which "[m]any questions were ... agitated: viz. about the Power of Imagination; and whether this Creature was endow'd with a humane Soul; and if not, what became of the Soul of the Embryo, that was five months old."[13]

What has been overlooked in the scholarly discussion of the early modern belief in the power of the mother's imagination to influence the developing fetus is that often this power not only was believed to play a role in the explanation of awkward cases of hypertrichosis or unsightly birthmarks but indeed was held by some to be a central factor in the normal or undefective reproduction of animal species.[14] And it was not just disreputable country doctors who thought as much, either.

---

[12] "Part of a Letter from Dr. Ashe, the Lord Bishop of Cloyne, dated March the 26th, 1687, concerning the Effects of Imagination the Vertues of Mackenboy, &c." *Philosophical Transactions* 20 (1698): 293 ff.

[13] "Extract of a Letter, Written from Paris, Containing an Account of Some Effects of the Transfusion of Bloud, And of Two Monstrous Births, &c." *Philosophical Transactions* 2 (1666–7): 479 ff.

[14] I mean to use the term "normal" in the most inoffensive way possible, to refer to those products of reproduction that would be considered in the history of embryology to represent their species in the same way their parents do. I recognize, however, that this is a notoriously problematic concept. For an extended meditation on just how problematic it is, see Georges Canguilhem, *Le normal et le pathologique* (Paris: Presses Universitaires de France, 2003).

We shall take mechanism as the view that all natural phenomena may be explained by appeal to matter, figure, and motion, without requiring recourse to final causes or ends. A mechanist physiology, for Descartes, would be one that, as he puts it, gives such a full account of the entire bodily machine that we will have no more reason to think that it is our soul which produces in it the movements which we know by experience are not controlled by our will than we have reason to think that there is a soul in a clock which makes it tell the time.[15]

It was in the domain of fetal development, among all natural phenomena, that mechanism showed itself most inadequate to the task it set itself: while mechanists could be reasonably confident that inorganic natural phenomena might soon be explained in terms of matter in motion alone, such limited means seemed utterly insufficient when attention was turned to the evidently infinitely complex phenomenon of generation, and Descartes's account of this phenomenon comes out looking like a fantastic exercise in wishful thinking.

As Dennis Des Chene has convincingly argued, Descartes sought in his discussion of animal bodies to radically reinvent the seemingly harmless language of functions.[16] As he writes, "The unswerving aim of [Descartes's] physiology is to show how the body is made – the structure and the processes – without ever mentioning what it is *for*." Des Chene notes that, for Descartes, "even the weakest hypothesis about mechanical causes is preferable to the ascription of ends."[17] For, on a strict mechanist understanding, "bodies can have no functional unity but only physical and dispositional unity."[18]

Descartes writes of the way the nerves "serve to move the exterior members" or "serve to turn the back and move the legs,"[19] always skirting any suggestion that the nerves may exist *for* something or other. In effect, he is striving to account for the formation of the fetus from the purely nonteleological mixing of the two parents' materials and the purely mechanical process this mixing sets in motion. While one might plausibly explain how blood

---

[15] AT XI, 226.

[16] Other scholars have argued that in his physiology Descartes is unable to do away entirely with the language of functions and ends and indeed that with respect to human physiology Descartes may not have wanted to abandon the idea that the body, considered in union with the mind, in fact does have an end: the preservation of functional integrity. See, in particular, Alison Simmons, "Sensible Ends: Latent Teleology in Descartes's Account of Sensation," *Journal of the History of Philosophy* 39 (2001): 49–75.

[17] Des Chene, *Spirits and Clocks: Machine and Organism in Descartes* (Ithaca: Cornell University Press. 2001), 45.

[18] Ibid., 121.

[19] Ibid.

congeals into tissue and organs in this way, it seems a much more difficult task to explain how this tissue and these organs eventually come to be the tissue and organs of a particular kind of entity, for example, of a human being rather than a pig. It is precisely this complaint that Malebranche will level against Descartes's embryology.

For roughly eighteen years, from 1630 to 1648, Descartes appears to have been frustrated by the evident intractability of animal generation and fetal development within the constraints of mechanism. In the *Discours de la méthode*, he acknowledges that he is simply too ignorant to explain these "in the same style as the rest, namely, by demonstrating effects from causes, and showing from what sort of seeds, and in what manner, nature must produce them."[20] Yet eventually Descartes nonetheless manages to produce something of a treatise on the formation of animals, even if incomplete, as a component of his *Description du corps humain*, first published posthumously in 1664. In this work, Descartes begins by describing the initial action of the two parental seeds upon one another, how they, as a result of their heat, "serve as a leaven to each other," which ultimately causes some particles "to gather toward some part of the space that contains them; and, expanding there, they press on the others that surround them, which begins to form the heart."[21]

In the *Description* Descartes has remarkably little to say about the subsequent acquisition of particular traits, save for a rather elaborate account of sexual differentiation, according to which every fetus sends forth its sexual organ in the direction of the mother's navel, and so if the child is facing toward her back, the genital organ will thrust inward and so result in a girl, while if the child is facing forward, it will thrust out and yield a boy.[22]

But what about more specific traits, such as the exact pigment of the skin, the modality of the earlobes, attached or unattached, or the color of the eyes? What about uncanny resemblances to great-great-aunts? Here, Descartes's account of the development of fetuses is at its most inadequate. There is, it would seem, only a hint of a solution, and it is to be found in scattered

---

[20] Ibid., 32.

[21] *Description* §28, AT II, 254. Des Chene describes this entire process in wonderful detail in *Spirits and Clocks*. I am only interested in communicating a basic sketch of Descartes's concerns and limitations in his account of fetal development, before moving on to what interests us here: the imagination theory and its use in mechanist embryology.

[22] The view that both sexes are equipped with a penis of sorts but that it only beomes visible in the male could rightly be seen as a continuation of the old Aristotelian perception of women as "men deformed." For a fascinating study of the history of the "one-sex" model of gender, see Thomas Laqueur, *Making Sex: Gender and the Body from the Greeks to Freud* (Cambridge, MA: Harvard University Press, 1992).

Justin E. H. Smith

comments, later more fully developed by other mechanist philosophers, concerning the formative power of the mother's imagination.

In a letter of January 29, 1640, Descartes writes to Meyssonier (whom he has accused of being misled by astrology, chiromancy, and other stupidities in a letter of the same day to Mersenne)[23] that he is suspicious of the fable according to which the urine of a man who has been bitten by a rabid dog contains "effigies of little dogs." But, he continues, if it actually does happen, this phenomenon should be no less difficult to explain than the causes of "those marks that children receive as a result of the desires [*envies*] of their mothers."[24] A few months later, he writes to Mersenne, who had worried that such marks might be damaging to Descartes's account of mind-body interaction, that there is nothing at all so strange about *marques d'envie*, and he now maintains that there is nothing impossible about the claim concerning the effects of a rabid dog bite.[25]

In his published works, there are three noteworthy references to the *marques d'envie* that interest Mersenne. One is in *La dioptrique*: "I could even demonstrate to you, moreover, how sometimes [the image] can pass from [the pineal gland] through the arteries of a pregnant woman into certain parts of the child that she carries in her womb, and how it forms its markings there, which cause such astonishment among the learned."[26] And again in *L'homme*, in almost exactly these words: "Here I could add something about how the traces of these ideas pass through the arteries to the heart, and thus radiate through all the blood; and about how certain actions of a mother may sometimes even cause such traces to be imprinted on the limbs of the child being formed in her womb."[27]

---

[23] This distinction between the legitimacy of some aspects of physiognomy and the fraudulence of chiromancy, astrology, and other disciplines seems widespread throughout the seventeenth century. Thorndike mentions a 1683 miscellaneous volume by Wilhelm ten Ryne that "upheld physiognomy but decried astrology and chiromancy and their association with physiognomy" (*History of Magic and Experimental Science*, 7:472). He also mentions François Bayle's *Dissertationes physicae*, in which the author "approaches the subject of physiognomy through the foetus. . . . By certain reactions certain parts of the body are nourished and strengthened more than others. The foetus is easily affected by the nervous juice. Other juices, too, excite reactions in the foetus, to which the mores and mode of life of the mother greatly contribute" (471). On this understanding, the mother is capable of transmitting moral as well as physiological traits to the fetus, and even this capacity is distinguished by its defenders from occult connections, such as that between the lines of the hand and a person's future.
[24] AT III, 21.
[25] AT III, 49.
[26] *La dioptrique* (Paris, 1637), Sixth Discourse.
[27] René Descartes, *Le monde, L'homme*, ed. Annie Bitbol-Hespériès and Jean-Pierre Verdet (Paris: Seuil, 1996), 152 ff.

The third and final allusion to the imagination theory, and certainly the most unequivocal of them, is found in the *Primae cogitationes circa generationem animalium*. Descartes has just finished explaining that the movement of the fetus's heart brings about sympathies between various parts of the fetus's body that assure its regular and uniform development. Thus whatever it sends to one leg, it will send to the other; if it sends material for the development of the head, it will send analogous material to the genitals, since these are corresponding body parts. The movement of the fetus's heart, he explains, is not governed by that of the mother; the fetus regulates the development of its own internal organs. However, it is the mother who determines, by way of the umbilical arteries, the form of all exterior parts. "For this reason," Descartes concludes, "if the imagination of the mother is harmed by some impression, the child's body parts will be monstrous." The mother can have no influence on the form or number of internal organs; the most her imagination can do is to sear a likeness to herself and other human beings, or, in the case of monstrosities, to sear a likeness to some other species, onto the visible surface of the developing fetus. Monstrosity, then, is only skin deep.[28]

This influence of the maternal imagination upon the development of external parts, in Descartes's view, is by no means a retreat into extramechanical causes to explain what would otherwise remain inexplicable. The imagination is not to be confused with some governing intelligence; it is a bodily process like any other, capable of being explained in terms of the laws that bind all of mechanical nature. Monsters are a product of the maternal imagination, a disruption of the normal development of the fetus, which, as we have seen, is governed primarily by a mechanical process kept in motion by the fetus's own heart. Yet the maternal imagination does not disrupt this normal process but is itself part of it; as Descartes puts it in the *Primae cogitationes*, the mother is the *"formatrix omnium membrorum exteriorum,"* regularly communicating images to the fetus through the umbilical arteries that serve to shape and imprint its visible body. This, however, is not, at least from the point of view of a mechanist like Descartes, the intervention of the mental in an otherwise physical process, for images, too, are entirely corporeal things. Thus Descartes hopes to be able to account for physiological capacities in the animal machine such as sensation and imagination (e.g., in the *Regulae*, AT X 412–17) without sacrificing its status as a mere machine.

---

[28] *Primae cogitationes circa generationem animalium* (Amsterdam, 1701), 11.

5. THE IMAGINATION IN CARTESIAN PHYSIOLOGY
AND PSYCHOLOGY

It is worth considering briefly here how well Descartes's evident adherence to the imagination theory fits within his broader theory of imagination. Does he, that is, really think that the imagination can be understood as producing "traces of ideas"? As is well known, in Meditation VI Descartes describes the faculty of imagination in some detail, with the aim of distinguishing it from what he calls "pure intellect":

> I remark, in the first place, the difference that subsists between imagination and pure intellection. For example, when I imagine a triangle I not only conceive [*intelligo*] that it is a figure comprehended by three lines, but at the same time also I look upon these three lines as present by the power and internal application of my mind, and this is what I call imagining. But if I desire to think of a chiliagon, I indeed rightly conceive that it is a figure composed of a thousand sides... but I cannot imagine the thousand sides of a chiliagon as I do the three sides of a triangle, nor, so to speak, view them as present.[29]

The imagination is the faculty that, as the etymology of the word suggests, is responsible for the rendering of images to the mind.

But are these images physical? Descartes does try to distance himself from the Scholastic view according to which "intentional forms" or some sort of picture of an object is transmitted to the perceiver and etched into the brain. Yet, as many scholars have noted, Descartes's adoption of the view that the pineal gland functions as a sort of transmitting station where images are converted into signals that will lead to the appropriate motor responses, such as the mechanical sheep's darting off in response to an incoming image of a wolf, made it impossible for him to fully abandon the view that the corporeal imagination is a mental faculty dealing in real images of external objects.[30] Indeed, his experimental work, reported in book V of the *Optics*, confirms this assessment. He explains there that, under the proper conditions, looking at "the white body [at the back of the ox's eye] you will see there, not perhaps

---

[29] *The Philosophical Writings of Descartes*, trans. and ed. John Cottingham, Robert Stoothoff, and Dugald Murdoch (New York: Cambridge University Press, 1992–3), vol. 20.

[30] See, e.g., John Cottingham, "Cartesian Dualism: Theology, Metaphysics, and Science," in *The Cambridge Companion to Descartes* (New York: Cambridge University Press, 1992), 236–57; and Gary Hatfield, "Descartes's Physiology and Its Relation to Psychology," in *The Cambridge Companion to Descartes*, 335–70. For thorough treatments of the role of imagination in Descartes's thought, see Dennis L. Sepper, *Descartes's Imagination: Proportion, Images, and the Activity of Thinking* (Berkeley: University of California Press, 1996); and Dominik Perler, *Repräsentation bei Descartes* (Klostermann, 1996).

without wonder and pleasure, a picture representing in natural perspective all the objects outside." In living creatures, whose eyes remain connected to the brain, "[t]he images of objects are not only formed in this way at the back of the eye but also pass beyond into the brain."[31] Recall, here, that the imagination in the animal machine serves to bring about appropriate motor responses to incoming images. Thus, the image of the wolf serves to trigger the flow of the animal spirits from the pineal gland, out through the nervous fibers, pushed along by the pumping of the heart, to the muscles, which brings about the inflations and contractions that will hopefully enable the sheep to evade its natural enemy.

If the animal spirits are like the air in the pipes of a church organ, the heart and arteries, Descartes says, are like its bellows,[32] pushing the spirits through and bringing about the appropriate bodily effects, which, evidently, are analogous to the organ's sounds. This account of the locomotion of the animal machine is fairly well known. What scholars have failed to note is that Descartes himself opens up the possibility of extending this account of the physiological effects of imagination – to speak with the Aristotelians – from locomotion to generation. In the passage from *L'homme* cited earlier, Descartes describes maternal influence on fetal development as the consequence of the radiation or an image from the heart out into the arteries and ultimately to the fetus; the difference between this example and that of the wolf and sheep is only the kind of effect brought about, not the physiological process that leads to the effect.

But why, one may easily ask, is it only the developing fetus that takes on the forms of images that enter the body rather than, say, any other internal organ of the mother, or, for that matter, of any animal, pregnant or not, male or female? The answer to this question would seem to have to do with the peculiar constitution of the fetus, with its softness and malleability. Thus in *Philosophia naturalis*, Pierre-Sylvain Régis offers a very detailed, and explicitly Cartesian, account of the way in which the mother's imagination influences the fetus while in a tender state:

> I am of the opinion that it happens in this way: the image of the thing seen, or otherwise thought, since it is truly depicted in the brain, is communicated by the power of the animal spirits present in the ventricles of the brain, to the conarium or the *sensorium commune*, and from here, through the mediation of the arterial blood, further on to the uterus, and at last through the umbilical vein to the fetus itself. And this happens in a manner in no way dissimilar to

---

[31] *The Philosophical Writings of Descartes*, 1:166.
[32] AT XI, 165–6.

that whereby images of visible things, by the power of intermediate ethereal globules, are depicted upon [a] dark wall . . . , or upon the retina of the eye; or whereby sounds are transported by the air to the most remote distances. The tender fetus, moreover, on account of its softness easily takes upon itself the image strongly impressed upon it from the imagination of the mother.[33]

This account relies, significantly, on the central principle of Descartes's embryology, that the development of the fetus proceeds through the steady hardening of its parts. Some years later, in the *Cours entier*, which was to serve as a textbook summary of Cartesianism, Régis describes this sort of influence as nearly universal: "[Since] there are scarcely any women who are not moved by some violent passion during their pregnancy, there can be few children that do not inherit from their mothers the beginnings of the same passion."[34] The malleability of the fetus suggests that, in a sense, it and the pineal gland – at least in humans – are analogous body parts (if the notion of "part" may be extended so far). Each straddles the boundary between the physiological and the psychological: the pineal gland in virtue of its unique capability to transduce psychological signals into somatic events; the fetus in virtue of its singular capacity to permanently take on the physical form of these signals.

Des Chene has characterized Descartes's highly speculative physiology as, in a certain sense, conservative: he seeks to explain as many traditionally accepted phenomena as possible by appeal to as few causal powers as possible; he does not, to the extent possible, have any interest in discerning new, previously undetected phenomena. Descartes "does not offend common opinion where it concerns the phenomena, and when he has no systematic reason for doing so."[35] It is fair to say that Des Chene's characterization of Descartes's embryology holds particularly true with regard to his appropriation of what we have now seen is a rather widely received idea about the fetus's development of individual traits. It is important to point out, moreover, that in retaining the theory of the maternal imagination's influence, Descartes is not forced, by his own lights, to retain any undesirable causal powers. Conservatism with respect to phenomena is not opposed to parsimony with respect to powers. The causal power of the mother's imagination is from his point of view physiological, not supernatural, just like the power to convert

[33] Pierre-Sylvain Régis, *Philosophia naturalis* (Amsterdam: Apud Ludovicum Elzevirium, 1654), 300.
[34] Pierre-Sylvain Régis, *Cours entier de philosophie ou Système général selon les principes de Descartes* (Lyon, 1691 [repr. New York and London: Johnson Reprint Corporation, 1970]), 3:314 ff.
[35] Des Chene, *Spirits and Clocks*, 50.

images into motor signals. While there may be a lingering problem about how exactly images are converted into physical signals via the pineal gland, this is to be resolved by physiological research, not by speculation on mind-body interaction. For the sort of images that are converted into motor signals are precisely the raw material of the *corporeal* imagination, the kind that even the soulless *bête-machine* is capable of having. The image that is seared on the fetus's body resides from its very beginnings in the eye and the brain of the mother. Of course, in human beings imaginings can also be present "in the internal application of my mind," and so it may be correct to say that in this one species it is mind that in part regulates the development of the fetus.

6. GASSENDI AND MALEBRANCHE ON SPECIES REPRODUCTION

Within a few years of Descartes's expressed interest in the imagination theory, Gassendi, an anti-Cartesian mechanist discussed in detail in chapter 5 of this volume, would extend the power of the imagination beyond its narrow teratological use to the explanation of any resemblance to parents. In the *Syntagma philosophiae Epicuri* of 1649, he observes that

> it seems that commonly the force of the imagination must be applied. . . . For the form of the image of an external object which has been impressed on the brain by the intervention of external sense and has moved the imaginative faculty residing in it seems so to set in motion the appetite and the spirits serving it that the spirits themselves also retain a trace of the impression that has been made and transport it with them through the body.

Like Aquinas[36] and some of the Muslim philosophers, Gassendi returns to a version of the imagination theory according to which both parents are capable of exercising this power in the moment of conception rather than the mother exercising it alone in the early stages of gestation. "It is . . . possible," he writes,

> for either a male or a female fetus to resemble the father if the imagination of the mother directed toward the father was more vehement and powerful than the imagination of the father; or conversely to resemble the mother if the imagination of the father directed toward the mother was more powerful than the imagination of the mother; or to resemble both parents promiscuously and confusedly if the imagination of both or of one directed toward the other was

---

[36] Aquinas notes in the fourth of his *Quaestiones de duodecim quodlibet* that married couples *in congressu* should be careful what they think about during the act, lest their imaginings inadvertently influence for the worse the baby they are in the process of creating.

of about equal force; or to resemble neither parent at all if the imagination of both was distracted elsewhere and in the male did not have the female herself as its object and in the female did not have the male himself as object.[37]

Malebranche for his part devotes an entire chapter of his *De la recherche de la vérité* to the question as to how like begets like.[38] Significantly, the title of the third section of this chapter is "Explanation of the generation of monstrous children and of the propagation of species." Like's begetting like and its failure to beget like, are now to be explained in the same way. As Malebranche explains:

> It is true that this communication between the brain of the mother and that of her child sometimes has bad results when the mother allows herself to be overwhelmed by some violent passion. Nevertheless, it seems to me that without this communication, women and animals could not easily bring forth young of the same species. For although one can give some explanation of the formation of the fetus in general, as Descartes has tried successfully enough, nevertheless it is very difficult, without this communication of the mother's brain with the child's, to explain why a mare does not give birth to a calf, or a chicken lay an egg containing a partridge or some bird of a new species.[39]

To be sure, Malebranche begins his discussion of the subject with the usual sensationalist anecdotes about defects; his chosen example is of a child born with bones broken in exactly the same places where a publicly executed criminal – whose execution the pregnant woman had the misfortune to witness – had had his own bones broken. He also tells us that women with excessive cravings give birth not only "to deformed infants but also fruits they have wanted to eat, such as apples, pears, grapes, and other similar things."[40] These things happen, he explains, because

> [i]nfants in their mothers' womb, whose bodies are not yet fully formed and who are, by themselves, in the most extreme state of weakness and need that can be conceived, must also be united with their mother in the closest imaginable way. And although their soul be separated from their mother's, their body is not at all detached from hers, and we should therefore conclude that they have the same sensations and passions.[41]

---

[37] Pierre Gassendi, *Opera omnia* (Lyon, 1698 [repr. Stuttgart-Bad Cannstatt: F. Frommann, 1964]), 2:284b–285a.

[38] Malebranche, *De la recherche de la vérité*, bk. II, pt. I., chap. 7.

[39] Nicolas Malebranche, *De la recherche de la vérité*, in *Oeuvres complètes de Malebranche*, ed. Geneviève Rodis-Lewis (Paris: Vrin, 1962), 1:243.

[40] Malebranche, *OCM* I, 241 f.

[41] Ibid., I, 234.

The "flow of spirits," Malebranche explains, "excited by the image of the desired fruit, expanding rapidly in a tiny body, is capable of changing its shape because of its softness."[42] The infants become like the things for which their mother has communicated to them a desire. But the mother, since her body is not soft enough to take on the figure of the things she imagines, remains unchanged.

7. TELEOLOGY, MECHANISM, AND THE SURVIVAL OF FUNCTIONS

Antoine Le Grand, to cite one more post-Cartesian mechanist, also extends the physiological effects of imagination beyond the account of trait acquisition to the regular course of species reproduction. In his *Entire Body of Philosophy According to the Principles of Renate Des Cartes*, translated into English by Richard Blome with assistance from the author himself (who had been based in England for a number of years), Le Grand relates some shocking hearsay tales of inherited abnormalities, such as that of "a certain Woman in France, being stabb'd with a Dagger," from whom

> an Embrio was taken out of her Body, mark'd with as many livid spots as she had received stabs, and in the very same parts of the Body. The Reason is, because the Embrio, carried in the Mothers Womb, makes up but one Body with the Mother; and as it is nourish'd with the same sort of Aliments, so it is vegetated with the same spirits according to all its parts. Wherefore no wonder if so tender a Body as that of the Embrio is subject to the same accidents as the Body of the Mother is subject; and any Mark or Brand whatsoever is the more easily imprest upon it, in regard the Animal Spirits, which convey the received Image, are directed by the Mother, whilst she touches any part of the Childs Body, and as it were marks it out.[43]

What Le Grand focuses on are acquired defects, but in the process of explaining how these come about, he reveals a picture of the relationship between the mother's body and the fetus little different from Aristotle's observation in the *Politics*. Significantly, Le Grand extends the power of the imagination beyond trait acquisition to the explanation of other physiological processes that today are understood to be triggered by hormonal changes.

---

[42] Ibid., I, 242.
[43] Antoine Le Grand, *An Entire Body of Philosophy according to the Principles of Renate Des Cartes* (London: Roycroft, 1694), 185.

He explains that a few days after giving birth women

> begin to think of Nutriment for the Child, and desire to quiet it from crying. From which affection the passages being loosned by a determinate influx of Animal Spirits, which before were carried to the Womb, the Chylie Juice is then converted to the Breasts. And this their strong intent and frequent rumination about their Milk, and the suckling of their Young Ones, may possibly cause the Chyle to be the better conveyed to the Breasts.

As anecdotal corroboration of this theory of lactation, Le Grand has occasion to attribute a *vis imaginativa* to men as well. "Santorellus relates of a Man," Le Grand tells us,

> who after the death of his Wife, not being able to hire a Nurse, one time above the rest, to still the Child, when it cryed, took it to his Breast, and gave it the Dug, which doubtless with great desire of satisfying its Appetite with Milk: and by this iterated application, together with an earnest intention of Mind, and the Childs sucking of the Teat from time to time, the Chyliferous passages were opened, and the Paps afforded plenty enough of Milk for the Childs nourishment.[44]

It is remarkable here to note how much of the work imagination is now doing in the explanation of biological phenomena that was once done by a teleology intrinsic to the new corporeal substance in the process of becoming. In mechanist embryology we see an old theory in a new guise: the power of the parent's mind has moved in to guide growth and development where otherwise the only causality to be discerned resides in the mechanics of heat and blood. Of course, as sketched out above, in animals the imagination is meant to be an entirely corporeal faculty. Yet it seems that, likely unwittingly, it is "the intention of Mind" of which humans are capable that is generally taken as the paradigm example of the causal power of the imagination in the triggering of somatic processes such as lactation or fetal growth. The fact that, strictly speaking, animals do not have minds may mean that we must limit to human beings the claim that the individual mother's mind moves in to take care of those aspects of fetal development once taken care of by the fetus's own immaterial formative faculty. But we may also wonder how present to mind the merely corporeal nature of animal imagination was for mechanist physiologists seeking – clearly very problematically – to account for the role of the imagination in fetal development.

[44] Ibid.

In Aristotle, the form or idea of a mammal includes lactation, and so the organism of a female mammal will arrange itself in such a way as to produce milk, thereby bringing to actuality the *function* of lactation, which on the Aristotelian understanding is ontologically more fundamental than the material organs that carry the lactation out. What is making things work, in both Aristotle and in mechanist physiology (bracketing the corporeality of animal imagination), is a mental model *of what the body ought to be doing*. What has changed, though, is the understanding of this mental model: it has been transformed from an objective blueprint that dictates the growth and development of entities according to their kind into something highly individualized, something produced in the minds of each individual subject to the formative power of the idea and exercising its causal power directly from the individual mind to that same individual's body. The fact that mechanists, evidently in spite of their best intentions, revert back to an explanation of biological phenomena so similar in this important respect to the old Aristotelian account illustrates just how great the challenge is to account for growth and development without recourse, in the case of trait acquisition, to genetic information, or, in the case of lactation, to some elementary understanding of endocrinology. The fact, moreover, that Descartes seeks to corporealize the imagination entirely, and deprives animals of any mental faculties at all, does not mean that the resort taken by Descartes and his followers to imagination in the explanation of trait acquisition was in no way continuous with premechanist embryological ideas about the role of a mindlike formative faculty in fetal development.

8. CONCLUSION

The imagination, for Descartes and his followers, was no occult force. It was an important component of Cartesian animal physiology, as many readers of Descartes have recognized. There is no good reason not to concede that it was an important part of mechanist embryology as well. As I hope to have begun to show, the imagination was of particular importance in mechanist embryology precisely because it had to be introduced to fill a vacuum left by the eradication of Aristotelian formative virtues that had once been so central to the account of how like regularly begets like, a vacuum that would not be adequately filled until the development of the science of genetics some centuries later.

III

The Gassendian Alternative

5

The Soul as Vehicle for Genetic Information

Gassendi's Account of Inheritance

SAUL FISHER

1. INTRODUCTION

Generation and heredity theories before early modern mechanist accounts might be faulted for numerous deficits. One might cite in this regard the failure to even attempt to explain how the inheritance of traits could occur, given what is known about the generation of new individuals. On the other hand, it would be hard to allow this as a true failure against the backdrop of a generation theory that poses form, and not matter, as the key to understanding the emergence of new structures in the offspring of two organisms. Hence the signal contribution of the mechanist accounts in this sphere was simply the suggestion that theories of generation and heredity might look to matter and its behavior in order to explain how new individuals are created and retain or discard features of the individuals whence they sprung. Among these accounts, one early important set of views was presented by Pierre Gassendi, eventually as a full-blown atomist conjecture but even beforehand, and early on, as a thoroughgoing materialist theory.

Gassendi's materialist mechanism for the transmission of traits is imperfect (and by the common sense of our day, implausible), not always or exclusively deployed in generation, and not clearly the *physical* bearer of information, as conformity to a materialist model would suggest. Nonetheless, his proposed mechanism represents a valuable attempt to provide an account of inheritance phenomena in materialist and ultimately atomist terms and rooted in an account of generation. Even his materialist account alone (without his specific commitment to atomism), however, is unique in his time in promoting a causal story that locates a special force in seminal matter which determines generation and, by extension, heredity phenomena. In this, his view marks an advance over – or at least a new challenge for – other corpuscularians and generation theorists of his day.

I begin with Aristotelian and early modern views of generation and heredity to provide an overview of the theories preceding and contemporary with Gassendi. Next, I introduce Gassendi's views on spontaneous and nonspontaneous generation and his notion of a "little soul," or *animula*, for which the seminal matter is a vehicle and which is responsible for the transmission of traits. The little soul is material in constitution and behaves no differently at root than any other bits of matter – which fact alone allows Gassendi a mechanist account of inheritance. This account is clearest in his views on animal generation. Yet his linkage of the main points of animal and plant generation, along with his global materialism, indicates that the bearer of inherited traits and developmental information is a common "flower" of matter. This commonality highlights a crucial distinction Gassendi wants to maintain between what may be passed on through natural processes and what elements of individuality require divine investment. As he suggests in an early exposition of his materialist account, only the generation of *material* aspects of individuals and their inherited *material* traits should be thought to flow from preexisting materially constituted individuals. Although Gassendi's scheme apparently preserves the Church view that the intellective soul is not passed on from parents to offspring, he does not shirk from suggesting that key elements of individuation among new creatures – including their first-order cognitive capacities – fall under the scope of his materialist account.

2. SETTING THE STAGE FOR PREFORMATIONISM: ARISTOTELIAN AND EARLY MODERN VIEWS

Aristotle's rejection of pangenesis, already discussed in the Introduction to this volume, is grounded in complaints about the explanatory capacity of a diffuse agent of heredity and in the failure of pangenesis to satisfy his causal scheme overall.[1] The available alternative in his view is that developmental information derives from only one bodily element – the semen – which in turn is borne by one sex only.[2] While this view sustains currency throughout the

---

[1] Aristotle's argument goes as follows: According to pangenesis, the seed would take information about development from the entire organism just in case it took such information from uniform parts, nonuniform parts, or both. None of these alternatives is satisfactory because resemblances of the progeny to the parents cannot be so explained, and because the arrangement of parts – in which such resemblances consist – is not itself a material cause, such as the seed should represent were this doctrine true; q.v. *GA* 1.17–18; G. E. R. Lloyd, "Empirical Research in Aristotle's Biology," in Allan Gotthelf and James G. Lennox, eds., *Philosophical Issues in Aristotle's Biology* (Cambridge: Cambridge University Press, 1987), 59–60.

[2] Robert Bolton suggests that Aristotle *intends* his criticism of pangenesis theory to undercut the view that both sexes contribute seminal matter to the generative process; q.v. *GA* 721b, 724a;

Aristotelian tradition, the notion that only one sex contributes seminal matter to the generative process or the inheritance of traits is rejected in a practical sense as early as Roman times, when animal husbandry and artificial selection are conducted on a sophisticated level. Further, the alternate ancient traditions of the Epicureans and Galenists offer a two-seed model.[3]

Yet these alternate models are not widely accepted as suggestive of distinctive theories of heredity until the early modern era. Even Harvey, who follows the dual *semina* view and suggests that heredity can be explained as the contribution of both sexes to the offspring, remains within the Aristotelian framework, viewing the *semina* as substance without form.[4] With the early mechanists we first see a general commitment to explaining generation – and in some cases heredity – by reference to a detailed story of matter in motion not essentially rooted in hylomorphism or a four-causes schema.

The first significant and thorough proponent of the new mechanical explanation is Descartes. This thoroughness is not, however, extended to his biological writings, and in his works there is no great single tract on generation, only two lesser works, *La formation du foetus* and *Primae cogitationes circa generationem animalium*. Neither account is well developed,[5] and, as discussed in chapter 4 of this volume, the focus is on embryonic development, with little suggestion as to how traits are transmitted. The *Formation* account offers an epigenetic proposal: the organism develops by heat and pressure on the embryo, leading it to ferment and expand outward in the direction of forming various organs and systems, first the heart, then the brain, blood, and

---

cf. Bolton, "Definition and Scientific Method in Aristotle's Posterior Analytics and Generation of Animals," in Gotthelf and Lennox, *Philosophical Issues in Aristotle's Biology*, 154.

[3] For Galen, q.v. "Utilités des parties des corps" and "Les organes genitaux" (bk. XIV, v II, chap. 7, 104–5, 1854–6). Descartes, as we see below, adopts this model, though in his view there is little to differentiate the two *semina* except for the sex of the parent.

[4] Harvey's primary contribution to the dual *semina* theory, then, is his emphasis on the importance of the communication between semen and egg, reflective of the sex-differentiated gametes.

[5] The former is a piece of *La description du corps humain* (2nd ed., 1677); the latter represents a set of posthumously published Latin notes (AT XI, 505–42). Descartes is at pains to explain why he feels ill-suited to fashion a deeper analysis of the reproductive role of seminal matter, falling back on the excuse that more empirical research – leading to experimental verification of his views – is required before a definitive account can be offered: "And although I have not wanted until now to undertake to write my views on this matter – because I have not yet done enough experiments to verify by those means all the thoughts that I have had about those views – nevertheless I cannot refuse to put here *en passant* something most general, and of which I hope that I will have less chance to retract afterwards, when such new experiments provide me with greater light" (pt. 4, §XXVII, *De la formation du foetus*, 125 [1677]; AT XI, 252–3). Descartes moved from suggesting that he indeed had sufficient data on such matters (cf. AT II, 525, February 1639 letter to Mersenne) to allowing, as in the cited passage here, that he lacked such data.

so forth. On the other hand, the *Primæ cogitationes* account has a weakly preformationist character: bits of embryonic matter split off to form organs as a consequence of the seed having subtler or thicker elements and being inclined accordingly to form different structures.[6] The hallmark of either view is a common particulate contribution to the structures developed at each stage – the seminal matter (from both parents), embryo, and organs of the new individual. This underlying particulate account allows Descartes to sketch a theory of "mathematical embryology," the notion that the organism's structural development follows a path that can be calculated geometrically from the spatial relations of the unfolding and growing new being and the environment in which it unfolds and grows.[7] Thus, from knowing the seed's parts for a particular species, one can deduce the shape and conformity of the parts it will yield, and from knowing the shape and conformity of those parts, one can deduce the seed's features. While this view of generation and development tells us little directly about heredity, there are some possible clues as to Descartes's view. Intriguingly, in the *Formation*, he suggests that the two parental seeds need not be very "diverse" themselves to produce diversity among parts of their progeny's body, as the seeds from each parent mix with and germinate one another.[8] By contrast, in the *Primae cogitationes*, we find the pangenetic view that seminal matter contains particles from around the body such that a sufficient diversity to account for the offspring's structures pertains before any union between the parents.[9] These sketchy and apparently conflicting accounts do not say how the transmission of traits occurs from one

---

[6] Andrew Pyle's *Atomism and Its Critics* (Bristol, England: Thoemmes Press, 1987) focuses on the earlier account and sees Descartes as an epigenesis proponent; Dennis Des Chene's *Spirits and Clocks: Machine and Organism in Descartes* (Ithaca and London: Cornell University Press, 2001) focuses on the latter account and sees Descartes as a preformationist.

[7] Pt. 5, §LXVI (146); the term "mathematical embryology" is from F. Duchesneau, *Les modèles du vivant de Descartes à Leibniz* (Paris: Vrin, 1998), 78.

[8] "I determine nothing regarding the shape and arrangement of the seminal particles. It suffices for me to say that those of plants are hard and solid, and may have their parts arranged and situated in a certain way, which could not be changed without rendering them useless – but that this is not the same with those of Animals, which are greatly fluid and are produced ordinarily by the conjunction of the two sexes, seemingly none other than a confused mixture of the two liquids, which serving to ferment one another, strongly reheat one another, such that some of their particles – acquiring the same agitation that fire has – dilate themselves and press the others, and by this means dispose them little by little in the way that is required to form the body parts. And these two liquids have no need at all to be very different [one from the other]. That is because, as one sees that the old pasta can enlarge the new, and that the foam that one beer throws off suffices to ferment another beer; thus it is easy to believe that the *semina* of the two sexes mix together, serving to ferment one another" (pt. 4, §XXVII, *De la formation du foetus*, 125–6 [1677]; AT XI, 253).

[9] AT XI, 534.

generation to the next. However, each of these views insists that a material vehicle is involved, writing out the role of the *anima* in the story of heredity and so markedly departing from the Aristotelian tradition.

From the perspective of generation theory, the Cartesian view (at least as presented in the *Formation*, though to a degree also as presented in the *Primae cogitationes*), resonates with that earlier tradition, in that both theories of embryo formation convey an epigenetic notion of gradually emerging and growing parts among the progeny. Development is stepwise and awaits the prior stage of insemination. For Aristotle, semen imposes form on the material of uterine blood; for Descartes, one sort of semen communicates with another, launching the embryo's developmental trajectory. From the perspective of heredity theory, though, Descartes offers the beginnings of a departure from Aristotelianism in his mechanical account of seed behavior – and by extension, trait transmission.

A more radical departure – centered on generation theory but with consequences for heredity theory – is the suggestion that the new individual's parts are all or partly formed at the inception of the embryo, with development less influenced by environment or the interaction of seminal matter than by the "simple" phenomenon of growth.[10] This diminished importance of development by stages – which I refer to here generically as "preformationism" (in his treatment of Malebranche in chapter 9, Andrew Pyle adopts a much narrower application of this term, as is now conventional in the scholarly literature) – could be understood as allied with the pangenesis promoted by Hippocrates and in later times by Descartes and Gassendi. Indeed, Aristotle seems to link the two views, suggesting that the parts of seminal matter gathered from parts of the parents' bodies could only be causally responsible for the development of organs in the progeny if the parts from the parents and their seminal matter resembled one another.[11] In other words,

---

[10] The main variant of this view, *preformationism*, suggests that the new individual's parts are fully formed while the seminal matter is still within the parent. Another variant is the *preexistence* stance, which says that all newly appearing individuals exist since Creation – in effect, there are no newly existing individuals. Hence such are transmitted from one generation to the next, originally in miniature form without needing to be formed within any given parent. A third variant is *metamorphosis*, which suggests that the parts of the new individual are formed all at once, though this happens in the embryo rather than prior to conception. A fourth variant is *spontaneous generation*. Bowler, following Roger, carefully distinguishes these views and sketches a picture of who among the major seventeenth-century naturalists subscribed to each of these perspectives.

[11] As Karen Detlefsen has pointed out to me, it matters what level of resemblance pertains here. Thus, pangenesis might account for morphological resemblance of particular traits found in parents and progeny though fail to explain the species-wise resemblance across individuals

the transmission mechanism for pangenesis on Aristotle's reading requires that the traits already appearing in seminal matter have a form found in the parents – per the preformationist doctrine. Historically, though, development of a mature preformationism awaits the formulations posed by Swammerdam and Malebranche.[12] At all events, one of Gassendi's innovations is to decouple those views, promoting a pangenesis that, while preformationist in some respects, is not entirely so. And the extent to which Gassendi's view *is* preformationist reflects his promotion of a mechanism for trait transmission not found in either the Cartesian or Aristotelian models. His proposed molecular structure of seminal matter allows the transmission of parental characteristics to the new organism through a material vehicle for organizational and developmental memory of body parts.

3. GASSENDI'S ACCOUNT

By contrast, the animula is an offshoot ("branch") of the parent soul but not a preformed miniature. This material soul yields a plan for the development of the new individual's material traits by passing on to that offspring information about the nature of the parents' physical traits. Thus, without specifically identifying the mechanism of transmission, Gassendi indicates a central role of trait inheritance in material models of animal generation.

Gassendi's considered views of generation and heredity, presented in the *Syntagma philosophicum*, represent an enthusiastic piece of his broader materialist and atomist doctrines.[13] This is also, to a significant degree, a long-standing view for Gassendi. As we will see, the roots of his materialism relative to these issues predate even his atomism.

First, however, let us consider his refined atomist account, which ranges over plant and animal generation and differs somewhat in each case. At the widest grain, Gassendi holds that generation may be either spontaneous, emerging from elements in the earth, or nonspontaneous, emerging from parent organisms, generally by sexual means – and involving what might be

which governs the likeness of members of a given species (e.g., bears or cows) or family of species (e.g., mammals).

[12] Q.v. Pyle, "Malebranche on Animal Generation: Pre-Existence and the Microscope," chap. 9 in this volume; and Karen Detlefsen, "Supernaturalism, Occasionalism, and Preformation in Malebranche," *Perspectives in Science* 11 (2004): 443–83. Further evidence for preexistence or preformationist views was thought to be found in other early modern microscopy – especially in the work of Leeuwenhoek and Malpighi.

[13] Q.v. *De ortu seu generatione plantarum* (pt. II, §3 [Prior], bk. IV, chap. 4) and *De generatione animalium* (pt. II, §3 [Posterior], bk. IV).

called "preorganized" seminal matter.[14] In plants, both types of generation entail development of a new organism from a single unfertilized seed bearing the new plant's soul. This soul is composed of a material yet *spirit-like* substance which directs the division, differentiation, and development of the corpuscles in the growing seed. This ontogenetic pathway produces the same results regardless of whether the seeds arise by spontaneous generation (SG) or preorganized generation (PG). Two distinct processes may yield similar results because seeds composed of elements either from the earth or from other plants may share an identical corpuscular composition. Gassendi writes,

> [N]othing forbids our saying that atoms or corpuscles were created by God in the beginning and given a certain mass, shape, and motion, and that while these are being variously moved, and while they are meeting, interweaving, mingling, unrolling, uniting, and being fitted together, molecules – or small structures similar to molecules – are created, from which the actual seeds are constructed and fashioned within the plant. In other words, the corpuscles which change into a seed within a plant are also attracted from the earth itself, and the only difference is that the fashioning of seeds can be more easily accomplished within a plant because a more select supply of corpuscles or similar principles has already been produced and is now gathering.[15]

Gassendi does not tell us, for either sort of generation, precisely how the right sorts of matter come together in just the appropriate configuration. Nor does he specify the subsequent behavior of any plant seeds so configured. However, following his general views on molecular formation, it is reasonable to surmise that the seeds giving rise to plants are formed from just the right combinations of only those atoms that may compose vegetative matter. For in Gassendi's matter theory, there are classes of atomic aggregates that come together using only a fixed range of atoms – as per their physical compatibility or tendencies to combine.[16] As a consequence, seeds resulting in

---

[14] For Gassendi, spontaneous generation entails generation of organisms from molecular aggregates that are predisposed to grouping in ways that yield the structures of organisms. This picture of things differs considerably from a more traditional picture that has it that generation is possible from elements that do not even bear the features of organic building blocks.

[15] *Opera Omnia* II 170b–171a (Lyon: Laurentius Anisson, 1658 [Facsimile reprint Stuttgart-Bad Cannstatt: Friedrich Frommann, 1964]); H. Adelmann, *Marcello Malpighi and the Evolution of Embryology*, 5 vols. (Ithaca: Cornell Univerity Press, 1966), 798.

[16] Cf. O. Bloch, *La Philosophie de Gassendi: Nominalisme, matérialisme et métaphysique* (The Hague: M. Nijhoff, 1971); Henk Kubbinga, *L'Histoire du concept de "molecule,"* 3 vols. (Paris: Springer, 2002). Margaret Osler, "How Mechanical Was the Mechanical Philosophy? Non-Epicurean Themes in Gassendi's Atoms," in *Late Medieval and Early Modern Corpuscular Matter Theory*, ed. Christoph Lüthy, John Murdoch, and William R. Newman (Leiden: Brill, 2002), and Duchesneau view the creation of atoms with dispositions to biological organization

SG and PG alike should comprise the same atomic combinations, given that those combinations are governed by the same compatibilities and tendencies – and produce like new individuals.[17] Yet only the occurrence of SG would depend on the chances that requisite but otherwise independent components are sufficiently abundant at the right place and time and that they tend toward the proper configurations. By contrast, the components of *semina* in non-spontaneous generation have already been brought within proximity of one another and fall into particular arrangements constitutive of *semina* by the time they constitute the characteristic seminal structures. Indeed, this is the sense in which they are "preorganized." Accordingly, PG should occur with greater ease than SG, *mutatis mutandis*, because the proper atomic mix for generation is already present within existing plants. As Gassendi puts it, PG occurs more easily in plants than SG does, since "a more select supply of corpuscles...has already been produced" in PG.

The distinction between PG and SG is not, in the end, richly detailed.[18] Yet Gassendi is clear in his view that PG accounts are more useful for generating biologically significant explanations than are SG accounts. One explanatory advantage of PG accounts over SG accounts is that only the former can tell us how a particular plant has a given set of traits as a consequence of any other plant having the same set. In other words, PG accounts can tell us about relations between individual plants and about inheritance phenomena, whereas SG accounts are rather more limited in what they can tell us about how a set of traits appears in a given plant. Gassendi draws this distinction by indicating that, as the product of parent organisms, only the seeds of PG bear

and function as a piece of teleology in Gassendi's theory. See Duchesneau (*Les modèles du vivant de Descartes à Leibniz*, 114) and Roger (*Les sciences de la vie dans la pensée française au XVIIIe siècle* [Paris: A Colin, 1963], 136), that *semina* in SG require an external impetus to start the process of self-organizing suggests that Gassendi needs teleology to explain SG. While Gassendi embraces final causation as a significant mode of explanation, in and out of biology (cf. *Opera Omnia* III 360b–361a; Bernard Rochot, *Les travaux de Gassendi sur Épicure et sur l'atomisme, 1619–1648* [Paris: J. Vrin, 1944], 406; Duchesneau, *Les modèles du vivant de Descartes à Leibniz*, 96), it is not clear that the divine endowment of atomic dispositions to organize as organic molecules counts as teleology, for it is not stated anywhere by Gassendi that such endowment (or any other) is done with any particular goal in mind (e.g., to meet a divine goal of populating the world with organic bodies).

[17] Indeed, as Duchesneau (*Les modèles du vivant de Descartes à Leibniz*, 114), following Roger (*Les sciences de la vie dans la pensée française au XVIIIe siècle*, 138), notes, if SG produces such new individuals as are comparable to those produced under PG, this is only possible given the common nature of the atomic structures and attendant processes underlying all organic phenomena.

[18] This is a natural consequence of Gassendi's view that a full account of why these processes unfold the way they do is unavailable to us and can be known only to God (*Opera Omnia* II 266b–267a, 274b; Duchesneau, *Les modèles du vivant de Descartes à Leibniz*, 109).

the special vehicles of miniature souls (*animulae*) that enable them to transmit their characteristics across generations and so allow for inheritance. Thus the contribution of the parent organism to the characteristics of the spawn is a function of the relation between two material souls: a soul spread throughout a plant's structure and generally animating the plant and an *animula* in the seed which is part of the larger soul until the seed falls from the plant.

Like the larger soul of the parent organism, the *animula* has an atomic composition. In addition to the inherent motive force found in all atoms, the molecular structure of *animulae* also features a special seminal force or virtue (*vis seminalis*) that guides development of the offspring (such a force is also found in crystals, guiding their uniform formation). The exact nature of this special force is left unclear, though it is apparently superadded to the inherent motive force and arises in the aggregate and not in the constituent atoms of the *animula* or seminal matter.[19] Yet as Duchesneau notes,[20] Gassendi's conception allows for a mechanical account of the *vis seminalis* as harnessing the motive force of the atoms in the *semina* to organize and stimulate the development of the embryonic matter. An intriguing possibility here is that, given that the *motion* of the *vis seminalis* is a function of those atoms' motive force, different compositions of atomic weights should give rise to distinctive patterns of organic development, depending on the weight distributions among the embryo's molecules.

The role of the *animula* is to transmit ontogenetic information from parent to new plant, and it does this by receiving, in concentrated form, "ideas" and "impressions" from all parts of the soul of the parent plant and communicating such information to the new organism which the seed produces.[21] Given the framework of Gassendi's matter theory, we find here micromerism, an atomist version of the classic pangenesis theory of Aristotle's opponents.[22]

---

[19] Q.v. *Opera Omnia* II 170b–172a, 260a–b.

[20] Duchesneau, *Les modèles du vivant de Descartes à Leibniz*, 100–3.

[21] When the *animulae* combine in the new embryo – whether in plants or animals – they form a new, single *anima* that guides development of the new individual (*Opera Omnia* II 275a–b).

[22] In animal pangenesis, Gassendi proposes, representative elements from different parts of the body travel through the nervous and circulatory systems, reaching the male or female genital parts, where the *semina* are formed, and these different elements combine to yield the *animula* (*Opera Omnia* II 270b–272b).

Roughly contemporary pangenesis theories were developed by Claude Perrault (*De la mecanique des animaux* [*Essais de physique ou Recueil de plusieurs traitez touchant les choses naturelles*, vol. III, 1680]) and John Ray (*The Wisdom of God*, 1692). The pangenesis notion of one small biological element (in Gassendi's picture, the *animula*) gathering and transmitting information from throughout the parent organism resembles the physical and chemical notion of *homeomerity*, or the representative mixture of all parts of a whole in the microstructure of matter (Gregory Vlastos, "The Physical Theory of Anaxagoras," shows that the early homeomerism

Saul Fisher

One noteworthy aspect of Gassendi's proposal is his notion that the *animula* is partly "transfused" from the root and accordingly may mimic the parent soul's directions and guidelines for development.[23]

By contrast, Gassendi assigns no such analogous structure or process to SG seeds, and it is hard to see how they could acquire anything like the same soul (*animula*) as PG seeds. Yet even if there were an *animula* in SG seeds, this miniature soul would of course have no relation to the soul of any parent plant because there is no parent. While there are no such sources of directions, ideas, or impressions, Gassendi supposes anyway that SG seeds somehow also bear a similar sort of developmental or organizational information. What makes this possible – and what also makes it possible that the two modes of generation yield roughly equivalent products – is that the atomic compositions of the types of seed do not differ:

> Hence when we have conceded that the seeds are equal, equal functions nec-
> essarily ensue from the equal dispositions we have assumed, equal stamina are
> formed, and equal plants arise from them.[24]

There is a hitch here: the seeds could not be equal in all respects, since the souls we find in PG seeds are related to their parent souls, which guarantees some continuity across individuals, whereas nothing in SG does (there being no *animulae*). It is conceivable that in Gassendi's model plants of the same species could develop from souls of either type – that "equal plants arise from them" – just in case the souls of PG seeds made a difference only for trait inheritance and not for generation *per se*. By this token, equivalent characteristics might arise in plants whether or not they resembled those possessed by members of previous generations; the distinction between resemblance relative to individuals and relative to species would thus collapse. But then on grounds of parsimony, Gassendi need not have introduced a distinct mechanism for trait inheritance altogether. In contrast to animal generation, he thinks plant

attributed to Anaxagoras explicitly embraces panspermism). Homeomeric mixtures were touted as holding the key to physical and chemical change by a number of iatrochemical writers and late Aristotelians who were Gassendi's contemporaries, such as Daniel Sennert; see Antonio Clericuzio, "Elements, Principles, and Corpuscles: A Study of Atomism and Chemistry in the Seventeenth Century," *Archives Internationales d'Histoire des Idées*, vol. 171 (Dordrecht: Kluwer, 2000); Roger Ariew, *Descartes and the Last Scholastics* (Ithaca: Cornell University Press, 1999); and Emily Michael, "Daniel Sennert on Matter and Form: At the Juncture of the Old and the New," *Early Science and Medicine* 2 (1997): 272–300.

[23] Lest we lose sight of the fact that Gassendi is talking about a *material* soul of plants, here he proposes that an identifiable *piece* of the parent's soul is responsible for development and is more pronounced in some parts of the organism than others are; q.v. *Opera Omnia* II 172b; Adelmann, *Marcello Malpighi and the Evolution of Embryology*, 800.

[24] *Opera Omnia* II 173a; Adelmann, *Marcello Malpighi and the Evolution of Embryology*, 801.

generation is a relatively simple affair, and so perhaps in a tidier version of his theory SG alone might have sufficed. Having introduced *animulae* along with SG, though, he has proposed two very different generative processes, whatever similarities may obtain. Hence, while he might well reason that seeds involved in the processes *could* have ontogenetic pathways with equivalent outcomes (producing species-wise resemblance across individuals) given their common atomic compositions, it is not clear why he thinks such commonality alone *guarantees* such an equivalence. In the absence of the organizing *animula*, nothing about SG in Gassendi's model provides such a guarantee. Nonetheless, this model signals an advance over prior SG theories, which do not even offer a mechanism to guarantee resemblance relative to species. Gassendi's version of SG tells an atomist story as to how such resemblance is possible at all.

In animal generation, we also find seeds (semen) that feature *animulae*. In a form of SG special to animals, one parent suffices, and the semen is created in the moment of generation. In animal PG, by contrast, two parents are required, and the two *semina* meet in the moment of generation. In both scenarios, the *semina* contain complete though folded-up and rudimentary forms of the animal offspring. Thus, in generation among animals, as among plants, new souls contain a plan for development. What is different is the idea – typically associated with preformationism – that the mature physical characteristics of new individuals are somehow contained in premature forms found in the *semina*. In Gassendi's view, as the *semina* are created, their tissue matter is differentiated into familiar body parts of the mature fetus. He states that "the semen both is manifestly heterogeneous and consists of the same parts from which gradually and in one series the organic parts will be perfected."[25] One distinctive feature of this account is that Gassendi thinks that animal generation, much more than plant generation, requires sexual communication to ensure that progeny receive the requisite matter and genetic information. Following the Epicurean view, he suggests that sexual generation involves the meeting of two seminal fluids and that this meeting is solely responsible for the creation of new fetuses in PG. To the new fetus the semen of the two parents contributes the material tissue (equally) and the *animula*-soul (less than equally, as one contribution dominates the other). Given that this fetal soul is responsible for the development of traits in a new individual, the particular mix of traits in the offspring results from the combinations of *animulae* borne by the two *semina* that meet in sexual reproduction.

---

[25] *Opera Omnia* II 280b; Adelmann, *Marcello Malpighi and the Evolution of Embryology*, 815.

The most unusual aspect of this view, though, is the proposal that, in addition to the two *animulae*, the semen from each parent contains its own new organism in miniature. Such a proposal appears to be a good candidate for a strong preformationism of a classic variety. Yet Gassendi's view does not quite fit this description. The argument for taking him to subscribe to a strong preformationism is centered on Gassendi's proposal that the new organism's soul – which directs the development of the seminal material from both parent organisms – itself comes from the parent organisms.[26] The notion is that a generation theory is preformationist if it entails that the organizing principle for the new individual exists intact in the parent organism prior to formation of the embryo. A "classic" preformationist view further entails that an actual miniature, the features of which are less developed versions of the features of the new individual, exists prior to conception. This classic view clearly upholds a standard that Gassendi's theory cannot meet, given that the organizing principles his theory describes – the *animulae* – do not bear the physical traits of the new individuals. This would be true, obviously, of immaterial souls, yet it is true of Gassendi's *animulae* as well, since they bear not the actual traits of the new individual but the organizational information (in physical form) which gives rise to those traits.[27] That the seminal matter *does* bear rudimentary forms of a new individual's traits must be considered irrelevant here since (1) the organizing principle is rather the distinctive *animulae* and (2) some selection and redistribution of parts from among the *semina*

---

[26] Q.v. Roger, *Les sciences de la vie dans la pensée française au XVIIIe siècle*; Peter J. Bowler, "Preformation and Pre-existence in the Seventeenth Century: A Brief Analysis," *Journal of the History of Biology* 4 (1971): 221–44; and Duchesneau, *Les modèles du vivant de Descartes à Leibniz*. Indeed, Duchesneau (114) suggests that, because Gassendi is a preformationist, he cannot accommodate a true SG since such generation cannot cull its seminal matter from any preexisting body. However, relative to PG, Gassendi is not a strong preformationist. Hence the epigenesis characteristic of Gassendi's SG is apparently also characteristic (albeit along different lines) of his PG.

[27] Bowler ("Preformation and Pre-existence in the Seventeenth Century," 228) and Roger (*Les sciences de la vie dans la pensée française au XVIIIe siècle*, 126–31) note that, for most generation theorists of Gassendi's era who also held that the parental soul gives rise to the soul of the offspring and guides its development, the new organism's soul was immaterial. These like-minded theorists included Fortunio Liceti (*De spontaneo viventium ortu: Libri quatuor in quibus de generatione animantium, quae vulgo ex putri exoriri dicuntur*, 1618), Aemylius Parisianus (*Aemylii Parisani ... nobilium exercitationum libri duodecim de subtilitate ... Accessit par et sanius judicium, de seminis a toto proventu, ac de stigmatibus*, 1623–43), Giuseppe degli Aromatari (*Epistola de generatione plantarum ex seminibus*, 1625), Nathaniel Highmore (*Exercitationes de generatione animalium*, 1651), and Honoré Fabri (*Tractatus duo, quorum prior est de plantis, et de generatione animalium, posterior de homine*, 1666). Gassendi – holding to his Epicurean tendencies – was alone among these writers in proposing a material soul.

must occur in the creation of the embryo given that not all parts come from any single parent and that there is an excess of parts from the two *semina*. Putting such classic preformationism aside, Gassendi could not have been a nonclassical strong preformationist either. The soul-like organizing principle contributed by one parent either makes the similar contribution from the other parent superfluous to the theory or else requires a theory of trait inheritance to explain how elements of the organizing principle from each parent make partial contributions to directing embryonic development.[28] In neither case do we see the robust determination of the offspring's traits by generative information, or actual traits, from a single source.

Instead, Gassendi offers a weak preformationism, according to which the new individual's parts are brought together – under the direction of the two *animulae* – from among those of the miniatures' parts extant in the two *semina*. This leaves a puzzle as to which semen the offspring will resemble and in what respects and how the *animulae* direct such combinations. In short, a theory of inheritance is required. Gassendi addresses this requirement by explaining how dominance is achieved among inherited traits and what characteristics of generation account for dominance relations. We might expect him, as a mechanist, to suggest that dominance results from physical interactions of tissues contributed by each parent's semen to the embryo. One could have it, for example, that bits of tissue from the two *semina* destroy, merge, or connect with one another so as to fashion particular mixtures of traits. Yet Gassendi thinks such physical interactions cannot account for dominance. He considers the scenario in which characteristics carried by each semen are joined in the fertilized embryo, such a joining process entails the total physical dominance of one set of seminal tissue over the other, and in particular the sex of the offspring is determined by the dominating set. In this case, he suggests, we should expect that a male child would exactly resemble the father and that a female child would exactly resemble the mother. Here he assumes that all the characteristics carried by a particular semen would be linked – that there would be no exchange of traits by the parents. Whatever the physical interaction of the two sets of *semina*, the semen of one parent would contribute all the tissue matter of the new individual whereas that of the other parent would contribute none. However, Gassendi notes, that scenario cannot be accurate, because male children sometimes resemble their mothers and female

---

[28] Bowler ("Preformation and Pre-existence in the Seventeenth Century," 228) makes a similar point relative to the "classic" preformationist view, noting that the existence of a miniature in one parent makes the seminal contribution of the other parent pointless. My suggestion is that the same superfluity applies to any two-*semina* theory that is taken to be strongly preformationist in character.

children sometimes resemble their fathers. From this he reasons that the key cause of trait dominance cannot be the physical domination of the parts from one semen over those of the other. Moreover, he proposes – without further justification – that trait dominance cannot be decided *at all* by interactions between bits of tissue from distinct *semina*.[29]

Gassendi suggests, rather, that the key causes of dominance are sets of mental impressions that bear information about the characteristic traits of the parents. The individual offspring inherits a given set of traits from one parent instead of another because that set develops from a bundle of dominant impressions transmitted to the fetus. We may find broad parallels between these suggestions and the modern account of genetic dominance. For each offspring, there are two potential sources of developmental information, so there is a surplus of both information and its sources, hence a contest between the two sources of developmental information. Gassendi's story of this contest is of course quite different from the modern account, and potentially even more complex. He holds that every *animula* contains a "précis" (*epitome*) of the corresponding parent soul, and the information each *animula* bears vies to direct development of the fetus. The contending bits of information consist of one parent's impressions of the other parent such that the dominating information passed on by the parent with that impression produces in the offspring more traits of the viewed parent than of the viewer parent.[30] He writes,

> [I]t seems that commonly the force of the imagination [*vis Imaginationis*] must be applied.... For the form of the image of an external object which has been

---

[29] *Opera Omnia* II 284b; Adelmann, *Marcello Malpighi and the Evolution of Embryology*, 815. This was not the only option open to Gassendi. In his view, each semen only contributes tissue matter to the fertilized embryo, whereas the development of particular characteristics in the new individual is directed by an *animula*-soul. Yet none of this blocks the possibility that, in the merger of underdifferentiated tissue matter from each semen, physical interactions of that matter could help decide which traits are dominant.

[30] Gassendi's role for parental impressions is in keeping with the long history, in science and medicine up through the modern era (and in medical traditions all around the world), of holding that maternal impressions may cause defects and diseases as well as features of "regular" development (cf. J. W. Ballantyne, *"Teratogenesis: An Inquiry into the Causes of Monstrosities,"* History of the Theories of the Past* [Edinburgh: Oliver and Boyd, 1897]). Such impressions were even thought to be a possible cause of "telegony," where offspring bear traits resembling those of a partner prior to the actual father, the notion being that the mother forms a sufficiently powerful impression of the prior partner to affect the character of the future offspring.

Such theories of maternal impressions are found as early as the story of Jacob in the book of Genesis (30:25–39) and appear throughout Greek and Roman thought (e.g., in Empedocles, Plato, Aristotle, Pliny, Plutarch, Hippocrates, and Galen) as well as in Avicenna and a host of late Renaissance and early modern medical and scientific writers, including Agrippa (*De occulta philosophia*, 1533), Van Helmont (*De injectis materialibus*, in the *Ortus medicinae*, 1643), Paracelsus (31–2 [*Jacobi*]; I/1 315, I/7 203 [Sudhoff & Mattiessen, *Sämtliche Werke*]),

impressed on the brain by the intervention of external sense and has moved the imaginative faculty residing in it seems so to set in motion the appetite and the spirits serving it, that the spirits themselves also retain a trace of the impression that has been made, and transport it with them through the body. Hence if the semen happens to be separating off, the spirits – which gather meanwhile and excite and variously pervade it – affect it [the semen] in accordance with their form (the whole mass of it and all its particles) and make them [the particles] partakers of their own impression; and thus while the particles are being suitably coordinated and are seeking their proper places in the fetus to be formed, they retain a trace of the impression or a resemblance to the image itself. It is therefore possible for either a male or female fetus to resemble the father if the imagination of the mother directed toward the father was more vehement [*vehementior*] and powerful than the imagination of the father; or conversely to resemble the mother if the imagination of the father directed toward the mother was more powerful than the imagination of the mother; or to resemble both parents promiscuously and confusedly if the imagination of both or of one directed toward the other was of about equal force; or to resemble neither parent at all if the imagination of both was distracted elsewhere and in the male did not have the female herself as its object and in the female did not have the male himself as object.[31]

Sennert (*Epitome naturalis scientiae*, 1632), and Thomas Feyens (*De viribus imaginationis*, 1608). Other early modern writers arguing for maternal impressions included Henry More (*The Immortality of the Soul*, 1639), Montaigne ("On the Power of Imagination," 1572–4), Descartes (correspondence, AT I, 113; II, 20, 49, 241; *Traité de l'homme*, AT XI, 176; *Passions de l'âme*, AT XI, 177, 518, 538, 606; cf. Justin E. H. Smith, "Imagination and the Problem of Heredity in Cartesian Embryology," chap. 4 of this volume), and Malebranche (*Recherches* II 1 Ch VII). Yet other authors defended the theory and reported putative cases up through the late nineteenth century.

The widespread appeal of maternal impressions theory, as Ross ("Occultism and Philosophy in the Seventeenth Century," paper presented at a conference of the Royal Institute of Philosophy, 1983) has noted, is undoubtedly due to its flexible nature, being variously explained by hylomorphist, materialist, and occult causal models. Among the early modern defenders of the theory, Thomas Feyens (Fienus) offers a particularly interesting antecedent to Gassendi's views, suggesting that maternal impressions shape the features of offspring when immaterial species of the imagination cause changes in the "emotions" that are transmitted, via the humors, to the fetus (q.v. L. J. Rather, "Thomas Fienus' [1567–1631] Dialectical Investigation of the Imagination as Cause and Cure of Bodily Disease," *Bulletin of the History of Medicine* 41 [1967]: 349–67). Gassendi's model partly resembles Feyens' hybrid of mental and physical causation (itself partly based on Galen's views). By adopting a physical notion of "species," though, Gassendi offers a wholly materialist account of maternal impressions. In this he holds closely to the Lucretian and Epicurean concept of imaginative powers as simply one more feature of a thoroughly atomist model of physiological phenomena.

[31] *Opera Omnia* II 284b–285a; Adelmann, *Marcello Malpighi and the Evolution of Embryology*, 815–16. According to Gassendi, there are other parties besides the parents who may be ideally represented in the *animulae*: if one parent's impressions of, for example, a statue are stronger

Gassendi does not suggest the nature of their struggle – specifically how such particles bearing information about one impression could contest another. He tells us that the stronger imagination from among the parents dominates, but the dominance as such cannot be transmitted. Instead, some medium must transmit to the *animulae* the relative strengths of the impressions. Although we are told little of the nature of that medium, it is undoubtedly just those impressions. Are such impressions material? While the *animulae* themselves are material, one could imagine that the bits of information they carry were not.[32] However, this is an instance where Gassendi's materialism is more thorough rather then less. In the passage cited here, he suggests that genetic information is carried by material impressions upon the *animulae*, just in the manner of his suggestion that folds and grooves in the brain are vestiges of neural activity from which we form memories. Yet none of this tells us how such impressions actually interact. Even with this account of the imagination's impressions as transmitted to *animulae*, we still lack a detailed explanation of dominance among inherited traits.[33]

Beyond his outlining an account of trait inheritance, the other principal novelty Gassendi brings to generation and heredity theories of his era is the suggestion of a force which, special to *semina*, is the underlying and direct cause of development in the offspring. Like a number of other early modern philosophers – including Descartes and Malebranche – Gassendi offers a mechanist account of animal generation. He approaches Malebranche's view in combining this mechanism with a preexistence theory of sorts, though he probably comes closer to Descartes in allowing that the definitive causal influence upon formation of the fetus is the genetic information carried by the embryo, entailing a combination of such information from the two *semina*.

---

than whatever the other parent's impressions are of, then the offspring will develop traits which match the impressions of the statue.

[32] While Gassendi holds that human beings also have immaterial souls, such are not equivalent to the *animulae*. As I note later, the intricacies and puzzles arising from this dualism do not weigh upon his generation theory.

[33] Duchesneau (*Les modèles du vivant de Descartes à Leibniz*, 110) has noted that the process of trait inheritance is not fixed rigidly by pangenesis, given the crucial "wild card" role of the parental imagination in determining traits of the offspring. Despite this randomizing factor, the process is nonetheless fixed in a broader mechanist sense. Gassendi's mechanical account of perception (and other cognitive faculties) allows him a seamless account of corporeal change resulting from mental impressions, consistent with his materialist mechanical philosophy. On the other hand, Gassendi's materialism is less easily reconciled with his additional proposal of a second, immaterial soul that is the seat of such higher-order cognitive capacities as abstract reasoning. That issue is addressed by, among others, Bloch (*La Philosophie de Gassendi*) and Emily Michael and Fred S. Michael ("Two Early Modern Concepts of Mind: Reflecting Substance vs. Thinking Substance," *Journal of the History of Philosophy* 27 [1989]: 29–48.

This view is characteristically mechanist in proposing that the development of new individuals results from interactions among atomic amalgams – in his view, constituting the *animulae* and tissue of the two *semina*. And it is characteristically (albeit weakly) preformationist in allowing that plant seeds and animal semen contain in a miniature, undeveloped form some or most of the actual physical characteristics of the maturing embryo and ultimately of the new individual. Yet Gassendi's view differs from the accounts offered by Descartes, Malebranche, and most other early modern generation theorists in its suggestion that a life force (*virtus seminalis*) inheres in those aggregates of atoms that constitute *semina*.[34] This view has the ring of the latter-day doctrine of vitalism: Gassendi's proposed life force – vested in the *animulae* – enables semen or seeds to direct ontogenesis simply on the basis of what they contain at the inception of the new individual, that is, without the benefit of interactions with any other external matter. Naturally, this cannot be true vitalism since his account has it that all the relevant players are composed of atoms, not spirit stuff. Further, in contrast to standard vitalist accounts, there is nothing emergent, strictly speaking, about the life force Gassendi proposes. Like the bud to a tree, the life force of the *animulae* is an offshoot of the parent organism's soul.

These various aspects of his account – so clearly in tension in their mature, doctrinal forms – somehow fit together in Gassendi's picture. His mechanism would be at odds with his talk of miniature souls if the latter were composed of soul stuff. They are not the stuff of souls, though, so it is hard to see any conflict here. Further, his wholly materialist account of souls makes it unreasonable to think of Gassendi as a true vitalist in any sense. His views about *animulae* would also sit poorly with his loose brand of preformationism if ontogenesis on his account did not require the organizing principle of the soul but instead required only that the preformed tissue matter of *semina* naturally unfolded in the course of generation and development. Yet here, too, Gassendi's views are compatible given that the material *semina* are responsible for physical unfolding and that the souls associated with *semina* are responsible for developmental direction. In his weak preformationist conception, an epigenesis-like development and refinement in the new organism follows the organizational principle established by the contributing seminal

---

[34] As Andrew Pyle details, a range of other early modern generation theorists – including the Cambridge Platonists, the late Scholastics, Boyle, and Malebranche – promoted causal factors that supplemented "pure" mechanist explanations of generation ("Animal Generation and Mechanical Philosophy: Some Light on the Role of Biology in the Scientific Revolution," *Journal for the History and Philosophy of the Life Sciences* 9 [1987]: 225–54). But none of these other authors located their proposed additional causal features in particles constituting seminal matter which gives rise to the embryo from which the fetus is formed.

structures.[35] Prior to combining to form the embryo, the mother's semen and the father's semen contain tissue bearing characteristics of the parents, per his panspermist-preexistence view. However, it is only after the cross-fertilization producing the embryo that the tissue combined from each of the *semina* expands and becomes differentiated, directed and managed by the *animula*-soul. In this vein, Gassendi suggests the task of the *animula*-soul is to "apply the given parts to parts . . . by replacing them in the position and in the arrangement with respect to one another [such] that they had to form a complete diminutive body."[36]

Gassendi's strategy for avoiding tensions in his view is to maintain a thoroughgoing materialist mechanism regarding all the agents of generation. This includes tissue matter with the potential for developing into a mature new individual, the material soul which lays out the plan for this development, and the seminal force which – though apparently superadded to the net quantity of force in the world – nonetheless counts in his scheme as a mechanically viable causal feature of the world (made necessary by the *de novo* nature of offspring organisms). It is not clear precisely how the different bits of matter must be related for generation to take place, or quite how the seminal force operates, and in the end all that Gassendi offers us is a gloss on what a viable account should look like. What is noteworthy about this gloss, however, is his insistence that any such account is viable only if it describes how traits are inherited and explains this in terms of the motions and interactions of the smallest bits of matter out of which the whole world, organic stuff or otherwise, is composed.

4. THE IMMATERIAL SOUL AND HEREDITY: THE LETTER TO FEYENS

Another noteworthy element of the accounts in the *Syntagma* is that Gassendi skirts the issue of how nonmaterial structures and traits of living individuals may be generated or, if such is appropriate, passed on. Gassendi is well aware of this issue, though, and treats it squarely in the letter to Thomas Feyens

---

[35] Bloch goes so far as to propose that Gassendi is a strong epigenesist, given the "perpetual changes" that the constant motion of organic molecules and their constituent atoms bring about as they yield the characteristic "form" of life (*La Philosophie de Gassendi*, 367). While epigenetic development clearly takes place in the Gassendist model, it is not clear that such development follows from these perpetual changes alone (i.e., in virtue of the constancy of motion).

[36] *Opera Omnia* II 275b; Adelmann, *Marcello Malpighi and the Evolution of Embryology*, 810.

of June 6, 1629,[37] where he attempts to reconcile a materialist account with Christian doctrine by distinguishing the generative role of the intellective soul from that of the material soul. His reasoning starts with the notion that, following the Aristotelian and Thomist tradition, we can distinguish material and immaterial (rational) souls. Then, we may stipulate that only material souls are generated from the souls of parent organisms, whereas immaterial souls are divinely invested in the individual, at the moment of (corporeal) individuation. Gassendi writes,

> Holy Faith taught us that the rational soul, if immaterial, is nonetheless an individual and likely to be produced only by God, from nothing. . . . [I]t occurred to me just now (some Catholics have also defended this idea) that the human soul was composed in fact of two parts, of which one is immaterial and intellectual, and the other corporeal and sensitive. The sensitive disposition is located in the body to attract the intellectual one – and to keep it bound, as if in chains. . . . I admit only that the sensitive part is derived from the parents. . . . In addition, I declared that there still remains the rational part, created by God and invested in the body – and that as soon as the rupture occurs, the infant or rational soul of the seed separated from the parent ceased giving form to the infant or seed, and that only the part of the sensitive soul remained. . . . [T]he soul of the son – spread throughout the whole of the body – though it comes from the parents, was present in the son's heart or brain, not as the flower of matter but as Divine energy, created by God, invested in the body and attached to the soul – whatever the part by which it is connected, whether the heart or brain.[38]

> [A]t the time the parents' *semina* arrived in the uterus, woven together and conjoined, God created the soul and transferred it into this seminal molecule (as the parents' soul had already left).[39]

Gassendi is careful to mention the exigencies of articles of faith in this context. It bears mentioning that, in the work to which he is responding, *De formatrice foetus liber* (1620), Feyens rejects the Aristotelian tripartite soul on the grounds that the continuity of the individual (in the embryo) is most simply explained by the unitary soul.[40] Gassendi carefully restores

---

[37] *Opera Omnia* VI 16b–19b; French translation in Sylvie Taussig, *Pierre Gassendi (1592–1655): Lettres Latines* (Turnhout: Brepols, 2003).

[38] *Opera Omnia* VI 19a–b.

[39] *Opera Omnia* VI 16b.

[40] In this letter Gassendi is endorsing Feyens' notion that the soul should appear early on after conception. Feyens bases this notion on the argument that, if there is a unitary soul which is the primary organizing principle of the body's structure, then only an early appearance of the soul accounts for the new body's form starting to take shape at that early point. Gassendi accepts Feyen's timetable without the underlying metaphysics.

the tripartite soul here, and yet his account is not clearly motivated by a need to reconcile faith and reason – at least not by any such need alone. This can be seen in his locating the seat of developmental information in the vegetative soul, which he holds – following classical and Church authors – is material in nature. As the "flower" of matter (*flos materiae*), this soul, on the *animulae* account, is transmitted by material means. As a consequence, Gassendi notes, the intellect owes its global plan to God, but all other particular features of the individual owe their global plan to the vegetative soul common to all organisms. "Also, one sees the son as a part and image of the father, and owes no less to his parents than what the stupid animal owes to his."[41] This view of the materialist source of generation and development contrasts neatly with the Cartesian view that, from a theological perspective, the fetus develops only as a result of divine force and not as a result of any force or power intrinsic to the body. In a physiological and natural-scientific light, Descartes sees the matter developing from the fetus as inert as all other matter, bearing no special "intelligence" regarding development and directed by no goals, except in the same sense that any other bits of matter have an ultimate cause in God.[42]

Gassendi, by contrast, assigns an extremely robust capacity to matter in this regard in virtue of the material soul's role in generation, heredity, and development – and the core status of corporeal bodies, including the sensory organs, for Gassendi. In his theory of cognition, our experience of the world is the ultimate source and arbiter of our knowledge. As a result, while his theory of the soul may allow that the intellect is divinely invested in the embryo, the nonintellective soul still plays a greatly important role in defining a person's individual mentality, and that soul is crafted by material means. Given this enduring importance of the material soul in constituting personal individuality, it is hard to see that Gassendi's account successfully reconciles a materialist generation and heredity account with Church teachings of the time – or that it was necessarily intended to do so.

The other critical point concerning the letter to Feyens is that Gassendi describes here the basic rudiments of his materialist account of generation,

---

[41] *Opera Omnia* VI 18b.

[42] Q.v. Stephen Gaukroger, "The Resources of a Mechanist Physiology and the Problem of Goal-Directed Processes," in *Descartes's Natural Philosophy*, ed. S. Gaukroger, J. Schuster, and J. Sutton (London and New York: Routledge, 2000), 387–8; Gaukroger follows here a general assessment of Cartesian force as divine outlined by Martial Gueroult, "The Metaphysics and Physics of Force in Descartes," in *Descartes: Philosophy, Mathematics and Physics*, ed. S. Gaukroger (Sussex: Harvester Press, 1980), and Gary Hatfield, "Force (God) in Descartes's Physics," *Studies in History and Philosophy of Science* 10 (1979): 335–70.

with implications for heredity, several years before he fully embraces atomism in the context of his Epicurean explorations:

> [T]he soul arrives already with instructions for the production of organs which it then initiates, beginning from the body generating it. This was because all parts of the soul existing in any living creature seemed to have the same function – so much so that all that the other parts comprise, any part could comprise that function insofar as that part contained the idea of all parts. I fancied taking the example of the cut branch of a willow or similar tree (as it is truer for a seed the soul of which is its principal concern?). After this branch was replanted, we recognized that, when one part was in the willow, its soul learned how to spread not only the branches and leaves but also the roots, which seemed to be the proper function of the other part of the soul. Thus, I concluded that a little soul existing within could, thanks to the seed that generated it, make the same thing as that which remained a whole and, once cut, carried with it the idea of all the parts to which the remaining part had given form. Thus, accustomed to the nutrition, conservation and continual conformity of a body which generates it, the little soul continues the work begun in the seminal particle. The little soul makes the same thing as what made it before – but initially in a weak way, according to its measure and with tenderness, according to a new way of acting and in order that other parts are entrusted to it. Then, gathering so to speak gradually and strengthening themselves, or because the little soul meets matter that is more ready, that soul accomplishes more suitable work and is much more energetic.[43]

There is other evidence that he had accepted a particulate matter theory this early, and his references to molecules and particles in this context are clearly consonant with an early corpuscularianism as well as his later atomism. One striking aspect of this timeline is that, by 1629, he had conceived of this account on what are perforce "generically" materialist grounds, without an explicit commitment to a particular particulate matter theory (save for his molecularism). The timeline is also striking because his later atomism changes nothing from 1629 and only adds the suggestion that seminal matter has a specifically atomist understructure and is endowed with a special force associated with atomic amalgams of that sort. Without diminishing the significance of his folding the generation and heredity accounts into his broad *atomist* program, this one insightful piece of the correspondence highlights Gassendi's even more central commitment to a broadly *materialist* story of the origins of individuals and the manner in which they inherit their features.

---

[43] *Opera Omnia* VI 19a–b.

6

Atoms and Minds in Walter Charleton's Theory of Animal Generation

ANDREAS BLANK

1. INTRODUCTION

The generation of animals, and especially the generation of human beings, is a recurrent theme in the work of the British physician and philosopher Walter Charleton (1619–1707). Based on an atomistic analysis of generation and corruption in his *Physiologia Epicuro-Gassendo-Charletoniana* (1654), he develops a purely mechanistic theory of animal generation in the *Natural History of Nutrition, Life, and Voluntary Motion* (1659). In later writings such as the *Natural History of the Passions* (1674), he expands the basic outlines of his views on animal generation into an atomistic account of emergent properties of higher animals. In addition, in works such as the *Dissertatio Epistolica de ortu animae humanae* (1659) and *The Immortality of the Human Soul, Demonstrated by the Light of Nature* (1659), he attempts to reconcile an atomistic view of the generation of the human organism with the Christian doctrine of an immortal human soul.

In recent years, two interpretations of the methodology behind Charleton's theory of animal generation have been influential. The first interpretation, put forth by Margaret Osler, ascribes to Charleton an empiricist methodology that, according to her view, is due to theological views that emphasize the role of God's will rather than the role of God's intellect.[1] According to Osler, Charleton's voluntarist theology leads to an empiricist theory of knowledge. For example, she ascribes to him the view that some of the primary qualities

[1] Margaret Osler, "Descartes and Charleton on Nature and God," *Journal of the History of Ideas* 40 (1979): 445–56, esp. 453–6. For an analogous interpretation of the theological background of Gassendi's philosophy of nature, see Margaret Osler, *Divine Will and the Mechanical Philosophy: Gassendi and Descartes on Contingency and Necessity in the Created World* (Cambridge and New York: Cambridge University Press, 1994). Charleton developed his theological views in *The Darknes of Atheism Expelled by the Light of Reason* (London: William Lee, 1652).

124

of atoms can be known only by empirical methods. Therefore, she thinks that according to Charleton primary qualities of atoms can be known at best with a good degree of probability. In fact, such an interpretation seems to be supported by the following passage from the *Physiologia*:

> Of the existence of *Bodies* in the World, no man can doubt, but He who dares indubitate the testimony of that first and grand Criterion, SENSE, in regard that all *Natural Concretions* fall under the perception of some of the Senses: and to stagger the Certitude of Sense, is to cause an Earthquake of the Mind, and upon consequences to subvert the Fundamentals of all Physical Science. Nor is Physiology, indeed, more then the larger Descant of Reason upon the short Text of Sense: or all our *Metaphysical* speculation (those only excluded, which concern the Existence and Attributes of the Supreme Being, the *Rational Soul* of man, and *Spirits*: the Cognition of the two former being desumed impressions implantate, or coessential to our mind; and the beliefs of the last being founded upon Revelation supernatural) other then Commentaries upon the Hints given by some one of our External senses.[2]

By contrast, the second interpretation reads Charleton's philosophy of nature as an example of early modern eclecticism. Michael Albrecht and Eric Lewis characterize this form of eclecticism as the attempt to select what is true in the work of ancient philosophers such as Empedocles, Plato, Aristotle, Anaxagoras, and Democritus and to combine this with what is true in modern mechanical philosophy.[3] In fact, talking about philosophers he labels "Electors," Charleton in the *Physiologia* confesses "[h]ere to declare ourselves of this Order."[4] As Albrecht points out, applying the notion of eclecticism to the *Physiologia* is problematic because this work is an expanded and reworked translation of passages from Pierre Gassendi's *Animadversiones in decimum librum diogenes Laertii*. As Albrecht argues, Charleton in the *Physiologia* does not select elements from various sources but rather adopts Gassendi's atomism as a whole.[5]

Although these interpretations shed light on interesting features of Charleton's thought, they do not adequately represent the extent to which

---

[2] Walter Charleton, *Physiologia Epicuro-Gassendo-Charletoniana: A Fabrick of Science natural Upon the Hypothesis of Atoms* (London: Thomas Heath, 1654), 18–19. Citations of passages from English works follow the original orthography; translations of passages from Latin works are my own; all italics are those of the original texts.

[3] Michael Albrecht, *Eklektik: Eine Begriffsgeschichte mit Hinweisen auf die Philosophie- und Wissenschaftsgeschichte* (Stuttgart: Frommann-Holzboog, 1994), 276–8; Eric Lewis, "Walter Charleton and Early Modern Eclecticism," *Journal of the History of Ideas* 62 (2001): 651–64.

[4] Charleton, *Physiologia*, 5.

[5] Albrecht, *Eklektik*, 276. For a full exposition of Gassendi's theory of animal generation, see François Duchesneau, *Les modèles du vivant de Descartes à Leibniz* (Paris: Vrin, 1998), chap. 3.

Charleton's views on animal generation are rooted in a kind of methodological pluralism which goes beyond empiricism and eclecticism. Already the *Physiologia* applies a range of argumentative strategies that, in Charleton's own view, do not have the status of proofs but rather are designed to show the "verisimility"[6] of the doctrines he defends. Thus, it is important to emphasize that these argumentative strategies do not amount to an aprioristic, axiomatic-deductive methodology. Nevertheless, he applies a variety of argumentative techniques that purport to provide rational grounds for preferring a version of an atomistic doctrine of animal generation to various alternative accounts. At various places, Charleton uses the Epicurean-Stoic theory of "common notions" for this purpose. In the *Physiologia*, he modifies Gassendi's views on the nature of common notions by interpreting these notions as something expressed in our everyday language. In Charleton's later writings, analogical arguments play a role in the explication of animal generation by means of its analogy with nutrition, thus filling in the conceptual framework provided by common notions. In particular, he uses this analogy to capture the role of vital heat and "vital spirits" – subtle parts of matter – in the process of embryo formation.

Moreover, as the passage from the *Physiologia* just cited indicates, Charleton already at this early stage makes two important restrictions to an empiricist program: according to his view, there are cognitive contents that can be known only by revelation, and, more significantly for the present context, there are cognitive contents that are in some sense innate. One of the aims of this essay is to spell out in what sense, according to Charleton, there is a realm of innate ideas and how this bears on his views on the origin of minds. In the *Immortality of the Human Soul*, he denies that common notions are all caused by sense perception and claims that some of them are known by means of reflection of the mind on its own operations. Contrary to Gassendi, he holds the view that through reflection these notions are not formed but rather made explicit. In this sense, he regards innate common notions as implicit knowledge always structuring our thought about the mind. Interestingly, throughout his work Charleton does not regard common notions as something that could be justified. Rather, like Gassendi he interprets them as criteria for judging the truth of given propositions. Whereas in the *Physiologia* common notions still are seen as something caused by sense perception and therefore as something revisable, in later writings common notions based on reflective knowledge are seen as something having certainty. Nevertheless, they are not introduced as an aprioristic starting point of deductive arguments; rather the process

---

[6] See Charleton, *Physiologia*, 419.

of transforming implicit knowledge of innate common notions into explicit knowledge is characterized as a process of analysis starting with a description of aspects of mental activity.

2. THE METAPHYSICS OF GENERATION AND CORRUPTION

Not only Charleton's version of atomism but also some of his most interesting methodological ideas are rooted in the philosophy of Gassendi. Thus, describing in what sense Charleton's methodology involves more than an early version of empiricism cannot be separated from the question in what sense Gassendi's methodology goes beyond empiricism. Let us begin with what Gassendi in the *Animadversiones* says about the nature of common notions or "anticipations":

> [A]n anticipation is at first some singular thing, or so to speak the idea of a singular thing, in so far as it is impressed by a singular thing and represents the singular thing by which it is created; but subsequently it is a universal, insofar as not only the thing by which it is created but also by means of its imitation several similar ones are imagined by the mind.[7]

Moreover, he gives to his view of the nature of anticipations the following canonical formulation: "All anticipation or precognition, which is in the mind, depends on the senses, and this either by means of incursion, or proportion, or similitude, or composition."[8] Interestingly, he deals with the issue of anticipations under the general heading of criteria of truth, and he explains that a criterion is "an organon or an instrument of judging."[9] Moreover, he explicitly identifies the Epicurean *prolepsis* with the Stoic "common notions," with Aristotle's "pre-existent cognition," and with Cicero's "presumption" and "information anticipated in the mind."[10] Especially using the juridical term "presumption" to characterize the nature of anticipations makes it clear that what Gassendi has in mind is not something like an empirical *justification* of common notions. If empiricism, in overly general terms, is characterized as the view that all cognitive contents can be justified in an inductive way, Gassendi's theory of common notions is not a form of empiricism. Rather, he

---

[7] Pierre Gassendi, *Animadversiones in Decimum Librum Diogenis Laertii, qvi est de vita, moribus, placitisque Epicuri*, 3rd ed. (Lyon: Barbier, 1675), 1: 80. For an exposition of Epicurus's theory of "proleptical" notions and Gassendi's adaptation of this theory, see Wolfgang Detel, *Scientia rerum natura occultarum: Methodologische Studien zur Physik Pierre Gassendis* (Berlin and New York: De Gruyter, 1978), esp. 33–8, 52–5.

[8] Gassendi, *Animadversiones*, 1:90.

[9] Ibid., 71.

[10] Ibid., 79. Gassendi refers the reader to Aristotle *An. post.* 1, 1, and to Cicero *De divin.* 2.

sees the relation of sense perception to common notions as initially a purely causal one. Nevertheless, these concepts subsequently acquire a new function as criteria for judgments about the truth of propositions. In particular, they acquire this function because they lay down standards of similarity. In the sense that particular things play a role in forming standards of similarity, Gassendi regards not only ideas representing particular things but also particular things themselves as something conceptual.

Much of what Charleton says in the *Physiologia* about the role of common notions in assessing the rational acceptability of various theories of animal generation can be seen as a modification of Gassendi's view of the nature of common notions. As in Gassendi, the connection between experience and common notions is not one of justification, and as in Gassendi, common notions are used as instruments of judgment. The modification Charleton introduces is twofold: First, he shifts the emphasis from the role of common notions as criteria of truth to their role as criteria of what he calls "verisimility." This concept clearly differs from an empiricist conception of inductive support and rather has to do with assessing the degree of rational acceptability as an indicator of closeness to truth. Second, he interprets common notions as something expressed in and therefore accessible through the analysis of everyday language. Thus, the conception of, for example, generation and corruption implicitly contained in our everyday language is seen as a criterion for judging the closeness to the truth of different metaphysical accounts of generation and corruption.

One of the applications of common notions can be found in Charleton's discussion of the question in the framework of which metaphysical theory of mixing parts the generation of complex objects (including animals) can be best understood. He formulates the following alternatives:

> [T]here are Two different kinds of *Commistion*, whereof the one is, by Aristotle (*de Generat. 1 cap. 10*) termed Σύνθεσις, *Composition*, and by others, παράθεσις, *Apposition*: the *other* is called, in the Dialect of the Stoicks, Σύγχυσις *Confusion*, and in that of *Galen*, πρᾶσις, *Coalition*, or *Temperation*. The *Former* is when those things, whether Elements, or others, that are mixed together, do not interchangeably penetrate each others parts, so as to be conjoined *by means of minima*; but either themselves in the whole, or their parts, onely touch each other superficially. . . . The *Latter*, when the things commixed, are so seemingly united, and concorporated, as that they may be conceived mutually and totally to pervade and penetrate each other, *by means of minimal parts*, so as that there is no one insensible particle of the whole mixture, which hath not a share of every ingredient.[11]

[11] Charleton, *Physiologia*, 418; see Gassendi, *Animadversiones*, 1:207.

Subsequently, he begins arguing for the adequacy of the first view of the nature of mixing parts and against the adequacy of the second view by using an ordinary language argument:

> If we look no further than the *Common Notion*, or what every man understand by the Terme, *Mistion*; it is most evident, that the things commixed ought to *Remain* in the *Mistum*; for if they do not remain, but Perish, both according to substance and Qualities, as *Aristotle* and the *Stoicks* hold, then is it no Mistion but a *Destruction*: and since the propriety of this Notion cannot be solved by any other reason, but that of the *Atomists*, that the particles of things are in commistion onely apposed each to other, without amission of their proper natures; what Consequence can be more naturall and clear than this, than that their opinion is most worthy our Assent and Assertion?[12]

This line of argument is developed further in the direction that not only a single common notion should be used as a criterion of judging the adequacy of a theory of mixing but a whole net of common notions involving not only the relation between the parts and a composite entity but also the structure of space and extension. This becomes clear in his discussion of an attempt he ascribes to Chrysippus to save the common notion by claiming that the particles of things mixed keep their substance and qualities but penetrate each other. Charleton objects that "from that his Position it necessary follows. (1) That two Bodies are at once in one and the same place, both mutually penetrating each others dimensions, or without reciprocall expulsion (2) That a pint of Water, and a pint of Wine commixed, must not fill a quart . . . (3) That a very small Body may be Coextensive, or Coequate to a very great one."[13] Thus, in Charleton's view Chrysippus's attempt to save the common notion of mixture violates other common notions concerning the location and extension of bodies. A similar strategy that uses a sorites argument to bring out everyday intuitions as a rational criterion for the adequacy of theoretical claims is expressed in a passage directly derived from Gassendi's *Animadversiones*:

> Nor, indeed, hath *Aristotle* Himself been more happy than *Chrysippus*, in his invention of a way, to remove or palliate the gross repugnancy of his opinion, to the proper importance of the term, Commistion, as may easily be evinced by a short adduction of it to the test of reason. . . . [W]e should only demand of him, if after the instillation of one single drop of Wine into 10000 Gallons of Water, a second drop should be superinfused, and after that a third, a fourth, and so more and more successively, till the mass of Water were augmented to

---

[12] Charleton, *Physiologia*, 419.
[13] Ibid.

ten, a hundred, thousandfold: *of what Nature would the whole mixture of Wine and Water be?*[14]

Charleton goes on again to invoke common notions. He argues that if Aristotle had in mind that the resulting middle thing arises from the destruction of both ingredients, the original parts would not remain, which, in Charleton's view, contradicts both Aristotle's own assumptions and our common notion. Alternatively, if Aristotle had in mind that the resulting middle thing participates in the properties of the ingredients, then the question Charleton (using the unusual term "mistile" for an ingredient of a mixture) poses is, "How, and in what respect, that Middle and Common thing comes to be participant of the Extremes of each Mistile?"[15] Following Gassendi, he claims that all answers that are possible from an Aristotelian viewpoint are contrary to common notions. In case the middle thing participates in the matter of the ingredients, Aristotle must admit that the whole matter of the parts is contained in the composite entity; and if, as the common notion demands, these portions of matter occupy different places in space, the parts can touch each other only at their surfaces. In case the middle thing participates in the forms of the ingredients, as Charleton believes, Aristotle would have to admit that the forms of the parts survive in the composite entities, because otherwise, contrary to the common notion, it would be a case of corruption of the ingredients. Finally, in Charleton's view something similar holds for the qualities of the ingredients: "neither ought *Aristotle* to deny the *permanence* of them: for, since in them consisteth the chief Capacity or Power of recovering the last Forms: if they perish, how can they be inservient to the recovery of the Forms?"[16]

Thus, a theory of the generation of composite entities in terms of composition of minimal parts that touch each other superficially but do not change their intrinsic qualities is seen as rationally preferable to the view Charleton ascribes to Aristotle and Galen. That the ingredients of composite entities have these structural properties and not those associated with the theory of the confusion of minimal parts is not presented as something capable of inductive support. At the same time, these concepts and intuitions also are not characterized as something providing support for logically conclusive arguments. Rather, thinking of composite entities in terms of a composition of unchangeable natural *minima* is described as something implied in our everyday concepts and intuitions used as rational criteria of judging the "verisimility" of theories. Charleton does not claim that an atomistic theory of the composition

---

[14] Ibid., 420; see Gassendi, *Animadversiones*, 1: 208.
[15] Charleton, *Physiologia*, 420–1; see Gassendi, *Animadversiones*, 1:208.
[16] Charleton, *Physiologia*, 421–2; see Gassendi, *Animadversiones*, 1:209–10.

of minimal parts is the only possible account of generation and corruption that is able to meet these criteria. Nevertheless, his use of common notions shows that an atomistic metaphysics of generation and corruption is compatible with the demands of reason, whereas existing alternative accounts are not.

3. VITAL HEAT, VITAL SPIRITS, AND ANIMAL GENERATION

To fill out the framework of generation as combination of minimal parts, Charleton in the *Natural History of Nutrition, Life, and Voluntary Motion* makes use of an analysis of the process of nutrition in order to arrive at a description of some more specific aspects of the process of animal generation by means of analogical reasoning. That an informative analogy between animal generation and nutrition holds is made plausible by the following argument:

> To forme, and nourish, are not only acts of one and the same soul; but so alike, that it is no easie matter to distinguish betwixt them. For, *Generation* and *Accretion* are not performed without *Nutrition*; nor *Nutrition*, or *Augmen-tation*, without *Generation*. To nourish, is to substitute such and so much of matter, as was decay'd in the parts. . . . In like manner, *Accretion* is not effected without Generation; for all natural bodies, upon the accession of new parts are augmented, and those new parts are such of which these bodies were first composed: and this is done, according to all the dimensions; so that, to speak properly, the parts of an Animal are encreased, distinguished, and organized all at once.[17]

In particular, the analogy between generation and nutrition adds to the general metaphysical theory of generation two features characteristic of animal generation. The first one concerns the idea of the homogeneity of parts entering into an organic body during the process of generation with those already components of the organism:

> Nature doth nourish and amplify all parts of an Animal with the same matter, or humour (not with a diverse) out of which she constituted or framed them at the first. Because, whatsoever is superadded to the parts, during their growth, ought to be of the same substance, with what was praeexistent, and so must consist of matter of the same genus: their Renovation as well as first Corporation

---

[17] Walter Charleton, *Natural History of Nutrition, Life, and Voluntary Motion* (London: Henry Herringman, 1659), 2. A Latin version was published under the title *Oeconomia animalis, novis in medicina hypothesibus superstructa & mechanicè explicata* (London: Daniel & Redmann, 1659).

being effected by épigénèsis, Aggeneration, or superstruction. So that we may
well conclude, that Nutrition is nothing else but continual Generation: and as
necessary to the Conservation of every individual nature, as Generation itself
is to the conservation of the Universe.[18]

The second feature of animal generation illuminated by means of the analogy
with nutrition is the role and nature of vital heat. In Charleton's view, the vital
heat at work both in nutrition and animal generation can be compared to a
flame constantly requiring fuel:

> That since the chief principle of life in every Animal, is a certain indigenary
> Heat (analogous to pure flame, such as the most rectified Spirit of Wine yields,
> upon accension) which by continuall motion and activity agitates the minute
> and exsoluble particles of the body, doth dissolve, and consume, or disperse
> them; of necessity, the whole Fabrick would soon be destroy'd, unless there
> were a continuall renovation or reparation of those decayes, by a substitution
> and assimilation of equivalent particles, in the room of those dispersed and
> absumed.[19]

This naturalistic conception of vital heat explicitly is seen as alternative to
Aristotelian accounts of vital heat involving a kind of celestial influence on
the sublunary world:

> [A]ll Fire whatever (that Elementary Fire, which the *Aristotelians* conceive to
> be so pure, as to need no *pabulum* or aliment, being a meer Chimera) doth
> conserve it selfe onely by the destruction of the matter, in which it is generated,
> So that, indeed, we have one and the same Cause both of our Life, and of our
> Death; or (to speak more properly) our Life is nothing but a continuall Death,
> and we live because we dye.[20]

In fact, an Aristotelian such as William Harvey holds the view that the conception of innate heat cannot be explicated without interpreting innate heat
as a "celestial substance." In this sense, Harvey characterizes innate heat as
the "instrument of God."[21] By contrast, Charleton uses the framework of an
atomistic ontology to formulate a thoroughly naturalistic theory of innate heat.
Nevertheless, this naturalistic conception is not the outcome of an inductive

---

[18] Charleton, *Natural History of Nutrition*, 2–3.
[19] Ibid., 3–4.
[20] Ibid., 4–5.
[21] William Harvey, *Exercitationes de generatione animalium: Quibus accedunt quaedam de partu: de membranis ac humoribus uteri: & de conceptione* (London: Octavianus Pulleyn, 1651), exercitatio 70.

methodology. Rather, the mode of operation of innate heat is characterized by means of an analysis of the concept of fire:

> Flame (as reason defineth it) is a substance luminous and heating, consisting in a perpetuall *Fieri, i.e.*, an indesinent accension of the particles of its pabulum, or combustible matter, and perishing as fast as it is generated: so that fire is made fire, and again ceaseth to be fire, in every, the shortest moment of time. . . . Continual *Dispersion*, therefore, being the proper effect of Fire; the matter or fewell, whereon it subsisteth, cannot but be in perpetuall flux or decay. In like manner . . . the *Lamp of life* consisting in a continuall accension of vital spirits in the blood, as that passeth through the heart; those vital spirits, transmitted by the arteries to the habit of the body, no sooner arrive there, but as they warme and vivifie the parts, so do they immediately fly away, and are dispersed into the air.[22]

The Stoic theory of common notions again comes into play, this time in the form of invoking a "common axiom," when Charleton writes about the causes of the renovation of parts in nutrition:

> The *Material*, or Constitutive principle, we take to be a certain sweet, mild and balsamical Liquor, analogous to the white of an egge, out of which the chicken is formed. For since all Animals are nourished with the same, out of which they were at first fabricated, according to that common Axiom, *we nourish ourselves of the same stuff we consist of;* . . . and since they have their origine *out of the Colliquamentum*: we may well conclude, that the *Nutritive Juice* is in all qualities correspondent to the Colliquamentum of the white of an Egge.[23]

He distinguishes three stages in the process of nutrition.[24] In the first stage, which he calls "apposition," parts of the nourishment are brought into contiguity with parts of the body. In the second stage, "agglutination," parts of the nourishment enter into a continuum with parts of the body. Given his general views about the nature of generation as the combination of minimal parts touching each other on their surfaces without internal change, such a process can take place only on the level of complex constituents of food and the organism. In the third stage, "assimilation" or "transmutation," parts of the nourishment are "made of the same substance with" parts of the body. Again, given the general account of generation and corruption, as well as the more specific view that food and the organism are built of parts of the same kind, the process Charleton may have had in mind here seems not to have

---

[22] Charleton, *Natural History of Nutrition*, 6–7.
[23] Ibid., 9.
[24] Ibid., 10.

been so much a change in the internal nature of the constituents of food but rather their acquiring a causal role within the organism.

In subsequent chapters of the *Natural History of Nutrition, Life, and Voluntary Motion*, he uses the analogy between animal generation and nutrition as the general framework of a physiological account of embryo formation. To develop this account, Charleton discusses whether there are "milky veins" (*venae lacteae*) connecting the stomach, the uterus, and the breasts, transporting nutritive juice or a substance analogous to it. As he points out, no anatomist ever has been able to discover these passages. Nevertheless, he thinks that the view that they exist is "highly probable."[25] In a first step, he argues for the claim that milk and the nutritive juice are "one and the same thing." In his view, this is apparent from the fact that milk and nutritive juice agree in "all their qualities" and, moreover, that they are convertible into each other. Among the qualitative resemblances, he lists the following:

(1) They both have a fatty substance: otherwise neither could be fit either to sustain the Lamp of life, or to instaurate the parts; nor can the bloud contain any such fatty substance in it, but what is derived from the Chyle. (2) As Milk doth consist of two parts, the *serum* and *crassamentum*; so likewise doth Chyle.... (3) As Milk, if kept over-long, especially in a warm place, or corrupted by any Acid juice, doth turn sowr; so also doth the Chyle.[26]

Moreover, he thinks that nursing a child proves that, in the woman, the nutritive juice is converted into milk and that, in the stomach of the infant, the milk is reconverted into nutritive juice. As he claims, considering these resemblances we can conclude that "they have much more of reason on their side, who conceive Milke to be nothing but meer Chyle brought from the stomach to the Paps, by peculiar passages; and therein promoted to somewhat more of perfection: than they, who think it to be made of bloud whitened in the glandules of the paps."[27]

In a second step, he argues for the "verisimilitude" of the claim that the nutritive juice plays a role in the formation of the embryo. To substantiate this claim, he invokes the authority of Hippocrates as well as the observations of Harvey. As he points out, according to Hippocrates, *"The foetus attracts what is most sweet in the blood, and at the same time benefits from a small portion of milk.* Where He hinteth the true cause, why it is unwholesome and dangerous for Infants to suck women with child, *viz.* because the best of the

---

[25] Ibid., 19–20.
[26] Ibid., 24.
[27] Ibid., 25.

milk is attracted by the Foetus, in the womb, and the worst is carried to the paps."[28] Moreover, he quotes Harvey's description of cavities (*cotyledones* or *acetabula*) found in the bellies of pregnant animals:

> [I]nto each of them penetrate deeply the most fine branches of the vessels of the umbilical cord: for in them the aliment of the foetus is contained, viz. not a bloody but a mucous one, of a very similar texture as the white substance in fat people. From which it is also manifest that the foetus of split-footed animals (as also all others) is not nourished by the mother's blood.[29]

To this, Charleton suggests to add "the consideration of that great *Sympathy* or consent betwixt the womb and paps, so frequently observed in women":

> Which Consent cannot be caused by nerves, nor by veins, nor by arteries, nor by similitude of substance ... and therefore most probably, by mediation of these presupposed Chyliferous vessels tending from the paps to the womb. (1) Not by *Nerves*; because the paps derive their nerves from the fourth intercostal pair, or the fifth pair of the thorax: and the womb is supplied with sense from the nerves of the *os sacrum*, and also from the sixth conjugation of the brain. (2) Not by *veins or arteries*; because they are, both, destitute of sense.... (3) Not by *Similitude of Substance*; because the paps consist mostly of Gland-ules, and the body of the womb is membraneous.[30]

Thus, in the two argumentative steps just considered, empirical observations are not used for giving a certain physiological claim inductive support. In this sense, the "verisimilitude" Charleton has in mind here is not one of a degree of inductive justification. Rather, observations are used to exclude alternative physiological views according to which both the production of milk and the formation of an embryo are due to the causal role of blood.

The view that animal generation can be understood through its analogy with nutrition also stands behind Charleton's account of the formation of blood. As in the case of the integration of material parts into the organism of an existing animal, he claims that blood is formed by the activity of vital spirits. Referring the reader to Harvey's description of the early stages of the development of the chicken in the egg, he writes about the transformation of the white of the egg into blood:

> Certain it is, this cannot be effected by any thing that was red before; because there is no part of the Egge of, or inclining to, that colour; and the yelk remains

---

[28] Ibid., 28; see Hippocrates *Liber de natura pueri* §21.
[29] Charleton, *Natural History of Nutrition*, 28–9; see Harvey, *Exercitatio de uteri membranis & humoribus*.
[30] Charleton, *Natural History of Nutrition*, 29–30.

intire a good while after there is bloud to be seen in the *punctum saliens*. Nor is it the Fleshy parts, that communicate this vermillion tincture to the bloud, because they remain white after the bloud is made out of the Colliquamentum: and it is much more reasonable, that the fleshy parts derive their rednesse wholly from the bloud, perpetually irrigating and washing them in its Circulation. . . . Again, nothing can have an activity, before it hath a being: and consequently the solid parts cannot give a rednesse to the bloud, because they are not in being, till after the bloud. Nothing, therefore, remains to be the Efficient of the Bloud, but the *Vital Spirit*, kindled originally in the purest part of the seminal matter, or *Colliquamentum* which we may well denominate the *Vital Liquor*.[31]

According to Charleton, this shows that the production of blood in the process of embryo formation is not the work of already formed organs. Rather, the production of blood, like the other stages in the formation of an embryo, is characterized as the activity of vital heat and vital sprits, which in a purely combinatorial way connect homogeneous parts into an organic whole. In this sense, Charleton characterizes the formation of blood as a process of "*simple Assimilation*" and consequently the activity of vital spirits as an "Action *similar*, not organical."[32] And, in his view, this is the type of action that takes place both in nutrition and animal generation.

4. EMERGENT PROPERTIES AND THE PROBLEM OF THE ORIGIN OF MINDS

His account of the role of vital heat and vital spirits in animal generation leads Charleton to a purely materialistic view of sensitive souls, including the sensitive part of human souls. In the *Natural History of the Passions*, he characterizes the sensitive soul as "*Corporeal*, and consequently *Divisible*, *Coextense* to the whole Body; of *Substance* either Fiery, or merely resembling Fire; of a *consistence* most *thin* and *subtile*, not much unlike the flame of pure spirit of Wine, burning in a paper Lantern."[33] For the coextension of the sensitive soul and the body, he there adduces an argument that shows that he regards passions as a result of corporeal and divisible vital spirits:

I am apt to suspect, that not only part of the Vipers Soul, but *Anger* and *Revenge* also survived in the divided head. For, it is well known, the bite of a Viper is never Venomous, but when he is enraged: the Chrystalline liquor contained in

[31] Ibid., 40–2.
[32] Ibid., 33.
[33] Walter Charleton, *Natural History of the Passions* (London: James Magnes, 1674), 5–6.

the two little Glandules at the roots of his fang teeth, being then by a copious afflux of Spirits from the Brain, and other brisk motions thereupon impress'd, in anger (of all passions the most violent and impetuous) so altered, and exalted, as to become highly active and venenate.[34]

But he ascribes to sensitive souls not only passions but also a kind of consciousness. In this, he draws on aspects of the theory of animal cognition developed in *De Anima Brutorum* (1672) by Thomas Willis (who in turn derives his atomistic analysis of the nature of sensitive souls from Charleton's earlier physiological works).[35] Already in the case of lower animals, Willis holds that the animal soul has the capacity "to moderate its own faculties." In his view, this capacity explains why the whole animal displays properties that go beyond the properties of its organic parts. Interestingly, he labels these properties as "emergent."[36] Moreover, he characterizes the capacities of higher animals as "the ability to modify and to combine [action] types in their souls." In particular, he ascribes to animal souls "the capacity to *know* about some necessary things & to be *active*."[37] Nevertheless, he limits this capacity to actions determined by instinct:

> Because in all these actions one thing is always performed without any variation and, what is more, in the same manner; this indicates that . . . it neither is initiated by external objects whose impulse is always varied & diverse, nor by an internal intention of the soul, which is mutable as the wind, but in fact by a more certain & fixed principle, which is always determined towards the same, which can only be *natural instinct*.[38]

Explicitly referring the reader to Willis's work, Charleton tries to understand animal consciousness in the framework of a theory of composite action:

> We are therefore to search for this *Power* of a Sensitive Soul, by which she is *conscious* of her own perception, only in *Matter* in a peculiar manner so, or so disposed or modified. But in what matter? this of the *Soul*, or that of the *Body*? Truly, if you shall distinctly examine either the Soul or the Body of a Brute, as not conjoyned and united into one *Compositum*; you will have a hard task of it, to find in either of them, or indeed in any other material subject whatever, any thing to which you may reasonably attribute such an *Energetic* and *self-moving Power*. But if you consider the *whole* Brute, as a *Body animated* . . . then you

---

[34] Ibid., 7–8.
[35] Thomas Willis, *De Anima Brutorum, quae hominis vitalis ac sensitiva est, exercitationes duae* (Oxford: Richard Davis, 1672). For Willis's atomistic theory of the sensitive soul, see esp. 86–90.
[36] Ibid., 95–9.
[37] Ibid., 99–100.
[38] Ibid., 102.

may safely conclude, that a Brute is... so comparated, as that from Soul and body united, such a confluence of Faculties should result, as are necessary to the ends and uses for which it was made.[39]

Thus, animal consciousness as the capacity to perform composite actions and thus to modify action types is analyzed in terms of emergent properties that are due to the compositional structure of the organic body of an animal. Nevertheless, in spite of the explanatory resources of an atomistic theory of sensitive souls, Charleton felt that there is something to human souls that cannot adequately be captured in terms of the causal role of vital heat and vital spirits. Recall that in the passage from the *Physiologia* concerning the role of sense perception in theory formation cited at the beginning of this essay, he claims that knowledge concerning God and the human soul is based on innate ideas.[40] Moreover, in *The Immortality of the Human Soul* one of the interlocutors in an imaginary dialogue, Lucretius, articulates Charleton's own theological concerns about the immortality of the soul: "You may remember, Sir: I told you in the beginning, that though I am an Epicurean, in many things concerning Bodies; yet, as a Christian, I detest and utterly renounce the doctrine of that Sect, concerning Mens Souls."[41] Although Charleton holds on to an atomistic account of the nature and operation of the sensitive soul (including the sensitive part of the human soul), he works out arguments for the existence of an immaterial rational soul. Again, he bases his views about the nature of the rational soul on the doctrine of common notions, thus connecting some nonempiricist aspects of his epistemology with his theological concerns:

> I presume it will not be accounted paradoxical in me to affirm, that Immaterial Objects are most genuine and natural to the Understanding; especially since *Des Cartes* hath irrefutably demonstrated, that the Knowledge we have of the existence of the Supreme Being, and of our own Souls, is not only Proleptical and Innate in the Mind of man, but also more certain, clear, and distinct.[42]

Here, Charleton adopts aspects of a Cartesian epistemology to defend the idea of a second kind of common notions, a kind that is not even causally

---

[39] Charleton, *Natural History of the Passions*, 33–4.

[40] See Charleton, *Physiologia*, 18–19.

[41] Walter Charleton, *The Immortality of the Human Soul, Demonstrated by the Light of Nature* (London: Henry Herringman, 1659), 185. For an account of the contemporary theological responses to atomistic theories of animal generation, see Matthew R. Goodrum, "Atomism, Atheism, and the Spontaneous Generation of Human Beings: The Debate over a Natural Origin of the First Humans in Seventeenth-Century Britain," *Journal of the History of Ideas* 63 (2002): 207–24.

[42] Charleton, *Immortality of the Human Soul*, 119.

dependent on sense perception. He uses this kind of innate common notions to defend the immaterial and, therefore, immortal nature of the intellectual part of human souls. Like some of Gassendi's arguments for a similar view in a much shorter passage from the *Animadversiones*,[43] some of Charleton's arguments draw on an analysis of the structure of reflection. However, he goes beyond Gassendi's interpretation of reflective knowledge by defending an innateness thesis:

> [T]he Common Notions, that are as it were engraven on our Minds, and that are not derived originally from the Observations of things by our selves, or the Tradition of them by others, do undeniably attest the contrary. Nor can any thing be more absurd, than to say, that all those Proleptical and Common Notions, which we have in our Mind, do arise only from impressions made upon the Organs of our Senses, by the incurse of External Objects, and that they cannot consist without them: Insomuch as all sensible Impressions are singular, but those Notions Universal, having no affinity with, no relation unto Corporeal motions or impressions. And... what kind of Corporeal impression that may be, which formes this one Common Notion in our Mind, *Things that are the same with a third thing, are the same among themselves.*[44]

Subsequently, the relation of external objects to innate ideas is described as one of an occasional cause. Although innate ideas are not derived from external objects, we would not think about them without our interaction with external objects: "Not that those Objects have immited those very Idea's into our Mind, by the Organs of the Senses; but because they have immited somewhat, which hath given occasion to the mind to form such Idea's, by its own Innate and proper Faculty, at this time rather than at any other."[45] He explicates his views about innateness further:

> [N]othing comes to the Mind, from External Objects, by the mediation of the Senses, besides certain Corporeal Impressions; and yet neither those Impressions, nor the Figures resulting from them, are such as we conceive in the Mind; as *Des Cartes* hath amply proved in his Diopticks: Whence it follows, that the Idea's of Motions and Figures are innate to the mind; that is, that the mind hath an essential power to form them: for, when I say that such an Idea is in the Mind, I intend that it is not alwaies actually there, but *Potentially.*[46]

In his view, self-referential mental activity serves as a criterion for distinguishing innate common concepts from other concepts. Whereas in

---

[43] See Gassendi, *Animadversiones*, 1:291–2.
[44] Charleton, *The Immortality of the Human Soul*, 92–3.
[45] Ibid., 94.
[46] Ibid.

imagination, the mind is directed toward an image of a particular thing, in pure intellection, where no image is involved in the cognitive process, the mind "converteth it self upon it Self."[47] He claims that the fact that there are acts of reflection "needs no other testimony but that of a mans own Experience; it being impossible for any person living not to know, that he knows what he knows."[48] And, as he argues, the intellectual part of human souls must be immaterial because, whereas no material object can move itself, and all apparent self-motion of composite objects results from the causal interaction between their constituent parts, minds genuinely can act upon themselves.[49] In this way, the analysis of the origin of innate common notions leads Charleton to a dualistic ontology of human beings, including a dualistic ontology of the human soul.

Naturally, this dualistic conception poses serious difficulties for Charleton's views about the generation of human beings. Whereas he thinks about the generation of the human organism in terms similar to the generation of other higher animals, he would like to allow for a diverging view as to the origin of the intellectual part of human souls. What he proposes, however, is not a specific theory about the origin of minds. Rather, in the *Dissertatio epistolica* he restricts himself to criticizing existing theories, thus leaving open the possibility of divine concurrence in the generation of human souls. Interestingly, he not only rejects the theory of transference of souls from parents to their seeds defended by medical authors but also the Scholastic theory of divine *inanimatio*. To refute the Scholastic view that the soul is implanted by divine agency into the embryo at some point in its development, Charleton refers to Harvey's observations of the early stages of embryo formation in eggs. What is relevant in his eyes is the fact that Harvey showed that vital functions such as the production of a "nutritive juice" and of blood are present before a visible organic body is formed. From this he concludes,

> What relates to the FIRST [the theory of divine *inanimatio*]; we say that from the most reliable observations of this great interpreter of nature, Harvey, . . . it is clear that the soul is present to the foetus from the very beginning of its conception, before any part of the body is formed, in fact even before the system of the embryo develops a plastic force of nature: therefore it is dissonant to imagine, that the soul is implanted by divine power into an already formed body.[50]

---

[47] Ibid., 98.

[48] Ibid., 100.

[49] Ibid., 100–1.

[50] Walter Charleton, *Dissertatio epistolica, de ortu animae humanae*, published as an appendix to Walter Charleton, *Oeconomia Animalis, novis in medicina hypothesibus superstructa &*

Similarly, the theory of transference is contrasted with Harvey's famous but misleading observation that in the uterus there cannot be found anything more after coitus than before.[51] However, Charleton adds a more convincing discussion of the version of the transference theory of the origin of souls defended by the German physician and philosopher Daniel Sennert:

> Sennert assumes that the soul is the principal efficient cause of the foetus . . . and among others he proposes the following axiom; *Whatever work or effect produces an effect that is more noble than itself, or an effect that is dissimilar to it; is not a principal efficient cause but only an instrumental one.*[52] But given that this is true, who would not see that the seed of animals is not the primary efficient cause in the work of generation, but only an instrumental cause? Because it is easy to see that the foetus is an effect that is much more noble than the seed, and it is plainly dissimilar to it: therefore, the soul cannot be the offspring of the [male] parent, and it cannot be transmitted together with the seed into the uterus of the female parent. That the foetus is an effect that is more perfect than the seed, is obvious from the fact that art, intellect, judgment, and highest providence are apparent from its fabric; but the seed is not of a kind that art, intellect, & c. can rightly be attributed to it.[53]

Charleton concludes that, due to the shortcomings of the transference theory of the origin of the soul, "one has to have recourse to a prior, higher, and more perfect efficient cause."[54] However, he does not try to specify further the nature of this prior efficient cause. As he points out, the requirement of an efficient cause in addition to the seed is compatible with various theories of secondary causation discussed in the sixteenth and seventeenth centuries, be it in the framework of Aristotelian theories of the ether, neo-Stoic hypotheses about an all-pervading *pneuma*, Averroistic conjectures about a universal active intellect, Neoplatonic accounts of emanative causation, or Christian ideas about divine concurrent causation.[55] Thus, he obviously, and intentionally, leaves ends loose at this point. Nevertheless, his arguments are intended to show that not only observation but also rational principles concerning the concepts of instrumental and primary cause require the existence of causal

---

mechanicè explicata, 3rd ed. (London: Roger Daniel, 1666), 288–290*. Charleton gives 1659 as the date of the *Dissertatio epistolica*; erroneous page numbers in the 1666 edition are marked with an asterix.

[51] Charleton, *Dissertatio epistolica*, 282*–284*.

[52] See Daniel Sennert, *Hypomnemata physica* (first edition, Frankfurt: Schleich, 1636), in *Opera Omnia*, editio novissima, vol. 1 (Lyon: Huguetan, 1676), 127.

[53] Charleton, *Dissertatio epistolica*, 285*.

[54] Ibid., 285*.

[55] Ibid., 287*–288*.

factors external to those inherent in the seed. And, in his view, this makes a nonnaturalistic account of the origin of minds compatible with an atomistic account of the generation of animals.

5. CONCLUSION

The interpretive strategy pursued in this essay has been to disentangle issues concerning the methodological basis of Charleton's theory of animal generation from issues connected with his voluntarist theology. What speaks in favor of such an interpretive approach is that it makes visible elements of Charleton's methodology that are significant but otherwise tend to get weighed down by his empiricist leanings. Even if there is a voluntarist theology in Charleton's early writing, and even if basic insights of British empiricism can be found throughout the development of his work on animal generation, his methodology cannot simply be reduced to an early version of empiricism. Rather, in his views on animal generation and the origin of minds, a kind of methodological pluralism is at work, some aspects of which the present essay tries to describe. Already in the *Physiologia*, the interpretation of common notions as concepts contained in our everyday way of thinking about the world goes beyond empiricism. As in the philosophy of Gassendi, these concepts are not seen as something justified by experience but rather as something first caused by the world of experience and afterwards – independently from the initial causal connection – used as a criterion of judging the truth and falsehood of propositions. Although some common notions can be said to "come from" the senses, their use as criteria of truth belongs to the realm of reason.

Charleton applies this insight to everyday concepts of generation and corruption, using them as criteria for judging the "verisimility" of certain philosophical accounts of animal generation. Another nonempiricist mode of reasoning can be found in Charleton's use of analogies. In this perspective, Charleton exploits the similarities between nutrition and animal generation to fill out the conceptual framework set up by common notions. As we have seen, this analogy leads him to a purely mechanistic account of the role of vital heat and vital spirits in the formation of the compositional structure of the organism.

Finally, although he explains emergent properties such as emotions and animal cognition through the compositional structure of the organism, Charleton connects the doctrine of common notions also with a version of a theory of reflection. Due to the capacity of a substance to act on itself involved in reflection, the existence of innate common notions that are not even causally derived from sense perception, in his view, cannot be explained in the same way as an emergent property of interacting material components. His dualistic ontology of human beings, including his dualistic view of the nature of human souls, is based on this argumentative strategy. The presence of these various types of arguments substantiates the claim that the methodology behind his views on the generation of animals and human beings could be characterized as pluralistic. In particular, these argumentative strategies add to the methodology behind Charleton's theory of animal generation a type of reasoning that – without being an instance of an axiomatic-deductive method – goes beyond empiricism and eclecticism.[56]

[56] I owe a huge debt of gratitude to two institutions that made work on the present paper possible. Initial research was conducted during my time as a Visiting Fellow at the Center for Philosophy of Science at the University of Pittsburgh in the academic year 2002–3. The final version was written during my time as a Visiting Fellow at the Cohn Institute for the History and Philosophy of Science and Ideas at Tel Aviv University in the academic year 2004–5. I would like to express heartfelt thanks to Adolf Grünbaum, Jim Lennox, Leo Corry, Marcelo Dascal, and Nicholas Rescher. Many thanks also to Stephanie Härtel and Justin E. H. Smith for their helpful comments on a penultimate version.

IV

Second-Wave Mechanism and The Return of Animal Souls, 1650–1700

7

Animal Generation and Substance in Sennert and Leibniz

RICHARD T. W. ARTHUR

1. INTRODUCTION

Gottfried Leibniz is well known for his claim to have "rehabilitated" the substantial forms of scholastic philosophy, forging a reconciliation of the New Philosophy of Descartes, Mersenne, and Gassendi with Aristotelian metaphysics (in his so-called *Discourse on Metaphysics*, 1686). Much less celebrated is the fact that fifty years earlier (in his *Hypomnemata physica*, 1636) the Bratislavan physician and natural philosopher Daniel Sennert had already argued for the indispensability to atomism of (suitably reinterpreted) Aristotelian forms, in explicit opposition to the rejection of substantial forms by his fellow atomist Sébastien Basson.[1]

In this essay I want to set Leibniz's philosophy in a novel perspective by comparing it with Sennert's anticipation of his reintroduction of substantial forms. And I shall be doing this by looking at motivations for his theory of substance in his views on biological generation. I shall be especially concerned with the genesis of Leibniz's views and therefore with the context of his youthful commitment to atomism in the period from roughly 1666 to 1678. And I shall argue here, by a comparison with the case of Sennert, that this

---

[1] For the seminal idea of this paper (so to speak) I am indebted to a remark of Antonio Clericuzio in his discussion of Sennert's atomism in his superb *Elements, Principles, and Corpuscles: A Study of Atomism and Chemistry in the Seventeenth Century*, Archives Internationales d'Histoire des Idées, vol. 171 (Dordrecht: Kluwer, 2000), 29: "Forms, however, do not disappear from Sennert's natural philosophy. He firmly opposes Basson's rejection of substantial forms." For the idea of writing a paper around it, I am indebted to the invitation of the editor of this volume, Justin E. H. Smith. I am also indebted to the insightful remarks of several colleagues at the joint meeting of the Central Canada Seminar for the Study of Early Modern Philosophy and the Midwest Seminar in Early Modern Philosophy, September 4–5, 2004, where I read a preliminary draft. All Latin translations are my own, except where noted.

Richard T. W. Arthur

atomism, rather than being irrelevant to his mature philosophy of substantial forms and monads, actually throws considerable light on it. For it turns out that, despite the huge differences in the contexts in which the young Leibniz and Sennert wrote and in the details of their views, there is a fair degree of commonality between the two in their motivations. Both Sennert and Leibniz in his youth were committed to the orthodox Lutheran doctrine of *traducianism*, the propagation of souls through the medium of parental seeds or *semina*; and both posited an indivisible core in individual substances in which such souls were contained. In each case, also, the soul or active principle within each substance was what accounted for the lawlike development from a seed to a full-grown animal, a fact that both authors believed was inexplicable on a purely materialist version of atomism.

To be clear, I am not claiming here that Leibniz was consciously developing the doctrines of Sennert. Whether or not Leibniz was familiar in his youth with the details of Sennert's views, many of the views they held in common were also shared by other writers of the period. Sennert himself acknowledges a debt to Fortunio Liceti for the idea of the preformation of the seed in an animated atom, and much of what he says about the soul, substance, and mixtion (the forming of chemical compounds) derives from the views of Julius Caesar Scaliger, whom the young Leibniz had certainly read. Again, similar views to Sennert's concerning atoms, mixtion, and generation could be found variously in contemporary authors such as Joachim Jung, Robert Boyle, and J. B. van Helmont, and also in Pierre Gassendi, whom Leibniz often mentions as having been influential on his early views. Indeed, in this connection Gassendi is of particular interest in that, although he opposed the doctrine of the immateriality of (all but rational) souls, he too was committed to traduction, to which he gave an atomistic interpretation indebted to Sennert's. Gassendi is also often cited as one of the earliest and most influential advocates of the biological theory of *preformation*, according to which the seeds of organisms preexist in invisibly small kernels in animals rather than forming by the accretion of homogeneous matter alone. Animal development is explained in terms of the unfolding of these seeds, organized around a soul that is active in organizing the matter, rather than by the action of some other cause upon a previously undifferentiated and homogeneous matter (epigenesis). While this develops the views expressed by Sennert, it also prefigures Leibniz's mature doctrine of *transformation*, which, I shall argue, is best seen as a development of preformation.

The term "preformation" has, of course, many shades of meaning. What I have in mind in ascribing it to these three thinkers is quite specific: Sennert, Gassendi, and the young Leibniz, I argue, all subscribe to the view that every

148

organism develops from a preexisting atomic or molecular seed;[2] this seed is an organic body whose organization and subsequent development is governed by its soul; and this soul is a principle of activity and a final cause.

My basic argument is that there is a substantial consilience between the doctrines of Sennert and the young Leibniz, not only in their motivations (e.g., in their shared commitment to traducianism), but also in many points of detail: their subscribing to atomism, to the idea of minds or souls vivifying these atoms, to the indestructibility of these minds or souls, and to a Scaligerian theory of mixtion. Perhaps the most striking consilience between them, however, is in the position upheld by Leibniz after he rejected atoms: that substantial forms are not to be rejected but to be reinterpreted in such a way that there is one immaterial form for each organic body, that such bodies contain other organic bodies with their own forms, that the form is identifiable with the soul in the case of living beings, and that all souls originated at Creation. The root idea is that soul or form is not a faculty that is material in plants and animals, nor an immaterial substance that is infused into human bodies by God at conception and thus superadded to them; rather it is an immaterial spirit that governs the development of each organic body and takes it through all its transformations. On this point Leibniz corrects Descartes and Gassendi in very much the same way that Sennert had earlier corrected Basson.

This supports, so I argue, a different perspective on Leibniz's theory of substance from that generally entertained. Standardly, it is held that in his later work Leibniz rejected the reality of corporeal substances in favor of the view that there are only immaterial simple substances – *monads* – and aggregates of these. I contend that Leibniz never abandoned the axiom that *in created things there can be no form that does not have an organic body it informs*, so that there is no created monad which is not manifested as a corporeal substance.[3] Monads are substances in the sense of immaterial substantial principles. But every such substantial principle has an organic body

---

[2] As Joseph Needham and Walter Pagel have observed, it was atomic theories like those of Liceti, Sennert, Kenelm Digby, Gassendi, and Nathaniel Highmore that Harvey's doctrine of epigenesis was designed to combat, not the later theory of *emboîtement*. See Walter Pagel, "Aristotle and Seventeenth-Century Biological Thought," in *From Paracelsus to Van Helmont: Studies in Renaissance Medicine and Science* (London: Variorum Reprints, 1986), 501.

[3] I committed myself to this interpretation in "Russell's Conundrum: On the Relation of Leibniz's Monads to the Continuum: An Intimate Relation," *Studies in the History and Philosophy of Science*, ed. James Robert Brown and Jürgen Mittelstraß (Dordrecht/, Boston/, and London: Kluwer, 1989). Glenn Hartz, in "Why Corporeal Substances Keep Popping up in Leibniz's Later Philosophy," *British Journal for the History of Philosophy* 6 (1998): 193–207, has given a detailed account of Leibniz's enduring commitment to corporeal substances, and the view that for Leibniz monads are always manifested as corporeal substances has been independently put forward by

Richard T. W. Arthur

as an immediate and necessary concomitant of its activity. This is consistent
with the view urged by Hans Poser and others, that Leibniz's fundamental
orientation is a Neoplatonist one, as has been argued eloquently by Justin
E. H. Smith.[4] He has provided the following succinct formulation of this line:
"On such a picture, it is neither the case that there is nothing but spirit, nor that
there is independently existing matter. Rather, all actual matter serves as the
body of some soul or soul-like form.... In Leibniz's variety of pananimism,
there is no *hyle* that is not a *soma*, no matter that is not the organic body of
some monad."[5]

2. ATOMS AND SOULS

Daniel Sennert (1572–1637) is one of those thinkers who was influential in his
own time but has since faded from the memory of all but a handful of specialist
scholars. His name was eclipsed, predominantly, by the thorough scholarship
and wide influence of Pierre Gassendi (1592–1655), so much so that we tend
to think of Gassendi as a lone voice for the rehabilitation of atomism in
the seventeenth century, thus doing a disservice to all of Gassendi's atomist
precursors: Thomas Harriot, Nicholas Hill, Walter Warner, Francis Bacon,
Isaac Beeckman, David van Goorle, Sébastien Basson, Kenelm Digby, and
many others. All of them claimed classical precedents, not only Democritus
and Epicurus, of course, but often Plato too. Sennert identified his atoms
with *minima naturalia*, enabling him to claim a further classical precedent in
Galen.[6] For him atoms were not indivisible in an absolute sense but rather
minimal particles of various natural kinds of bodies:

Pauline Phemister in "Leibniz and the Elements of Compound Bodies," *British Journal for the
History of Philosophy* 7, nos. 1–2 (1999): 57–78.

[4] In his "Are Leibniz's Monads Corporeal Substances? Old Answers to a New Problem" (paper presented at the workshop on "Corporeal Substances and the Labyrinth of the Continuum in Leibniz" at the Florence Center for the History and Philosophy of Science, Florence Italy, November 23–25, 2000), Hans Poser argued that a placing of Leibniz's philosophy in the context of its Neoplatonist foundations helps explain his commitment to corporeal substances. Christia Mercer's *Leibniz's Metaphysics: Its Origins and Development* (Cambridge and New York: Cambridge University Press, 2001) provides an analysis of the Neoplatonist foundations of Leibniz's thought.

[5] Justin E. H. Smith, "Christian Platonism and the Metaphysics of Body in Leibniz," *British Journal for the History of* Philosophy 12, no. 1 (2004): 48 ff.

[6] Daniel Sennert, *Hypomnemata physica* (1636), Hyp. III, *On Atoms and Mixtion*, chap. 1, "On Atoms," p. 17, c. 2. For my own original research I consulted the Venice edition (Francis Baba, 1651) in the Fisher Rare Book Library at the University of Toronto. In the late stages of composition of this essay, Roger Ariew drew my attention to the existence of an online photo reproduction of a contemporary English translation of Sennert by Nicholas Culpeper, *Thirteen Books of Natural Philosophy* (Sennert, 1659), of which the last five are the *Hypomnemata*. The translation is excellent, although unfortunately Culpeper does not include the preface.

It seems, however, that the doctrine of atoms can be expounded in the following way. In natural things, which are liable to generation and corruption, because there is a perpetual alternation of generations and corruptions, it is necessary that there be certain simple bodies of their own kind, from which composites are generated and into which they pass away again.... These minimal natural bodies are called atoms, atomic corpuscles indivisible bodies; and they are called this because of the fact that in a natural resolution no further progress can be made to smaller ones, and vice versa; and from them natural bodies have their origin. And these indeed really are in their nature so small that they are not accessible to the senses. (1651, 17, c. 2; 1659, 446)

Sennert infers these atoms from chemical operations such as sublimation and distillation, particularly through a process of *reductio ad pristinum statum*: the fact that metals such as gold and silver can be recovered from a homogeneous alloy by applying *aqua fortis*, he claimed, is evidence that their constituent atoms remain unchanged in the process.[7] Likewise mercury keeps its nature in all chemical changes and can be recovered easily because atoms of the same metal tend to unite together. But Sennert also used atoms to explain a wide range of other natural phenomena, including the action of magnetism, poisons, contagion, and natural generation. It is the latter that is of most interest to us here. Central to his views on generation is the idea of atoms that are animated:

Atoms of Animate Bodies: Indeed, there are atoms not only of inanimate bodies, but of certain animate bodies too: and meanwhile the soul itself can remain whole in such minimal bodies and conserve itself: as will be said afterwards, *concerning mixtion and the spontaneous birth of living creatures*: and the Very Learned Fortunio Liceti built upon this doctrine of atoms almost the whole of his opinion on the spontaneous birth of living creatures. (1651, 21, c. 2; 1659, 453)

As Sennert goes on to argue, the apparent phenomenon of spontaneous generation is in fact a transformation of matter due to putrefaction, and the new forms of life are produced by *semina* (seeds; but also semen) latent in the rotting matter, each of which is animated by a soul. This is in keeping with his whole philosophy of generation, for which he argues clearly and methodically, beginning with the generation of plants.

---

[7] See (1659), 452. See also Clericuzio, *Elements, Principles, and Corpuscles*, 28–9, and the references contained therein. See also Emily Michael, "Sennert's Sea Change: Atoms and Causes," in *Late Medieval and Early Modern Corpuscular Matter Theories*, ed. Christoph Lüthy, John E. Murdoch, and William R. Newman (Leiden, Boston, and Cologne: Brill, 2001).

A plant produces a seed, which contains a vegetative soul, and this can only be an emanation from the mother plant.[8] Likewise each animal seed, male and female alike, contains a sensitive soul that emanates from the soul of the father or mother. As Roger explains, "[T]he two souls are united at the moment of conception as easily as they propagate themselves, and despite its double origin, the soul of the embryo is simple: 'a fire which burns with different flames is not called composite' (*Hypomnemata* [287])." This soul, moreover, is the only soul the animal has; Sennert rejects the Aristotelian orthodoxy of vegetative, animal, and (in the case of humans) rational souls, all existing as entities or essences in the same being. "It is denied that vegetative, sensitive and rational are three different entities [esse] working from different forms; instead they are powers of the same soul."[9] In the same way, "There is only one soul in man, that is the rational soul, which is however equipped with the virtues of all the inferior souls, and endowed with the faculty of vegetating and feeling."[10]

Sennert is explicit about where this leads in connection with the controversial matter of the propagation or origin of the human soul. Concerning plants and animals, he takes himself to "have demonstrated on firm grounds that plants and brute animals are generated from seed [semen], that the seed contains a soul communicated from the thing generating, which, lying hidden in the seed, makes a home for itself suitable for a plant or animal of the same species from which it was split off" (1651, 54, c. 1; 1659, 509). He then argues that it is only reasonable to suggest that the same account applies to human generation, and he declares himself on three counts:

> First, it is more probable and more consonant with the truth that the soul propagates from the parents into the offspring by dint of a divine blessing. Second, whatever else is established, whether it is the soul's creation and infusion or its propagation and traduction, it must nevertheless be held that the soul is present in the seed at first conception, when the seed of the father and mother are conjoined in the womb and retained, and as soon as the work of formation

---

[8] Jacques Roger has already charted this progression in his *Les sciences de la vie dans la pensée française au XVIIIe siècle* (Paris: A. Colin, 1963), 108. Unfortunately, he had a different edition of Sennert's *Hypomnemata* than mine, and I was not always able to find the corresponding passages. I have set his page references to his edition of *Hypomnemata* in square brackets (e.g., 108 [250–8]).

[9] Sennert 1651, 59, c. 2. Cf. Roger *Les sciences de la vie dans la pensée française au XVIIIe siècle*, 110, citing [75] in his edition: "The nature of a plant is not defined by a mere vegetative faculty, but each plant has a specific form, without doubt endowed with the vegetative faculty, but also with other faculties."

[10] Roger, *Les sciences de la vie dans la pensée française au XVIIIe siècle*, 109 [315].

of the human body begins; nor can there be any other efficient cause which could accomplish the work of conformation than the human soul itself. Third, even though I know great men disagree here, still, if their reasons alone are considered in isolation from their authority, I will show that scarcely any solid reason can be brought against the propagation of the human soul through the seed. (Ibid.)

Here we see a firm commitment to the doctrine of traduction; indeed, Sennert defines traduction as a truth: "What traduction [*tradux*] is: The truth is namely that the seed (I mean the body of the seed) is the vehicle by which the soul is communicated from the parents to the offspring: which action is called traduction [*traductio*], by a word taken from plants." The analogy, indeed the root concept, is taken from grafting:

For example, in trees, in which the seminal force is diffused throughout the whole body, if some part is cut off, and that of some other tree is inserted, the soul of the tree from which the shoot was cut is communicated into the other tree: in the same way, when the soul is conveyed with the semen of animals into the female uterus, so finally from this semen endowed with its own soul the complete animal emerges. (1651, 54, c. 1)

Sennert firmly rebuts the objection that traduction is incompatible with the nobility of the human soul:

Nor does the nobility of the intellect, with respect to both essence and operations, show anything else. For it pleases the Creator thus to associate the noblest substance with the body. And so if the immortal rational soul can spend a long time in the mortal body, why may it not also be propagated with the seed? . . . The nobility of the human soul is consistent with traduction. (1651, 57, c. 2; 1659, 515)

In a manner consistent with these premises, Sennert upholds the axiom that *omnis forma est sui multipicativa*, that is, that every form multiplies itself; in particular, in the generation of living things, every soul multiplies itself. This is something he extrapolates from Zabarella, who held it to be the case for sensible species. But "if accidental forms have the force to multiply themselves," Sennert argues, "how much more will substantial forms have it?" (1651, 42, c. 2; 1659, 488). Thus it is more proper to say that the soul is multiplied than that it is divided, since it has no quantity except what it gets by being predicated to something extended; but when the seed is cut off, it is extended to the whole of that body too.

153

Richard T. W. Arthur

Although traducianism was Lutheran orthodoxy in the seventeenth century, this is hardly to say that it was uncontroversial, and indeed it still is controversial.[11] After the publication of the *Hypomnemata*, Sennert was accused of impiety in a barrage of attacks by Johann Freytag (1581–1641), who denounced him "to the Academies of the Christian world."[12] The nub of Freytag's denunciations is precisely "the doctrine of the transmission of souls from the parents *per traducem*, which," according to Freytag, "would imply the transmigration of souls as well as the immortality of the souls of beasts."[13] The transmigration of souls would follow from the fact that, according to Sennert's theory, all souls were created by God *ex nihilo* in the Creation and since then have simply propagated from one generation to the next, budding at each conception, "organizing, animating and defining each being individually," as Roger puts it. Sennert's answer, of course, is that the alternative – having souls created *ex nihilo* at each generation, as in the doctrine of eduction – contradicts what is implicit in God's command in Genesis: "Increase and multiply!"[14] For "since man consists of a body and soul, if the soul does not emanate from the parents, man would not engender man" (1651, 54, c. 1; Roger, 109 [292]; 1659, 509–10). Still, even though Sennert's motives in defending traduction are pious enough, his defence of the self-multiplying of the soul seems only to reinforce Freytag's charges. For example, appealing to the Scaligerian doctrine that "our intellect comprehends itself as a species of substances," he also endorses Scaliger's claim that "[t]he form is so divine a thing that, since it is a substance, it fills another whole substance with itself,

---

[11] The *Catholic Encyclopedia* on the Web comes out firmly against traducianism, claiming that "the organic process of generation cannot give rise to a spiritual substance, and to say that the soul is transmitted in the corporeal semen is to make it intrinsically dependent on matter. Since the soul is immaterial and indivisible, no spiritual germ can be detached from the parental soul" (http://www.newadvent.org/cathen/15014a.htm).

[12] See Clericuzio, *Elements, Principles, and Corpuscles*, 31, for a listing of the various works by Freytag attacking Sennert, one published in 1630, two in 1632, two in 1633, and one in 1636. In the preface to his *Hypomnemata*, Sennert reports Freytag as describing him as "substituting [for Aristotelian principles] his truly absurd, false, rancid, heretical and blasphemous paradoxes, . . . and putting forward as principles his truly inept, colder than January, impious, false, heretical paradoxes and dogmas brimming with contradictions that have either been previously condemned or recently hatched in the brains of sycophants"! See Michael, "Sennert's Sea Change," 350.

[13] Clericuzio, *Elements, Principles, and Corpuscles*, 32.

[14] Sennert quotes this passage from Genesis 1 many times: "And God blessed the fishes and the birds he had created, saying: increase and multiply; and fill the water of the seas, and let the birds be multiplied upon the earth" (1651, 33, c.1); again, noting how a single tobacco seed can engender 300,000 in one year so that by the third year it will produce "100,000 myriads" (i.e., 10^9), he comments, "so powerful was that Divine Blessing: Increase and multiply" (1651, 48, c.1; 1659, 498). Cf. also Sennert 1659, 489, 510.

154

in such a way that out of the two one is made" (1651, 41, c. 2; 1659, 486). But such claims seem only to support Freytag's charge that Sennert makes animal souls immortal.

Turning now to Leibniz, we find in his early work the same commitment to traduction, and also to the idea of atoms containing within them minds, interpreted as principles of activity and individuation. This is particularly evident in 1671, when he was working out a neo-Hobbesian account of the cohesion of atoms or naturally indissoluble bodies, and together with this a projected book, *Elementa de mente* (A II i 114). One of the governing ideas is that the mind is contained in a mathematical point and is therefore necessarily indivisible.[15] As he later confides to Des Bosses, "I thought that the multiplication of souls could be explained through traduction, since many points can be made out of a single point, as the vertices of many triangles can be made through division from the vertex of one."[16] This punctual soul is in turn encased within a physical point, variously described as an atom, a globule, or a *bulla*, the latter being a hollow but unbreakable bubble. Another leading idea is that mind is associated with a vortical motion and that this is responsible for the cohesion of the globules and *bullae* and their resistance to being broken up.

The book on mind never came to fruition, although Leibniz did publish his novel views on cohesion in the *Hypothesis physica nova* (*HPN*) in 1672, of which the "concrete" part was sent to the Royal Society of England and the "abstract" part, *Theoria motus abstracti*, to the Academie Française. In the latter, Hobbesian endeavors (*conatus*) are used to found a neo-Scaligerian theory of mixtion, according to which bodies cohere when one endeavors to penetrate the space of the other.[17] These endeavors, momentary in bodies, are sustained in mind, where they are proto-thoughts or emotions that can be compared and contrasted.[18] The *bullae*, Leibniz explains in *HPN*, were

---

[15] "For I shall demonstrate that mind consists in a point. Whence it will follow that mind can no more be destroyed than a point. For a point is indivisible, and therefore cannot be destroyed. Therefore body is obliterated, and dispersed to all corners of the earth. Mind endures forever, safe and sound in its point. For who can obliterate a point?" (A II i 113).

[16] Leibniz to Des Bosses, April 30, 1709, quoted from *Gottfried Wilhelm Leibniz: Philosophical Papers and Letters*, ed. and trans. Leroy E. Loemker (Dordrecht and Boston: D. Reidel, 1976), 599.

[17] See the extracts from these two works in my *G. W. Leibniz: The Labyrinth of the Continuum* (New Haven and London: Yale University Press, 2001), along with my comments on the theory of cohesion in the introductory essay there. See also my "The Enigma of Leibniz's Atomism," *Oxford Studies in Early Modern Philosophy* 1 (2003): 243–302.

[18] As Leibniz writes to Oldenburg, March 11, 1671, "Every body is a momentaneous mind, and therefore without consciousness, sense, or recollection. Indeed, if two contrary endeavors could persevere in the same body for longer than a moment, every body would be a true mind. But wherever this does come about, minds are produced, and these are naturally indestructible,

155

formed shortly after Creation by the rectilinear action of the sun's rays on the rotating matter of the Earth, like the beads in glassmakers' workshops. From then on they are "the seeds of things, the warps of species" (*HPN, Theoria motus concreti,* §§6–11). Mind, he speculates in a related piece from the same time, can be generated in the place where the actions of stars, and possibly suns, collide.[19]

In early May 1671 Leibniz wrote an enthusiastic letter to Lambert van Velthuysen extolling the virtues of his *HPN* and drawing together some of these themes:

> I judge, however, that just as *corporalia* are to be explained by spaces and motions, so *mentalia* are to be explained by points and endeavors. I explain the way in which God could construct a body which is agitated by motions of such a kind that it would from then on be naturally indissoluble...; indeed, what is more, I explain by means of this body, in which a mind is implanted, that mind can multiply itself, without new creation, *per traducem*, with no mention of incorporeality, which no one has hitherto been able to explain clearly and distinctly. Nor do I think Traduction [*Traducem*] is despised by certain theologians except insofar as it seems to imply corporeality and divisibility, and thus mortality. That it does not is shown by the very nature of indivisibles with as much clarity as sunshine. Once this is supposed, it is at least more rational to concede human propagation to be natural than needlessly to invoke God to perform the perpetual miracle of new creation, not to speak of other difficulties. (A II i 97–8)

What is noteworthy for my purposes in these speculations is that Leibniz's minds, just like Sennert's souls, are implanted in bodies that are naturally indissoluble; that each is able to multiply itself *per traducem*; and that the indivisibility of minds is intended to support their immortality, with the consequence that traduction is protected from the charge that it implies the materiality of souls.

> because, as I shall demonstrate in the appropriate place, once two contrary endeavors in the same point of a body are compatible beyond a moment, no other bodies can slip between them, nor can they be prized apart by any force for all eternity" (A II i 90).

[19] Leibniz writes, "It is possible that a mind is born in the place of a rectilinear collision of all the actions of suns; certainly of all the actions of stars. It is not at all necessary that stars themselves should have minds.... These things [are inferred] from the establishment of a Globular Hypothesis." This is the fourth fragment of a set titled *De conatu et motu, sensu et cogitatione* (A VI ii 281–2). It continues: "An Essay on Traduction [*Traduce*]. What if congenital eggs or minds are determined in them? But what about ones that are not given birth to? Do these minds not then perish, and are they to be excluded from serving their time with the other things? Must we say that the mind alone is able to impress equally perfectly an equilateral triangle onto another body touching it in a certain way, and thereby produce a new mind? And this is to be explained by means of the establishment of globules."

These themes are also evident in a short treatise *De usu et necessitate demonstrationum immortalitatis animae* ("On the Use and Necessity for Demonstrations of the Immortality of the Soul"), written for Duke Johann Friedrich in 1671 and enclosed in a letter to Oldenburg only a couple of weeks after the letter to van Velthuysen quoted earlier. Here Leibniz writes of "how mind can produce mind from its substance without being diminished, or, 'On Traduction' (for I do not see why it is necessary to posit a new creation of minds as often as a man is generated, by always calling on God for perpetual miracles, like a *Deus ex machina* ...)." (In another piece from the same period, perhaps connected with his project of reconciling Catholicism with Protestantism, Leibniz reveals one of the religious motivations for upholding traduction when he observes that Augustine inclined toward traduction since he could not otherwise explain original sin.)[20] Again, whereas in his letter to van Velthuysen Leibniz had referred to "the nature of indivisibles," without specifying what kind, in the treatise for Johann Friedrich, in a discussion of cannibalism, Leibniz explicitly calls the physical point that contains the soul or mind an *atom*: "even if not even an atom (other than that point in which the mind is implanted) is now left of me" (A II i 115).[21]

But as I said earlier, I am not claiming that Leibniz came to adopt either atomism or traducianism under the direct influence of Sennert's views. In fact, Leibniz makes no reference to Sennert as an early influence, whereas he does several times claim to have followed Gassendi in his youth. The relevance of this is that Gassendi, having already followed Sennert in distinguishing certain concretions of atoms (which he calls *molecules*) as the principles of most chemical reactions, also followed him in holding that generation occurs in plants, animals, and humans alike by the passing down of invisibly small seeds which are indivisible – not atoms of the elements but still indivisible molecules or larger particles of the organism in question:[22] "Of course the seed is woven together from several molecules, which are not all always broken

---

[20] "Concerning traduction. Augustine approved inclining toward a certain causal Traduction, as he called it, since he could not explain the propagation of Original Sin if the soul were not an offshoot of Adam's [*decisa ab Adamo*]" (A VI ii 144). The Akademie editors date this 1670–beg. 1672 (?).

[21] For an account of Leibniz's early atomism, see my "Enigma of Leibniz's Atomism." Concerning the uniting of the soul with an atom, Parkinson observes (in G. W. Leibniz, *De Summa Rerum: Metaphysical Papers 1675–1676*, ed. and trans. G. H. R. Parkinson [New Haven and London: Yale University Press, 1992], 130) that Leibniz had stated in an earlier paper of 1669–70 (A VI i 533) that the soul is substantially united, not with all the corpuscles of the body, but with one in the center of the cerebellum.

[22] For a good recent account of Gassendi's views, see Clericuzio, *Elements, Principles, and Corpuscles*, 63–74. Leibniz could also have been influenced by Gassendi through his reading of

down into atoms, nor into anything close to them in primary or subsequent generation."[23]

Each of these seeds contains a little soul that is an offshoot of its parents' souls, in accordance with traduction, as Saul Fisher explains in chapter 5 of this volume. But this is not to say Gassendi's position is identical to Sennert's. In fact, he reverts to the orthodox Thomistic position that the sensitive soul, responsible for the particular physical traits of the individual, is material in nature, since he cannot otherwise see how corporeal faculties could inhere in it. But this means that, when he comes to the topic of human generation, he must give a different account of the traduction of the human soul than that cherished by Sennert, since the rational soul is immaterial. Thus, despite subscribing to the traduction of the sensitive soul, Gassendi has the rational soul divinely implanted in the individual at conception:

> For at whatever time God finally creates and infuses the rational soul we can understand there to pre-exist [*præesse*] in the seed or embryo an Irrational or sensitive soul derived from the parents, to which He unites it. That is, when the seed has broken away, it must be understood that what is broken off with the matter of the seed is only a portion of the irrational soul of the parent (for the intellective or rational soul is undivided [*individua*]), and immediately afterwards the rational soul is united with it in exactly the same way as the rational soul of the parent was united with the whole sensitive soul of the same parent.[24]

Still, Gassendi does follow Sennert's teaching that there are not two distinct souls in animals, vegetative and sentient, but only different faculties of one soul:

> But when I say the sensitive soul alone, I certainly do not exclude the vegetative faculties, such as the nutritive, augmentative and generative faculties; but I do this in order not to suggest that there is in the seed, embryo or new-born human two distinct souls, one vegetative and one sentient. (Ibid.)

Here Sennert's position appears more consistent. If there are not distinct vegetative and sentient souls in any individual animal but only distinct faculties, all of them faculties of an individual soul, why should there be two distinct souls in humans? Why should there be an immaterial rational soul as opposed to just an intellective faculty of the sentient soul? Or if, as Gassendi

Boyle, who followed Gassendi's identification of primary concretions of particles with seminal principles.
[23] Gassendi, *Syntagma* (1658), sect. III, bk. III, 285, c. 2.
[24] Ibid., 256, c. 2.

suggests, the rational soul can be immaterial so long as it is accompanied by an organic body, why shouldn't that solution apply to all souls?[25] Clearly Gassendi was compelled to adopt this position as a result of his materialist construal of vegetative and sentient souls:

> It seems to me therefore that the soul is rather a certain very tenuous substance, as it were a flower of matter [*flos materiæ*], with a special disposition or arrangement [*habitudine*] and symmetry of the parts passing within the denser mass itself; in fact, because of its mobility, it can be the principle of acting, just as is a substance (for we have deduced in its own place that the primary matter from which the substance of bodies comes is not something otiose, or devoid of motion and action). . . . By this argument the soul is also a body, but it is very tenuous, and with respect to the rest of the mass it is as if it were incorporeal.[26]

The point is that, having avoided the charge of heresy that Freytag leveled at Sennert by making the souls of animals and plants material and therefore mortal, Gassendi could not have made the rational soul a mere faculty of the sensitive soul without committing the opposite heresy of making human souls material and therefore mortal. Nevertheless, Gassendi was sensitive to the fact that one still had to account for the generation of highly ordered and well-designed creatures, especially in the case of spontaneous generation. On this he noted Liceti's solution, that there must be souls everywhere in living or dead matter where such generation occurs. But how could one believe "that there would be a form in matter without its informing the matter?" (261, c. 2; Roger, 138). To resolve this he follows Sennert (without attribution) in holding that "the cause of generation of animals of this kind is nothing other than the seed itself, or the little soul [*animula*] contained in it and prepared for its service" (262, c. 1). This in turn leads him to assert that these seeds must be already formed and lying dormant in matter:

> Thus one can also say that even now the seeds of animals are formed, whether out of atoms or out of other principles, which God created in the beginning, and that he willed them to be endowed with forms and movements of such a kind that, on coming together, mixing, combining and remaining attached to

---

[25] Cf. Gassendi's explanation of how the rational soul can elicit corporeal vital actions: "Since it is universally unclear how in the rational soul, an incorporeal subject, there can appear corporeal faculties which are organic, inasmuch as through them a corporeal action, internal and by its own nature vital, is elicited; or how there can be received in them (as the Philosophers claim) vital actions which are corporeal; . . . nothing prevents us from saying that corporeal faculties are received in the body, since the soul is the eliciting principle" (ibid.).

[26] Ibid., 250, c. 2.

Richard T. W. Arthur

one another in such a manner and in such order, they made such seeds, and such animals. But can one not say equally plausibly that seeds of this nature were made at the beginning of the world by the Supreme Creator of things, and dispersed variously through the Earth and the Water? [Here Gassendi quotes Genesis 1.] And it seems on other grounds that this fecundity of the Earth and the Water should not be understood to come from the first principles of which these seeds are composed, but rather from the seeds themselves, which God himself has composed.[27]

Gassendi concludes, just like Sennert, that this enables one to understand what is said in the Scriptures, that God created all things together (*creasse Deum omnia simul*):"insofar as God can have created the seeds of all things in the beginning, all things can be said to have been produced in that first primeval generation" (262, c. 2). Thus, on the one hand Gassendi's account of generation recapitulates Sennert's theory of spontaneous generation; on the other, it anticipates Leibniz's mature view that all forms were created by God in the beginning and lie dormant everywhere in creation as invisibly small, as yet undeveloped seeds.

3.   ATOMISM AND MECHANISM

I agree entirely with the followers of those excellent gentlemen, *Descartes* and *Gassendi*, and with whomever else teaches that in the end all variety in bodies is to be explained in terms of *size, shape* and *motion*.[28]

Because of Leibniz's trenchant opposition to materialism, some modern scholars have had difficulty accepting his claims that he was ever much influenced by Gassendi. From his earliest writings, they point out, he was concerned to argue the inadequacy of a purely mechanical account of body. Thus as early as 1668, Leibniz argued in his *Catholic Demonstrations* that Descarte's *extended substance* could not contain the basis for the activity of a body and that, being purely passive and unable to act, it would not qualify as a substance. But as I have argued elsewhere,[29] these criticisms of Descartes by Leibniz are echoes of Gassendi's criticisms in his *Disquisitio metaphysica*: "Whoever says a thing is only extension says, among other things, that it

---

[27] Ibid., 262, c. 2; also quoted in Roger, *Les sciences de la vie*, 139.

[28] Leibniz, *HPN* §57; A VI ii 248; cf. Paniel Garber, "Leibniz: Physics and Philosophy," in *The Cambridge Companion to Leibniz* (Cambridge: Cambridge University Press, 2000), 275. Leibniz continues the passage by saying that he nevertheless finds that the hypotheses of Epicurean atoms and Cartesian globules are "rather remote from the simplicity of nature and from any experiments" and that his hypothesis of *bullae* is much to be preferred.

[29] Arthur, "The Enigma of Leibniz's Atomism," 215.

160

is not active. Therefore there will be no action, and no faculty of acting, in bodies."[30] Thus, Gassendi claims in his *Syntagma*,

[I]n natural things there is an Agent operating inside them, which is indeed distinguished from matter in part, but not from matter as a whole . . . since in everything there is a principle of action and of motion, . . . and as it were the flower of the whole of the matter, which is also the very thing that is usually called Form.[31]

Now what Gassendi here calls *form* is very much like what Sennert meant by the *spirit* that was the causal means for the soul's operations. Sennert had spoken of the spirit as an "innate heat [*calidum innatum*]" that is the agent or principle of activity, acting as a vehicle for the soul. Gassendi's description of the soul as a "sort of little flame" is also evocative of Sennert's comparison of the soul derived from the two parental souls as "a fire which burns with different flames":

It seems therefore that this soul is a sort of little flame [*quandam flammulam*] or very tenuous kind of fire, which thrives or keeps burning for as long as the animal lives; when it no longer thrives but is extinguished, the animal dies.

Thus Gassendi can be seen as (perhaps implicitly) assimilating Sennert's immaterial soul to the semimaterial spirit that assists it. Leibniz, on the other hand, followed Sennert in insisting on the *immateriality* of all souls. Nevertheless, in his early work he defended this in terms of the unextendedness of minds due to their consisting in points, a fact which also accounted for their indestructibility. With minds so very much implanted in matter, and souls governing bodies by means of vortices, we can say that perhaps Leibniz was not after all exaggerating when he referred to this earlier phase of thought by saying, "when I was more materialistic." At any rate, a Gassendian strain in his early thought seems quite conceivable.

In this connection it is hard to resist seeing something of Gassendi's *flos materiae* (flower of matter) in Leibniz's talk of a *flos substantiae* (flower of

---

[30] Pierre Gassendi, *Disquisitio metaphysica*, vol. III 305b in (1658); quoted by O. Bloch, *La Philosophie de Gassendi: Nominalisme, matérialisme et métaphysique* (The Hague: M. Nijhoff, 1971), 207. As I have noted elsewhere ("The Enigma of Leibniz's Atomism," 215 n. 73), Leibniz acknowledges his debt to Gassendi on this point in a letter to Honoré Fabri written in December 1676: "Truly, I hold for certain that there are incorporeal substances, that motion does not come from body but from outside; . . . Nonetheless I agree with Gassendi rather than Descartes that the essence of body does not consist in extension" (A II i 289).

[31] Gassendi, *Opera*, I 336a, 337a; cf. Bloch, *La Philosophie de Gassendi*, 216.

Richard T. W. Arthur

substance).[32] In the treatise on resurrection he composed for Duke Johann Friedrich (May 21, 1671), he writes, "Even the Jews relate the story that in a certain little bone they call the *Luz*, the soul survives intact in all cases together with the flower of substance" (116). Leibniz summarizes some of the main points of the treatise more pithily in German in the body of the letter accompanying the treatise:

> In particular, I hold fast to the opinion that every body – people as well as animals, herbal plants and minerals – has a kernel of its substance which, distinct from the *Caput Mortuum*, is understood to be just like what the chymists take "from the damned and phlegmatic earth." This kernel is so *subtle* that it even remains left over in the ashes of things that have been burnt, and can, as it were, draw itself together into an invisible *centre*. As one can, to a certain extent, use the ashes of plants as seeds, and as the salient point in the foetus or fruit of a plant or animal already encompasses in itself the kernel of the whole body. Now as this kernel of substance consisting in a physical point (the proximate instrument, and as it were the vehicle of the soul constituted in a mathematical point) remains always, so it is of little consequence whether all gross matter is so in us.[33]

Leibniz's use of the "damned and phlegmatic earth" is certainly a reference to the alchemist literature.[34] But the passage about the kernel of substance remaining intact in the ashes of things that have been burnt ("lying dormant," to use his later phrase) is evocative of what Catherine Wilson describes as "a beautiful passage"[35] in Gassendi's *De plantis*, where he talks of the souls of grains of wheat that may apparently be lost by boiling, roasting, or crushing possibly surviving these ordeals "to be preserved like trees that are dormant through the winter and those swallows of northern climates, and in particular of Muscovy, that grow stiff in the ice," before awakening to the warmth of spring.

---

[32] This is hardly conclusive evidence of influence, of course. The alchemical literature abounds with talk of the *flos* of one substance or another, and Leibniz himself mentions the "chymists" in talking about the kernel of a body's substance. But so far as I can determine, the term *flos materiae* is original with Gassendi, and likewise *flos substantiae* with Leibniz.

[33] Leibniz, letter to Duke Johann Friedrich, May 21, 1671; A II 1 58, p. 108.

[34] Kenelm Digby (*Two Treatises* [Paris: Blaizot, 1644], "*Of Bodies,*" chap. 16) explains that "water, the third instrument to dissolve bodies, dissolveth calx into salt, and so into *Terra Damnata*" (135): "the more grosse and heavy earthy partes (having nothing in them to make a present combination between them and the water) do fall downe to the bottome, and settle under the water in dust. . . . Which ordinary Alchymistes looke not after: and therefore call it *Terra damnata*: but others, find a fixing quality in it, by which they perform very admirable operations" (136).

[35] Catherine Wilson, *The Invisible World* (Princeton, N.J.: Princeton University Press, 1995), 120.

But whatever may have been the sources from which Leibniz developed this view, it certainly opens a fascinating window on the development of his thought and in particular on his motivations for atomism. This is especially relevant for his writings in the spring of 1676, where he asserts atomism with increasing vigor despite the fact that the theoretical foundation provided by the theory of cohesion in the *Theoria motus abstracti* has foundered on the reefs of the infinite. By then he has concluded that there are after all no infinitely small things and also that "endeavours are true motions, not infinitely small ones" (A VI iii 492). Thus the theories of cohesion and of indivisibles he had proposed in the *Theoria motus abstacti* are both overturned. But instead of rejecting atoms, he assumes they are necessary *because* there are minds: "There seem to be elements, i.e. indestructible bodies, precisely because there is a mind in them" (A VI iii 521). One reason Leibniz gives for this is that mind alone can give a criterion for something's remaining self-identical; another is that there is nothing in purely material bodies to ground the laws of motion. But the indissectible bodies Leibniz assumes also have a biological flavor: "Body is as incorruptible as mind, but the various organs around it are changed in various ways."[36] What this indicates is a continuing source of motivation for his atomism in the theory of animal generation and its connection with theological doctrine. Not only did mind-containing atoms solve the problem of the self-identity of substance, but the indestructibility of mind entailed the immortality of the soul and solved the problem of the origin of forms.

This biological-cum-theological connection is confirmed by Leibniz's sustained commitment to the doctrine of the flower of substance he had outlined in his thesis for Duke Johann Friedrich, as evidenced in a piece written in February, 1676, "On the Seat of the Soul." There, commenting on Boyle's work *On the Possibility of Resurrection* (1675), he refers to the paper he had been instructed to prepare for the Duke six years before, reaffirming that "the argument based on cannibals is demonstrative, and proves that a flower of substance must be understood" (A VI iii 478). "Add to this," he says, "what Borel has in his *Microscopic Observations* about the shape of the cherry tree enclosed in the shell of the seed or of the wild fruit." In this rendition of the doctrine, however, Leibniz makes no reference to the multiplication of minds, perhaps because he no longer subscribes to his earlier view that the soul is situated in a mathematical point.[37] Here the "flower of substance is our

---

[36] A VI iii 510. Again: "Every mind is organic. The transmigration of souls is sufficiently refuted by new experiments concerning the already preformed foetus" (A VI ii 394).

[37] However, Leibniz still claims that minds do not obstruct matter because they take up no volume: cf. "Immortalitas mentis mea methodo statim probata habetur, quia . . . rerum cursum non imminuit. Quia mentes n'ont point de volume" (A VI iii 581–2).

[whole] body, subsisting perennially in all changes," or at least "is diffused throughout the whole body, and in a way contains only form" (A VI iii 478). This brings him more in line with Sennert's view of the way the soul informs the body: the soul is implanted in the body, which is invisibly small prior to conception, and it occupies all of the body as it grows.

Indeed, as Christia Mercer has observed, a fascinating aspect of this flower of substance doctrine is the way it anticipates the mature doctrine of the dominant monad.[38] The soul or form implanted in the flower of substance operates or has influence through the whole organic body and is wholly responsible for the self-identity of the individual. Thus despite the dramatic change in death, where the volume of matter organized by the soul shrinks to some minute portion, at resurrection the kernel of substance can diffuse itself through a quantity of matter equal to what it did before death and thus reconstitute the same individual. This prefigures Leibniz's later doctrine of *transformationism*, according to which death is merely a transformation of the organism in such a way that the domain of influence of its dominant monad shrinks to a physical point.[39]

This raises the question, why did Leibniz abandon atomism in favor of his doctrine of transformationism, where corporeal substances, unlike atoms, do not retain the same material casing, size, and shape? A related question is, when did this change in his views occur? It is hard to be precise. As we have seen, the atoms of 1676 are already conceived as cores of organic bodies, and by November of that year (in his dialogue *Pacidius Philalethi*) Leibniz has rejected Gassendi's perfectly hard atoms, arguing for an infinite enfolding of matter, with one fold inside another and no last fold, in keeping with his rejection of a categorematic infinite division. A month later he visits the Dutch microscopists Leeuwenhoek and Swammerdam, who would have shown him the fantastically minute living organisms in apparently inert matter which they saw as dazzling visual evidence for preformationism. It is

---

[38] Christia Mercer, *Leibniz's Metaphysics: Its Origins and Development* (Cambridge: Cambridge University Press, 2001), 282. In this book (200 ff., 223 ff.) Mercer traces Leibniz's sympathy to the *rationes seminales* of Plotinus and Ficino to the influence of his teacher Jakob Thomasius.

[39] A representative statement of this mature doctrine is given in Leibniz's *Specimen Inventorum* of ca. 1686: "Indeed, just as some people have proposed that every generation of an animal is a mere transformation of the same animal now living, and a kind of accretion that renders it sensible, so by parity of reason it seems defensible to hold that every death is a transformation of the living animal into another smaller animal, and is a kind of diminution by which it is rendered insensible" (A VI.iv 1623–4; Richard T. W. Arthur, ed. and trans., *G. W. Leibniz: The Labyrinth of the Continuum* [New Haven and London: Yale University Press, 2001], 317). Cf. Leibniz's letter to Arnauld of October 9, 1687, where he cites both Leeuwenhoek and Swammerdam in support of his belief in transformation.

tempting, in light of Leibniz's repeated later references to this evidence, to see it as a profound factor in his adopting transformationism. But although it can hardly be discounted as an influence on his subsequent views, it is hard to see it as decisive for his rejection of atoms. On the one hand, to judge from the similarity of his "kernel of substance" to the views of Gassendi and Sennert, there is no reason to think he would see atomism as in conflict with preformation. And given his explicit appeal to Pierre Borel and the preformation of the tree in a cherry stone, the Dutch scholars' evidence for preformation must have been marvellously confirming but not revelatory.[40] On the other, the rejection of atoms may have been conditioned more by developments internal to his thought on the infinite and the collapse of his endeavor theory of cohesion.[41] In any case, by 1678 we find him rejecting atoms and advocating the necessity of forms in almost the same breath:

> It must also be demonstrated that every body is actually divided into further parts, i.e. that there are no atoms. Here therefore the soul must be treated, and it must be shown that all things are animated. Unless there were a soul, i.e. a kind of form, a body would not be an entity, since no part of it can be assigned which would not again consist of further parts, and so nothing could be assigned in body which could be called *this something*, or *some one thing*. That it is the nature of a soul or form to have some perception and appetite, and why.[42]

And in another piece written not long after this:

> Every body is organic, i.e. is actually divided into smaller parts endowed with their own particular motions, so that there are no atoms. Every body is animate, i.e. has sensation and appetite.
> Substantial form, or soul, is the principle of unity and of duration, matter is that of multiplicity and change. Every form is in a way a *soul*, i.e. capable of sensation and appetite.[43]

What is interesting about these passages is the link they show in Leibniz's own mind between the organic nature of bodies and the falsity of atomism. But equally, this change in Leibniz's philosophy can be seen in terms of

---

[40] One might add that Swammerdam himself subscribed to "living atoms" in the sense of smallest living organisms, so Leibniz's exposure to his thought is unlikely to have been a cause of his abandoning atomism.

[41] Just how Leibniz could have simultaneously maintained a commitment to both atomism and the actually infinite division of matter is the subject of Arthur, "The Enigma of Leibniz's Atomism."

[42] "Conspectus for a Little Book on the Elements of Physics" (Summer 1678–Winter 1678–79?), in Arthur, *G.W. Leibniz*, 233–5.

[43] "Metaphysical Definitions and Reflections" (Summer 1678–Winter 1680/81), in Arthur, *G. W. Leibniz*, 245–7.

an abandonment of the essentially Cartesian program to understand the way forms "inform" matter in terms solely of the laws of motion. As Malebranche expressed that program, "[A]ll the parts essential for the machine of animals and plants are so wisely disposed in their germs that they will, in time and as a consequence of the general laws of motion, assume the shape and form which we observe in them."[44] Leibniz's change of heart concerning mechanism is connected with his critique of its inability to account for individuation. The laws of motion (as he conceived them prior to 1678) do not constitute form in such a way as to provide the basis for the self-identity of an individual substance. But let me attempt to make this clearer by comparing Leibniz's reintroduction of substantial forms with Sennert's.

4. REINTRODUCING SUBSTANTIAL FORMS

Sébastien Basson (1592–1655) was in his late twenties when he published his *Philosophiae naturalis* (1621), a vigorous championing of atomism. According to Basson, atoms were created by God (Amsterdam edition, 14), and they "originated after the first creation, and are free from all destruction" (126). Following Plato, he claims that "these primary particles are so minute that, unless very many of them come together into one mass, they do not affect the senses." Various orders of compound particles are formed through composition: "From these primary particles secondary ones are composed, and from the secondary ones tertiary ones, and from the tertiary, particles of fourth, fifth and sixth degree, etc." The primary particles are those of the elements and of the five chemical principles (earth, salt, mercury, sulphur, and phlegm), and these "conserve their own nature in a mixture, and act otherwise than through the force of some form." The form of a mixture is characterized as "a composition or singular mixture." "Hence through the whole variety of composition, things can be varied *in infinitum*." In other words, the generation of forms is explained entirely in terms of addition or subtraction of particles and change of their mutual situation:

> Furthermore, parts varying to infinity can be forged in this way from those very different prime particles; and through either the subtraction or addition of some particles, or the variation of the situation of parts, it is not difficult to understand that some pass over into the nature of others.[45]

---

[44] Malebranche [1688] 1997, 176; quoted from Andrew Pyle, *Malebranche* (London and New York: Routledge, 2003), 171.

[45] Translated from the quotation given in Clericuzio, *Elements, Principles, and Corpuscles*, 40; Clericuzio cites it as p. 72 in the Amsterdam edition of *Philosophiae naturalis*.

From this description, Basson perhaps sounds more of a Democritean than he was. In fact, for him atoms do not combine by pure chance, as in materialist atomism, but through appetite, attractions, and sympathies, with which they are endowed by the universal spirit or world soul, depending ultimately on the will of God:

> We have said that that universal spirit endows those elements with an appetite doubly conjoined. [I]t excites two motions, the primary one being that by which like attracts like ... the secondary motion being that by which there is a connection between them, as long as like has an appetite for like.[46]

Nevertheless, despite appealing to such nonmechanistic causes of combination, Basson's explanation of the natural phenomena produced by them is entirely in terms of corpuscular combination, dissociation, and rearrangement of the elements. The forms are mechanistically explained, and the only teleology resides in the actions of the universal spirit acting under God's direction.

For Sennert this was quite unacceptable, but his reasons have nothing to do with a rejection of spirit as a causal agent. What was absurd for him was the notion that forms could be reduced to mere combinations and situations of elements. He makes this criticism in the *Hypomnemata*, where he explicitly criticizes Basson for rejecting substantial forms. Noting the difficulty Zabarella has in explaining how, as an animal acquires parts in growing, the soul manages to extend to the accruing parts, he comments that this "is so obscure as to make Sébastian Basson, otherwise perspicacious enough in Physics, propose that substantial form be denied, and hold, absurdly, that form is nothing but a certain aggregation of parts of the same kind optimally concentrated" (1651, 41, c. 1; 1659, 486). Sennert follows Zabarella and Scaliger in claiming that the soul does not itself move, nor does it make sense for it to change by an accretion of matter; rather "the soul remains numerically the same throughout a whole life, and as new matter arrives it is extended to inform it" (ibid.). It does this "not by a motion to inform new matter, but in a certain way that is occult and hidden from us; and even if Basson ridicules these things, they are nonetheless very true when rightly explained" (1651, 41, c. 1; 1659, 487). The trouble people have had stems from this one mistake: "that they attribute to form and soul things which belong to bodies, namely quantity, divisibility and parts" (ibid.).

---

[46] Again, this is translated from the quotation given by Clericuzio, *Elements, Principles, and Corpuscles*, 40 n. 21, cited as p. 391 in the Amsterdam edition of *Philosophiae naturalis.*

But Sennert's reasons for rejecting the sufficiency of Basson's construal of form as an aggregation of elements are perhaps more transparent in his *De chymicorum* (1619), written two years before Basson's *Philosophiae naturalis*:

> The elements give the matter, but not the form. That mixture, *qua* mixture, is informed by elementary forms, I do not contest. Yet I deny that a specific form for each thing, which gives the essence and name of the thing, is produced from the elements. For there is in every natural thing, and in the parts of body, besides the matter that the elements supply, a certain divine principle and fifth nature, by which they are what they are, and are reduced to a certain family of a natural species. For elements are material, and so are not capable of giving rise to action. (358)

There is quite a lot packed into this succinct passage. First there is the question of the forms of mixtures. As Emily Michael has explained, Sennert's own position had undergone a wholesale change from his earlier writings, in which he had upheld the doctrine of the refraction of forms advocated by Averroes and Zabarella, according to which the elements lose their individuality on becoming mixed and there emerges a form of the compound substance. But by 1619 Sennert had gone over to "the opinion of Avicenna that the Elements not only remain in the mixtures, but also retain their forms perfectly and integrally." The same experiments that had convinced him of the existence of atoms of matter by 1619 had also convinced him that they retained their forms.

But according to Basson's atomism, the forms of the mixtures that the atoms combined to make are nothing more than combinations and situations of the constituent atoms. It seemed impossible to Sennert that the individual natures of different natural kinds, especially living things, could be accounted for in this way. "Even if maybe Democritus thought so, natural things do not come about through the fortuitous collision of atoms."[47] On the contrary, "all qualities and properties of things flow from their forms."[48] Even if in every natural thing there is a semimaterial spirit or innate heat (*calidum innatum*) that is the agent or principle of activity, the spirit is subordinate to the form or formative power: "natural things come about under the direction of a superior form which by the instrument of heat attracts, restrains, mixes and organizes everything as a work in progress."[49] It is the form, the soul in living things,

---

[47] Sennert 1676, vol. III, 780; translated from the Latin quotation in Michael, "Sennert's Sea Change," 357 n. 91.

[48] Sennert 1676, vol. I, 153; translated from the Latin quotation in Michael, "Sennert's Sea Change," 356 n. 87.

[49] Sennert 1676, vol. III, 780; translated from the Latin quotation in Michael, "Sennert's Sea Change," 357 n. 91).

which is responsible for generation.[50] These immaterial forms determine all the actions and passions of any natural thing:

> Today too in the generations of things, mixtion does not occur by the Elements running together of their own accord to constitute each thing, but every one is directed by a form. For forms are the divine and immutable principle that determines all the actions and passions of a natural thing; and they are, as it were, the instrument and hand of the most wise Creator and Workman God, who in creation freely bestowed this force and efficacy onto these his instruments, things than which nothing more marvellous can be thought. (353)

But if the atoms retain their specific forms while in a compound, and the compound has a form of its own, then there must be more than one form in a compound. Sennert did not shrink from this implication, but rather developed it into a hierarchical analysis of substances. Taking inspiration from both Jacopo Zabarella and Jacob Schegk (1511–87), he wrote, "Nor does it imply any absurdity that besides the specific form there should be still other subordinate forms" and "There is no absurdity in saying that in one substance there are many forms."[51] As Emily Michael explains, not only do elemental atoms (e.g., of sulphur, salt, or mercury) combine into compound atoms, which can in turn be "subordinate to a higher supervening compound form, e.g. of gold or silver," even "corpuscles that are substances can be combined to form inanimate bodies that are not substances, e.g. water and wine mixed" – these are the same *entia per aggregationem* that will figure prominently in Leibniz's writings about substance. But this also implies that the bodies of substances, living things, contain a plurality of subordinate forms, each one governing one of the compounds from which the body is composed:

> To me it seems more consonant with the truth that in living bodies there are many auxiliary and subordinate forms, yet such that one form is principal and ruler [*domina*] and informs the living thing, and is that from which the living thing gets its name.[52]

That is, the dominant form is the substantial form of a living thing, identifiable with its soul. The substantial forms of the various substances making up its

---

[50] Sennert 1619, 353. Cf. Clericuzio: "*Semina* organize matter, which is completely inert. The fortuitous concourse of atoms cannot explain the generation of natural bodies. Even in the smallest atoms there are seminal principles which inform and give instructions to the passive matter" (*Elements, Principles, and Corpuscles*, 30; Sennert [Amsterdam ed.], 420).

[51] Sennert 1676, vol. I, 218; translated from the Latin quotation in Michael, "Sennert's Sea Change," 347 n. 64.

[52] Sennert 1676, vol. I, 218; translated from the Latin quotation in Michael, "Sennert's Sea Change," 347 n. 64.

body are subordinate to it so long as it is alive. But on death, says Sennert, again following Zabarella, the dominant form is extinguished, and the body is reduced to the next lower grade of forms making up the substances that compose it.

Turning now to Leibniz's version of the doctrine of substantial forms, we find him boldly taking the next step at which Sennert had hesitated: that of ascribing immortal souls to animals. Thus in "Wonders concerning Corporeal Substance," jotted down on the back of a bill in 1683, Leibniz writes,

> Every created thing has matter and form, i.e. is corporeal. Every substance is immortal. Every corporeal substance has a soul. Every soul is immortal. It is probable that every soul, indeed every corporeal substance, has always existed from the beginning of things. A pile or entity by aggregation such as a heap of stones should not be called a corporeal substance, but only a phenomenon. . . .
>
> There are as many souls as there are substantial atoms or corporeal substances. This puts an end to the inextricable difficulties concerning the origin of things and forms, because they have no origin and there is no generation of substances.[53]

Thus Leibniz adopts a position not unlike that recommended by Gassendi to account for spontaneous generation: the souls of such animals have existed as the forms for their substantial atoms since the Creation. Only this is now extended to all animals, and indeed to all substances: for Leibniz, there are no substantial forms that are not the forms of living beings of some order or other, no matter how lowly. But there is no traduction in Sennert's sense, that is, no multiplication of souls. Each substantial form has existed from the Creation and is just as impervious to destruction as he had previously held the human soul to be.

Sennert had contended that his living atoms could, through the faculty of the soul they contained, extend to a growing body under the direction of the soul and its nutritive faculty, in which case they would be a dominant form, with the subordinate form-matter compounds acting as the matter for that body. On his "flower of substance" view, Leibniz had maintained that the human soul would not die but would shrink back down to a physical point. Now he has extended this analysis to all forms: although animals evidently die, their souls are not destroyed, but only their realm of influence. Typically, Leibniz writes in the *Monadology*, "There are no *souls* which are completely *detached* from matter, and no *spirits* without bodies. . . . [B]ecause of this there is never either complete generation nor total death in the strict sense,

---

[53] "Mira de natura substantiae corporeae," A VI iv 1465: Arthur, *G. W. Leibniz*, 262–5.

which consists in the detaching of the soul. [W]hat we call *death* is enfolding and diminution"⁵⁴ (278). This is true even of the spermatozoa identified by Leeuwenhoek:

> §75. These *animals*, of which some come to be raised to the level of larger animals through the process of conception, we call spermatic. But even those among them which remain within their own kind, i.e. the majority of them, are born, reproduce themselves, and are destroyed, just like the larger animals. It is only a small number of the elect who move up onto a larger stage. (278–9)

Sennert, of course, knew nothing of these spermatic animals. There is, nevertheless, a parallel to Sennert's view. Whereas Gassendi (rather uneasily) submitted to the orthodoxy that the rational soul is superadded to the material human soul at conception, Leibniz has an "elevation" of certain immaterial souls to the "elect" or "society of minds," reminiscent of Sennert's idea that the Creator "associates the noblest substance" with such fertilized seeds at the moment of conception.

Meanwhile, we find a strong (even if unwitting) echo of Sennert in Leibniz's doctrine of dominant and subservient forms:

> §70. We can see from this that every living body has a dominant entelechy, which in an animal is its soul; but the parts of that living body are full of other living things – plants, animals – each one of which also has its entelechy or dominant soul. (278)

5. CONCLUSION

In this essay I have tried to establish that there is a strong consilience between the views on generation of Sennert and Leibniz. Both authors are committed to the origin of forms in Creation itself, as indeed is Gassendi (although they take the position that forms multiply throughout history, a view held by Leibniz even when young). Both Sennert and the mature Leibniz are committed to the view that every substance has its own substantial form, conceived as an immaterial architectonic principle that governs its development and "determines all the actions and passions of a natural thing," and that this form is the soul in living things. Both hold that the forms of substances that are components of the body of another substance remain intact, even though subordinate to the dominant form governing that body, so that within any complex organism like an animal there is a hierarchy of forms. But at the

---

⁵⁴ *G. W. Leibniz: Philosophical Texts*, ed. and trans. R. S. Woodhouse and Richard Franks (Oxford: Oxford University Press, 1998), 278.

foundation of their views in common is a principle also shared by Gassendi, that *there can be no form in matter without its informing the matter*. This is the foundation of the preformationism upheld by all three authors. Granted that all souls were created at the beginning of the world, whether they multiply at every conception, as in Sennert, Gassendi, and the young Leibniz, or remain intact through all time, as in Leibniz's mature view, they must always have been accompanied by an organic body. Gassendi allowed teleology a role in generation and animal development; Leibniz, reminiscent of Sennert in his criticisms of Basson, could not see how a purely material form or soul could perform that function. On their way of conceiving it, a form is responsible for the individuation of a substance and thus can neither be educed from mere matter nor superadded to it.

Sennert and Gassendi (at least in the case of spontaneous generation) inferred that a soul can never come into being where there was not already organic matter organized by a soul, and they therefore held that every existing soul has a continuous history traceable back to the Creation itself. Leibniz, with admirable consistency, inferred that the converse has an equal claim to truth: just as each soul has a continuous history from the beginning of time, so it will continue to exist until the end. But there are no disembodied souls in the whole of creation. As he writes in 1705,

> There must be machines in the parts of a natural machine to infinity, so many enveloping structures and so many organic bodies enveloped, one within the other, that one can never produce any organic body entirely anew and without any preformation, nor destroy entirely an animal which already exists.... For since animals are never formed naturally from an inorganic mass, mechanism, although incapable of producing their infinitely varied organs anew, can at least draw them out of pre-existing organic bodies by a process of development and transformation.... I have said here that there is no part of matter that is not actually divided and does not contain organic bodies, that there are also souls everywhere as there are bodies everywhere, that the souls and even the animals subsist always, that organic bodies are never without souls and that souls are never separated from organic bodies.[55]

This comparison with Sennert and Gassendi casts a different light, I believe, on Leibniz's ontology. Granted, Leibniz was no dogmatist, and it certainly would not have been atypical of him to have changed his position on corporeal substances later in life. This is precisely what most of Leibniz's recent interpreters have asserted, claiming that, if not already in his middle period,

---

[55] "Considerations on Vital Principles and Plastic Natures" (1705), in Loemker, *Gottfried Wilhelm Leibniz*, 589–90.

then certainly in his mature work, Leibniz came to abandon his earlier commitment to corporeal substances in favor of an immaterialist monism where monads are the only real constituents of the created world.

But I believe this immaterialist monism is undercut by the development of his position I have outlined here. Granted it is beyond question that for Leibniz forms have a more fundamental reality than bodies. It is also plausible that he was sympathetic to the Neoplatonist theory that bodies emanate from these forms, as the forms themselves do from God, and that (especially later in life) he applied himself to giving a construal of the former kind of emanation in terms of the perceptions of monads. But none of this detracts from the axiom that *there can be no form that does not have an organic body it informs*, which Leibniz appears to have upheld consistently throughout his career. For without a body and organs of perception, a monad cannot perceive. As Leibniz wrote in 1676, "Since in us there is nothing but mind, it is a wonder how so many different things are perceived in it. Actually, though, it is added to matter, and without matter it would not perceive as it does."[56] It is by means of the body that the rest of the universe is represented in the soul; the fact that the perceptions we receive are of various things is the ground of the argument for plurality, so matter is the *principium multitudinis*, the principle of multiplicity; again, without a body, a substance could not be acted upon, so matter is the *principium passionis*, the principle of being acted upon. In a nutshell, if there were only bodyless monads, there would be only one monad, namely, God.

In conclusion, it does not seem that Leibniz ever abandoned corporeal substances in later life. For even in the *Monadology*, which commentators promote as a prime example of his late idealism, Leibniz upholds the necessity of an organic body to each monad. Even though the name "monad" now denotes immaterial entelechies rather than body-form composites, bodies are their vehicles of perception, and thus bodies are necessarily organic, on the one hand, and essential to the monads' being able to represent the universe, on the other:

§63. The body belonging to a monad, the monad being either its entelechy or its soul, makes up together with that entelechy what we can call a *living thing*, and together with a soul what we call an *animal*. Now that body of a living thing or animal is always organic, because, since every monad is a mirror of the universe in its own way, and since the universe is regulated in a perfectly orderly manner, there must also be order within that which represents it, i.e. in

---

[56] A VI iii 518; *G. W. Leibniz: De Summa Rerum: Metaphysical Papers 1675–1676*, ed. and trans. G. H. R. Parkinson (New Haven and London: Yale University Press, 1992), 77.

the perceptions of the soul, and therefore also in the body by means of which the universe is represented in the soul.[57]

No doubt this interpretation of Leibniz's monads will be disputed. But if I have been successful in this essay, Leibniz's earlier atomism will no longer seem irrelevant to the development of his mature philosophy of substance, and his reintroduction of substantial forms, as well as his vaunted solution to the problem of the origin of forms, will be seen to belong to the same context of biological and theological concerns that confronted his atomist predecessors, Sennert and Gassendi, in their theories of animal generation.

[57] *Monadology*, in *G. W. Leibniz: Philosophical Texts*, 277.

8

Spontaneous and Sexual Generation
in Conway's *Principles*

DEBORAH BOYLE

1. INTRODUCTION

To philosophers such as Descartes, Hobbes, Hooke, and Boyle, mechanism seemed to have great promise for explaining all manner of natural phenomena. Yet later seventeenth-century philosophers were critical of mechanism and often appealed in their explanations to what a thoroughgoing mechanist would consider "occult" forces. This was particularly true in explanations of the generation of animals, where satisfying mechanical explanations were not forthcoming; for example, despite promises to explain animal generation, Descartes never did provide a worked-out account. One fascinating figure in the history of later seventeenth-century responses to mechanism is Anne Conway (1631–79). In her *Principles of the Most Ancient and Modern Philosophy*, Conway raises objections to the materialism of Hobbes and Spinoza and to the dualism of Descartes and Henry More. In the course of arguing for a monistic vitalism, she presents interesting views on animal generation. While Conway's monistic vitalism has recently received increasing attention in the secondary literature,[1] the details of her views on generation have not been explored.

---

[1] Jacqueline Broad, *Women Philosophers of the Seventeenth Century* (Cambridge: Cambridge University Press, 2002), 70–9; Alison P. Coudert, "Anne Conway: Kabbalist and Quaker," in *The Impact of the Kabbalah in the Seventeenth Century* (Leiden: Brill, 1999), 177–219; Jane Duran, "Anne Viscountess Conway: A Seventeenth Century Rationalist," *Hypatia* 4, no. 1 (1989): 64–79; Sarah Hutton, "Anne Conway critique d'Henry More: L'esprit et la matière," *Archives de philosophie* 58 (1995): 371–84; Sarah Hutton, "Anne Conway, Margaret Cavendish and Seventeenth-Century Scientific Thought," in *Women, Science and Medicine 1500–1700*, ed. Lynette Hunter and Sarah Hutton (Thrupp, England: Sutton Publishing, 1997), 218–34; Sarah Hutton, "Henry More and Anne Conway on Preexistence and Universal Salvation," in *"Mind senior to the world"*: *Stoicismo e origenismo nella filosofia platonica del Seicento inglese*, ed. Marialuisa Baldi (Milan:

Deborah Boyle

2. CONWAY'S ONTOLOGY

Anne Conway, née Finch, had gained an introduction to her half-brother's Cambridge tutor, Henry More, with whom she began corresponding in 1650.[2] More guided her through Descartes' *Principles of Philosophy*; after Anne's marriage to Edward Conway in 1651, More and other Cambridge philosophers frequently visited the Conways' home, and Conway and More continued their correspondence. In 1670, More introduced Conway to Francis Mercury van Helmont, hoping that van Helmont could cure her chronic and very severe headaches. Although van Helmont failed to cure her, he and Conway became close friends. Through van Helmont, Conway learned about the Kabbalah and the radical views of the Quakers, then an unpopular sect, which Conway herself joined in 1677, two years before her death.

Sometime during the 1670s, Conway wrote the book that came to be called *Principles of the Most Ancient and Modern Philosophy*.[3] She probably did not intend to publish the writings, which were, More writes, "abruptly and scatteredly, I may add also obscurely, written in a Paper-Book, with a Black-lead Pen, towards the latter end of her long and tedious Pains and Sickness; which She never had Opportunity to revise, correct, or perfect."[4] After Conway's death, van Helmont and More arranged for the work to be translated from English into Latin; van Helmont took the book to Holland, where it was published in 1690.[5] In 1692, it was retranslated into English and published as *Principles of the Most Ancient and Modern Philosophy*;

Francoangeli, 1996), 113–25; Jennifer McRobert, "Anne Conway's Vitalism and Her Critique of Descartes," *International Philosophical Quarterly* 40, no. 1 (2000): 21–35; Carolyn Merchant, "The Vitalism of Anne Conway: Its Impact on Leibniz's Concept of the Monad," *Journal of the History of Philosophy* 17 (July 1979): 255–69.

[2] These biographical details are to be found in Marjorie Hope Nicolson, ed., *The Conway Letters*, rev. ed. by Sarah Hutton (Oxford: Clarendon Press, 1992).

[3] Nicolson suggests the period 1671–4 as the most likely date of composition, (*The Conway Letters*, 453). Peter Loptson argues for a later date of composition, suggesting that Conway's references to Spinoza's critics are very likely to a 1676 work by Franciscus Cuperus with which Conway and More were both familiar (introduction to *The Principles of the Most Ancient and Modern Philosophy*, by Anne Conway [The Hague: Martinus Nijhoff, 1982], 8–9).

[4] Anne Conway, *The Principles of the Most Ancient and Modern Philosophy*, ed. Allison P. Coudert and Taylor Corse (Cambridge: Cambridge University Press, 1996), 3. The ridicule heaped on Margaret Cavendish, who published numerous works under her own name in the 1650s and 1660s, may well have deterred Conway from publication.

[5] The work was among three treatises published together as *Opuscula philosophica quibus continentur principia philosophiae antiquissimae et recentissimae ac philosophiae vulgaris refutatio*.

176

the original English version was apparently lost.[6] The book had at least one important (and appreciative) reader: Leibniz, who learned of it from van Helmont and who mentions Conway approvingly in several letters and in the *New Essays on Human Understanding*.[7]

Because Conway's remarks on animal generation depend on her ontological views, it is essential to understand her account of substance. Conway is typically labeled a monist, but she actually identifies three fundamental species of substance, which differ according to their mutability. She writes,

> There are therefore three kinds of being. The first is altogether immutable. The second can only change toward the good, so that which is good by its very nature can become better. The third kind is that which, although it was good by its very nature, is nevertheless able to change from good to good as well as from good to evil.[8]

These three types are, respectively, God, Christ, and creatures. Because the three substances differ in their essential attributes, Conway's view would more accurately be described as "trialism" than monism; still, she is indeed a monist at the level of created substance. Every creature is composed of spirit and body,[9] but these differ only in degree, not in kind.[10] What we perceive as matter is simply a "fixed and condensed" form of spirit, and spirit is "nothing but volatile body or body made subtle."[11] Under the right conditions, matter can be transformed into spirit, and spirit can be hardened such that we call it "matter." Because spirit and body are essentially the same, every created thing has "spirit or life and light, . . . the capacity for every kind of feeling, perception, or knowledge, even love, power and virtue, joy and fruition."[12] The Cartesian notion of a nonthinking extended substance utterly distinct from

---

6 Allison P. Coudert and Taylor Corse, introduction to *Principles of the Most Ancient and Modern Philosophy*, xxxviii. The 1690 Latin version and 1692 retranslation into English were published in 1982 in a volume edited by Peter Loptson (see n. 37). Because the modern translation by Coudert and Corse is clearer, all references are to that edition.

7 In a 1697 letter, Leibniz writes, "My philosophical views approach somewhat closely those of the late Countess of Conway" (Leibniz, *Die philosophischen Schriften*, ed. C. I. Gerhardt, 7 vols. [Berlin, 1875–90] 3: 217; cited in Merchant, 258). See also Gottfried Leibniz, *New Essays on Human Understanding*, trans. and ed. Peter Remnant and Jonathan Bennett (Cambridge: Cambridge University Press, 1996), 72.

8 Conway, *Principles*, chap. 5, sect. 3, 24.

9 Ibid., chap. 6, sect. 11, 38.

10 Ibid., 40.

11 Ibid., chap. 8, sect. 4, 61.

12 Ibid., chap. 9, sect. 6, 66.

spirit is thus a fiction.[13] Nonetheless, Conway attributes to spirit many of the qualities that materialists assigned to body: extension, infinite divisibility, and impenetrability.[14]

According to Conway, each created thing is composed of a multitude of bodies and a multitude of spirits. These multitudes are ordered in a certain way, with one ruling spirit which determines what type of creature that thing is.[15] For example, a tree contains the ruling spirit of a tree, which is a less noble spirit than that which rules a horse, which, again, is less noble than that of a human. All spirits, the lesser as well as the nobler, are contained in a creature, albeit only potentially.[16] Thus, the matter of which a creature is composed has the potential to be refined into the spirit of any other sort of creature.[17] In other words, matter and spirit are essentially the same substance.

If all created things are made of the same basic stuff, then what we perceive to be different objects in fact differ only in "certain modes or attributes,"[18] entailing that created beings can be transformed into other types of created beings. Conway writes,

> For example, water does not change but stays the same, although when cold it freezes, where it was fluid before. When water turns to stone, there is no reason to suppose that a greater change of substance has occurred than in the earlier example when it changed from water to ice. And when a stone changes back into softer and more pliant earth, this too is no change of substance. Thus, in all other changes which can be observed the substance or essence always remains the same. There is merely a change of form inasmuch as the substance relinquishes one form and takes on another.[19]

Created substance is, in its nature, mutable for either better or worse, but Conway believes that "the divine power, goodness, and wisdom has created good creatures so that they may continually and infinitely move towards the good through their own mutability."[20] Thus their natural tendency is to strive to be better. For this to be possible, a creature of a certain type which becomes as perfect an instance of that type as possible must be changed into a nobler,

---

[13] Ibid., sect. 2, 63.
[14] Ibid., chap. 7, sect. 4, 50–3; chap. 7, sect. 3, 49. See Broad, *Women Philosophers of the Seventeenth Century*, 70–9.
[15] Conway, *Principles*, chap. 6, sect. 11, 39; chap. 7, sect. 3, 53.
[16] Ibid., chap. 8, sect. 4, 60.
[17] Ibid.
[18] Ibid., chap. 6, sect. 3, 29.
[19] Ibid., 29–30.
[20] Ibid., sect. 6, 32.

better type of being; otherwise, it would stagnate at the lower stage of goodness and would no longer be striving for the better.[21] Indeed, Conway holds that such transformations are evidence of God's justice, in which creatures are rewarded or punished for their actions.[22]

According to Conway, transformations involve changes of both shape and ruling spirit. All spirits are potentially present in all creatures, but different spirits can become actual and govern the others;[23] then the spirit of the creature can be said to have changed. The presence of a certain type of spirit causes the construction of a certain type of body for that spirit. For example, through its "formative power," a brute spirit will direct the construction of a brute body, while a human spirit will direct the construction of a human body.[24] Why is this so? Conway maintains that no spirit can exist without a body, whether that body be "terrestrial, aerial, or etherial."[25] The body is needed, she says, to "receive," "reflect," and "retain" the image of the spirit.[26] Conway's views here echo Henry More's views that souls must at all times be joined with a body;[27] that souls can join with terrestrial, aerial, or aetherial "vehicles";[28] and that a soul can organize matter into an appropriate vehicle.[29]

In chapters 7 and 8 of the *Principles*, Conway presents six arguments for her matter-spirit monism. While the general thrust of the *Principles* was against both dualism and materialism, Conway's six arguments do not, for the most part, engage with thoroughgoing materialism; as Sarah Hutton has shown, Conway's work was in large measure a critique of Henry More's dualist system.[30] Conway's assumption is always that spirits do exist; the question she pursues is whether spirit is distinct from body (as dualists like More claim) or whether there is a single substance which shares extension and the "more excellent attributes" of "spirit or life or light."[31] Her objection

---

[21] Ibid., 32–3.
[22] Ibid., sect. 7, 35.
[23] Ibid., chap. 7, sect. 4, 55.
[24] Ibid., chap. 6, sect. 7, 36.
[25] Ibid., chap. 5, sect. 6, 27.
[26] Ibid., sect. 11, 38. See also chap. 6, sect. 7, 36, where Conway writes that a spirit's "formative power is governed by its imagination, which imagines and conceives as strongly as possible its own image, according to which the external body must take shape."
[27] Henry More, *The Immortality of the Soul*, in *A Collection of Several Philosophical Writings* (London: James Flesher, 1662; repr., New York: Garland Publishing, 1978), 146–7.
[28] Ibid., 119.
[29] Ibid. See also p. 150.
[30] Hutton, "Anne Conway critique d'Henry More."
[31] Conway, *Principles*, chap. 9, sect. 6, 66.

to Hobbes and other materialist monists is that they simply *ignore* these latter attributes.[32] Conway's arguments are designed not to persuade the materialist that spirit exists but rather to persuade the dualist that spirit and matter are actually a single substance.

Conway's first argument derives from her trialism: if God, Christ, and creation are the only substances, then the realm of creation cannot have more than one nature or essence, for if it did, there would be more than just these three substances.[33] Clearly, this argument is only as good as her arguments for trialism. Chapter 6 had contained three reasons in support of trialism. The first is that positing a fourth substance is "superfluous," since all natural phenomena can be satisfactorily explained in terms of just the three that she posits.[34] Second, the three substances exhaust the possibilities for definition: God is immutable, Christ is partly mutable (changeable only for the good), and creation is altogether mutable (changeable for better or worse).[35] Third, positing a fourth kind of substance would destroy "the most excellent order which we find in the universe, since there would be not only one mediator between God and creatures but two, three, four, five, six, or however many can be imagined between the first and the last."[36] Conway also thinks it fitting that created substance have only a single essence, since God and Christ themselves have only one essence each.[37]

In Conway's second argument for matter-spirit monism, she appeals to the divine attributes, particularly God's goodness. Being infinitely good, God shares his goodness with his creations.[38] There are, of course, attributes that are unique to God, which he cannot confer on anything else, such as "subsisting by himself, [being] independent, immutable, absolutely infinite, and most perfect."[39] But God is also "spirit, light, life ... good, holy, just, wise, etc.," and whatever God creates must possess at least some of these "communicable attributes"; God cannot make something which shares none of his qualities.[40] Now, Conway argues, if the mechanists' "dead matter" existed, then it could share none of God's attributes, because neither being dead nor being corporeal are among God's attributes. Thus if God creates anything corporeal, it

---

[32] Ibid.
[33] Ibid., chap. 7, sect. 1, 41.
[34] Ibid., chap. 6, sect. 4, 30.
[35] Ibid.
[36] Ibid.
[37] Ibid., 30–1.
[38] Ibid., chap. 7, sect. 2, 45.
[39] Ibid.
[40] Ibid.

must share some of God's communicable attributes. Dead matter is therefore impossible;[41] matter must be living and perceiving, and thus it must be spirit as well as body.

Conway's third argument is that spirit and matter must be of the same basic nature because of the great love that souls have for bodies. As evidence that souls do in fact love bodies, she writes that "as is widely known, the souls of certain people remain with their bodies and in their power after the death of the body and until it decomposes and turns to dust,"[42] though her grounds for accepting that claim are unclear. She maintains that any instance of love must be based on one of three possible bases: being of the same nature and substance, being like each other, or one having its being from another.[43] Since body and soul cannot give being to each other (creation being the bailiwick of God alone), and since (Conway evidently assumes) similar attributes can only occur between things with a shared nature, body and soul must share the same nature.

Conway's fourth argument is reminiscent of Princess Elisabeth's objection to Descartes' claim that mind and body have distinct essences: namely, that if this is so, then their union and interaction are inexplicable. Conway's questions, however, are directed not at Cartesian dualism but at More's dualism. What could unite spirit and body if they are so radically different that they have no affinity for each other? And if there were an affinity between them, then that affinity should be present regardless of the condition of the body; thus it is unclear why mind and body should separate at death.[44] If one hopes to answer this by claiming that spirit needs an "organized body" to function properly, Conway points out that this conflicts with the dualist's original claim, that mind and body are entirely different in nature.[45] Conway then turns to interaction problems: How can the soul produce bodily motion, given that the soul can supposedly penetrate all things and that bodily motion involves the resistance of impenetrable objects against each other?[46] And if the soul, being penetrable, can simply pass through the body, how could the body ever cause pain or happiness in the soul?[47]

---

[41] Ibid., 45–6.

[42] Ibid., sect. 3, 48.

[43] Ibid., 46.

[44] Ibid., chap. 8, sect. 1, 57.

[45] Ibid.

[46] Ibid. This objection is especially pertinent to More's version of dualism, since More emphasizes the soul's ability to penetrate things. See Hutton, "Anne Conway critique d'Henry More," 376–7.

[47] Conway, *Principles*, chap. 8, sect. 2, 58.

Jumping ahead for the moment to Conway's sixth argument for her monism, she maintains simply that Scripture supports her theory.[48] She cites various passages which refer to all things being "living"; the implication is that the mechanists' "dead matter" contradicts such passages.

3. SPONTANEOUS GENERATION AS AN ARGUMENT FOR MONISM

Conway's fifth argument for monism, presented in chapter 8, bears most on her biological views. There she argues that if mind and body were distinct substances, the spontaneous generation of animals would not be possible:

> I take the fifth argument from what we observe in all visible bodies such as earth, water, stones, wood, etc. What an abundance of spirits are in all these things! For earth and water continually produce animals, as they did in the beginning. Hence a pool full of water produces fish, although there were no fish put there to propagate. Since all other things come from earth and water originally, it necessarily follows that the spirits of all animals are in the water.[49]

Later in the same section, she asks, "How, finally, does it happen that when a body putrefies, other species are generated from this putrefaction? Thus animals come forth from putrefying water or earth. Even rocks, when they putrefy, turn into animals. Thus mud or other putrefying matter generates animals, all of which have spirits."[50] She had said much the same thing in chapter 6, asking, "Does not rotting matter, or body of earth and water, produce animals without any previous seed of those animals?"

Conway herself does not use the phrase "spontaneous generation," or the synonymous phrase "equivocal generation," but her reference to propagation without parents of the same type makes it clear that this is what she means.[51] Conway's belief in such phenomena may seem quaint, but such belief was common from antiquity through the first half of the seventeenth century (and even beyond). In the *Historia animalium,* Aristotle had catalogued various cases of supposed spontaneous generation, both plant and animal.[52] Aristotle alleged that different types of mud could produce different

---

[48] Ibid., sect. 7, 62.
[49] Ibid., sect. 4, 60.
[50] Ibid., 61.
[51] For discussion of the senses of "spontaneous generation," see D. M. Balme, "Development of Biology in Aristotle and Theophrastus: Theory of Spontaneous Generation," *Phronesis* 7 (1962): 91–104.
[52] See also Aristotle, *Generation of Animals,* trans. A. L. Peck (Cambridge, MA: Harvard University Press, 1953), 721a8–10, 758a30–b28, 762a19–763b15. On Aristotle's account of spontaneous

varieties of shellfish;[53] that certain insects could be generated from dew, putrefying mud or dung, wood, hair, flesh, animal residues,[54] old snow,[55] fire,[56] wool clothes, old cheese, and books;[57] that some fish could be generated from mud or sand;[58] and that eels are produced "neither by copulation nor out of eggs" but rather from the water in marshy pools.[59] St. Augustine also mentioned cases of supposed spontaneous generation,[60] and Albert the Great reiterated Aristotle's views.[61] Medieval Islamic philosophers also developed Aristotelian theories of spontaneous generation, even occasionally suggesting that humans could be formed spontaneously.[62]

Spontaneous generation continued to be accepted by many thinkers in the first half of the seventeenth century.[63] Believers included Francis Bacon,[64] Descartes,[65] Pierre Gassendi,[66] William Harvey (Conway's physician,

---

generation, see Balme, "Development of Biology in Aristotle and Theophrastus"; Allan Gotthelf, "Teleology and Spontaneous Generation in Aristotle: A Discussion," *Apeiron* 22 (1989): 181–93; J. G. Lennox, "Teleology, Chance, and Aristotle's Theory of Spontaneous Generation," *Journal of the History of Philosophy* 20 (1982): 219–38. G. E. R. Lloyd, *Aristotelian Explorations* (Cambridge: Cambridge University Press, 1996), 104–25.

[53] Aristotle, *Historia animalium*, 3 vols., trans. A. L Peck (Cambridge, MA: Harvard University Press, 1970), 547b18–22.

[54] Ibid., 551a1–10.

[55] Ibid., 552b6–8.

[56] Ibid., 552b10–13.

[57] Ibid., 557b1–13.

[58] Ibid., 569a10–b9.

[59] Ibid., 570a3–14.

[60] St. Augustine, *De Trinitate*; quoted in Howard B. Adelmann, *Marcello Malpighi and the Evolution of Embryology* (Ithaca: Cornell University Press, 1966), 2: 750.

[61] Adelmann, *Marcello Malpighi*, 2:751.

[62] Remke Kruk, "A Frothy Bubble: Spontaneous Generation in the Medieval Islamic Tradition," *Journal of Semitic Studies* 35, no. 2 (1990): 265–82.

[63] Everett Mendelsohn points out that the believers in spontaneous generation were often strange bedfellows, including orthodox Aristotelians, Paracelsians, and those whom Mendelsohn terms "the philosophical atomists" ("Philosophical Biology versus Experimental Biology: Spontaneous Generation in the Seventeenth Century," in *Topics in the Philosophy of Biology*, ed. Marjorie Grene and Everett Mendelsohn [Dordrecht: D. Reidel, 1976], 43, 52–9). For discussion of other, less well known writers who believed in spontaneous generation, see Lynn Thorndike, *History of Magic and Experimental Science* (New York: Columbia University Press, 1958), vols. 7–8.

[64] On Bacon's views, see Thorndike, *History of Magic and Experimental Science*, 7:75.

[65] René Descartes, *Primae cogitationes*, in *Oeuvres de Descartes*, ed. Charles Adam and Paul Tannery (Paris: Vrin/C.N.R.S., 1957–76 [repr., Paris: Vrin, 1996]), 11:506.

[66] On Gassendi, see Adelmann, *Marcello Malpighi*, 2:776–7. According to John Farley, however, Gassendi "refuted the concept of spontaneous generation" (*The Spontaneous Generation Controversy from Descartes to Oparin* [Baltimore: The Johns Hopkins University Press, 1974], 12).

briefly),[67] Nathaniel Highmore,[68] Robert Hooke,[69] and Daniel Sennert.[70] Francesco Redi's 1688 *Experiments on the Generation of Insects* offered evidence against the theory of spontaneous generation but did not entirely silence the theory's proponents.[71]

The passages already quoted show that Conway believed that decaying matter could produce animals, although she does not specify which ones. Conway maintains that her vitalist monism has little difficulty explaining such cases. On her view, because all spirits are contained potentially in matter, spirits can "flow out or fly away"[72] from putrefying matter and through their "formative power" direct the construction of new bodies in which to reside.[73] A particular kind of spirit will direct the construction of a particular kind of body, for the body reflects and retains the image of a spirit just as a mirror reflects and retains an image of a face.[74] Thus, for example, if the spirit of a fish is released from water, that spirit will direct the construction of a fish body using the matter of the water.

Conway clearly thinks that the ability of her metaphysical theory to explain spontaneous generation is a reason to accept her vitalist monism over dualism. That her target is dualism is clear from her phrasing of a possible objection to her argument:

> One might say that this argument does not prove that all spirits are bodies but only that all bodies have in themselves the spirits of all animals; hence every body has a spirit in it, and although spirit and body are united, they always remain different from each other in their natures and cannot therefore be changed into each other.[75]

[67] William Harvey, *Anatomical Exercises on the Generation of Animals*, in *The Works of William Harvey*, trans. Robert Willis (London: Sydenham Society, 1847), 456–7. However, commentators do not agree on how to understand Harvey's position; see Elizabeth Gasking, *Investigations into Generation 1651–1828* (Baltimore: The Johns Hopkins University Press, n.d.), 18–19; Lois N. Magner, *A History of the Life Sciences* (New York: Marcel Dekker, 1994), 185–6; Mendelsohn, "Philosophical Biology versus Experimental Biology," 46; Catherine Wilson, *The Invisible World: Early Modern Philosophy and the Invention of the Microscope* (Princeton: Princeton University Press, 1995), 198.

[68] Nathaniel Highmore, *History of Generation* (London: John Martin, 1651), 57–9. On Highmore, see Adelmann, *Marcello Malpighi*, 2:777–8; Mendelsohn, "Philosophical Biology versus Experimental Biology," 58–60.

[69] This is Catherine Wilson's interpretation (*Invisible World*, 199), but not all scholars share it. See Mendelsohn, "Philosophical Biology versus Experimental Biology," 50.

[70] On Sennert, see Mendelsohn, "Philosophical Biology versus Experimental Biology," 53–4.

[71] On responses to Redi's work, see Farley, *Spontaneous Generation Controversy*, 1–7; Wilson, *Invisible World*, 199–205; Magner, *History of the Life Sciences*, 243–6.

[72] Conway, *Principles*, chap. 8, sect. 4, 61.

[73] Ibid., chap. 6, sect. 7, 36.

[74] Ibid., sect. 11, 38.

[75] Ibid., chap. 8, sect. 4, 60.

This is just the kind of response that a dualist might make. That is, a dualist could seek to explain spontaneous generation by maintaining that every body contains multiple types of spirit; thus, when a body decays, the spirit of a different animal might be released. Conway's reply is that the dualist cannot, in fact, hold that all bodies contain the spirits of all types of animals. The contained spirits, she claims, would have to be either *actual* spirits or *potential* spirits. If they are actual, the dualist is faced with two problems. First, the dualist must explain how so many actual spirits could be contained in a very small body.[76] The only possible explanation, according to Conway, would be through "intimate presence," in which the spirits mingle in the body "in a most perfect way, while not adding to its weight or extension"; however, according to Conway, existing as intimate presence is one of God's incommunicable attributes, which cannot be possessed by creatures.[77] Conway's second objection to the suggestion that a body might contain the actual spirits of all kinds of animals is that the dualist has no way to explain why the decaying matter gives rise merely to one kind of animal and not to any of the other kinds whose spirits are contained.[78] What happens to all the other actual spirits, which are presumably also released when the matter containing them decays?[79]

The remaining option, if multiple actual spirits cannot be contained in a body, is that they must be contained merely potentially. But, argues Conway, if body can potentially be changed into any spirit, then that is just to say that body and spirit are the same basic stuff, contrary to the dualist's view.[80] In sum, explaining spontaneous generation requires abandoning dualism in favor of Conway's monism.

One might think that Conway's target here is Descartes. However, although some dualists might make the objection Conway frames in the quotation above, Descartes himself would not. Conway's argument assumes that the animals created from water, earth, or decaying matter have spirit. But Descartes would reject this assumption, as he held that only humans have souls. Someone who does not think that animals have souls will not be faced with the

---

[76] Ibid.

[77] Ibid., chap. 7, sect. 4, 50.

[78] Ibid., chap. 8, sect. 4, 60.

[79] The Aristotelian account of spontaneous generation also had to explain why a given chunk of matter turned into, say, an eel rather than a flea. Commentators debate how Aristotle and his later followers answered this question; see Balme, "Development of Biology in Aristotle and Theophrastus," 99; Gotthelf, "Teleology and Spontaneous Generation in Aristotle," 187–9; Kruk, "Frothy Bubble," 269.

[80] Conway, *Principles*, chap. 8, sect. 4, 60.

puzzle of explaining how animal souls can be generated from matter. Indeed, as the *Primae cogitationes* make clear, Descartes thought he *could* explain spontaneous generation, relying on purely mechanical principles.

Why doesn't Conway argue against a purely mechanical account of spontaneous generation? It might seem important that she do so. After all, dualism was not the only competitor to her vitalistic monism; another contender was Hobbesian materialism, which purported to explain all natural phenomena in strictly mechanical, materialist terms.[81] And since Conway presents her discussion of spontaneous generation as the fifth argument for her monism, one might accuse her of begging the question in assuming, in this argument, that spirits do exist. However, this difficulty becomes less pressing once we note, again, that Conway's target in this part of the *Principles* is Henry More. Conway's argument is aimed at someone who endorses dualism, believes that humans and nonhumans alike have souls, and accepts the existence of cases of spontaneous generation; More fits the bill in these respects.[82] Conway also specifically mentions "plastic natures," a key feature of More's account of spirit, asking,

> How does the corruption or dissolution of the body lead to the new generation of animals? If one should say that spirits of these animals are, as it were, released from their chains and set free by this dissolution, and they then form and shape new bodies for themselves from the aforesaid matter by means of their plastic natures [*plastica*], I reply, how did the first body hold those spirits captive to such a degree?[83]

Conway seems to be responding directly to book II, chapter 15 of More's *Immortality of the Soul*, where More claims to answer the puzzle of "how the Soul can get out of the Body, being imprisoned and lockt up in so close a Castle," saying that it is because spirit can penetrate all things. Later in the same chapter, he relies on his theory to explain stories of "the suddain ingendring of Frogs upon the fall of rain, whole swarms thereof, that had no Being before."[84] Still apparently responding to More, Conway goes on to argue that if spirit can penetrate all things, then it is unclear how spirit could ever be imprisoned in a body such that only the death of the body could release it. She concludes, "This captivity of spirits in certain hard bodies, and their

---

[81] Whether or not Hobbes really held such a view matters less than the fact that Conway interpreted Hobbes in this way, as is clear from her discussion in chap. 9 (pp. 64–5).

[82] More, *Immortality*, 34–5.

[83] Conway, *Principles*, chap. 8, sect. 4, 61.

[84] More, *Immortality*, 122–5.

liberation when the bodies become soft, offers a clear argument that spirit and body are of one original nature and substance, and that body is nothing but fixed and condensed spirit, and spirit nothing but volatile body or body made subtle."[85]

Thus, Conway's claim to be able to provide a superior account of spontaneous generation is aimed squarely at More. This helps to explain why Conway does not address mechanical, materialist explanations of spontaneous generation. More had argued against just such explanations in the *Antidote against Atheism*.[86] Thus it may be that Conway was presupposing the force of More's arguments against a materialist account of spontaneous generation, yet trying to show that More, too, failed to provide a satisfactory account of the phenomenon.

In addition to accepting the standard case of spontaneous generation from decaying matter, Conway suggests a more radical thesis. In both chapters 6 and 8, she suggests that God's original creation of living beings was a form of spontaneous generation, insofar as those original creatures lacked parents. In chapter 6, she asks, "And in the creation of this world did not the waters produce fish and birds at God's command? Did the earth not also at the same command bring forth reptiles and beasts, which were, on this account, real parts of earth and water?"[87] She goes on to note that the first humans, too, were "made from earth," though she notes that, while all creatures received their spirits from the earth, humans received an additional, dominant spirit "from above and not from the earth."[88] In the passage quoted earlier from chapter 8, Conway suggests that such spontaneous generation still occurs, writing that "earth and water continually produce animals, as they did in the beginning."[89]

---

[85] Conway, *Principles*, chap. 8, sect. 4, 61.

[86] More describes atomists' attempts to give mechanistic explanations of the creation of animals and plants, reporting that some proponents of mechanistic atomism have agreed that nature may have originally produced "ill-favoured and ill-appointed Monsters" but have maintained that the more perfect creatures destroyed the imperfect ones. More's objection appeals to spontaneous generation: "all Creatures born of Putrefaction, as Mice and Frogs and the like, as those many hundreds of Insects, as Grasshoppers, Flies, Spiders and such other, that these also have a most accurate contrivance of parts, and that there is nothing fram'd rashly or ineptly in any of them" (*An Antidote against Atheism*, in *A Collection of Several Philosophical Writings* [London: James Flesher, 1662; repr., New York: Garland Publishing, 1978], 136). The suggestion is that since mechanistic production often results in mistakes, and spontaneously generated creatures show no evidence of such mistakes, spontaneous generation does not occur mechanistically.

[87] Conway, *Principles*, chap. 6, sect. 6, 34.

[88] Ibid.

[89] Ibid., chap. 8, sect. 4, 60.

The idea that the act of creation described in Genesis should be understood as a kind of spontaneous generation was not new.[90] In the *Summa Theologiae*, Aquinas allows that creatures can be produced from putrefying matter through the influence of a "formative power" deriving from some "heavenly body";[91] analogously, he says that "in the case of creation . . . the active principle was the Word of God, producing animals from elementary matter."[92] Pierre Gassendi, too, treated the original generation of creatures in Genesis and the ongoing spontaneous generation of certain creatures in the same way.[93] And in his 1653 *Conjectura Cabbalistica*, Henry More had described God's original creation of living beings in the same terms he used in his 1659 *Immortality of the Soul* to describe the spontaneous generation of frogs from droplets of water.[94] Thus, although Conway describes these processes in terms quite different from those of any of these earlier thinkers, her analogous treatment of God's original creation and of cases of spontaneous generation echoes the approach of her predecessors.

4. CONWAY'S ACCOUNT OF SEXUAL GENERATION

Though belief in spontaneous generation was widespread at the time Conway was writing, much of the seventeenth-century writing on and research into animal generation focused on sexual generation. While Conway's views on sexual generation share some affinities with Galen's views, she ultimately understands sexual generation in the same terms as spontaneous generation: both

---

[90] Even those who objected to this idea typically accepted that the original act of creation was through spontaneous generation; what they rejected was the claim that such spontaneous generations still occur. The grounds for this rejection were not scientific but religious; spontaneous generation is a form of creation, but according to Genesis, God ceased creating on the sixth day (Mendelsohn, "Philosophical Biology versus Experimental Biology," 47–8).

[91] Thomas Aquinas, *Summa Theologiae*, vol. 10: *Cosmogony*, trans. William A. Wallace, O.P. (Blackfriars, 1967), 1a q.71.129.

[92] Ibid.

[93] In Gassendi's account, God's original creation of plants and animals occurred through seeds scattered "variously over the earth, each in the place more suitable for it" (*Syntagma philosophicum*, in Adelmann, *Marcello Malpighi*, 2:798). Spontaneous generation, too, occurs from seeds (ibid., 804). For further discussion, see Matthew R. Goodrum, "Atomism, Atheism, and the Spontaneous Generation of Human Beings," *Journal of the History of Ideas* 63 (2002): 209–12. Of course, Gassendi's account, being atomistic, was radically different from Aquinas's.

[94] Henry More, *Conjectura Cabbalistica*, in *A Collection of Several Philosophical Writings*, 16; More, *Immortality*, 125.

involve the transmutations of matter and ruling spirit from one kind of organism to another.[95]

Conway's remarks on sexual generation occur at the end of chapter 6, in the context of a discussion of how creatures are composed. As we have seen, Conway holds that every creature is composed of spirit and body, where body is itself a form of spirit. She now interprets spirit and body in traditional Aristotelian terms, equating spirit with an active "male" principle and body with a "more passive" female principle.[96] Aristotle had maintained that the male's contribution to the development of a fetus was the form, transmitted through the male semen, while the female contributed only matter – specifically, the menstrual blood.[97] Galen objected that the female must also influence an animal's form, or offspring would never resemble their mothers. Thus he posited the presence of female semen (which he also claimed could be established empirically). According to Galen, both types of semen are formed when blood, lingering in the "spermatic duct," "undergoes coction and clotting,"[98] and both male and female semen contain a kind of formative power, although he considered female semen to be weaker than male semen.[99] Moreover, Galen held that male semen joined with the menstrual blood to become part of the matter of the embryo and that the menstrual blood itself contained certain powers.[100] Thus menstrual blood, female semen, and male semen each contribute both matter and form to the embryo.

Conway's association of spirit with the male and body with the female may suggest a more Aristotelian account, but in fact her view of the parents' contributions more closely resembles Galen's. Her claim that male semen is the "spirit and image of the male"[101] suggests that she shares their view that male semen contains a nonmaterial formative power (remembering, of course, that for Conway even spirit, the "nonmaterial," is really a form of matter). She does not say whether she thinks male semen contributes to the matter of a fetus, though her vitalist monism would seem to imply that it does; after all, the way that a potential formative power comes to be actual is

---

[95] Conway was not alone in treating spontaneous and sexual generation in essentially the same way. As Mendelsohn points out, Paracelsus and Daniel Sennert also took such a view ("Philosophical Biology versus Experimental Biology," 52–4).

[96] Conway, *Principles*, chap. 6, sect. 11, 38.

[97] Aristotle, *Generation of Animals*, 727b31–33, 729b19–23.

[98] Galen, *On Semen*, ed. and trans. Phillip de Lacy (Berlin: Akademie Verlag, 1992), 117–23.

[99] His argument is that the female, being wetter and colder than the male, "lacks something for the precise perfection of semen" (ibid., 177).

[100] Ibid., 165.

[101] Conway, *Principles*, chap. 6, sect. 11, 39.

through its "release" from the matter in which it is potential. If this is right, then Conway shares with Galen the view that the female is the primary, but not the only, source for the matter of the fetus. Unlike both Aristotle and Galen, however, she does not assign this role to the menstrual blood, of which she says nothing. Rather, she focuses on "female semen." She describes female semen as a "perfect extract of the whole body,"[102] echoing Galen's view that it is produced from throughout the body. Female semen has a "a remarkable power of retention";[103] this has affinities with Galen's view that the uterus has a special retentive faculty.[104] Given her view that spirits require matter to "retain" an image of the spirits, describing the female semen as "retentive" suggests that it constitutes the matter which retains an image of the spirit which acts on it.

But the spirit which acts on the female semen is not necessarily from the male semen. The male semen indeed contains spirit, but Conway holds that the female contains spirits, too (though, curiously, she does not specify that these spirits are in the female *semen*).[105] Which parent provides the spirit to guide fetal formation? Conway's account includes an element of struggle. She writes,

> In this [female] semen, as in the body, the masculine semen, which is the spirit and image of the male, is received and retained together with the other spirits which are in the woman. And whatever spirit is strongest and has the strongest image or idea in the woman, whether male or female, or any other spirit received from outside of one or the other of them, that spirit predominates in the semen and forms a body as similar as possible to its image.[106]

In other words, the spirit contributed by the masculine semen has to *compete* with other spirits contained in the feminine semen. Only the strongest wins. Thus the male might contribute *nothing* to the formation of the fetus if the spirit in his semen does not win this contest.

Here, too, there is an echo of Galen, for Galen had written that male or female semen might "dominate" in directing the creation of different parts of the fetus, resulting in a child whose eyes (for example) resemble her father's eyes but whose mouth resembles her mother's mouth.[107] The difference is that

[102] Ibid.
[103] Ibid.
[104] Galen maintains that the uterus provides a paradigmatic example of a "retentive faculty," which ensures that something remains for a prolonged period in one place (*Galen on the Natural Faculties*, trans. Arthur John Brock [Cambridge, MA: Harvard University Press, 1963], 225, 229).
[105] Conway, *Principles*, chap. 6, sect. 11, p. 39.
[106] Ibid.
[107] Galen, *On Semen*, 181.

for Conway the spirit which wins out over the other spirits evidently directs the entire construction of the fetus.

One might ask, though, how other spirits could be "received from outside" the male or female and come to be present in the female semen. As we have seen, Conway's view is that a dead creature can be transformed into a different creature when a spirit is released from the recently deceased creature and forms for itself a new body. In spontaneous generation, it creates this new body by shaping the very matter which it previously inhabited; thus when an insect is spontaneously generated from mud, both its spirit and its matter come from the mud. Conway sees sexual generation, too, as involving transmutation, but in sexual generation, the newly released ruling spirit presumably uses and shapes the matter of another creature altogether: the mother's semen. How does a spirit get from the recently deceased body that previously contained it and into the mother of the soon-to-be-formed embryo? Conway herself provides only the barest hints of how to answer this question.

One possibility, though an odd one, is suggested by the Lurianic Kabbalah, which we know to have influenced Conway:[108] a human might come to contain the spirit of another creature through consuming and digesting that creature.[109] And, indeed, Conway's remarks seem to suggest such a view:

> In the human body ... food and drink are first changed into chyle and then into blood, and afterwards into spirits, which are nothing but blood brought to perfection. ... From these spirits also come the semen, through which the race propagates, and especially the human voice and speech, which is full of those good or bad spirits made and formed in the heart.[110]

Conway apparently holds that a creature can be transformed into a human through being consumed by a human. Indeed, some of her examples of

---

[108] In a 1671 letter to Conway, More writes, "I am glad your Ladiship takes so much pleasure in reading Peganius" (Hutton, *Conway Letters*, 515). "Peganius" was the pseudonym of Christian Knorr von Rosenroth, editor of various Kabbalistic texts, which he published as the *Kabbala denudata*; Gershom Scholem says that Knorr von Rosenroth's book "served as the principal source for all non-Jewish literature on Kabbalah until the end of the 19th century" (Gershom Scholem, *Kabbalah* [New York: Dorset Press, 1974], 416). For more discussion of Conway, More, and the Lurianic Kabbalah, see Allison Coudert, "A Cambridge Platonist's Kabbalist Nightmare," *Journal of the History of Ideas* 36 (1975): 633–52.

[109] Allison Coudert writes that "the souls in food could be elevated through the gustative attentions of wise and pious men." She quotes from the *Kabbala denudata*: "he who is a wise disciple and eats his food with proper attention is able to elevate and restore many revolving souls. Whoever is not attentive will not only restore nothing, but he will sustain damages" (Coudert, *Impact of the Kabbalah in the Seventeenth Century*, 123).

[110] Conway, *Principles*, chap. 8, sect. 5, 61–2.

non-spontaneously-generated transmutations are cases in which one creature consumes another: grass becomes sheep, sheep becomes human.[111] So, Conway may think that the first step in sexual generation is that the would-be parents consume various creatures, whose matter and spirits become part of the male and female semen, by way of digestion and then "perfection" of the blood. At some point in this process, of course, there must be a change of the ruling spirits in the semen; that is, a human spirit, which was already potentially present in the matter of the consumed creature, has to be released from that matter, and the ruling spirit of the consumed creature must be demoted, as it were. For the next step is copulation, causing male and female semen to be mixed in the female; and, as we saw in the passage quoted earlier, the spirits in the semen compete to be the ruling spirit of the soon-to-be-formed embryo. If human spirits had not already been made actual in the male and female semen, then the winning spirit could be the spirit of some animal the parents had recently consumed; thus, a woman could give birth to, say, a duck or a cow. Galen had explained why offspring are always of the same type as the parents by appealing to the type of matter generating the creature;[112] for Conway, a fetus is of a certain species because of the ruling spirit which directed the construction of its body.

However, requiring that creatures eat other creatures in order for sexual generation to occur would not only be odd, it would also severely constrain Conway's account, insofar as none of the creatures humans *do not* eat could ever be transformed into humans. This seems at odds with her belief that transmutation is an instrument of God's justice. Might there be other ways in which Conway could claim that spirits can come to be in semen? Another possibility is suggested by her distinction between mechanical motion and vital motion, "the motion of life."[113] Conway says that "when a creature attains a more noble kind or degree of life, it acquires greater power and ability to move itself and transmit its vital motions to the greatest distance,"[114] suggesting that a spirit could use its vital power to move into an appropriate location for formation of a new body to occur. Nonetheless, she also stresses that vital motion is "inseparable from" mechanical motion.[115] Thus, although spirits have the power of self-motion, they presumably have to move from place to place in the same way as other objects and cannot simply flit out of existence in one location and into existence at the new location. Do spirits

[111] Ibid., chap. 9, sect. 5, 65.
[112] Galen, *On Semen*, 163.
[113] Conway, *Principles*, chap. 9, sect. 9, 69.
[114] Ibid.
[115] Ibid., 67.

fly through the air, to be breathed in by the soon-to-be mother? Henry More held that a spirit can inhabit the air after death, using an "aiery vehicle,"[116] and then rejoin with matter to be born as a new creature.[117] Van Helmont, too, maintained that air contains a spirit which is drawn in by the lungs and circulated throughout the body; having absorbed an individual's essence, this spirit is emitted either through ejaculation (in semen) or through speech (in ideas).[118] Conway may well have shared such a view; it is striking that immediately after giving her account of sexual generation, she discusses how thoughts are produced, writing that "in the same way, the internal productions of the mind . . . are generated."[119]

Even if we can successfully fill in the gaps in Conway's account of sexual generation, Conway may seem to face a problem with explaining personal identity. It is important to Conway that personal identity be preserved through transmutations, for if one individual is reborn as an entirely different individual, then there is no sense in which the first individual has been rewarded or punished for its actions. It seems that continuity of matter cannot be the source of personal identity across transmutations, for in sexual generation the matter of the developing fetus is not the same as the matter of the virtuous animal whose spirit is being rewarded. Yet the dominant spirit of the new fetus is also different than the dominant spirit of the now-deceased virtuous animal. If it were not, then, as we saw earlier, the fetus's spirit would direct the construction of a body just like the body it left rather than a body like the one of the mother, so that a human mother could give birth to a horse or sheep. But if both the matter and the spirit are different, in what sense is the developing fetus the same individual as the previous creature?

Conway may have the resources to respond to such a worry, however. She could point out that the new dominant spirit was in fact potentially present in the now-deceased virtuous creature. Indeed, it was present because all matter is potentially changeable into spirit (since they are the same substance). Thus, what is now a dominant spirit was previously part of the matter of the creature. In a sense, then, it is continuity of *both* matter and spirit which accounts for personal identity, insofar as the factor underlying this identity was once matter but is now spirit. But this is just what we would expect from a monist who holds that matter and spirit are one.

---

[116] More, *Immortality*, 119.
[117] Ibid., 121.
[118] Coudert, *Impact of the Kabbalah in the Seventeenth Century*, 67.
[119] Conway, *Principles*, Ibid., chap. 6, sect. 11, 39.

9

Malebranche on Animal Generation

Preexistence and the Microscope

ANDREW PYLE

INTRODUCTION

The theory of the preexistence of germs, either ovist or animalculist, was biological orthodoxy from the late seventeenth century to the end of the eighteenth. Although there were always dissenting voices,[1] biologists of the calibre of Réaumur, Haller, Bonnet and Spallanzani lent the weight of their considerable authority to the theory – usually in its ovist version, which was generally considered to have superior empirical credentials, founded on a mass of observations of the development of eggs of various kinds.[2] The rival theory of epigenesis seemed, to the orthodox biologists of the period, to require mysterious and occult powers or faculties and thus to be ruled out by the austere metaphysics of the mechanical philosophy.

In more recent times, the theory of preexistence has had a very bad press. It is generally derided as the product of speculative philosophy rather than of sober observation of and reflection on nature. In popular accounts of the history of biology, the 'Russian dolls' caricature of the theory of *emboîtement*[3] is

---

[1] See, e.g., Maupertuis, who in his *Vénus physique* and *Système de la nature* subjected the theory of preexistence to some searching questions and advocated instead a return to the rejected theory of epigenesis. The other famous defender of epigenesis was of course Caspar Wolff in his *Theoria generationis* (1759).

[2] For details of the debate between ovists and animalculists, see Clara Pinto-Correia, *The Ovary of Eve: Egg and Sperm and Preformation* (Chicago: University of Chicago Press, 1997). Although her work is often deficient in analytical clarity (she fails to make the crucial distinction between preformation and preexistence), it remains a wonderfully vivid account – informed by genuine biological insight and understanding – of the work of the central figures in the debate.

[3] The term 'emboîtement' was not used by Malebranche himself, although it is an apt label for his theory. According to Pinto-Correia (*The Ovary of Eve*, 32–3), the term was introduced by Nicolas Andry.

often introduced as a joke for the amusement of students. But if the theory was little more than speculative philosophy masquerading as empirical science, why did so many serious and sober biologists endorse it? The biologists of the period were all signed up – at least in principle – to the Baconian orthodoxy of the primacy of observation over theory. They must have been persuaded that the theory of preexistence had sound empirical credentials.

The most natural derivation of the theory of preexistence takes its origin from the principles of the mechanical philosophy. Organisms are likened to machines, and their parts to wheels, cogs, levers, pulleys, hydraulic systems, sieves, hawsers, etc. But machines do not (cannot) self-assemble. Given a sufficiently complex machine, one might in principle imagine how it could grow by adding and arranging new materials, but one still cannot explain how such a machine could assemble itself. One can only explain its current operations by reference to its contrivance and arrangement of parts; but if generation is our topic, it is precisely this arrangement and contrivance that demands explanation. The origin of a complex machine seems to transcend mechanism and to demand a non-mechanical agency of some kind.

It is precisely to avoid this implication that mechanists turned to the theory of preexistence. Here it is essential to distinguish clearly between the entirely distinct theories of preexistence and preformation.[4] On the theory of preformation, the new organism is 'elaborated' in the testes of the father, or the ovaries of the mother, in a process governed by the soul of the respective parent.[5] (To admit a genetic role for both parents, i.e., a double preformation, raises obvious difficulties.) To the mechanists, this process of elaboration seemed occult and unintelligible, and the role of the parental soul was obviously objectionable, especially to supporters of Descarte's theory of the *bête-machine*. On the theory of preexistence, the new organism was always present either in the egg (ovism) or the sperm (animalculism). On the ovist theory, the male provides only a trigger or stimulus for growth and development; on the animalculist theory, the female provides only protection and nourishment. In an important sense, preexistence is not a theory or account of generation at all, since it explicitly denies the very existence of plant and animal generation in nature. Every plant and every animal is the product of the original supernatural act of creation, not of a natural process of generation.

---

[4] I owe this important distinction to Jacques Roger, *Les sciences de la vie dans la pensée française au XVIIIe siècle* (Paris: A. Colin, 1963), 2nd ed., 325–6, and Peter Bowler, 'Preformation and Pre-existence in the Seventeenth Century: A Brief Analysis', *Journal of the History of Ideas* 4 (1971): 221–44.

[5] For the preformation theories of Liceti and others, see Roger, *Les sciences de la vie*, 125–30.

Andrew Pyle

The most prominent philosophical advocate of (Ovist) pre-existence in the late seventeenth century was Nicolas Malebranche. In his *Recherche*[6] (1674) and *Entretiens*[7] (1688) he set out and defended the central philosophical argument for the theory of preexistence. This argument resists easy categorization as either a priori or empirical, since it rests on considerations of both kinds. Two central pillars of the argument may be identified. There is the mechanical philosophy, which Malebranche takes to have been established by Descartes and which requires natural philosophers to provide intelligible (i.e., mechanistic) accounts of all the phenomena of nature and thus eliminate all references to the mysterious agency of substantial forms, vegetable or animal souls, and their supposed powers and faculties. And there are some very deep principles of biology,[8] notably functional complexity and functional interdependence of parts, which do rest on detailed investigation of the subtlety of nature.[9] The mechanical philosophy is fundamentally a priori – driven by considerations of intelligibility and the demand for clear and distinct ideas – but might also derive empirical support from its explanatory power.[10] The biological principles of functional complexity and functional interdependence of parts are clearly based on observation but also draw support from the conception of organisms as products of divine design and craft and thus from religion. It would therefore be an oversimplification to describe the first premise of Malebranche's argument (mechanism) as a priori and the second premise (functional complexity and functional interdependence of parts) as empirical. Both premises rest on considerations of both kinds. From these two premises Malebranche draws the conclusion that preexistence provides the only intelligible account of the (apparent) origin of plants and animals.

---

[6] *Recherche de la vérité, ou l'on traite de la nature de l'esprit de l'homme et de l'usage qu'il en doit faire pour éviter l'erreur dans les sciences* (1st ed. 1674). References will be made both to the standard French edition, the *Oeuvres complètes de Malebranche* (OCM), ed. André Robinet, 20 vols. (Paris: Vrin, 1958–66), and to the English translation of Lennon and Olscamp (LO), *The Search after Truth*. Thomas M. Lennon & Paul J. Olscamp. Columbus: OSU Press, 1980.

[7] *Entretiens sur la métaphysique et sur la Religion* (1688), in *Oeuvres complètes de Malebranche*, vol. 12. References will be made both to the French of the OCM and to the English translation of Nicholas Jolley and David Scott (JS), *Dialogues on Metaphysics and on Religion* (Cambridge: Cambridge University Press, 1997).

[8] I trust the mild anachronism of my use of this term will be permitted.

[9] This is a theme, prominent in much seventeenth-century discussion of plants and animals, that seems to have made little impact on Descartes, who endorses spontaneous generation (abiogenesis) with the astonishingly casual remark that 'il faut si peu de choses pour faire un animal' (see AT XI 50). For more on 'the subtlety of Nature', see Catherine Wilson, *The Invisible World* (Princeton: Princeton University Press, 1995), 39–69.

[10] Some Cartesians emphasize the a priori argument for the mechanical philosophy; others (e.g., Rohault) lay more stress on its role as providing a research program for the empirical investigation of nature, with the justification of the program largely by its fruit (i.e., a posteriori).

In addition to this main argument for the theory of preexistence, Malebranche also claims – in a number of places – that preexistence can draw empirical support from the observations of microscopy. Malebranche took a keen interest in the work of microscopists such as Marcello Malpighi and Jan Swammerdam and argued that their observations not only refuted earlier, mistaken theories of generation but could also serve to provide powerful support for preexistence. This essay both documents this claim and subjects it to critical scrutiny. Is there direct empirical evidence from microscopy for preexistence? Or (more weakly) does the evidence of the microscope count against rival theories of generation and thus lend some support to preexistence by means of an indirect eliminative argument? Or is the available microscopic evidence evidentially neutral, consistent with all the rival theories?

This essay divides naturally into three sections. Section 1 sketches Descartes' mechanistic cosmogony and his tentative attempts to extend this program to the origin of animals.[11] It also highlights the main difficulties that led later mechanical philosophers to abandon this part of Descarte's ambitious program. Section 2 sets out Malebranche's central argument – based on the functional interdependence of all the parts of an organism – for the impossibility in principle of all naturalistic accounts of generation and thus for the conclusion that preexistence is the only intelligible account of the (apparent) origin of plants and animals. Section 3 addresses the precise role of the evidence of the microscopists in lending empirical support (direct or indirect) to the theory of preexistence.

1. THE CARTESIAN PROGRAM

From its beginnings in *Le monde*,[12] Descarte's natural philosophy took the form of a cosmogony, officially presented in the guise of a fiction or an account of the genesis of a merely hypothetical world. Although the fictional nature of his account is stressed both in *Le monde* (AT XI 31, CSM I 90) and again in the *Principles* (AT IX 123–4, CSM I 255–7), it is clear that he regards the account as having explanatory power in the real world. Even if God created everything in its perfect state, we will understand more about a cosmos – or an organism – if we see how such a thing could have emerged from the play

---

[11] See Aucante's essay in this volume (chap. 3) for much more detail. Descartes clearly saw animal generation as a legitimate field for mechanical explanation but was never fully satisfied with any of his various attempts to provide such an account, which remained unpublished at his death.

[12] References to Descartes will be to Adam and Tannery (AT) and, where possible, to the English translations of Cottingham, Stoothoff and Murdoch (CSM) and (for vol. 3) Kenny (CSMK). For *Le monde*, see AT XI 3–118 and CSM I 81–98 (only part translated).

of corpuscles moving and interacting in accordance with merely mechanical laws (AT IX 201, CSM I 267). Matter, we are told in article 47 of part 3 of the *Principles*, takes on all the forms of which it is capable, so an organized world can emerge out of chaos by the laws of nature alone (AT IX 126, CSM I 258).

The genetic character of Descarte's natural philosophy is one of its most striking features: he only felt he had explained something when he had 'deduced' it from its causes, that is, had shown how it could have come to be. That he intended this method to be extended to biology is evident both from his published works and from his correspondence. In part of the famous *Discours* (1637), Descartes admits that the generation of animals had been omitted from *Le monde* because 'I did not yet have sufficient knowledge of them to speak of them in the same manner as I did of the other things – that is, by demonstrating effects from causes and showing from what seeds and in what manner nature must produce them' (AT VI 45, CSM I 134).

In his correspondence, Descartes was rather bolder than in his published works. In a letter to Mersenne of 1639 he insists that no part of the animal body is inexplicable by natural – that is, purely mechanical – causes. If, he writes, *Le monde* were to be rewritten, 'I should undertake to include also the causes of its formation and birth' (AT II 525, CSMK III 134–5).[13] The problem did not even strike him as particularly difficult. In the misleadingly named unpublished manuscript *Primae cogitationes circa generationem animalium*,[14] he accepts 'spontaneous' or 'equivocal' generation (abiogenesis) without a qualm: 'Since therefore so few things are necessary to make an animal, it is assuredly not surprising to see so many animals, so many worms, so many insects form spontaneously in all putrefying matter' (AT XI 50).[15]

Although animal generation was omitted again from the published *Principles* (1644), it remained a topic of intense interest for Descartes: he seemed forever on the point of publishing his great work on the subject. As Jacques Roger insists, embryology was to be the touchstone of the entire mechanical philosophy:[16] if matter and motion can account for a frog or a chicken, they can account for anything; conversely, a check on this point threatens the whole system. On this view, the little-known and posthumously

---

[13] For commentary, see Aram Vartanian, *Diderot and Descartes: A Study of Scientific Naturalism in the Enlightenment* (Princeton: Princeton University Press, 1953), 247.

[14] According to Aucante (chap. 3 of this volume), some of these 'first thoughts' actually date to 1648, i.e., towards the end of Descarte's life.

[15] See also Roger, *Les sciences de la vie*, 79.

[16] Ibid., 141–2.

published *De la formation de l'animal*[17] assumes a pivotal importance as a central text for the whole Cartesian program.

In this work the generation of the higher animals is said to occur, as in Galenic physiology, by a mixture of male and female semen, under the action of heat alone, without the traditional array of faculties. It is an advantage of his theory, Descartes boasts, that 'it is not necessary to assume . . . any strange or unknown faculties' (AT XI 244). Whereas the seeds of plants are solid and may therefore possess preformed structure, the *semina* of animals are fluid, and the embryo 'seems merely to be a confused mixture of two liquids, which serving to leaven one another, are warmed in such a way that some of their particles, agitated as if by a fire, become dilated and press on the others, and thus dispose them gradually in the manner required to form the members' (AT XI 253).

The process is a true epigenesis, with the heart (in this final version of the theory) the first organ to be formed.[18] The heart in turn produces a sort of distillation or reflux of the blood, from which the other organs condense out in due order. Blood currents to and from the heart become veins and arteries respectively; streams of blood passing through pores in the walls of blood vessels deposit muscle fibres; a secondary reflux of 'spirits', centred in the brain, produces the nervous system. The details need not detain us long: it suffices to see that we have here an immensely ambitious research program for embryology, explaining epigenesis (the successive appearance of organs out of a homogeneous mixture of fluids) in purely mechanistic terms.

Although Descartes was confident that his theory was broadly correct – if perhaps in places in need of more detailed observational support – the posthumous publication in 1664 of this key work by his disciple Clerselier did not impress his contemporaries.[19] Descarte's account of generation, according to his immediate successors, fails to account for (1) specificity, (2) functional adaptation and (3) the structural complexity of even lowly organisms. With regard to the first point, it is clear that, by abandoning the substantial forms of the Aristotelians, Descartes had imposed upon himself an additional and weighty explanatory burden. All he can do is to sketch the beginnings of a reply. Admitting that his account was perfectly general, he added that 'if one

---

[17] AT XI 223 ff. The more familiar title *De la formation du foetus* was given to it by Clerselier in his 1664 edition of previously unpublished works of Descartes.

[18] In earlier versions of the account, the brain and the liver had been given priority. See chapter 3 of this volume for more details.

[19] The publication of *De la formation du foetus*, according to Elizabeth Gasking, did nothing to promote Descartes's theory but rather served 'to provide an argument against it' (*Investigations into Generation: 1651–1828* [Baltimore: John Hopkins University Press, 1967], 67).

were familiar with all the parts of the semen of some particular species of animal, for example of man, one could deduce from that alone, by entirely certain mathematical reasoning, the whole figure and confirmation of each of its members' (AT XI 277).

This hand-waving failed to impress the critics. From the many who raised the specificity of generation as an objection, let us cite only two: Joseph Glanvill and Ralph Cudworth. In his *Scepsis scientifica* (1665), Glanvill insists that mere mechanism cannot explain many obvious features of generation, notably its specificity:

> [W]hy the *wood-cock* doth not sometimes borrow colours of the *Mag-pye*; why the *Lilly* doth not exchange with the *Daysie*; or why it is not sometime painted with the blush of the *Rose*. Can unguided matter keep it self to such exact conformities, as not in the least spot to vary from the species?[20]

In his *True Intellectual System* (1678), Cudworth endorses this criticism. There is, he notes, a piece 'imputed to Cartes', entitled *De la formation du foetus*, which pretends to explain generation in terms of 'fortuitous mechanism'. But, he retorts, '[I]t is all along precarious and exceptional; nor does it extend at all to the differences that are in several Animals, nor offer the least reason why an Animal of one Species might not be formed out of the Seed of another.'[21]

With regard to functional adaptation and the question of teleology, Descartes is on still shakier ground. He explains, for example, the formation of the cardiac valves in terms of the deposition of a 'skin' at the interface of opposed blood currents (AT XI 279), yet offers no account of the functional significance of their location and structure. In his *Rebuttals* of the *Meditations*, Descarte's great rival Gassendi sprang to the defence of final causes.[22] You, Descartes, say that we should investigate efficient rather than final causes, yet 'no mortal can understand, much less explain, what agent forms and disposes those valves which are placed in the cavity of the heart'. If you know the efficient cause, 'you are no man, no hero, no demigod, but truly some God'.[23] The utility of the valves is, however, perfectly clear, once Harvey's discovery of the circulation of the blood has been grasped. Why then should we not marvel at 'this most extraordinary and ineffable providence which fashioned

---

[20] *Scepsis Scientifia: or, Confest Ignorance, the Way to Science: in an Essay on the Varity of Dogmatizing and confident Opinion* (London, 1665; repr., New York: Garland, 1978), 33–4.

[21] *The True Intellectual System of the Universe*, 3 vols. (London: Tegg, 1845), 2:615.

[22] Gassendi, 226.

[23] Ibid., 237.

these valves so beautifully for their function?' Many of the second-generation mechanists took the side of Gassendi against Descartes in this controversy.[24]

The *coup de grâce* was struck by the microscopists. After the revelations of Hooke and Malpighi, Leeuwenhoek and Swammerdam, who could ever again believe that 'il faut si peu de choses pour faire un animal'? Where the naked eye discerns only homogeneous tissues, the microscope uncovers a dazzling array of complex, beautiful and (no doubt) highly functional structures.[25] The publication of Hooke's *Micrographia* (1665), Malpighi's *De bombyce* (1669) and Swammerdam's *Historia insectorum generalis* (1669) showed that even such lowly organisms as fleas and lice are extremely complex in their structures, which are (no doubt) exquisitely adapted to their particular lifestyles.[26] The idea that such highly contrived structures could simply have condensed out of a mixture of fluids was dismissed as incredible. Crystals, says Boyle in his *Origin of Forms and Qualities*, might arise by mere condensation out of a fluid, but 'the stupendous and incomparably more elaborate fabrick and structure of animals' could not conceivably do so.[27] To the eyes of the pious Jan Swammerdam, microscopic observations of insects seemed to provide direct evidence of divine craftsmanship, as the striking lines of a letter to Thevenot bear eloquent witness. 'Herewith', he writes, 'I offer you the Omnipotent finger of God in the anatomy of a louse.'[28] After the revelations of the microscopists, it was impossible to believe any longer that the mere 'play of corpuscles' could fashion a plant or an animal. The second-generation disciples of Descartes chose not to pursue this part of his ambitious research program, opting either for a discreet silence on the subject (Rohault) or for the theory of preexistence (Régis).[29]

---

[24] See, e.g., Robert Boyle's *Disquisition about the Final Causes of Natural Things* in his *Works*, vol. 5, 392 ff., esp. 427.

[25] For more details on this theme, see Marian Fournier, *The Fabric of Life: Microscopy in the Seventeenth Century* (Baltimore and London: Johns Hopkins Press, 1996), esp. chap. 5, 'The Fabric of Living Beings', 135–77, and Catherine Wilson, *The Invisible World*, esp. chap. 2, 'The Subtlety of Nature', 39–69.

[26] Robert Hooke, *Micrographia*, in R. T. Gunther, ed., *Early Science in Oxford*, 15 vols. (Oxford: Oxford University Press), vol. 13. For Malpighi's comments on the microanatomy of insects, see his letter to Oldenburg of 22 March 1668, in *Henry Oldenburg: Correspondence*. Ed. & trans. A. Rupert Hall & Marie Boas Hall. Madison; University of Wisconsin Press, 1965–86, IV 92.

[27] Robert Boyle, *Origin of Forms and Qualities*, in *Works*, ed. T. Birch, 6 vols. (London, 1772), 3:55.

[28] Swammerdam to Thevenot, April 1678, from *The Letters of Jan Swammerdam to Melchisedec Thévenot*. Ed. G. A. Lindeboom (Amsterdam: Swets & Zeitlinger, 1975), 105.

[29] Part 4 of Jacques Rohault's *System of Natural Philosophy* (1671) discusses a variety of physiological topics but conspicuously omits animal generation. For Régis's conversion to the theory of preexistence, see Roger, *Les sciences de la vie*, 348.

2. MALEBRANCHE'S MAIN ARGUMENT FOR PREEXISTENCE

In book 6 of the *Recherche*, Malebranche presents a beautifully lucid exposition of Cartesian cosmogony and a measured defence of the program against its critics (OCM II 321ff, LO 453 ff). Descarte's theory of vortices is, he insists, the world system that emerges if one simply follows the light of reason. As for the objection that Descarte's cosmogony contradicts Scripture, one might either reinterpret the notoriously obscure text of Genesis or opt, more cautiously, for Descarte's own avowed fictionalism. Malebranche agrees with Descartes that an explicitly fictional account can still have explanatory value, that is, that we will understand our world better if we see it as the product of a fictional evolutionary history. Consider, he says, the stability of our solar system. If God had chosen a different arrangement of Sun and planets than one that could have arisen from the unaided laws of motion, the system would not be stable (OCM II 341–2, LO 464). If, for example, God had created a cubic Sun, displaced from the centre of its vortex, the laws of motion would wear away its corners and shift it back to its 'proper' place in the center of the vortex.

Even the natures of living things, Malebranche continues, can be illuminated by a genetic approach. Suppose a man wants to discover the nature of the chicken:

> To do this, every day he opens all the eggs it lays. In them he notes a vesicle that contains the embryo of the chicken, and in this vesicle a projecting point that he discovers to be its heart; that from there, blood-carrying conduits that are the arteries go out in all directions, and that this blood returns toward the heart via the veins; that the brain also appears at an early stage, and that the bones are the last parts to be formed. By this means he is delivered from a great many errors, and he even draws many conclusions of considerable use in the understanding of animals from these observations. What can one find to criticize in the conduct of this man? (OCM II 342, LO 464)

At this point, however, Malebranche makes a crucial departure from Descartes. The solar system, he thinks, could have emerged out of a mere chaos by the unaided laws of motion; a plant or an animal could not have done so. Why not? The reason is the functional interdependence of parts:

> An organized body contains an infinity of parts that mutually depend upon one another in relation to particular ends, all of which must be actually formed in order to work as a whole. For it need not be imagined with Aristotle that the heart is the first part to live and the last part to die. The heart cannot beat without the influence of the animal spirits, nor these be spread throughout the heart without the nerves, and the nerves originate in the brain, from which they receive the

spirits. Moreover, the heart cannot beat and pump the blood through the arteries unless they as well as the veins that return the blood to it are already complete. In short, it is clear that a machine can only work when it is finished, and that hence the heart cannot live alone. Thus, from the time this projecting point that is the heart of the chicken appears in a setting egg, the chicken is alive; and for the same reason, it is well to note, a woman's child is alive from the moment it is conceived, because life begins when spirits cause the organs to work, which cannot occur unless they are actually formed and connected. It would be wrong then to pretend to explain the formation of animals and their plants and their parts, one after the other, on the basis of the simple and general laws governing the communication of motion; for they are differently connected to one another by virtue of different ends and different uses in the different species. But such is not the case with the formation of vortices; they are naturally born from general laws, as I have just in part explained. (OCM II 343–4, LO 464)

For Malebranche, in contrast to Descartes, many things are necessary to form an animal, and these many things are both functionally specific and functionally interdependent.[30] If no part can exist and function without the others, an organism cannot come into existence gradually, part by part, as theories of epigenesis require. In *De la formation de l'animal*, Descartes started with the heart, but how can the heart function without arteries, veins, nerves and brain? (We might add lungs, liver, kidneys, etc.) The heart could presumably function in the absence of arms and legs, so the argument does not strictly require that the fetus be fully formed, a minute replica of the eventual adult, but it does entail that all the vital organs, all the organs necessary for life itself, must be present *ab initio*. The quoted passage tells against Descarte's theory in particular, but its implied moral is perfectly general: no theory of this kind could possibly account for the type of organized complexity that is characteristic of plants and animals.

It is worth pausing here to reflect on this argument. It is clearly founded on genuine biological insight, but its conclusion is much too strong to be credible. So where does the argument go wrong? The problem seems to be that it presupposes both that there is a strict (one-to-one) correspondence between organs and functions and that each vital function must always be served by its own particular specialist organ. But neither assumption is strictly true. Suppose a modern biologist were to argue as follows: The heart cannot function without a supply of oxygen. The function of the lungs is to supply oxygen to the heart (and other organs). Therefore, the heart cannot function without the lungs.

---

[30] According to Catherine Wilson (*The Invisible World*, 120), this key argument was anticipated by Gassendi. But Malebranche drew more radical conclusions from the argument than Gassendi.

The premises are true, the conclusion false, so the argument is invalid. Why is it invalid? Because it presupposes that only the lungs can serve the function of providing oxygen. But in mammals, another system (placenta and umbilical cord) provides oxygen to the developing foetus; this function is transferred to the lungs at birth. And in its very earliest beginnings, the embryo may be so small that mere diffusion will do the job, without any need for a specialist organ or system. So even though the argument is based on genuine biological insight, its conclusion is an exaggeration, representing a difficulty as an impossibility.[31] But let us return to Malebranche.

It is organisms, not worlds, we are told, that provide evidence of divine design and craftsmanship. 'The smallest fly' provides better evidence of the power and wisdom of God than 'everything the astronomers know about the heavens' (OCM II 61, LO 296–7). Every organism is a particular product of divine craft, something that could not have arisen from any other cause. It could not have arisen from the unaided laws of motion, for the reason just given, and there are no subordinate but non-mechanical natural agencies (animal and vegetable souls, 'plastic nature', the *Anima Mundi*) available for the task. Every cabbage and mushroom, every beetle and mosquito, is the product of a particular providence, a special act of creation. But if we posit a supernatural origin for every individual plant and animal, are we not abandoning *natural* philosophy altogether?[32] It is to meet this objection that Malebranche has recourse to the theory of preexistence. Every organism is the product of creation (supernatural)[33] rather than of generation (natural), but God has packed all His creative acts into the one great miracle of the Creation; at present, He is 'resting', that is, acting only by the *volontes generales* of his ordinary providence, namely, the laws of nature. These laws of nature suffice to account for growth and development but not for generation.

---

[31] It is striking that very similar objections were raised by Cuvier against early theories of evolution and by modern critics of Darwinism who cite so-called irreducible complexity as evidence against the theory of evolution by natural selection. In all three cases, the argument starts with genuine biological insight but concludes with exaggeration and overstatement.

[32] Precisely this objection would later be pressed by Maupertuis in his *Vénus physique* (*Oeuvres*, 4 vols. Hildesheim: Olms, 1965–74), who argued that there was no explanatory gain in packing all the miracles into the first moment of creation.

[33] For the striking claim that the mechanical philosophy, by reducing the powers of matter, led to 'Supernaturalism', see Keith Hutchison, 'Supernaturalism and the Mechanical Philosophy', *History of Science* 21 (1983): 297–333. For a qualified endorsement of this claim, see Andrew Pyle, 'Animal Generation and Mechanical Philosophy: Some Light on the Role of Biology in the Scientific Revolution', *Journal for the History and Philosophy of the Life Sciences* 9, no. 2 (1987): 225–54.

The theory of preexistence is first articulated in book 1 of the *Recherche*, ostensibly devoted to errors of vision. Malebranche begins with the observation of micro-organisms under the microscope and with the natural inference from function to structure, from a vital activity (e.g., locomotion) to the structures (organs, arrangements of parts) needed to sustain that activity (OCM I 81, LO 26). Given the infinite divisibility of matter (which can be demonstrated from our clear and distinct idea of it as three-dimensional extension), nothing prevents the existence of smaller and smaller organisms *ad infinitum*. If our imagination boggles at the prospect of an endless descending hierarchy of smaller and smaller micro-organisms, so much the worse for the imagination. Reason shows that such a hierarchy is possible; microscopes reveal to us the existence of some of these micro-organisms; better and more powerful microscopes will doubtless reveal still more and even more minute ones.

So far, this emphasis on the infinite divisibility of matter, and hence on the possibility of ever-smaller organisms without end, has just been a softening-up exercise on Malebranche's part, designed to shake his readers' confidence in the testimony of their (unaided) senses. Now, however, he begins to present his main argument for preexistence:

> When one examines the seed of a tulip bulb in the dead of winter with a simple magnifying glass, or even merely with the naked eye, one easily discovers in this seed the leaves that are to become green, those that are to make up the flower or tulip, that tiny triangular part which contains the seed, and the six little columns that surround it at the base of the flower. Thus it cannot be doubted that the seed of a tulip bulb contains an entire tulip. (OCM I 81, LO 26)

The next step is a simple generalization. What is visibly true of the tulip is also true of all other plant seeds, even if the parts of the eventual plant are not readily discernible in the seed. What drives this hasty generalization? Not the evidence of the senses. Why must what is visibly true in one case also be true in all the others? Because complex functional structures cannot simply emerge out of nothing. The oak tree must be present in the acorn as the tulip is present in its bulb. But this, of course, only displaces our search for explanations. We can now explain how the oak tree arises from the acorn (it is just unpacking and growth), but how was the acorn formed in the first place? Does the oak tree have the utterly mysterious power to create mini-replicas of itself? That theory (a version of preformation) is for Malebranche simply inconceivable. But the only remaining alternative is preexistence:

> Nor does it seem unreasonable to believe even that there is an infinite number of trees in a single seed, since it contains not only the tree of which it is the seed but also a great number of other seeds that might contain other trees and

other seeds, which will perhaps have on an incomprehensibly small scale other trees and other seeds and so to infinity. So that according to this view, which will appear strange and incongruous only to those who measure the marvels of God's infinite power by the ideas of sense and imagination, it might be said (1) that in a single apple seed there are apple trees, apples, and apple seeds, standing in the proportion of a fully grown tree to the tree in its seed, for an infinite, or nearly infinite number of centuries; (2) that nature's role is only to unfold these tiny trees by providing perceptible growth for that outside its seed, and imperceptible yet very real growth in proportion to their size, for those thought to be in their seed. (OCM I 82, LO 27)

Here we have the classic statement of the notorious doctrine of *emboîtement*, promptly extended from plants to animals. The argument starts once again with observation, but reason soon outstrips the relatively modest evidence provided by the senses:

Likewise, a chicken that is perhaps entirely formed is seen in the seed of a fresh egg that has not been hatched. Frogs are to be seen in frogs' eggs, and still other animals will be seen in their seed when we have sufficient skill and experience to discover them. But the mind need not stop with the eyes, for the mind's vision is much more extensive than the body's. We ought to accept, in addition, that the body of every man and beast born till the end of time was perhaps produced at the creation of the world. My thought is that the females of the original animals may have been created along with all those of the same species that they have begotten and that are to be begotten in the future. (OCM I 82–3, LO 27)

Although the argument here starts with observations of eggs, it is clearly not meant to be inductive. Reason shows (given the infinite divisibility of matter) that *emboîtement* is at least a possible account of the (apparent) generation of plants and animals. Experience shows that (in some cases at least) we can see preexistent structures in the egg or seed. This lends, however, only the weakest of empirical support to the doctrine of preexistence. Malebranche must have been aware of this fact, insisting as he often does that the intellect must go beyond the evidence of the senses and demand an intelligible account of the origin of plants and animals. The argument for preexistence can then be formulated as an argument from elimination: once all accounts that fail to satisfy the intellect have been rejected, only the theory of preexistence remains.

The same mixture of weak empirical evidence and strong a priori constraints reappears in the tenth of the *Entretiens*, in which Malebranche's spokesman Theodore begins by citing empirical evidence for the existence of micro-organisms (OCM XII 228, JS 174). There are, Theodore tells us, an untold number of animals even more minute than those discovered by Leeuwenhoek. But what is the origin of all these minute organisms? The mere laws of motion, Theodore argues, can explain growth but not generation:

> We see . . . that, unless we wish to have recourse to an extraordinary providence, we must believe that the seed of a plant contains in miniature the plant which it engenders, and that in its womb an animal contains the animal which should come from it. (OCM XII 229, JS 175)

Either God is forever intervening in the course of nature, fashioning every new mosquito and mushroom, or each new organism was preexistent in its germ. But where did this germ in turn come from? The only conceivable answer, Theodore concludes, is that the organisms have been packed inside one another, generation inside generation, from the first moment of creation:

> We even understand it to be necessary that each seed contains the entire species it can conserve; that every grain of wheat, for instance, contains in miniature the ear it germinates, each grain of which contains in turn its own ear, all the grains of which can always be as fertile as those of the first ear. (OCM XII 229, JS 175)

Here we find a clear statement of the doctrine of *emboîtement*, supported once again by an argument that is eliminative in form. Merely mechanical causes ('matter and motion') cannot, *pace* Descartes, account for generation. There are no subordinate but non-mechanical agencies (animal and vegetable souls, 'plastic nature', etc.). It follows that the only possible cause of animals and plants is God himself. Faced with a forced choice between continual interventions and preexistence, the natural philosopher must opt for the latter. Follow this line of argument to its logical conclusion and one has the full theory of *emboîtement*. The course of nature is entirely mechanical; the ultimate origin of all things is supernatural. Anyone who wishes to resist this conclusion must explain which of its premises he is prepared to reject. Is he going to defend the discredited Cartesian project of mechanistic epigenesis? Or abandon the mechanical philosophy? Or admit incessant divine interference in the course of nature?

Theodore goes on to make an important qualification of the doctrine of preexistence. The miniature plants and animals contained in their seeds need not, he tells us, resemble fully-formed adults:

> Nevertheless, it is not the case that the tiny animal or the germ of the plant has precisely the same proportion of size, solidity, and figure among all its parts, as the animals and plants. But it is the case that all the parts essential of the

machine of the animals and plants are so wisely disposed in their germs, that they will, in time and as a consequence of the general laws of motion, assume the shape and form which we observe in them. (OCM XII 229, JS 176)

The theory of preexistence requires only the existence of a determinate (and functional) arrangement of the vital organs, the parts essential for life. It does not require that the preexistent germs be precise miniatures of their parents. Here Malebranche does rely on empirical evidence. When Malpighi dissects embryonic chickens, he finds that their heads are relatively large and their bones not fully formed. When Swammerdam dissects the grubs of insects, he does not find every detail of the imago present in the larva. As Malebranche explains, the preexistent germs of insects such as bees need not be iconic, need not be mini-replicas of the adult form:

> I simply claim that all the organic parts of bees are formed in their larvae, and are so well proportioned to the laws of motion that they can grow through their own construction and through the efficacy of those laws, and can assume the shape suitable to their condition, without God intervening anew through extraordinary providence. (OCM XII 253, JS 195–6)

Since preexistence does not require that the mini-organisms in the germs or seeds be precise replicas of their eventual adult forms, the theory does not exclude a causal role for environmental factors in development. One such factor that Malebranche explicitly accepts is the maternal imagination, routinely invoked at the time to account for monstrous births. He even suggests at one point in the *Recherche* that the maternal imagination may explain the specificity of animal generation (OCM I 242–3, LO 117). If this is right, the miniature germ of a future animal is not yet determinately a mini-dog or a mini-cat until acted on by the maternal imagination. But this is surely just an aberration on his part. The theory of preexistence has no need to ascribe such a crucial role to the maternal imagination: God can pack cats within cats and dogs within dogs *ad infinitum*. Divine foresight and preplanning can account for any degree of resemblance between parents and offspring. Malebranche sees this perfectly clearly (OCM II 242, LO 117–8). Worse still, no remotely plausible mechanism is postulated for transferring corporeal images from the brain of the mother to the developing foetus. The alleged empirical evidence is merely anecdotal and is contradicted by some plain facts. Plants breed true without any imagination at all. When a hen sits on duck eggs, it is ducklings, not chickens, that hatch. A blind mother bears sons who resemble their father. Given these facts, it is puzzling that Malebranche accepts, on the basis of the usual anecdotal evidence (which in other contexts he would despise), the importance of the maternal imagination in the development of the germs of animals and humans.[34]

3. PREEXISTENCE AND MICROSCOPY

Malebranche's main argument for preexistence, as here presented, is eliminative in form and (largely) a priori, based on a rigorous and exhaustive search for an intelligible account of the origin of plants and animals. It is true, of course, that one of the key premises (the functional complexity and functional interdependence of the parts of organisms) is informed by detailed observations of nature, so the argument as a whole cannot strictly be labelled a priori. It remains the case that the argument, as we have presented it, has the strongly abstract air of a rational exercise in natural philosophy rather than an inductive argument from close and detailed observations of nature.

Malebranche was nevertheless convinced that the new discoveries of the microscopists – Hooke, Malpighi, Swammerdam, Leeuwenhoek and Grew[35] – provided additional and more direct observational support for preexistence. In the *Entretiens*, for example, we are referred to Leeuwenhoek's discovery of micro-organisms (OCM XII 227, JS 174),[36] to Swammerdam's famous dissections of insects (OCM XII 229, JS 176)[37] and to Malpighi's observations of the development of the chicken in its egg (OCM XII 260, JS 201).[38] There can be no doubt both that Malebranche took a keen interest in the work of contemporary microscopists and that he firmly believed that

---

[34] I have elsewhere suggested (see my *Malebranche* [London and New York: Routledge, 2003], 182) that on this occasion Malebranche's habitual rationalism gave way to his desire to accommodate an article of Christian dogma – i.e., to find room within his natural philosophy for the Christian doctrine of original sin. For an alternative account of the role of the maternal imagination in early modern embryology, including Malebranche, see Justin E. H. Smith's essay in this volume (chap. 4).

[35] Marian Fournier lists these five men as the chief scientific microscopists of the seventeenth century (see *The Fabric of Life*, chap. 2, 49–91).

[36] The reference is to Leeuwenhoek's letter to Sir Christopher Wren, for the Royal Society, of 25 July 1684. For Leeuwenhoek, see Clifford Dobrell, *Antony van Leeuwenhoek and His 'Little Animals'* (London: Bale, 1932), and Fournier, *The Fabric of Life*, 79–91.

[37] For commentary on Swammerdam, see A. Schierbeck, *Jan Swammerdam, 1637–1680: His Life and Works* (Amsterdam: Swets Zeitlinger, 1967); Matthew Cobb, 'Reading and Writing the Book of Nature: Jan Swammerdam (1637–1680)', *Endeavour* 24 (2000): 122–8; id., 'Malpighi, Swammerdam and the Colourful Silkworm: Replication and Visual Representation in Early Modern Science', *Annals of Science* 59 (2002): 111–47; Fournier, *The Fabric of Life*, 62–72. The reference in the *Entretiens* is to the *Historia insectorum generalis* of 1669.

[38] The reference is to the *Dissertatio epistolica de formatione pulli in ovo* (1673). For commentary, see Howard B. Adelmann, *Marcello Malpighi and the Evolution of Embryology*. 5 vols. (Ithaca: Cornell University Press, 1966), and Fournier, *The Fabric of Life*, 55–62.

their work supported his own conception of nature. But what sort of evidence does he have in mind? Is it direct ocular evidence of the preexistence of the eventual plant or animal in its seed or egg before fertilization? Or is it indirect evidence, supporting certain implications of the theory of preexistence or counting against rival theories? If one is a Bayesian, one partitions one's subjective or personal probabilities between the competing hypotheses. Any evidence that counts against one hypothesis will thus tend to support – in this negative and roundabout manner – its rivals. On the face of it, this supposition of indirect support only for preexistence seems more plausible. But if the support is only indirect, is it strong enough to convince an open-minded observer?

One use for the evidence of the microscopists is perfectly clear and straightforward. Their observations can be used to lend additional support to the second premise of what I have labelled the main argument for preexistence. If even lowly organisms like insects turn out – when closely observed – to be staggeringly complex, miniature masterpieces of design and contrivance, the explanatory burden of the rival theories (epigenesis, metamorphosis) becomes so much greater. No one could admire the illustrations of Hooke's *Micrographia* (1665) or Swammerdam's *Historia insectorum generalis* (1669) and think it an easy matter to explain how such organisms could simply condense out of a mixture of fluids. Such works may thus have served to shift the balance of probabilities away from epigenesis and towards preexistence. But we are looking for more than this. Let us see what we can find in Malebranche's explicit references to Leeuwenhoek, Swammerdam and Malpighi.

Preexistence requires the existence of organisms within organisms *ad infinitum* – or at least, for countless generations.[39] The imagination may boggle at the thought, but the imagination is no guide to truth, as Malebranche has told us in the *Recherche* (OCM I 81, LO 26). Given the (demonstrable) infinite divisibility of matter, there can be nothing impossible a priori about the postulation of ever-smaller organisms with the same basic structure and arrangement of parts. But showing that a given hypothesis is not inconceivable – in the relevant sense of *rational conceivability* – does little or nothing to render it positively credible. It is precisely here that Leeuwenhoek's micro-organisms can play an important role. In *Entretien* X, Malebranche's spokesman Theodore has been admiring insects but continues as follows:

> After having for some time admired this tiny creature so unjustly scorned . . . I began reading a book I had with me, and I found something quite amazing there:

---

[39] God could, of course, intend the world to come to an end after some large but finite number of generations. The theory of preexistence could easily be combined with speculations about the end of the world.

in the world there are an infinite number of insects at least a million times smaller than the one I had just considered, and ten thousand times smaller than a grain of sand. (OCM XII 227, JS 174)

Leeuwenhoek's micro-organisms are not 'insects' in the precise modern sense, of course, but the term was used much more broadly in the seventeenth century. What his observations show is that the existence of a descending hierarchy of ever-smaller organisms is not merely possible but actual. Objections to preexistence based on incredulity about its implication of a descending hierarchy of ever-smaller organisms within organisms must therefore give way to the plain facts of observation. The evidence of Leeuwenhoek is not direct evidence for preexistence (which there is no reason to believe he accepted),[40] but it is important indirect evidence in the sense that it provides direct support for what might otherwise have been a highly contentious implication of the theory.

A little later in the same dialogue, the pupil Ariste shows us that he has learned his lesson:

As matter is infinitely divisible, I understand quite well that God was able to produce in miniature everything we see on a larger scale. I have heard that a Dutch scientist discovered the secret of showing, in the cocoons of caterpillars, the butterflies which issue from them. (OCM XII 230, JS 176)

The "Dutch scientist" is Jan Swammerdam, whom Malebranche had met in the late 1660s and with whom he had discussed the problems of animal generation.[41] Swammerdam had shown by careful dissection the existence in the pupa of the parts of the butterfly or moth that was to emerge from it. Strictly speaking, we observe not a fully-formed miniature moth or butterfly but only the 'imaginal discs' or markers, clusters of cells that await the stimulus to develop into legs, wings, antennae, etc.[42] But the absence of strict resemblance here does not threaten the theory of preexistence: no advocate of that theory ever believed that it required strict or exact resemblance. Swammerdam found some preexistent structure in the pupa and – continuing

---

[40] Leeuwenhoek was more of an observer of nature than a theorist, but the evidence of his letters suggests that he did not believe in preexistence. After his famous discovery of the spermatozoa in the semen, he favored animalculism against ovism but seems to have believed that the spermatozoa are generated in the testicles. See Roger, *Les sciences de la vie*, 309; Gasking, *Investigations into Generation*, 51 ff.; and Fournier, *The Fabric of Life*, 90.

[41] Schierbeck, *Jan Swammerdam*, 115–16. The theory of *emboîtement*, says Swammerdam, was first proposed to him by 'a learned friend', since identified as Malebranche. See also Cobb, 'Reading and Writing the Book of Nature', and Roger, *Les sciences de la vie*, 335–7.

[42] Pinto-Correia, *The Ovary of Eve: Egg and Sperm and Preformation*, 29–32.

the same reasoning further back in the line of development – postulated only a 'mystical simulacrum' in the egg, a 'scenario for the individual's future development, imprinted on the matter within the egg'.[43]

Preexistence requires only that the final structure of the adult insect emerges from a determinate preexistent structure, not that the latter exactly resembles the former. Swammerdam's observations support this by refuting the rival theory of metamorphosis. On that theory, favored by Harvey, the structure of the caterpillar is completely broken down in the pupa, and the matter is then subjected to a new form, as if it had been placed in a mould. On this theory, Swammerdam contends, every metamorphosis of every insect would be 'death and resurrection', not organic growth and development. He succeeded in showing that this account was quite mistaken and that the transitions from caterpillar to pupa and from pupa to butterfly are transitions in the life story of a single abiding organism.[44] This provides no direct support for preexistence – he does not claim to have observed the structure of the eventual butterfly in an unfertilized egg[45] – but does count strongly against one of the rival theories. If the central claim of the theory of preexistence is 'structure comes from structure', or 'no structure without preexistent structure', then Swammerdam's observations of insect larvae can be cited in its support, even without direct observation of preexistent organisms in eggs.[46] We see the rudiments of the butterfly in the pupa, trace a single continuous life from caterpillar to pupa to imago and are led to postulate some preexistent structure (albeit perhaps very remote from that of the imago) in the egg.

When we turn our attention from invertebrates to vertebrates, it is the eggs of frogs and birds that offer the best opportunities for close microscopic observation. Here the work of Marcello Malpighi takes pride of place. In his famous study *De formatione pulli in ovo* (1673), he claims to have detected the presence of structure – of future organs – in the egg even before incubation. Strictly speaking, this is irrelevant to the issue of preexistence, since such observations are perfectly consistent with preformation or even epigenesis. If one seeks to trace earlier and earlier structures, the microscope will eventually fail us. In an unfertilized egg, Malpighi tells us, preexistent structure cannot be clearly detected: 'Everything is so mucous, white and transparent that the

---

[43] Fournier, *The Fabric of Life*, 71.
[44] Ibid, 68–9.
[45] See Bowler, 'Preformation and Pre-existence in the Seventeenth Century': A Brief Analysis," *Journal of the History of Biology*, 1971, 4 (2): 221–234.
[46] The microscope, according to Catherine Wilson (*The Invisible World*, 103), supports those philosophers (mechanists) who believe in the actuality of form against those (Aristotelians) who believe in mere potentialities not grounded in actuality.

eye, with whatever instrument it may be fortified, is unable clearly to detect the contexture of the parts.'[47]

There were claims made in the seventeenth century to have observed pre-existent structure in an unfertilized or 'wind' egg, but such claims were not made by the most authoritative and respected of the microscopists, were not corroborated by contemporaries and were not accepted into the received wisdom of the scientific community.[48] So what is the evidential relevance of Malpighi's observations? Clearly, they provide no direct empirical support for the theory of preexistence: that would require precisely what Malpighi says he cannot clearly observe, that is, preexistent structure prior to fertilization.[49] But they do provide some support (albeit fairly weak) for the maxim that structure comes from structure, refuting not epigenesis as such but some epigenetic accounts and timetables for development. In this negative and indirect manner, these observations too can support preexistence.

The answer to our original question – about the nature and extent of the support given by microscopy to the theory of preexistence – is gradually becoming clear. There is scarcely any direct positive evidence:[50] none of the great microscopists ever claimed to observe fully-formed organs in unfertilized eggs, still less the organisms within organisms within organisms required by *emboîtement*. But there is a significant body of indirect evidence, of the following kinds:

1. Evidence of the extreme complexity of all animals and plants (even the microscopically small) and for the 'contrivance' or functional interdependence of their parts.
2. Empirical proof of the real existence of organisms whole orders of magnitude smaller than anything observed with the naked eye (Leeuwenhoek).
3. Refutation of the theory that insects undergo a complete metamorphosis (abrupt remoulding) in their changes from larva to pupa and from pupa to imago (Swammerdam).

---

[47] Quoted from Fournier, *The Fabric of Life*, 59. Malpighi dismissed the theory of ovist *emboîtement* as 'strange and erroneous'. See his letter to Torti of January 1691 in Adelmann, *Marcello Malpighi*, vol. 4.

[48] See Pyle, 'Animal Generation and Mechanical Philosophy', 249.

[49] Since, as Malpighi notes, one can only observe structures that are present to be observed, there is a reason of principle why generation as such must escape not just the naked eye but even the microscope. For some perceptive remarks, see Wilson, *The Invisible World*, 122–3.

[50] The one minor qualification here is the existence of parthenogenesis in some species of aphids, in which females can give birth to a new generation of females already pregnant with the eggs of the next generation. This does seem to provide the sort of direct positive evidence that the theory of preexistence needs.

4. Support for the maxim that – at least as far as our observation extends – structure always emerges out of structure, never out of a homogeneous fluid (Malpighi, Swammerdam).

To be fair to Malebranche, he never explicitly claims more than this. He is confident that his theory of preexistence does gain support from the work of the microscopists, but he never cites any claim to have actually observed organisms within organisms within organisms. Guided by certain strict a priori constraints, we seek an intelligible (i.e., mechanistic) account of the generation of plants and animals. We soon become convinced that – given the complexity and functional interdependence of parts that characterize all organisms – such beings could not have come to be as a result of unaided matter and motion. Preexistence now offers itself as a plausible hypothesis to the mechanical philosopher. On this theory, there is no generation in nature, so no problem of accounting for it – there is simply the one great miracle of the Creation. But preexistence offends against the prejudices both of ordinary people and of those philosophers (e.g., Aristotelians) who place too much reliance on the senses. Such people are inclined to dismiss preexistence as fantastic and wildly speculative, even as quite literally incredible. It is here that the evidence of the microscopists comes into play, providing not direct proof of preexistence but rather a sustained assault on the prejudices that would prevent us from taking it seriously. This is all that the actual observations of Leeuwenhoek, Swammerdam and Malpighi can do for Malebranche, and this is all he explicitly claims for them. Before reading the microscopists, he thinks, one would find the theory of preexistence wildly speculative; after reading them, one would find objections met and doubts put to rest. This could aid acceptance of the conclusion of the main argument: that only the theory of preexistence provides an intelligible account of the emergence (*not* generation) of plants and animals.

10

Animal as Category

Bayle's "Rorarius"

DENNIS DES CHENE

Sometime before 1539 in Velletri, after dinner, Cardinal Bernard of Cles, a "man of all hours," had a dog perform for his guests. Not only did it perform the usual tricks, jumping through hoops and so forth, but when its trainer brought out a book of music, the dog, "jumping up on his knee," began to sing, "now with a high voice, now with a low, sometimes drawn out with continuous breath, sometimes varied and modulated." Girolamo Rorario, better known by his Latin name "Rorarius," was quite taken with the animal's uncanny performance. The cardinal, for his part, reflecting on astrology and fate, asked his guest about the power of the stars to govern human acts. Rorarius answered that although the stars may incline us this way or that, still "by reason, which rules most powerfully, man may abstain from those things he knows will lead to unhappiness" (Rorarius 1647, 18). "But why", the cardinal asked, "have you said that reason rules most powerfully in man? Do you not believe that reason likewise occurs in animals?" Rorarius said that he had long been troubled by that very thought – that indeed "reason oftentimes is found to be better in brutes than in men" (19).

Out of that conversation came the manuscript of a work that was published only a century later. In 1648 Gabriel Naudé dedicated Rorarius's manuscript, with the slightly altered title *Quod animalia bruta ratione utantur melius homine*,[1] to the brothers Dupuy as an illustration of the *libertas philosophandi*.[2] A half-century later it drew the attention of Pierre Bayle, who devoted an article of the *Dictionnaire historique et critique* to Rorarius

---

[1] It is likely that Naudé omitted the qualifier *saepe* (often) from the original title.
[2] *Quod animalia utantur* was published again, without alteration, in 1654 and 1666. Naudé's 1648 edition has been recently reprinted in facsimile in the series Aurofodina Philosophica, with a preface by Maria Teresa Marcialis; that reprint is the edition used here. On the life and works

and his work. The ostensible subject of the article, almost vanishing at the top of a page nine-tenths of which is devoted to footnotes, serves mostly to provide an occasion for Bayle to engage in controversy. Descartes and those of his persuasion had argued that animals, being machines crafted by God, have no souls. Against them were aligned not only the Schools but also many among the new philosophers, notably Leibniz. Bayle's argument issues in a dilemma. The Cartesian position is most favorable to religion. But it is incredible. What remains is to grant that animals have souls; it then appears that either animal souls are immortal or human souls are not. Neither position is entirely palatable. But the second is far more offensive to religion than the first. Hence Bayle's evident sympathy for the Leibnizian view, a "third way," which nevertheless he rejects because he cannot stomach preestablished harmony.

The concept *animal* is charged not only with designating a class of creatures, real and imagined, but also with supplying a contrast to the human.[3] In Christian anthropology, the animal represents material nature in the most perfect condition it can attain without the intervention of spiritual nature. It designates that condition not only in nonhuman species but in us, by virtue of what we share with those species, physically and morally. The complexity of the concept remains even when the theological implications fade into the background. Questions about "the animal" are not strictly natural-philosophical

of Rorarius, see Aidée Scala, *Girolamo Rorario: Un umanista diplomatico del Cinquecento e i suoi "Dialoghi"* (Florence: Olschki, 2004). Rorarius's work was one of a number of works from the Italian cinquecento brought to light by Naudé during his travels to Italy on behalf of his employer Mazarin and published in the 1640s; among the others were works by Bruni, Campanella, Cardano, and Nifo. As Marcialis notes, in this group Rorarius stands out as distinctly second-rate, and Naudé's motives in publishing *Quod animalia utantur* are unclear; but in the preface Naudé writes of Rorarius as resembling himself in conversation, and it may well have been that he found the "naturalism" of the Italian congenial (7; see Marcialis's introduction, pp. x–xi).

[3] The *animal*, in philosophical usage, almost always denotes "whatever is neither plant nor human among living things" generally and is opposed to an equally general *human*. The animal-machine is *any* animal, the human contrasted with it *any* human. In general, discourse on the animal-machine has as its subject not only the animal which is said to be a machine but the human which is said to be other or more than a machine. "All discourse on the animal is a discourse on man, resting on conscious or unconscious choices: the choice of the 'fact' that constitutes animal action, the choice of a 'cipher' permitting one to interpret this action at the level of a psychic faculty, and to generalize from what is affirmed of an individual or a species to a discourse on *animal nature* generally – defined in terms of its proximity or otherness in relation to *human nature*. . . . The human is therefore not only the *key* to the interpretation of the actions of animals: it is also the *real object* of discourse on the animal" (Thierry Gontier, *De L'homme à l'animal* [Paris: Vrin, 1998], 14).

or biological questions. It is not surprising that Bayle, in his treatment of animal souls, should devote only a small portion of his argument to anatomy or physiology. Indeed it is characteristic of the controversy that its factual content should consist mostly in anecdotes. With no systematic relations between sensation and action to appeal to, what remained were *experientia*: illustrations of animal sagacity and foresight, on the one hand, debunkings or claims of explicability in principle, on the other.

For the Schools,[4] the human soul shares two of its three "parts" with animals. Only the intellectual soul is distinctively human, and only by virtue of that third and highest part does the human soul attain to immateriality. The distinction between human and animal rests upon demonstrating that the rational powers of the mind – discursive reasoning and free will – are distinct from the sensitive powers and absent from animals. The immateriality of the soul follows from that distinction. In Cartesianism, discursive reasoning remains the criterion; but the distinction is now between creatures in whom an immaterial soul is united with a body and creatures which are body alone. Sense and sensation become powers and acts of an immaterial substance which has no counterpart in the animal. It no longer senses or feels, strictly speaking, though its body undergoes changes similar to those we experience in our bodies. No longer is the animal, now a machine, a distinctive *kind* of thing in nature; it is distinguished from the machines we build only by the size, number, and intricacy of its parts. The notorious implication is that animals have no feelings – neither sensations nor passions. The soul is a unity, its every operation, active or passive, a thought: there can be no granting to animals of *some* but not *all* its powers, and thus there can be, if the human soul is immaterial, no "material," hence perishable, soul like that which the Schools attributed to animals.

Though he accepts neither, the School philosophy and Cartesianism set for Bayle the terms of the problem. Together with the ancient sources they draw upon, they determine the field of force to which any philosophical solution must respond: the claims and methods of natural philosophy, in particular, the predominance of mechanism; the theological necessity of preserving for humans their unique role in the destiny of the created world; and the moral position customarily assigned to animals, to whom no rights are granted and on whom no duties imposed.

---

[4] I use this term, customary in the period, without pejorative connotation, to denote briefly the late Aristotelian philosophers whose textbooks were, until the second half of the seventeenth century, the basis of university teaching.

Dennis Des Chene

1. AGAINST THE SCHOOLS

After touching on Rorarius and his work, Bayle turns to the real topic: the controversy of his own day concerning animal souls. The Schools[5] and Descartes: those are the alternatives, it would seem, both of them unsatisfactory.

Bayle's article has the form, loosely, of a disputation, a disputation without a thesis. The opinions of the Schools and of the Cartesians are presented and rejected: "The facts concerning the abilities of animals are an embarrassment both to the sectaries of M. Descartes and to those of Aristotle" (note B, *OD* Suppl. 2:970). The opinion of Leibniz is then taken up. The reader might expect that in Leibniz a resolution of the question is to be found. But it is not. Leibniz, too, is rejected, with regret. *If* the system of Leibniz were satisfactory, his account of animal souls, which solves a great many problems, would be welcome. But preestablished harmony, despite its benefits, is too much to swallow.

Two boundaries are at issue, one scientific, one moral. The scientific is between that which admits of a mechanical explanation and that which does not. The implicit finding of Bayle is that either sensation and reason *both* admit of such an explanation, in which case the Cartesians, though correct in their opinion about animals, would have gone too far, or else *neither* admits of such an explanation – an outcome that presents dangers of its own. The moral boundary is between those creatures whom we or God treat as subject to law and thus as punishable for their crimes and those which have neither right nor obligation under the law and to whom our only relations are relations of power. The customary basis upon which the boundary is fixed is the presence or absence of reason and free will, which though distinct (for all the philosophers discussed here except Spinoza) are inseparable. The argument of "Rorarius"

---

[5] Bayle taught philosophy at the Protestant university in Sedan from 1675 to 1677. The *Système* is "a conscientious mosaic" assembled from notes by colleagues (among them Pierre Maignan); only a few "isolated doctrines" of the new philosophers could be included in what was otherwise a traditional four-part *cursus*. As Élisabeth Labrousse notes, the *Système* is "interesting especially because it is so typical and, in a sense, so banal," even though Bayle, like many other professors at the time, sought to introduce bits of the new philosophy into his courses (Labrousse in Pierre Elisabeth Bayle, *Oeuvres diverses*, ed. Elisabeth Labrousse [Hildesheim: Olms, 1982], 4: xiv–xvi [hereafter cited in the text as *OD*]; Labrousse *Pierre Bayle* vol. 1, *Du Pays de Foix à la cité d'Erasme*, 2nd ed. [Dordrecht: Martinus Nijhoff, 1985], c. 6). But the framework and the mode of argument remain as before. The Aristotelian cursus, built on the model of the disputation and based on arguments ideally syllogistic in form, aims not at certainty but at probability, "confronting one thesis with another, and deciding in favor of one or the other by means of argumentation purely logical in form" (Labrousse, "Introduction," in Bayle, *Oeuvres diverses*, 4:xv). The eclecticism of the late Aristotelian manner persisted in Bayle's work, even when, exiled, he had no university curriculum to conform to, nor institutional inertion to contend with.

implies that animals cannot be denied the use of reason; nor can the usual grounds be adduced to show that their souls are mortal; only free will remains to distinguish them, and even that is dubious. Bayle does not, all the same, draw the evident conclusion: the departure from custom would, it seems, be too extreme, and the moral consequences for us, if we do not change our ways of acting toward animals, too dire.

Rorarius's work contains a mass of "singular facts on the industry of animals and the malice of men," facts which embarrass Schoolmen and Cartesians alike. It is not immediately clear, however, why Aristotelian philosophers *should* be embarrassed. Illustrations of animal sagacity – nestbuilding by birds, the dog's recognition of its master, the social ways of ants – were well known to them and were happily put to use in demonstrating that nature acts toward ends.[6] Those feats, however marvelous, do not exceed what can be ascribed to instinct. Although, for example, "some animals are in a certain way teachable, this does not exceed the grade of the sensitive [powers], because it can occur by memory together with a natural instinct (Suárez, *De anima* 1c5no2, *Opera* 3:500). Even the recognition of benefit and harm is not beyond the capacities of a creature without reason. The sheep has a natural capacity, the *vis aestimativa*, by which it recognizes the wolf not only as wolf but as inimical to it; in exercising that capacity it does not subsume the wolf under a concept of badness, it simply *recognizes* the wolf as bad (Suárez, *Disp. meta.* 23§10no14, *Opera* 25:889).

In Aristotelian psychology, the soul, defined as the form of living things, is regarded as having three functional "parts." To the vegetative or nutritive part belong the powers of generation, growth, and nutrition. Of these, generation is the most natural in the sense of following most immediately from the essence of living things, whose forms are distinguished among material forms by their capacity to reproduce themselves in new matter. Growth and nutrition have as their end the preservation of the organism and its preparation for the task of generation. Among Aristotelians there was general agreement that even the vegetative powers of the soul, the least perfect of its powers, could not be found among the capacities of inanimate matter. But no one doubted that the vegetative soul is material, "immersed in matter." It was thought, for example, to be divisible; more importantly, unlike the senses, its powers in no way rise above their material basis.

Little attention was paid to the vegetative powers in controversies on the mechanistic explanation of living things. It was instead the sensitive part of

---

[6] See, e.g., Coimbra *In phys.* 2c9q1a1, 1:323 (educating the young, building, seeking food, fighting enemies, and so forth).

the soul that proved to be pivotal. In the Aristotelian science of the soul, the sensitive part, common to humans and animals, distinguishes animals from plants; the rational part is found in humans alone. Thus the burden of explaining those operations of animals that are not due to their matter, nor merely vegetative, rests entirely on the sensitive soul, however ingenious they seem to be. Moreover, it is incumbent on the philosopher, if the distinction between humans and animals is to be maintained so that humans but not animals have immortal souls, to show that having sensation (in all its aspects, including the internal senses of memory, imagination, and the *vis aestimativa* mentioned earlier) does not entail having reason.

That way of putting the issue is not the Aristotelians' own but a way that arose only after Descartes had insisted that for a thing to have genuine sensations – or to stand in any *intentional* relation to other things, including teleological – it must have a Cartesian mind, which is an all-or-nothing affair. In Aristotelian texts the issue instead arises in arguments showing that the human soul is immaterial. The principal argument is that the human soul is capable of operations no merely material thing, even a living thing, can perform. Those operations are assigned to the rational part of the soul; and to the sensitive all those operations a material thing can perform. It then follows that having sensation does not entail having reason.

Bayle's first sally against the Schools consists in turning against them the very argument they use to show that animal souls are not reducible to the natural capacities or forms of the material constituents of their bodies. The Aristotelian points to instances of animal industry and demands of the opponent (here the Cartesian) that he show how mere inanimate stuff could perform such operations.

Every Peripatetic who wants to say the animals are not just automata objects first of all that a dog when it is beaten for having thrown itself upon a plate of meat no longer touches it when it sees its master menacing it with a staff. But to show that this phenomenon cannot be explained by the one who proposes it, it suffices to say that if the action of the dog is accompanied by knowledge, then necessarily the dog must reason: the dog must compare the present with the past and draw a conclusion; it must remember the blows given to it and why it received them; it must know that if it pounced upon the plate of meat that impresses itself upon its senses, then that action would be the same as the action it was beaten for; and it must conclude that to avoid new strokes of the staff it must abstain from the meat. Is this not a true reasoning (Bayle, "Rorarius" rem. B, *OD* Suppl. 2:970a)?

There is, in short, no halfway house between the wholly inanimate – the automaton – and the fully rational.

That argument, however, is insufficient. Ignace Pardies sets forth very eloquently a version of it. If you grant, he says, that animals can act toward ends, foresee the future, remember the past, and so forth, then "why do you not say that men are capable of exercising their functions without a spiritual soul?" (Pardies, *Discours* 49, 100; see also Cureau de la Chambre, *Traité* pt4c1, 222). That humans are capable, but animals not, of thinking of universals, the infinite, and spiritual things suffices to show, indeed, that we have a rational soul. But it does not suffice to show that animals do not. "Those operations you find so extraordinary differ only in degree from the operations you attribute to animals" (103). The singing dog who acts to please his master does all a thing must do to exhibit reason.[7]

To this Pardies has an answer. Animals have "sensible" but not "intelligible" knowledge. The difference between animal and human reason is not, therefore, merely one of degree. Spiritual knowledge is "a perception that carries with it essentially a species of reflection that it makes indivisibly on itself, so that we know full well that we know" (78, 150). Sensible knowledge is perception with reflection. To see that our knowledge is sometimes spiritual requires only that we "consider what goes on in us" when, for example, "after having considered the admirable arrangement of this world," we conclude that God exists. At the same time we know "intimately" that we are thinking that very thought – without any *further* act of understanding.

It may happen, on the other hand, that we perceive *without* perceiving that we perceive:

> For example, it often happens that, when the mind is extremely occupied in the consideration of some object that pleases us greatly, we are so absorbed in this consideration that there remains to us no means, almost, to think of anything else. And so, having our eyes open, we do not perceive the objects that are before us, and one of our friends could have passed without our taking note of it (80, 154).

It cannot be denied that we have seen that person. After all, we did not suddenly go blind. It is true that we did not *attend* to our friend. But seeing with attention is reflexive, while in the case at hand we saw but did not attend, which is to say, we had sensible but not spiritual knowledge of what we saw.

---

[7] The singing dog is the dog in Rorarius's story. Pardies cites (incorrectly, as Bayle notes) "Horarium oratione peculiari de ratione brutor[um]" – which is either a printer's misreading or a secondhand citation (Pardies, *Discours* ¶57, 113). Rorarius's oration *is* peculiar, but that is not what it is called.

Bayle rejects that distinction. It is "chimerical" to suppose that animals can know all they know without having reason:

> It is evident to anyone who knows how to judge, that every substance which has a sensation knows that it senses; and it would not be more absurd to maintain that the soul of man knows at this moment an object without knowing that it knows it, as it is to say that the soul of a dog sees a bird without seeing that it sees. This shows that all the acts of the sensitive faculties are by nature and by essence reflexive upon themselves. ("Rorarius" rem. E, *OD* Suppl. 2:973a)

Since this is the *point fort* of Bayle's argument, it is worth asking where he takes it from. He cites here the *Philosophia naturae* of Emmanuel Maignan, like Mersenne a Minim very much taken with the new science but not a wholehearted Cartesian. Maignan, though he retains the Aristotelian doctrine on the souls of animals, nevertheless argues that "what we call sensing is not without a cognition of the act which occurs in us as we sense," precisely because to sense is to recognize (*agnoscere*) the action of some external thing upon the senses (*Philosophia naturae* c24no2, 527). Before Maignan, Suárez had already argued that in the sense as it senses there is always at least something akin to reflection:

> Every sense perceives in a certain way its own action not by reflection, but imperfectly, and as in the actual exercise. . . . This is proved by the fact that every cognition occurs by a vital attention and immutation of the power itself: and therefore when the power senses an extrinsic object, it is changed [*immutatur*] by that sensation, and in the actual exercise in some way experiences itself sensing. (*De anima* 3c12no5, *Opera* 3:654)

Suárez, of course, holds that the human soul differs essentially from that of animals, but on the more traditional grounds that it has the capacity to grasp universals and to consider immaterial things and that it is capable of "formal" reflection on its own thoughts.

Descartes too held that sensing entails reasoning. In the Sixth Replies, he notes among the errors of his youth that of judging weight to be a *tendency* of heavy things to move toward the center of the world. But for those things to have that tendency would be for them somehow to know or to sense where the center of the world is. But that entails thought of a sort; and whatever has thoughts of any sort is in every essential respect a *mind*; thus the error of his youth amounted to the attribution of "little souls" to stones and other nonhuman things, including animals. If, as Pardies urges, there is "sensing" without awareness – sensing from which the *cogito* cannot be extracted – then that must be put entirely on the side of the body.

The Cartesian cast of Bayle's thought can be discerned already in his *Système*. The proof there of the immortality of the soul rests entirely on its indivisibility, not on the character of its operations. Bayle notes that "the Philosophers" ordinarily prove the immateriality of the soul by reference to its powers to "conceive being in general" and so forth – the usual list, including the power to perform "reflexive acts" (*actes réfléchis*). He adds that "every thought is essentially a reflexive act, that is, known to itself," so that to know a thing is to know that one knows it (*Système* pt2tr2c3, *OD* 4:457). From this alone it would follow, even if Pardies' distinction were granted (Bayle does acknowledge a difference between the reflection essential to every thought and that by which "the soul examines its acts, in such a way that one thought is the object of another"), that human thought and the supposed thought of animals have nothing in common – save if matter were capable of reflection.

At this point, Bayle's argument takes an odd turn. He takes it to be proved that animals think and that thinking is essentially reflexive, from which it follows that if animal souls are material, then matter is capable of reflection; if not, then animal souls are immaterial (and so, by the usual arguments, immortal). Elsewhere Bayle observes that among the anti-Cartesians some have been reduced to bestowing an immortal soul on animals: "Vanquished by the purest ideas of Theology, and forced into all sorts of retrenchment, they whisper in one another's ears that the soul of animals does not perish" (*Nouvelles* mars 1684, art. II, *OD* 1:10).

Rather than take up that unorthodox possibility, Bayle instead refutes a position which he does not attest by citation.[8] Since the soul of an animal is capable of judging, discerning, pursuing the beneficial, and so forth, one must conclude that if (as the Schoolmen say) "it produces no further acts as noble as those of our soul, or if it is of a nature less perfect than the soul of man, it can only be because the organs it animates do not resemble ours" ("Rorarius" rem. E, *OD* Suppl. 2:973b). Now the cognitive capacities of people do differ, because of the condition of their organs. "The same soul, which makes us admire its reasoning and its wit in a great man would only dote in an old man, extravagate in a madman, sense in a child." Nevertheless there are no *essential* distinctions among human souls.

"The soul is . . . a thinking substance, it is therefore capable of thought in general." It cannot be that animal souls should differ *essentially* from the human simply by virtue of being in bodies not like ours. Interestingly enough, the Coimbrans acknowledge that if the body of an oyster were (miraculously)

---

[8]   In the article "Pereira," Diogenes (*apud* Pliny) is said to hold the view.

to be joined with a human soul, it would be capable, so far as external acts were concerned, only of oysterish acts. But they do not entertain the thought that it would cease to be human: for them too different bodies do not make for different souls except in the ordinary course of nature.

As against the Aristotelians, this part of Bayle's argument succeeds at best in pinpointing the real issue (supposing that the notion of "reason" is not in dispute). "It would be necessary first that you [the Aristotelian] should prove that the defect of reasoning in animals proceeds from a real and interior imperfection of their soul, and not from the organic dispositions on which it depends" (*OD* Suppl. 2:974a). The Peripatetic cannot hold (against the Cartesians) that animals have reason and cannot be automata and at the same time hold (against Pythagoreans and the like) that animal reason differs essentially from ours, because the reasons offered for the difference – the capacity of humans to grasp general concepts and so forth – suffice to show only that a certain kind of *body* is needed to support those sorts of thought. The challenge is a skeptical challenge: show us, in the face of all the evidence you yourself have brought forward in defense of animals, that they do not have souls like ours – and conversely that if a corporeal soul can produce all the acts of an ape, it could not also produce all the acts of *un gros lourdaud paisan*, a fat stupid peasant.

2. AGAINST THE CARTESIANS

"It is too bad that the sentiments of M. Descartes are so difficult to uphold, and so far from verisimilitude; for they are otherwise very advantageous to the true faith" (Bayle, "Rorarius" main text, *OD* Suppl. 2:970; compare *Nouvelles mars 1684*, art. II, *OD* 1:8b). Chief among those advantages is that animals, having no soul, certainly cannot have an immortal soul and that, having not even sensation, let alone thought, they cannot suffer. If, as Augustine holds, where there is no sin there can be no suffering, then the Aristotelian doctrine would entail that animals must be capable of sin. Moreover, proofs of original sin that depend on the principle of Augustine would "fall to the ground": but the maladies of infants, otherwise sinless, can be explained only on that principle.

The basis of the Cartesian view is familiar enough. In the *Treatise on Man*, written in the 1630s but not published until thirty years later, Descartes attempts to demonstrate that, in a world consisting only of extended things in motion and therefore lacking the forms and qualities of Aristotelian natural philosophy, including animal souls, there could be machines that exactly imitate those actions of the human body that do not require thought (and thus *all* the actions of thoughtless brutes). It follows that, if animal souls were intro-duced simply to explain those actions – the vital operations of animal bodies – then in the world of the *Treatise* they are superfluous. Descarte's systematic proscription against forms and qualities in natural philosophy generally is thus reinforced by showing that even in the study of animals and plants they are not required, provided that we agree that the only ground for introducing them was to explain how animals respond to things around them, pursue what will benefit them (i.e., conserve their machines), and so forth. We attribute souls to them only because the similarity of some of their acts to ours has misled us from our first years into thinking that the cause of those acts must resemble the cause of our own, which we know to be a soul.

The one major lacuna in the theory of the *Treatise* was generation. Descartes attempted to supply the missing pieces when, in the 1640s, he returned to the study of living things. In the *Description of the human body*, he attempted to show how from the seed of the parents the body of their off-spring can be formed in purely mechanical fashion. With that the last reason for attributing souls to animals was refuted, and with that refutation a host of problems surrounding generation, which in the Schools were discussed under the heading of the "eduction of forms," and which had been solved only by appealing to the acts of celestial intelligences or God himself, could be dismissed.

Nevertheless, the Cartesian opinion must be abandoned. Bayle argues at length, mostly by abundant citations, in both the article "Rorarius" and the article "Pereira," that animals have not only sense but reason. On this point he agrees with the anti-Cartesians. Descartes and his sectaries had not solved this problem. Indeed they had not even solved the problem of generation. In the article on Daniel Sennert, Bayle, considering that problem, rejects both the Aristotelian appeal to substantial forms, the hypothesis of Sennert and More that the cause of fetal organization is a soul within the seed, and – not surprisingly – the appeal of some philosophers to celestial intelligences:

> I know able men who vaunt themselves for understanding that the general laws of the communication of movements, however simple, however few in number they are, suffice to make a fetus grow, supposing it to be organized. But I admit to weakness in this respect: I cannot understand it. It seems to me that in order for a little organized atom to become a chicken, a dog, a calf, etc., it is necessary that an intelligent Cause should direct the movement of the matter that makes it grow. . . . I find it therefore rather closer to the truth that the growth of the fetus, organized if you will since the beginning of the World [an allusion to the views of Leibniz and other preformationists], is directed by a particular cause that has an idea of the work [i.e., the already organized seed] and the means of

making it larger, as an Architect has the idea of a building and of the means for making it larger. ("Sennert" rem. G, *OD* Suppl. 2:1040b)

The argument against the Cartesians is developed furthest here. Its conclusion is not quite that animals have souls (for that, it would seem the innumerable examples drawn from the ancients and cited earlier already suffice) but that mechanism is *in general* incompetent to explain the operations of life, a conclusion from which it would follow that not only are animal souls to be retained but they are something like the Aristotelian notion of a soul in general, a nonmechanical principle by which the operations of living things are explained.

More problematically: mechanism is either insufficient, as it seems to be in generation, or else it explains *too much*. The Peripatetic takes his revenge by turning the tables: you, the Cartesian, argue that on my principles not only humans but animals must have immortal souls; I argue in return that on your principles the human being has no soul at all – or at least no other than yourself.

> The Cartesian has no sooner overturned, ruined, and annihilated the opinion of the Scholastics on the soul of animals, than he realizes that one can defeat him with his own arms, and show that he proves too much, and that if he reasons consequently, he will renounce his opinions, which he cannot hold on to without exposing himself to ridicule and admitting obvious absurdities: where is the man who would dare to say that only he thinks, and that all others are machines. (Bayle, "Rorarius" rem. G, *OD* Suppl. 2:975b)

Bayle's argument is taken from the *Voyage du monde de Descartes* of Father Gabriel Daniel, who elaborates his version of it with no small relish. Even the very certain criterion for distinguishing people from automata put forward by Descartes in the *Discourse* – the use of language – will hardly do:

> But to consider things well and without preconceptions, as you eternally advise Philosophers to do, is there, in your opinion, much more mystery in the coherent discourse of men than in an infinity of very coherent actions of animals? What, after all, is a "coherent discourse"? Let us see what is comprised in it. In coherent discourse there is movement; the lips move, the tongue, the jaws, and by this movement the air is pressed, fragmented, reflected in various ways. Clearly a thinking principle is not needed for that. The diverse modifications of all those movements that make certain sounds rather than others, [sounds] which form the words "French," "Latin," "Spanish"; all this is as yet nothing – parakeets, crows, and magpies form those sounds; and yet they do not think. And so when one disputes with a Cartesian one has only to bring in proofs of similar things. If, therefore, coherent discourses require us to suppose a principle that thinks

in the machine that pronounces them, it is because they are coherent. Let us examine now what "coherent" means. (*Voyage* 476–7)

Father Daniel goes on to show that the marks of coherence, whether sounds similar to those one hears, or the use of sounds to coordinate action, or whatever, can be found also in the actions of soulless animals and that the Cartesian is therefore committed to holding that other people are automata. He adds that "wherever there is order, subordination, and a constant and regular usage, it is a necessity that there should be a knowing Principle, fully rational." But the "immediate principle" of the movements of things so ordered can be either a rational soul or the "disposition of a machine." In particular, the production of coherent discourse by a thing does not entail that the principle of its actions is an *internal* principle (485–6).

The debate between the Cartesians and their Peripatetic opponents ends in stalemate, each group vanquishing the other with its own weapons. The argument of Bayle to this point can be put in the form of a complex dilemma:

(i) If animals have no souls, then humans do not have souls (Daniel's argument; or parity of reasoning, given that animals perform acts indicative of the possession of reason);

(ii) if animals have souls, then

   (ii.1) if their souls are mortal, and if animal reason is indistinguishable in kind from human reason, then human souls are mortal;

   (ii.2) if human souls are immortal, then (because there is no essential difference between human and animal reason) animal souls are immortal, and if (what seems plausible) animals also have free will, they must enjoy equal title with us to the status of moral agents.

None of these outcomes is acceptable. Bayle, invoking the argument of Darmanson's *La bête transformée en machine*, concludes that if the last holds, then much of our behavior toward animals is unjust.

> There is no Casuist who believes that anyone sins in making bulls fight dogs, etc. or in making use of a thousand ruses and violences in hunting and fishing in order to destroy animals, or in diverting himself by killing flies as did Domitian. But is there not cruelty and injustice in subjecting an innocent soul to so many misfortunes? (Bayle, "Rorarius", rem. C, *OD* Suppl. 2:971a [cf. Darmanson *La bête transformée* 25–26])

It is not clear from the immediate context how Bayle would answer that question. It is clear that a great revision in our practices and moral judgments would be entailed by attributing rational souls, endowed with free will, to animals; that in itself may suffice to make the proposition doubtful.

3. AGAINST LEIBNIZ

It is with Leibniz's system (taken from the *Système nouveau de la communi-cation des substances*, published in 1695) as with Descartes. Leibniz's system solves nicely a host of problems, notably that of the generation of animals and the origin of souls. But it cannot be adopted without reservations, because preestablished harmony "presents impossibilities" whose resolution Bayle cannot conceive ("Rorarius" rem. L, viii, *OD* Suppl. 2:981b). In particular, Bayle finds it difficult to understand how a simple substance, such as the souls of animals are supposed to be, can spontaneously give rise to a variety of phenomena if (as Leibniz holds) it is never acted upon by any other substance.

The *Système nouveau* is a précis of Leibniz's principles in metaphysics and physics, principles arrived at, he says, some ten years earlier (the period of the *Discours de la métaphysique* and the beginning of his correspondence with Arnauld). Leibniz's nature is mechanist, he says; indeed it is not only mechanist but hypermechanist: an infinity of machines is contained in every visible portion of matter. But although he rules out the *archaei* of More or plastic natures of Cudworth, his mechanism is, by Cartesian standards, heterodox. The motions of bodies and the laws that govern them are not the fundamental laws of creation; above them are the laws that govern what Leibniz calls "force" or "potency" (*puissance*). Force is a "medium between power (*pouvoir*) and action, that envelops an effort, an act, an entelechy" (*Philos. Schriften* 4:472), and it is constitutive of substance, resembling in that respect the *conatus* of Spinoza. The laws that govern substances so conceived are not the laws of motion appealed to in physics but the laws of the order and perfection of nature by which God created the world.

The principle of unity of substances is not to be found in matter as the Cartesians understand it. "We will never be able to find a corporeal mass or portion of matter which would be a true substance. It will always be a collection, since matter is actually infinitely divided so that the least particle envelops a truly infinite world of creatures, and perhaps of animals" (*Philos. Schriften* 4:473; cf. 4:482). In the "true unities" of which bodies are composed is found "something that answers to what is called *me* in us," indivisible and without parts. In animals that something is what we call the soul, or what the Schools call a substantial form. Leibniz cites here with approval the opinion of St. Thomas according to which the souls of *bruta* are not divisible (4:479).[9]

Animal souls, being substances, cannot be generated or corrupted by natural forces: "true unity is absolutely indissoluble" (4:474). They can only be created or annihilated – acts reserved to God. The vexed question of the origin of animal souls is thus not so much answered as mooted. In a passage quoted by Bayle, Leibniz writes,

> [T]he duration that must be attributed [to animal souls], in place of that which had been attributed to atoms, might make one doubt whether they do not go from body to body, which would be Metempsychosis.... But this imagining is very far from the nature of things. There is no such passage, and it is here that the *transformations* of Messieurs Swammerdam, Malpighi and Leeuwenhoek, the most excellent observers of our times, came to my aid, and led me to admit more easily that the animal, and every other organized substance, has no beginning, in the way we believe, and that its apparent generation is only a development, a species of augmentation. (4:480; "Rorarius" rem. H, *OD* Suppl. 2:976b)

The immortality of souls entails the immortality for each soul of a kind of body. Bayle recalls here the opinions of certain Scholastics who, according to the anonymous author of a *Philosophia vulgaris refutata*, hold that the soul is composed of two substances, "one of which, immaterial, is created by God the other of which, material, is born *ex traduce*" (*OD* Suppl. 2:977a, note 87; *ex traduce* is a Scholastic phrase used to explain why parental characteristics are transmitted to their offspring). Bayle adds to this the recent opinion of Poiret, according to whom the body of Moses, which will appear on the day of transfiguration, included a "portion of internal matter, more spiritual, more subtle, and purer" than the gross matter of the cadaver that was destroyed, a portion that remained united with the soul of Moses after his death.

It is worth stepping back a moment to understand how such a view, some version of which Leibniz credits to Malebranche, Régis, and Hartsoeker, could come to be resurrected. The late Aristotelians who for the seventeenth century represented the opinions of the Schools hold that animals do have souls but that, because the operations of their souls require the assistance of a suitably disposed matter, their souls cannot exist naturally apart from matter. In generation a soul must be made (since it cannot preexist), and in the corruption of the body the soul perishes (since there is no transmission of souls from matter to matter). Human souls, on the other hand, have operations that

---

[9] On the divisibility of souls in the late Aristotelian science of the soul, see Dennis Des Chene, *Life's Form* (Ithaca: Cornell University Press, 2000).

require no material organ, chief among which is reason. They can therefore exist apart from all matter.

Descartes makes an analogous distinction (but an even more radical one, denying to animals even the sensitive powers). It is by virtue of possessing reason and free will (but above all reason, since for that we have the evidence of the use of language) that the human is not a machine, not an arrangement of parts of matter whose existence even as a collection ends when those parts are separated. Like the Aristotelians, Descartes holds that death, in humans and in animals, is the destruction of the body (even if its matter is concerned in other configurations) and that it therefore entails the entire separation of soul from matter.

In Bayle's "Rorarius," the distinction upon which the separability of the soul relies is rejected. Either animals have reason of the same sort we have, and so, by the usual arguments, immortal souls, or else human reason is not superior to that credited to animals; thus, if animal souls are mortal, the human soul is too, or if animals are perishable machines, then humans are too.

What cannot be maintained, then, is the *separability* of souls from matter. It follows that if the soul is indestructible, the body or some portion of it is indestructible too. Bayle exhibits no hostility to that consequence of Leibniz's view. What remains problematic is the doctrine of preestablished harmony. Bayle does not reject it outright – he does not go so far as to assert that the "impossibilities" entailed by it cannot be removed – but he cannot see that it has any advantage over the doctrine of occasional causes. To examine the bulk of his criticisms of Leibniz, and the responses of Leibniz to which Bayle in turn replies, would take us into matters only distantly relevant to animal souls. But one aspect of their exchange is worth noting: that which pertains to the notions of *automaton* and *machine*.

Bayle's objection consists in doubts, variously expressed, that a Leibnizian soul could exhibit the phenomena we normally associate with living things and humans – notably, responsiveness to things around it – if, as Leibniz says, it is neither acted upon nor acts upon anything else. It would be like a miraculous ship that somehow, through storms and diverse currents, magically reached its destination without its captain at any point observing the winds or even the current location of the ship. Leibniz has little difficulty answering the queries of Bayle: the greater part of his effort consists in trying to get Bayle to understand his view. That is the point of interest.

The view that animals are automata – that is, self-movers *of a sort* – is the basis of Aristotle's definition of living things, and so also of the soul as the form proper to living things. As Bayle notes, Aristotle even makes use of analogies between animals and the artificial "automata" of his time – puppets and wind-up toys. The Cartesian animal-*machine* is also an automaton *of a sort*. It carries within it the heat by which its movements are caused: that heat, or the blood that transports it, is what in animals Descartes calls their principle of life or their soul. But the machine is not *autonomous*; it is, rather, automatic in the sense that its actions have their remote, but not always their proximate, causes within. The Leibnizian "machine," the animal soul, is not only automatic, it is autonomous. The *law* of its transformations is within; it has as it were its own law (hence Leibniz's recollection of the Thomistic doctrine that each angel is its own species, a doctrine referred to by Bayle at the end of "Rorarius"). Bayle's difficulty seems to lie in making the transition from the automaton-of-a-sort proposed in Cartesian physiology to what might be called the metaphysical automata of Leibniz, metaphysical because the basis of their automatism is their individual essence and not a contingent mediation of external causes by internal causal mechanisms.

Cartesian mechanism met with wholehearted approval among only a minority of seventeenth-century philosophers. Mechanism was, almost from the start, thought to be insufficient. Animal souls were perhaps the most glaring example of that insufficiency. In Leibniz we see the dissolution of the temporary bond between two ways of conceiving the animal: as a machine and as a self-mover, an automaton. A radical move, but all the same one that confers many advantages on the philosopher willing to make it. The Cartesian view, on the other hand – the view that the insufficiencies of mechanism are a matter of feasibility, not of principle, proved to be the more fruitful for science.

V

Between Epigenesis and Preexistence:
The Debate Intensifies, 1700–1770

11

Explanation and Demonstration
in the Haller-Wolff Debate

KAREN DETLEFSEN

1. THE HALLER-WOLFF DEBATE: PREEXISTENCE
VERSUS EPIGENESIS?

One issue that surfaces with regularity in studies on the problem of organic
generation in the early modern period is the battle waged between preexistence theorists and advocates of epigenesis. Roughly, by preexistence, I mean
the theory that, at the Creation, God preformed (to some degree at least)
every living organism that would ever exist.[1] One especially distinct form of
preexistence is the *emboîtement* theory, according to which each organic individual is encased within the reproductive organs of one of its parents (either
the mother on the ovist theory or the father on the spermist or animalculist theory), its parent is encased within the reproductive organs of one of its parents,
and so forth. This accounts for all organic individuals of every generation – all
future members of a given species are found encased within the first member
of that species upon creation. And roughly, by epigenesis, I mean the theory
that posits a truly new development of organic form.[2] Upon coitus, matter

---

[1] I use the term "preexistence" to differentiate this theory from preformation. As Jacques Roger
and Peter Bowler have both noted, there are two distinct theories to be considered, one of which
specifies God as the creator of organic forms, and the other of which simply states that the
organism is formed before conception, but by a natural agent (Jacques Roger, *The Life Sciences
in Eighteenth-Century French Thought*, trans. Robert Ellrich, ed. Keith R. Benson [Stanford:
Stanford University Press, 1963 (1997)], 259–60; Peter J. Bowler, "Preformation and Preexistence
in the Seventeenth Century: A Brief Analysis," *Journal of the History of Biology* 4 [1971], 221–
2). As Bowler has shown, early modern authors themselves often used the term "evolution" to
describe preformation, but with a pre-Darwinian meaning of that term (Bowler, "The Changing
Meaning of 'Evolution,'" *Journal of the History of Ideas* 36 [1975]: 95–114.

[2] See Jacques Roger, "La notion de développement chez les naturalistes du XVIIIe siècle," in *Entre
forme et histoire: La formation de la notion de développement à l'âge classique* (Paris: Meridiens
Klincksieck, 1988), on the notion of "development." In that article, Roger notes that the term was

235

Karen Detlefsen

that seemed previously to be homogenous, undifferentiated, noncomplex, unorganized, and nonunified becomes heterogeneous, differentiated, complex, organized, and unified into a living, functional individual. The form not only develops anew but is actually brought into existence as this process continues, and the process is often considered one of *self*-development, even self-creation. Malebranche is frequently credited with the revival of preexistence in the seventeenth century,[3] while William Harvey coined the term "epigenesis" and developed that theory in the seventeenth century.[4]

frequently used in reference to preexistence, not epigenesis, but when thus used it does not mean creation of new organic parts but rather the growth and filling out of existing essential organic parts. Within the class of preexistence theorists, Roger also notes that the concept of development is more readily applied to Leibniz, for example, than Malebranche, due to the degree of organic development permitted on his theory of preexistence – a degree that I believe places his theory partway between preexistence and epigenesis.

[3] For example, he writes in the *Entretiens sur la métaphysique et sur la religion*, "At the time of creation, God constructed animals and plants for future centuries. He established the laws of motion necessary for making them grow. Now he is at rest because he does no more than follow these laws" (OC XII, 253–4/DMR XI, 196). There is some dispute as to whether Malebranche or Swammerdam (or even Malpighi) is to be credited with the revival of the theory in its seventeenth-century form. See Howard B. Adelmann, *Marcello Malpighi and the Evolution of Embryology*, 5 vols. (Ithaca: Cornell University Press, 1966), 2:869–70; Richard Aulie, "Caspar Friedrich Wolff and His 'Theoria Generationis', 1759," *Journal of the History of Medicine* 16 (1961): 124–44; Peter J. Bowler, "Preformation and Preexistence in the Seventeenth Century: A Brief Analysis," *Journal of the History of Biology* 4 (1971): 234 n. 34, 237; Bowler, "The Changing Meaning of 'Evolution'," 233; Daniel C. Fouke, "Mechanical and 'Organical' Models in Seventeenth-Century Explanations of Biological Reproduction," *Science in Context* 3 (1989): 365–81; Edward Ruestow, "Piety and the Defense of Natural Order: Swammerdam on Generation," in *Religion, Science and Worldview*, ed. Margaret Osler and Paul Farber (Cambridge: Cambridge University Press, 1985), 231 ff. For studies of Malebranche's theory of generation see, e.g., Andrew Pyle's essay in this volume (chap. 9) and Karen Detlefsen, "Supernaturalism, Occasionalism, and Preformation in Malebranche," *Perspectives on Science* 11(2003): 443–83; Pyle, *Malebranche* (London: Routledge, 2003); André Robinet, *Malebranche de l'académie des sciences* (Paris: Vrin, 1970); and Paul Schrecker, "Malebranche et le préformisme biologique," *Revue internationale de Philosophie* 1, no. 1 (1938): 77–97.

[4] For example, in his 1651 *Disputations Touching the Generation of Animals*, he writes that "an animal which is procreated by epigenesis draws in the material and at the same time prepares and concocts and uses it; at the same time that the material is formed, it grows.... The formative power of the chick takes the material to itself and prepares it, rather than finds it ready prepared, and the chick seems less to be made or given increase by another then by its own self" (William Harvey, *Disputations Touching the Generation of Animals*, translated and introduction by Gweneth Witteridge [Oxford: Blackwell Scientific, 1981 (orig. pub. 1651)], 204). Some commentators believe that epigenesis as a biological theory can be traced back to Aristotle (e.g., A. L. Peck, introduction to Aristotle, *On the Generation of Animals* [Cambridge, MA: Harvard University Press, 1963]; Anthony Preus, "Science and Philosophy in Aristotle's *Generation of Animals*," *Journal of the History of Biology* 3 [1970]:1–52). 1970). For studies of Harvey's theory of generation, see, e.g., James G. Lennox's essay in this volume (chap. 1) and Don Bates, "Machina Ex Deo: William Harvey and the Meaning of Instrument," *Journal of the*

The debate surrounding organic generation between the Swiss poet, botanist, anatomist, and politician Albrecht von Haller (1707–77) and the German doctor and professor of anatomy and physiology Caspar Friedrich Wolff (1734–94) is often considered a classic version of the preexistence-epigenesis debate in the eighteenth century. The crucial details of that debate as it is generally understood are as follows. Early in his career, Haller wavered on whether to accept preexistence or epigenesis, initially supporting animal-culist preexistence, as had his teacher Boerhaave, then converting to epigen-esis in order to account for experiments in grafting and the recent discovery that polyps regenerate themselves when severed[5] and his own observations that organic parts appear to develop gradually. He eventually lost faith in epigenesis at least in part due to his own extensive experiments on chicken eggs at various stages of their development, which convinced him that certain organic parts are preformed in the egg. Moreover, he believed that the func-tional organization of living beings seems to require an intelligent builder absent in epigenesis but most certainly present in preexistence, given that God is responsible for organic formation on that theory.[6] In his late work

---

History of Ideas 61 (2000): 577–93; Edward T. Foote, "Harvey: Spontaneous Generation and the Egg," *Annals of Science* 25 (1969): 139–63; and Roger French, "Two Natural Philosophies," in *William Harvey's Natural Philosophy* (Cambridge: Cambridge University Press, 1995). For a general historical account of the rise of epigenesis a century and a half after Harvey coined the term, see Helmut Müller-Sievers, *Self-Generation: Biology, Philosophy, and Literature around 1800* (Stanford: Stanford University Press, 1997).

[5] Abraham Trembley, *Mémoires pour servir à l'histoire d'un genre de polypes d'eau douce, en bras en forme de cornes* (Leiden: Jean & Herman Verbeek, 1744), 26; Albrecht von Haller, ed., *Praelectiones academicae in proprias institutiones rei medicae*, authored by Hermann Boerhaave, notes added by Albrecht von Haller, 6 vols. (Göttingen: A. Vandenhoeck, 1739–44), 5:504–6.

[6] There is some dispute among commentators about the role that Buffon's theory of generation played in Haller's conversion to preexistence after having endorsed epigenesis for a few years. Haller wrote a preface to the German translation of the second volume of *Histoire naturelle* (Haller 1752), the text of which had appeared the year before in French as *Réflexions sur le système de la génération de M. de Buffon* (Haller 1751). Some commentators believe that Haller's criticism of Buffon in this piece spurred his return to preexistence. For various approaches to this issue see Amor Cherni, "Haller et Buffon: À propos des Réflexions," *Revue d'Histoire des Sciences* 48 (1995): 267–305; and Cherni, *L'Épistémologie de la transparence: Sur l'embryologie de A. von Haller* (Paris: Vrin, 1998); Francis J. Cole, *Early Thories of Sexual Generation* (Oxford: Clarendon Press, 1930); François Duchesneau, "Haller et les theories de Buffon et C. F. Wolff sur l'épigénèse," *History and Philosophy of the Life Sciences* 1 (1979): 65–100; Duchesneau *La physiologie des lumières: Empirisme, modèles et théories* (The Hague: Martinus Nijhoff, 1982), 281 ff.; Elizabeth, Gasking, *Investigations into Generation: 1651–1828* (Baltimore: Johns Hopkins University Press, 1967), 108 ff.; Shirley Roe, "The Development of Albrecht von Haller's Views on Embryology," *Journal of the History of Biology* 8 (1975):167–90; Roe, *Matter, Life, and Generation: Eighteenth-Century Embryology and the Haller-Wolff Debate* (Cambridge: Cambridge University Press, 1981), 26–32; Roger, *The Life Sciences in Eighteenth-Century French Thought*, 708 ff. For an account of the relation between regeneration and generation, see

Elementa physiologiae corporis humani (hereafter *Elementa*), he notes that while a snowflake could be produced by forces alone, an organism must be produced by forces and wisdom (EP 8:117–8). In his 1758 *Sur la formation du coeur dans le poulet* (hereafter *Formation*), Haller writes in support of preexistence:

> It seems to me very probable that, at all times, the essential parts of the fetus exist formed; true, not in the way that they appear in the adult animal: they are arranged in a way that allows certain prepared causes to hasten the growth of some of the parts, to delay the growth of other parts, to change positions, to render organs that were transparent visible, to give consistency to the fluid and mucous, and thus to end up forming an animal that is very different from the embryo, and yet in which there is no part that did not essentially exist in the embryo. (Haller 1758, 2:186; cf. EP 8:148–9)

And in his *Elementa*, he explicitly identifies God as the creator of organisms and the mother as the parent who hosts the germ (EP 8:143).[7]

In 1759, one year after the publication of Haller's *Formation*, Wolff published his dissertation, *Theoria generationis*, in which he defends an epigenetic account of generation according to which plant and animal fluids are secreted from the developing organism and are solidified into parts. The process of secretion and solidification is accomplished by means of the *vis essentialis*. Just as this force accounts for the absorption of nutrients from the earth and the distribution of them throughout plants in both generation and self-maintenance, so too is it responsible for the development of animal bodies during generation: "At the start of its development, the chicken embryo takes food from the substance of the egg. It is absorbed by a force that is not the heart's contraction, and neither the arteries nor the pressure caused by them in the neighboring veins nor their compressions by the activity of muscles.... This force is called the *vis essentialis*" in animals just as it is in plants (Wolff 1759, §168, 73). Wolff sent his dissertation to Haller and thus

---

Charles W. Bodemer, "Regeneration and the Decline of Preformationism in Eighteenth-Century Embryology," *Bulletin of the History of Medicine* 38, no. 1 (1964): 20–31.

[7] For sustained investigations of Haller's changing ideas on generation, see Cherni, *L'épistemologie de la transparence*, 29–68; Duchesneau, "Haller et les théories de Buffon et C. F. Wolff sur l'épigénèse"; Duchesneau, *La physiologie des lumières*, 277–90; Maria Teresa Monti, "Théologie physique et mécanisme dans la physiologie de Haller," in *Science and Religion/Wissenschaft und Religion: Proceedings of the Symposium of the XVIIIth International Congress of History of Science* (Bochum: Universitätsverlag Dr. N. Brockmeyer, 1989); Roe, "The Development of Albrecht von Haller's Views on Embryology"; Roe, *Matter, Life, and Generation*, chap. 2; and Richard Toellner, *Albrecht von Haller: Über die Einheit im Denken des letzen Universalgelehrten* (Wiesbaden: Franz Steiner Verlag, 1971).

began a direct and extended debate which lasted until 1777, the year Haller died. The exchange included a series of letters (Wolff's survive, Haller's are lost), along with two reviews by Haller of Wolff's work and a German version (much revised due to the intervening polemic) of Wolff's dissertation under the title *Theorie von der Generation* (1764).[8]

In her impressive study of the dispute, Shirley Roe indicates that the Haller-Wolff controversy is especially interesting because they interpret so differently the apparently shared observational data culled from their dissection of chickens during their formation and development.[9] Two points stand out from their dispute over the meaning of the evidence. First, Haller attributes the phenomenon of the gradual formation of parts to the transparency of those parts before they grow and gain the solidity which makes them visible, while Wolff attributes this phenomenon to the prior nonexistence of those parts. Second, Haller responds to the fact that the heart appears later than other organs by attributing this to its tiny size and, again, transparency, while, once again, Wolff attributes this phenomenon to the prior nonexistence of the heart. It is crucial for Haller to suppose the preexistence of the heart because he takes the stimulation of this organ by the semen as that which starts the heartbeat, and the heartbeat is that which starts the circulation of fluid through the transparent, collapsed body, thus bringing it to life and to a visible state. What might account for Haller's and Wolff's divergence on the issue of generation in light of the shared empirical data? Roe's approach is to turn to "a whole host of 'extrascientific' assumptions and expectations, which fundamentally colored the observational level of their debate."[10] She sketches three broad areas of disagreement between the two which together account for their different theories of generation. First, she claims that Haller holds a "mechanical view of physiological explanations" (90), while Wolff holds a view of nature closer to vitalism and does not believe that a reduction of life phenomena to mechanism

---

[8] For sustained accounts of Wolff's theory of generation and its development, partly Under the pressures of Haller's criticisms, see Duchesneau, *La physiologie des lumières*, 312–40; A. E. Gaissinovitch, "C. F. Wolff on Variability and Heredity," *History and Philosophy of the Life Sciences* 12 (1990): 179–201; Reinhard Mocek, "Caspar Friedrich Wolffs Epigenesis-Konzept: Ein Problem im Wandel der Zeit," *Biologisches Zentralblatt* 114 (1995): 179–90; Garhard H. Müller, "La conception de l'épigénèse chez Caspar Friedrich Wolff (1734–1794)," *Revista di Biologia* 77 (1984): 343–62; Shirley Roe, "Rationalism and Embryology: Caspar Friedrich Wolff's Theory of Epigenesis," *Journal of the History of Biology* 12 (1979): 1–43; and Roe, *Matter, Life, and Generation*, passim. For Wolff's influence upon the latter half of the eighteenth century, see A. E. Gaissinovitch, "Influence des travaux de Caspar Friedrich Wolff sur la biologie du XVIIIe siècle," in *Lazzaro Spallanzani e la biologia del settecento* ed. Giuseppe Montalenti and Paolo Rossi (Florence: Leo S. Olschki Editore, 1982).

[9] Roe, *Matter, Life, and Generation*, chap. 3.

[10] Ibid., ix.

is possible (108).[11] Second, Haller conceives of God as a providential God, and the beauty in structure and use in function found in living beings only serves to prove that God alone could have been directly responsible for the existence of each one (90–2, 119). Wolff conceives of God as less directly interventionist and so believes that he created the world such that it could generate living beings from its own sufficient power (111–12). Third, Roe believes that Haller is an empiricist in the Newtonian tradition, with strong antirationalist leanings, while Wolff is a rationalist in the tradition which he inherited from Leibniz through Christian Wolff, and this rationalism demands that Wolff give a sufficient reason for generation, whereas Haller feels no such compulsion (95, 105).

Roe's approach is surely valuable, and we learn a great deal about Haller and Wolff as natural philosophers from her study. While I have misgivings about some of the details of her analysis, I proceed not by directly dealing with those misgivings but by asking two more fundamental questions. First, are Haller and Wolff really seeing the same thing when they conduct experiments on living beings? This is a question about the techniques used in their experiments and how these techniques condition the way they see the empirical data. Second, what exactly *are* the theories of preexistence and epigenesis, and are they so clearly distinct and mutually exclusive as they are often portrayed to be? Frederick B. Churchill is surely right to note that in discussing this "awkward dichotomy" between preexistence and epigenesis, "a thorough analysis must tangle with those intractable questions about the meaning of 'novelty,' 'emergence,' 'coming-to-be,' and 'form.'"[12] In fact, some commentators have questioned (implicitly or explicitly) the strict division of these theories. Marjorie Grene and David Depew argue that Buffon's theory of generation, based on the hypotheses of organic molecules and the

---

[11] On this score, Roe is in line with many commentators who have interpreted the preexistence-epigenesis debate as a debate between mechanism and vitalism. According to this account, with the rise of mechanism, organic generation became impossible to explain, and so it was explained away by saying that God created all organic beings. While mechanism could explain organic growth, it could not explain organic generation. If one wishes to explain generation by appeal to epigenesis, then one had to abandon mechanism for vitalism so as to be able to explain organic formation. For examples of this general approach, see Theodore Brown, "From Mechanism to Vitalism in Eighteenth-Century English Physiology," *Journal of the History of Biology* 7 (1974): 176–216; Hans Driesch, *The History and Theory of Vitalism*, trans. C. K. Ogden (London: MacMillan & Co., 1914), 12; Stephen Jay Gould, forward to Clara Pinto-Correia, *The Ovary of Eve: Egg, Sperm and Preformation* (Chicago: Chicago University Press, 1997), xiv–xv; and E. S. Russell, *The Interpretation of Development and Heredity* (Oxford: Clarendon Press, 1930), 132–4.

[12] Frederick B. Churchhill, "The History of Embryology as Intellectual History," *Journal of the History of Biology* 3 (1970): 171.

internal mold (*moule intérieur*), brings together elements of both epigenesis and preexistence despite the belief of Buffon's contemporaries that he is an epigenesist.[13] Similarly, in the late nineteenth century, the German biologist Oscar Hertwig compared his theory of embryological development with that of August Weissmann and determined that the debate between them was essentially the old epigenesis-preformation debate.[14] Yet Jane Maienschein believes that their views "were neither as extreme nor were they as distinctly separated as previous preformationist or epigenetic views."[15] Such examples alert us to the caution we ought to take in referring to the "divide" between preexistence and epigenesis. It is helpful to retain skepticism on this matter when examining the details of the Haller-Wolff debate. François Duchesneau, for example, notes Haller's accommodation of some aspects of epigenesis.[16] While Haller and Wolff adopt elements from the rival generation theory, there is a salient difference between their theories, and recognizing this is helpful, both for gaining a better understanding of epigenesis and preexistence and for gaining a clearer understanding of Wolff's and Haller's different scientific methodologies.

I devote the next section of this essay to showing that while Wolff's theory of generation during the years of his dispute with Haller is surely quite different from Haller's, an argument could be made that this difference comes about because Haller and Wolff are engaged in different sorts of projects. According to this argument, Wolff simply describes the visible, sequential development of embryos whereas Haller is more interested in explaining what causal source could realistically give rise to the fetus. When we turn to Wolff's later work on heredity and variation, together with his paper on the nature of the *vis essentialis*, this interpretation gains further support since he seems to move closer to a preexistence stance when attempting to properly explain the process of generation. Still, I think this interpretation of Wolff is lacking, and so, in sections 3 and 4, I investigate the nature of experiment and explanation in

---

[13] Marjorie Grene and David Depew, *The Philosophy of Biology: An Episodic History* (Cambridge: Cambridge University Press, 2004), 82–7. Anna Tymieniecka argues that Leibniz's theory of generation also bears marks of both preexistence and epigenesis (*Leibniz's Cosmological Synthesis* [Assen, The Netherlands: Van Gorcum & Co., 1965], 142–51).

[14] Cf. August Weissmann, *Germplasm: Walter Scott's Contemporary Science Series*, 14, 1893.

[15] Jane Maienschein, "Preformation or New Formation – Or Neither or Both?" in *The Eighth Symposium of the British Society for Developmental Biology: A History of Embryology*, ed. T. J. Horder, J. A. Witkowski, and C. C. Wylie (Cambridge: Cambridge University Press, 1985), 79. If Weissmann's theory was a preformationist theory, it was, indeed, pre*formation* rather than preexistence, but this fact does not alter the point under consideration, and in fact I argue later that there can be a blurring of the line between epigenesis and preexistence as well.

[16] Duchesneau, *La physiologie des lumières*, 311.

their work, first detailing their reactions to Descarte's methodology in natural investigations and then turning to how their reactions to Descartes can help us accurately interpret their modes of experimentation. The latter work shows both how the interpretation of Wolff in section 2 is lacking and that Haller and Wolff actually see quite different things in their experimental manipulations of living beings. Throughout this essay, then, I aim to gain a clearer understanding of what essentially divides Haller's theory of generation from Wolff's, to show the defining impact of the nature of experimentation upon their theories of generation, and to indicate my points of departure from Roe's account of the conceptual underpinnings of these two thinkers' embryological theories.

2. EXPLANATION AND DESCRIPTION IN THEORIES OF GENERATION

J. S. Wilkie suggests that some versions of epigenesis are mere *descriptions* of embryonic formation and growth but that preexistence is a theory truly meant to *explain* that phenomenon,[17] thus implying that the two doctrines are not two different answers to the same problem but are rather different answers to different issues: while the epigenesist is not interested in providing an ultimate explanation for the observations, the preexistence theorist is, and if the epigenesist should shift emphasis to explanation, he too might arrive at something like preexistence. As a generalization, this suggestion cannot hold (though Wilkie does not present it as a general rule). One need only examine the theories of Harvey and Descartes to see examples that falsify this position.[18] But might one view the Haller-Wolff debate in these terms? If so, then it may be argued that, while Wolff's epigenetic theory of generation is indeed different from Haller's preexistence theory, this difference comes about because the two have distinct concerns. There are reasons, found in Wolff's early writings, for believing this to be true. Moreover, this possibility gains strength when we note that Wolff himself proposes something much closer to preexistence later in his life when he tries to explain elements of his generation theory that he had earlier left unexplained. While I eventually argue against the interpretation of Wolff offered in this section, it is nonetheless helpful to present it here in order to detail crucial elements of the two theories

---

[17] J. S. Wilkie, "Preformation and Epigenesis: New Historical Treatment," *History of Science* 6 (1967): 142–3.

[18] Descartes is occasionally taken to have posited mechanical epigenesis (e.g., Phillip R. Sloan, "Preforming the Categories: Eighteenth Century Generation Theory and the Biological Roots of Kant's A Priori," *Journal of the History of Philosophy* 40 [2000]: 233–4).

of generation, including the ways in which Wolff's later account both bears marks of preexistence and is still quite different from Haller's theory.

On one understanding, preexistence simply cannot be considered an explanation of generation because it transforms what ought to be explained naturally into a supernatural event.[19] Wolff himself believes that preexistence cannot be taken as an explanation of generation because it in fact denies generation (Wolff 1759, *Expositio et ratio instituti*, §3, 5). But under one common understanding of preexistence,[20] the theory comes about because it seems that generation could not be explained any other way given the premise of the mechanical philosophy. Descarte's own theory, according to which organic forms would emerge part after part from bits of matter moving according to the laws of nature, establishes the extreme improbability of an epigenetic account within the confines of a nascent mechanism. Given that there is no possible natural explanation of generation (according to the present interpretation), generation could occur *only* by God's forming each organic individual. So in the case of generation, turning to the supernatural provides, in fact, a *true* account of the origin of organic beings and to this extent actually explains how they come to be. Thus, the sequential emergence of parts from undifferentiated material may be an accurate description of what we can see during fetal formation, but we cannot conceive how the finished product could ever arise from matter in motion, and so epigenesis can be only a description of events seen, not an explanation of how they actually occur.

Wolff himself believes that in positing the twin suppositions of secretion-solidification and the *vis essentialis*, he is providing an explanation and not a mere description of an epigenetic generation theory because these two principles are the sufficient reason for generation (Wolff 1759, §242, 115). He draws upon a distinction adopted from Christian Wolff between historical and philosophical knowledge (a third kind of knowledge, mathematical, is not salient to the current discussion). Historical knowledge merely lays out the facts, and in the case of generation it simply provides a description of changing appearances of the developing fetus. Philosophical knowledge provides an explanation for the changing appearances, and it does so by providing the sufficient reasons for the appearances (§§5–7, 5). Wolff believes that, with the exception of Descartes before him (to be dealt with in the next section), no one had provided a sufficient reason to explain the phenomena of generation.

---

[19] Andrew Pyle, "Animal Generation and Mechanical Philosophy: Some Light on the Role of Biology in the Scientific Revolution," *Journal for the History and Philosophy of the Life Sciences* 9 (1987): 246.

[20] Roger, *The Life Sciences in Eighteenth-Century French Thought*, chap. 3; Bowler, "Preformation and Preexistence in the Seventeenth Century," 236.

But by positing the *vis essentialis* acting along with the secretion and solidification of nutritive fluids, he believes that he has provided a proper explanation
(§242, 115) and that he has thus produced philosophical knowledge.[21]

While Wolff is justified in this claim, it is not *prima facie* clear why he is
justified, and a proper account of this justification will be given in section 4.
At this juncture, however, one may argue against Wolff's assertion that he
has produced philosophical knowledge by pointing out that, while he has
identified sufficient reasons for the emergence of organized form, he has
left these reasons themselves entirely unexplained. The process of secretion
and solidification of parts (one of two sufficient reasons identified by Wolff)
is a mere description of the phenomena seen, namely, the fact that from
"budding points" in plants, for example, globules of fluid are secreted and
then gradually solidify into organic structures. The second of Wolff's sufficient
reasons for generation is the *vis essentialis*, the force which is responsible for
the secretion-solidification process unfolding as it does. The *vis essentialis* is
meant to be the cause that brings about the effects experienced in generation.
His denial that the heart is preformed is crucial for his theory because without
a beating heart some other principle must be identified as the efficient cause
of fetal development, thus encouraging the supposition that the *vis essentialis*
is that principle (Wolff 1759, §§167–8, 72–3). Yet one might argue that it is
unclear in his early works that this "cause" is anything other than a name that
stands for the described progression of organic formation. This suspicion is
vindicated by Wolff's discussion of the *vis essentialis* in a late essay intended
to deflect criticisms of his reliance upon it:

> One could have eliminated it [the *vis essentialis*], and attributed the motion of
> the [nutritive] juices to other causes. Or one could have accepted no cause for
> the motion, and left it unexplained. Nonetheless, the motion of the juices could
> not be denied, and the way in which the parts are produced and formed – the
> main issue of a theory of generation – would always remain identical. (Wolff
> 1789, 50n)

Yet once Wolff admits that the supposed "sufficient reason" of the *vis essentialis* (which was meant to explain why organic development proceeds as
it does) can be eliminated from our "explanations" and that we would still
depict the production and formation of the organic parts exactly as we do when
supposing the existence of the *vis essentialis*, the explanatory worth of the

[21] Roe, *Matter, Life, and Generation*, 103–5; Joan Steigerwald, "Instruments of Judgment: Inscribing Organic Processes in Late Eighteenth-Century Germany," *Studies in History of Philosophy
of Biological and Biomedical Sciences* 33 (2002): 86–7.

concept is thrown into doubt. So it seems (1) that this "sufficient reason" does not provide much of an explanation at all but rather is an unexplained cause named as the reason for why generation proceeds as observed and described and (2) that it is the accurate description of effects that matters.

One way of bolstering the interpretation that Wolff's epigenesis is merely descriptive while also showing the divergence of Haller and Wolff in their theories of generation is to pay regard to the use each makes of forces in their theories of generation. Newton had a clear impact on the life sciences in the eighteenth century, and this is certainly true in the cases of both Haller and Wolff. As Thomas Hall points out, physiologists consciously adopted Newtonian paradigms to their own investigations of living beings. Hall calls these "physiological unknowns" or the "inexplicable explicative devices" used to explain organic phenomena which, like Newton's gravity, may be unknown as causes but are well known by their constant and predictable effects.[22] There are two elements to this adoption of forces. First, ontologically, life scientists appeal to forces as causes that explain the effects studied. And second, epistemically and methodologically, we do not know what the natures of these causes are, but we can still rely upon them in our explanations because we can study the actions of the causes and thus the effects they bring about.[23]

Haller posits two such forces – the force of irritability and that of sensibility – and he presents a detailed account of his discovery of these forces in his 1752 "Dissertation on the Sensible and Irritable Parts of Animals" (hereafter "Dissertation"). The force of irritability is the tendency, found only in muscle fiber and semen, to contract, and Haller hypothesizes that this force resides in the gluten of the irritable fiber (Haller [1752] 1936, 675). The force of sensibility, on the other hand, is the tendency, found only in nerves, to feel or sense (658–9). Haller discovered these forces through a series of vivisections that underscore the central import of experiment in his studies:

> I took living animals of different kinds, and different ages, and laying bare that part which I wanted to examine, I waited till the animal ceased to struggle or complain; after which I irritated the part, by blowing, heat, spirits of wine, the scalpel, *lapis infinalis*, oil of vitriol, and butter of antimony.... The repeated

---

[22] T. S. Hall, "On Biological Analogs of Newtonian Paradigms," *Philosophy of Science* 35 (1968): 6–7.

[23] See A. E. Gaissinovitch, "Le role du newtonianisme dans la renaissance des idées épigénétiques en embryologie du XVIIIe siècle," *Actes du XIe Congrès International d'Histoire des Sciences* 5 (1968): 105–10, for an historical account of the relationship between Newtonianism and theories of epigenesis (including Wolff's) in the eighteenth century.

events of those experiments I wrote down faithfully, whatever I found them to be. (659–60)

The results of all these experiments have given place to a new division of the parts of the human body . . . by distinguishing those which are susceptible of Irritability and Sensibility, from those which are not (657–8).

These multiple experiments lead Haller to the conclusion that the forces of irritability and sensibility surely exist – we witness the actions and effects of them – even though we are no more familiar with the nature of these forces than we are familiar with the nature of gravity (692). The force of irritability is useful for Haller, as he employs it to explain the onset of fetal growth on a preexistence theory. Given that the heart is the most irritable of all muscles (686–8), it is the first to be stimulated by the inherent irritability of the semen when the semen contacts the embryo in the uterus upon coitus. This starts the beating of the heart, the life of the fetus, and the filling out of shriveled and transparent but preexisting body parts.

Haller's force of irritability is somewhat akin to Newtonian gravity. Certainly, epistemically and methodologically it is precisely Newtonian, and Haller claims exactly this. We do not know its nature as cause, but we know it by its effects. Those effects are simple in that they are a single type of motion – the effect is simply to contract. While Haller advocates a suspension of judgment on the nature of forces, he does not suggest a suspension of research upon them.[24] Haller is also not averse to allowing that organic actions, including the effects of irritability, may well one day be subject to calculation.[25] But ontologically Haller's forces, both irritability and sensibility, are somewhat different from Newtonian gravity in that the latter is universal but the former are confined to specific sorts of organic parts within an already organized body.[26] The importance of this departure from Newton will come clear in due course.

---

[24] Cf. Maria Teresa Monti, "Les dynamismes du corps et les forces du vivant dans la physiologie de Haller," in *Vitalisms from Haller to the Cell Theory*, ed. Guido Cimino and F. Duchesneau (Florence: Leo S. Olschki Editore, 1997), 59–60.

[25] See Magarete Hochdoerfer, *The Conflict between the Religious and the Scientific Views of Albrecht von Haller (1708–1777)* [1932], reprinted in Shirley Roe, *The Natural History of Albrecht von Haller* (New York: Arno Press, 1981), 11 ff., and Maria Teresa Monti, "Les dynamismes du corps et les forces du vivant dans la physiologie de Haller," 46ff. and 64, for Haller's equivocal attitude to the use of mathematics in investigations of life phenomena.

[26] Duchesneau, "Haller et les theories de Buffon et C. F. Wolff sur l'épigénèse," 77–8; Duchesneau, *La physiologie des lumières*, 284 ff.; Brigitte Lohff, "The Concept of Vital Forces as a Research Program: From Mid-XVIIIth Century to Johannes Müller," in Cimino and Duchesneau, *Vitalisms from Haller to the Cell Theory*, 127ff. For a discussion of whether or not Haller's irritability is

Wolff also relies upon the concept of force – his *vis essentialis* – throughout his discussion of the generation of living beings (animals and plants) and their self-maintenance (plants). His force is also akin to Newton's force of gravity epistemically in the sense that we cannot know its nature, but its existence as a cause of living processes is clear to us due to the regular effects it brings about: "It suffices that we know it is there, and to identify it by its effects, as it is required purely and simply so as to explain the development of parts" (Wolff 1764, 160). Just as Haller's forces are nonuniversal and confined to the body of organic structures, so too is Wolff's *vis essentialis* unique to living beings. Moreover, no less than Haller's forces, Wolff's force is not connected with a soul. But Wolff's *vis essentialis* is different from Haller's forces in at least two ways. First, while it is unique to living beings, it does not require an already existing organic structure in order to be brought forth. While Haller's forces of irritability and sensibility can only emerge once such a structure is in place, Wolff's *vis essentialis* is responsible for bringing such a structure into existence. And so, second, its effects are not simple linear contractions but the quite complex development of a heterogeneous, integrated form from homogeneous matter. Haller's insistence that his force of irritability has no shaping or forming capacity at all (EP 4:64; EP 8:112) thus sets his force squarely apart from Wolff's *vis essentialis*, which, at least in Wolff's early work, would seem to have to accomplish exactly this.

These last differences might well be the locus of the distinction Roe makes between Haller as a mechanist and Wolff as more of a vitalist. Some commentators believe that even Haller himself might be taken as a vitalist, since, after all, he no less than Wolff identifies forces that only living beings possess – life forces of some sort.[27] But Haller is determined to disavow vitalism: "Some famous men have recently called this [irritable force] a 'vital force,' a name I do not like at all.... I prefer to call this the muscle's implanted or proper force" (EP 4:64). Wolff is similarly determined to reject "mechanical medicine" precisely because one cannot explain what an animal does in terms of what a machine might do; only the former can build, grow, and maintain itself (Wolff 1759, §255, scholium 1, 125–6). So it may be possible to make a distinction between Haller as mechanist and Wolff as vitalist in this way: both

a force in the Newtonian tradition, and for a general discussion of force and vitalism in Haller, see Maria Teresa Monti, "Les dynamismes du corps et les forces du vivant dans la physiologie de Haller," 56ff.

[27] Dietrich von Engelhardt, "Vitalism between Science and Philosophy in Germany around 1800," in Cimino and Duchesneau, *Vitalisms from Haller to the Cell Theory*, 161; Richard Toellner, "Principles and Forces of Life in Haller," in Cimino and Duchesneau, *Vitalisms from Haller to the Cell Theory*, 31.

may conceive of the product of generation as unique – living beings are not reducible to machines (nor to other naturally organized forms such as crystals) because they include a force unique to living beings – yet they conceive of the process by which this product comes to be in distinct ways. Precisely because Haller's living forces have simple, linear effects, they cannot form a precisely organized, complex living body. Nor can such a nonpurposive force form a living body that manifests purposive functionality. So these natural, living forces cannot explain generation. In contrast, Wolff's force is posited precisely as the source of organic formation and so cannot be merely simple and linear. Therefore, since Haller's force requires the existence of a living structure, it is parasitic upon and emerges from that structure. And since Wolff's force is posited as the source of living structures, it is primitive and fundamental. Haller's living forces cannot account for all living functions because generation is precluded, but Wolff's living force can account for all living functions. Hans Driesch takes vitalism to be a theory about the unanalyzable autonomy of living beings, but this must mean that they are autonomous in their generation as well as in every other function.[28] So taking a basically Drieschian theoretical approach, the divide between Haller as mechanist and Wolff as vitalist can be reached by noting the capacities of the latter's living force to affect all organic phenomena including generation.[29]

By focusing on the way in which Wolff employs the *vis essentialis* to account for generation, we see how one might charge him with simply describing (and not explaining) generation by claiming it is accomplished

---

[28] Driesch, *The History and Theory of Vitalism*, 1914, 1–6.

[29] While this may provide an internally nonarbitrary way of distinguishing between vitalism and mechanism, and of categorizing Haller and Wolff, it must be borne in mind that there are many other equally legitimate ways of distinguishing between vitalism and mechanism. See, e.g., François Duchesneau, "Vitalism in Late Eighteenth-Century Physiology: The Cases of Barthez, Blumenbach and John Hunter," in *William Hunter and the Eighteenth-Century Medical World*, ed. W. F. Bynum and Roy Porter (Cambridge: Cambridge Univerity Press, 1985), and Hilde Hein, "The Endurance of the Mechanism-Vitalism Controversy," *Journal of the History of Biology* 5 (1972): 159–88. In this paper, I leave aside the broader question of whether there is any meaningful way of achieving such a distinction, especially given the myriad ways in which vitalism and mechanism are defined. For an example of the difficulty, see Maria Teresa Monti, "Les dynamismes du corps et les forces du vivant dans la physiologie de Haller," including her struggle to come to terms with the question whether Haller's physiology is mechanist or vitalist and the arguments that might be launched in favor of either approach. See also Duchesneau, *La physiologie des lumières*, 338, for a similar discussion regarding Wolff. See also Bernard Balan, *Génération, organisation, développement: L'enjeu de l'épigénèse: Entre forme et histoire: La formation de la notion de développement à l'âge classique* (Paris: Meridiens Klinksieck, 1988), 115, for his claim that Wolff shared a mechanistic style of explanation with Descartes. But see the next section of this paper for my contrasting view of the methodological elements Wolff shares with Descartes.

through epigenesis. This is because of the Newtonian epistemology that grounds his theory of the *vis essentialis*: we may not know what the force is as a cause – we may not know what its nature is and how it accomplishes what it accomplishes – but we know that it exists by studying the effects that it surely does have, such as the process of generation. Yet studying these effects seems to amount to no more than describing the constant and regular sequential development of parts, a development not even subject to calculation. Contrast this with Haller's approach, which admittedly starts with a description of the effects of living forces. But because these forces are simple and linear, they cannot satisfactorily explain generation. Only God, believes Haller, is up to the task, and so acknowledging the truth of God's role in generation is an accurate explanation of that process. It is more than mere description.

The conclusion that Wolff describes but does not explain generation through his theory of epigenesis is encouraged by his later works. In his 1789 paper *Von der eigenthümlichen und wesentlichen Kraft der vegetabilischen sowohl als auch der animalischen Substanz*, Wolff writes that "through this present treatise, this essential force, which I posited at that time [in earlier works] as the foundation which I also proved existed, but which I in no way explained, now will be explained" (50n). In doing so, he contrasts his *vis essentialis* with Blumenbach's *Bildungstrieb*, or formative power, since the *vis essentialis* "exists in nothing further than a particularly defined kind of attractive and repulsive force" which draws like substances together and drives unlike substances apart (42).[30] Not only is the *vis essentialis* not to be equated with the soul, but it is not to be understood as selective and purposive, capable of accomplishing different things from moment to moment, as Blumenbach's force would seem to do. This is because there would be no sufficient reason why it would act in these different ways precisely as it does and in no other way (66n). Wolff thus echos Haller's own earlier stated concern regarding the *vis essentialis*: there is no reason why a hen should produce a chicken while a peacock produces a peacock (EP 8:117). Wolff concludes that the principle of sufficient reason, as he understands it, requires that we reject the idea of a purposeful, determining building force that can produce myriad effects because there would be no necessary connection between such a force and its actual very precise effects. Because there is no necessary connection, in reality it should thus be unpredictable in what it does (Wolff 1759, 67). In his criticism of Blumenbach, and in his definition of the *vis essentialis* as a nonselective attractive or repulsive force, Wolff places himself squarely in

---

[30] For one discussion of Wolff's denial that his *vis essentialis* is like Blumenbach's force, see Müller, "La conception de l'épigénèse chez Caspar Friedrich Wolff (1734–1794)," 350 f.

Haller's camp with regard to the simplicity of effects brought about by the force found in living beings. And just as Haller was skeptical about being able to account for the creation of the complexity of the living body on the basis of such a simple force, so too is Wolff. Part of his solution to this new obstacle is to postulate not just one *vis essentialis* but a number of nonselective attractive forces as causes that bring about the activity of generation (66).[31] Further, and crucial for our concerns, there are "countless other concurring causes" which guide the formation of the embryo (67).

One of these concurring causes comes clear in his unpublished notes, *Objecta meditationum pro theoria monstrorum*, which are concerned with explaining the generation of monsters in light of the facts that like usually generates like (inheritance of specific traits), that children often resemble their parents or other relatives (inheritance of familial traits), and that individuals nonetheless are usually unique (variability). Wolff here suggests the new hypothesis that the initial matter of generation is already conditioned in certain ways, for he now believes there must exist various forms of "qualified vegetable matter" (*materia qualificata vegetabilis*), or vegetable substance (*substantia vegetabilis*), each of which vegetates in its own fashion. In every vegetative body, writes Wolff, there are three aspects to vegetation. First, there is the act of vegetation itself, which produces the animal or plant as such and is thus responsible for generation in general; second, there is the mode of vegetation, which produces the species; and third, there is the degree of vegetation, which produces individual variety within animal and plant species (Wolff 1973, 168).[32] Qualified vegetable matter is relevant to the first two aspects of the vegetative process. It is in the essence of qualified vegetative matter to vegetate, and it therefore contributes to generation *per se*, and each sort of qualified vegetable matter vegetates in its own manner or mode, thus contributing to species production (Wolff 1973, 158, 171). As Wolff writes: "[I]n any soil and any climate they [plants] have their special structure which they create *for themselves* [emphasis added]. Therefore . . . neither the climate nor soil but the plant in its diversity, its growing qualified matter in the process of growth, creates the structure and form." Further, it is because of their specific qualified vegetable matter that plants "preserve not only the ability of vegetation [which produces organisms as such] but also their qualification . . . maintaining the species" (Wolff 1973, 181, 186; trans. Gaissinovitch 1990, 193–4). This already somewhat predetermined material

---

[31] For an alternative account of the evolution of Wolff's concept of the *vis essentialis*, see Mocek, "Caspar Friedrich Wolffs épigénèsis-Konzept."

[32] Cf. Roe, *Matter, Life, and Generation*, 129–30.

which determines species is passed from generation to generation, from parent to offspring, while no organic structure is so transmitted because the species consists in the cause of the structure (and that cause is the qualified vegetable matter); the species does not consist in the structure itself:

> [I]n vegetation and generation there exists something that is associated with the specific structure observed, and the basis of this structure. The latter – all that could be called *the inner* – is hidden from us. . . . Only this is transmitted from the parents to the offspring by way of generation; the former [the specific structure] emerges in the offspring as the result of that which is transmitted. (Wolff 1973, 155; variation of translation in Gaissinovitch 1990, 189)

Moreover, this transmitted qualified vegetable matter has its ultimate origin in God, for this alone can explain the replication of species from generation to generation. Species, then, are eternal and have existed, as we now know, from the Creation. The supposition of qualified vegetable matter as the bearer of specific traits "altogether proves [the existence] of the primary reasonable acting cause," as it depends upon "God's providence" (Wolff 1973, 187; trans. Gaissinovitch 1990, 194).

Recall Roe's belief that one of the "extrascientific" factors motivating the divergence in generation theories between Haller and Wolff is the greater role for a providential God in Haller's natural philosophy.[33] In light of Wolff's later philosophy, it is far from clear that this belief can be sustained. Once Wolff starts to ask how the *vis essentialis* produces a fetus, and once he starts to investigate why species produce like kinds – that is, once he turns to seeking explanations for *how* generation occurs as it does – he must turn to God in order to answer these sorts of questions. According to Haller, at the Creation, God produced organic structures and endowed matter with the forces of irritability and sensibility (EP 7:xii), as this alone could explain organic processes, including generation. According to Wolff, at the Creation, God created various types of qualified vegetable matter, each with its mode of vegetation, as this alone could explain generation of species from like kinds. Though they differ in what they believe God created at the Creation, Haller and Wolff both do appeal to God's providence and foresight as an essential element in their theories of organic generation, a point Wolff draws to Haller's attention in his letter of April 17, 1767 (Wolff [1759–77] 1981, 168).[34] This

---

[33] Shirley Roe, *Matter, Life, and Generation: Eighteenth-Century Embryology and the Haller-Wolff Debate*, 90–2 and 119.

[34] Margarete Hochdoerfer believes that Haller maintains a basic belief in the divine order of the universe, whether that order is expressed in terms of preexisting germs formed directly by God or laws of epigenesis that will produce organic order (*The Conflict between the Religious and*

will not, however, be the last word on the role played by God in these two naturalists' theories.

So Wolff abandons the theory of epigenesis as he had previously understood it. Once he queries the way in which the *vis essentialis* operates, and concludes that it is simple, nonselective force, he posits a variety of special substances created by God and passed from generation to generation in order to explain the results of generation. Once Wolff turns from describing the process of generation to explaining how this process is brought about, he moves decisively in the direction of preexistence in the sense that he turns to God's having predetermined to a considerable degree the matter of generation. Moreover, Haller's belief, as one example, that the preexisting structure created by God is merely an organized fluid that, upon conception, develops boundaries and a consistency that can resist pressure (Haller 1758, 2:175) indicates the strong epigenetic elements to be found in his theory: there is, in an anatomically significant sense, notable organic development. If the preexistence-epigenesis divide is to be meaningfully retained in the case of Haller and Wolff, we must provide a more careful explication of the distinction between the two theories than has thus far been given.

3. HALLER AND WOLFF ON DESCARTES

Roe is surely correct when she writes,

> Admittedly, embryos on Wolff's theory do not start out in a state of absolute homogeneity. Yet one must be careful not to define epigenesis so narrowly that clearly epigenetic systems like Wolff's are excluded. Gradual development of complex heterogeneity from simple heterogeneity can provide a valid epigenetic viewpoint. In Wolff's system, the embryo's initial heterogeneity is of a potential nature, based only on physical factors like solidification and attraction and repulsion, which produce the structures of the organism through a gradual, but automatic, sequence of events. This is a far cry from preformation, especially in its eighteenth-century *emboîtement* form.[35]

Haller insists that what God created and what is passed on from generation to generation is a fairly complete structure of well-integrated essential organs (EP 8:148–9). Wolff emphatically denies that physical structure is what is passed from generation to generation since what is passed on – the species – is

---

the *Scientific Views of Albrecht von Haller (1708–1777)*, 9). If she is right, then this would reinforce Wolff's point in his letter to Haller, that God need not work in the world by creating all organisms at the Creation in order to work providentially in the world.

[35] Roe, *Matter, Life, and Generation*, 147.

simply a mode of vegetation (Wolff 1973, 154): "the thing which is, in the beginning, excreted from or produced by the maternal ovary is none other than the drop of liquid located in the egg in which there is no structure similar to the structure of the parent" (Wolff 1973, 149; trans. Gaissinovitch 1990, 188). So if we define, reasonably, preexistence as the passing on of the essential structure from generation to generation, and if we define epigenesis as the development of the essential structure where there was no such organic form before, even if the matter of generation is differentiated in a nonstructural way so as to necessitate a specific physical form, then the portrayal of Haller as preexistence theorist and Wolff as epigenesist holds.

But this, then, raises a crucial question. Why does Wolff posit matter that is somehow informed but not structurally formed, contrary to Haller's view? There are many viable answers to this question, but the one I will pursue requires that we reconsider Wolff's earlier account of generation in order to determine why he thinks that it is an explanatory account of organic formation and development when it seems to be a mere description of the process. This brings us to the other fundamental issue of this essay: how experimental techniques, in the case of Haller and Wolff, condition the way they experience the data and consequently influence the theories of generation that they adopt. To set the stage for this work, I deal first with Haller's and Wolff's different reactions to Descartes in order to clarify their epistemologies and thus their methodologies in natural investigations.

Haller is emphatic about the need to place experiment and observation above rational speculation and theory building. This should be clear by the fact of his extensive research on chicken eggs and his vivisections of 190 animals during his investigations into sensibility and irritability, a "species of cruelty for which [he] felt such a reluctance, as could only overcome by the desire of contributing to the benefit of mankind" (Haller [1752] 1963, 657). In the preface to the German translation of Buffon's first volume of the *Histoire naturelle*, he writes, "[M]ore convenient telescopes, rounder glass drops, more accurate divisions of the yardstick, syringes, and scalpels have contributed more to the enlargement of the domain of science than the creative spirit of Descartes, the father of classification, Aristotle, and the erudite Gassendi" (Haller 1750, x). Descartes, in Haller's estimation, misuses hypotheses because he starts from them rather than starting from observation and experiment. According to Haller, Descartes consequently speculates in the complete absence of empirical data. Haller is wrong, of course, that Descartes failed to conduct adequate experiments. In fact, Descartes is reluctant to put pen to paper to tackle the problem of generation because he has not had enough opportunity to conduct enough experiments (AT 5:261), even though, in a letter to Mersenne of February 20, 1639, he reports having spent eleven years doing dissections in order to further his knowledge (AT 2:525, CSMK 134–5). Still, Haller is correct that rational theorizing comes first for Descartes in the sense that his theory of matter as extension is arrived at through the use of the pure intellect and that his matter theory sets limits upon any scientific explanation he can give of the empirical data.

Wolff, conversely, praises Descartes. Descartes "showed what a proper explanation must look like, and he taught how one must philosophize" (Wolff 1764, 6). The reason Wolff thinks that Descarte's attempt at explanation is so successful is that he follows the method which Wolff lauds: "the only clear demonstration is to prove that if laws and principles are assumed an organic body necessarily follows, or to show the sufficient connection between principles and laws and the generated organic body" (Wolff 1759, §§6–7, 5). This is what Wolff believes he himself is doing when he posits the secretion and solidification of nutritive juices together with the *vis essentialis* as the sufficient reasons for the organic body. Descartes goes wrong, Wolff believes, in his theory, but at least he followed the correct method.

Roe takes Haller's rejection of Descartes and Wolff's praise of him as evidence of Haller's empiricism and Wolff's rationalism.[36] But this is an odd claim to make in light of these facts: that the theory of preexistence is surely as divorced from experimental proof as a theory could be (there is no possible way of observing God creating all living beings at the Creation); that Wolff himself is an avid experimentalist and relies heavily upon experimentation in developing his theory of epigenesis; that Wolff also expresses hostility to hypothesis in a November 16, 1765, letter to Haller (Wolff [1759–77] 1981, 165);[37] and that of the two of them it is Haller who propounds nonempirical theories such as the supposed transparency of parts (Wolff 1764, 126, 188). A closer examination of Haller's and Wolff's reactions to Descartes establishes that what Haller criticizes in Descartes is not what Wolff praises and that Wolff can agree with Haller's criticism of Descartes while still lauding Descartes for providing the correct kind of explanation. Despite Descarte's own reluctance to speculate on generation in the absence of any relevant experiments, and despite the fact that he eventually did observe fetuses in various stages of growth, Haller seems to believe that Descarte's system of generation is built upon speculative causes considered completely in isolation from effects realized. This is what he criticizes in Descartes. Wolff praises the ability to

[36] Roe, *Matter, Life, and Generation*, 105.
[37] Duchesneau, *La physiologie des lumières*, 294.

give an account of causes sufficient to show that a given effect is the necessary outcome. But this is compatible with a rejection of what Haller finds objectionable, specifically theorizing about causes in the complete absence of experience. That is, Wolff can criticize Descartes for providing the wrong principles and laws – such as the assertion that all organisms are generated from the motion of matter of certain sizes, shapes, and so forth – because these are arrived at through speculation. But Wolff can still commend Descartes for providing the right *kind* of explanation (i.e., for showing how these [erroneous] causes and principles necessarily yield an organic body). The fact that Wolff thinks Descartes gets the details of the explanation of generation wrong suggests he might accept Haller's criticism of Descartes while also believing that Descartes nonetheless proceeded in an appropriate explanatory manner by positing principles and laws and proceeding to show how these would (at least in Descarte's view) necessarily produce an organism. We need, therefore, to reconsider Roe's claim about Haller's empiricism and Wolff's rationalism to make sense of their epistemologies and methodologies. This will entail in turn a rejection of the depiction developed in section 2 of Wolff's early theory of epigenesis as descriptive rather than explanatory.

4. EXPLANATION, EXPERIMENT, CAUSES

One of Haller's criticisms of Wolff during their dispute over their observations of chicken eggs is that Wolff (according to Haller) erroneously assumes that what is not seen is not there (Haller 1760, 1226–9). It does, indeed, seem to be the case that this is the basis of Wolff's theory of epigenesis (encouraging all the more the supposition that the "theory" is mere description of the observed):

> No one has yet seen parts with the aid of a stronger lens that he could not also see through weaker magnification. These parts either have not been seen at all or have appeared of sufficient size. It is a fable, then, to suppose that parts may remain concealed on account of their infinitely small size and then gradually emerge. (Wolff 1759, §166, 72)

In response to Haller, however, Wolff says that he is not asserting that what is not seen is not there. Rather, Wolff believes that he has based his theory of generation by epigenesis on having positively seen the development of organic parts that previously did not exist (Wolff 1764, 87–8). What justifies Wolff's claim?

Gerhard Rudolph draws our attention to Haller's "Baconian" method:

[E]xperimentation is not mere observation but the willful and thoughtful pro-voking of an experience.[38] This is true too of Wolff, but the way in which the two naturalists manipulate and intervene in their experiments is critically different. Throughout the first twenty-four sections of the *Theoria generationis*, Wolff describes how he experimentally demonstrates the actions of the *vis essentialis*. By using instruments such as scalpels and needles, he actively takes on the role of that force to imitate its actions upon the nutritive fluids found in plants, thereby showing how the effects of generation follow from that cause. For example, he uses a needle to jiggle the nutritive juices in plant roots and stems in order to show how the *vis essentialis* would redistribute the fluid and even how it would create new vessels through which the fluid would be secreted (Wolff 1759, §§1–24, 12–18). This is highly interventionist and manipulative in a particular way: Wolff's instruments *become* the *vis essentialis* for the purposes of demonstration and thus for the purposes of giving a proper explanation. Wolff does not explain the *vis essentialis* in the sense of explaining its nature, nor why it exists and acts as it does. But he does explain the *vis essentialis* by claiming that it does exist (a claim based on its effects such as the generation of living forms) and, crucially, by demonstrating that it operates in such-and-such a manner.[39] This gives us a way of revisiting Wolff's reaction to Descarte's mode of explanation. According to Wolff, explanations require that we expli-cate causes from which the effects can be deduced, and Descartes provided an explanation of generation by satisfying this requirement. But an explanation is a mere "proposition that lacks a demonstration," and a demonstration must be added to any explanation (§255 scholium 2, 127); Descartes did not do this (and he could not provide a demonstration because the causes he posited were subsensible sizes, shapes, and speeds of bits of matter). Wolff, once again, could accept Haller's criticism of Descartes – that his theories are too specu-lative – while still praising Descartes for providing adequate explanation, yet also faulting Descartes for not providing a demonstration of the explanation. Wolff, then, demonstrates experimentally how a sequence of actions brought about by the *vis essentialis* would necessarily produce the effects he details in his explanation of generation.

---

[38] Gerhard Rudolph, "La méthode hallérienne en physiologie," *Dix-Huitième Siècle* 23 (1991): 78.

[39] Duchesneau alerts us to another element of Wolff's experimental demonstration that is not at odds with the depiction here (*La physiologie des lumières*, 330). Wolff has a double scheme of demonstration, first providing an account of vegetation "in abstract," by laying out the order of generation in general according to the supposed law of epigenesis. Second, Wolff examines the question of generation in "inverse" order, by starting with an examination of the nutritive actions of the *vis essentialis* in the fully grown adult (e.g., in digestion), then deriving the probability that there are laws governing these actions, and finally arguing by analogy that there must be a law governing generation by epigenesis.

Haller, too, is willfully interventionist and manipulative in his experimentation. Dissections will not suffice to give the experimenter the information she needs about the living organism, and so she must perform vivisections so that the living tissue can be manipulated in various ways.[40] We get multiple examples of this manipulation in his "Dissertation," such as the experiment of pricking the bladder of a nearly dead dog to see whether it would still contract and expel urine (Haller [1752] 1936, 682). As another example, in a letter to Bonnet of August 25, 1765, Haller describes experiments he performed in an attempt to determine the nature of the changes that happen to the blood vessels in the membrane surrounding the embryo in an egg in the first few days after fertilization. Specifically, he wants to determine whether these changes indicate a new formation or merely the becoming visible of the previously invisible, and to determine this he uses a scalpel to poke at the vessels both in an early stage when they are still yellow and at the later stage when they become red with blood. By poking and moving these lines around, the color in them does not spill out of the pathways as would be expected if the vessels were not yet formed and closed, and the lines of color return to their original positions once the scalpel is removed, thus indicating that these are, indeed, fully formed and closed vessels even at the earliest stage of organic life. (Haller [1754–77] 1983, 436–7)

In the experiments meant to elucidate the actions of the forces of irritability and sensibility, however, Haller is interventionist in a different way than is Wolff in the latter's attempts to elucidate the actions of the *vis essentialis*. He manipulates the environmental conditions in order to bring forth the causal actions of the forces of irritability and sensibility, but it is the force itself that is causally active in Haller's experiments. He does not imitate or mimic or take on the role of the force in order to show how (by what series of events or by what process) it produces its actions and effects. In this way, by Wolff's lights, Haller would not have produced an adequate demonstration of those forces. Both Haller and Wolff claim ignorance regarding the nature of the forces they posit. But Wolff believes that, through his demonstrations, he can claim knowledge of at least some of the steps or part of the process by which the *vis essentialis* brings about its effects, and Haller can make no such claims about the process by which the forces of irritability and sensibility produce their actions. Wolff, then, believes that he can answer the question of how (through what steps, by which process) the *vis essentialis* produces its result, while the best Haller can do is to explain under what circumstances the forces might act and how the effects they produce appear to us (e.g., as nonselective

---

[40] Rudolph, "La méthode hallérienne en physiologie," 79.

contracting forces); he cannot explain, however, how the forces produce these effects.[41]

Haller's experiments to identify the forces of irritability and sensibility presuppose the existence of a specific organic structure made up out of organic material.[42] He is looking for the parts of the human body that have the capacity to be stimulated to exhibit irritability and sensibility, and so the parts are already assumed. At least insofar as his investigation into the forces of living bodies is concerned, then, Haller's experiments are premised on the assumption that living functions and living structure are inseparable and mutually interdependent,[43] indeed, so much so that without the structure the functions cannot occur, and without the functions the structure cannot endure. Unless there is an organic being that has at least its essential parts fully integrated and connected, there will be no living function (EP 8:278). This is why generation (the function of bringing forth an integrated organic form) by epigenesis (the gradual formation and integration of organic structure, part after part) is epistemically impossible.[44] In claiming this, he is echoing a misgiving that Harvey himself has about his own theory of generation by epigenesis. It is a "paradox," Harvey admits, that "the body is nourished and increased before the organs dedicated to concoction, namely, the stomach and the viscera, are formed" (Harvey [1651] 1981, 295). And, again, "it seems a paradox to say that the blood is created and made to move and imbued with vital spirit before any organs for making it or giving it movement exist" (294). One way of solving the paradox is Haller's way: presuppose the essential structure by claiming that God created it because nothing natural could *function* prior to the structure in order to bring the structure into existence.[45] A second way of solving the paradox is Wolff's way: ask a different question. Do not ask how the *vis essentialis* could possibility function outside of an organic structure, but rather demonstrate that it does. Ask how (as a matter of fact) it produces

---

[41] For a general account of Haller's methodology, including his experimental approach, see Maria Teresa Monti, "Difficultés et arguments de l'embryologie d'Albrecht von Haller: La reconversion des catégories de l'anatome animate," *Revue des Sciences, Philosophiques et Théologiques* 72 (1988): 301–12.

[42] Engelhardt, "Vitalism between Science and Philosophy in Germany around 1800," 166.

[43] Duchesneau, *La physiologie des lumières*, 296–7.

[44] Duchesneau, "Haller et les théories de Buffon et C. F. Wolff sur l'épigénèse," 88 ff. Amor Cherni, "Haller et Buffon," 278 ff.

[45] See Duchesneau, *La physiologie des lumières*, 289 ff. for a treatment of preexistence as more probably true of the world than epigenesis. See also Duchesneau, *La physiologie des lumières*, 297, for an account of epigenesis as epistemically impossible. Two elements of organic formation that contribute to its epistemic impossibility are the facts of the production of extreme complexity from relative simplicity and the production of a functioning structure from a natural cause that lacks intelligence.

the results that it does, and then reproduce the process by way of a demonstrated explanation. Duchesneau perceptively identifies this divide as Wolff's departure from Haller's philosophy of organic functions in favor of a new such philosophy.[46]

This different philosophy of functions brings us back to theology, and we can now see the weight behind Roe's claim that God plays a much more significant role in Haller's theory than he does in Wolff's. For both, God created something essential for generation at the Creation. According to Haller, God created the essential structure of each organic form together with the forces found within some of the matter of these structures (EP 7:xii). For Wolff, God created the various forms of vegetative matter which account for the continuity of species from generation to generation. For both, God must be involved to explain the regularity, order, and reasonable variation that we experience in the living world. But for Haller, the irreducibly teleological nature of the structure requires an intentional, intelligent builder as the source of the usefulness of the structure. That is, his philosophy of function as one that attributes usefulness for achieving certain functional goals of various structures implies the presence of an intellect, the source of which must be God, for it cannot be found in matter or forces. For Wolff, there is no such assumption about the need for structure in order to permit useful, goal-directed functions. Rather, he starts with the functioning *vis essentialis* and demonstrates how (in terms of process) this force can bring forth a structure without worrying about how the developing organism can do this minus a preexisting structure. It is enough for explanation that he has demonstrated that it does so.[47]

How, then, are we to think about Roe's claim about Haller's empiricism and Wolff's rationalism? It is true that Wolff is looking for the sufficient reason by which to explain generation, but for him this amounts to a demonstration of the proximate causes of generation only. And even then it amounts to a demonstration only of how the proximate causes accomplish the task and not an investigation into their nature nor why they proceed as they do. Perhaps for this reason, Wolff claims to be searching a posteriori for the principles and laws of generation (Wolff 1759 §71 scholium 2, 38). It is true that Wolff turns to God's initial creation of the various forms of qualified vegetable

---

[46] Duchesneau, *La physiologie des lumières*, 315.

[47] Hochdoerfer says that Haller was also influenced by the Biblical story of Creation in upholding preexistence (*The Conflict between the Religious and the Scientific Views of Albrecht von Haller (1708–1777)*, 31). This story provided him with a more direct reason for believing that God is needed for generation than the indirect reason suggested in the text of this paper, namely, that observations of functional organic structures indicate that there must exist an intellect responsible for that structure, an intellect for which God could be posited as the probable source.

matter to account for the continuity of species and as a necessary element of generation, but we still need a philosophical explanation for the process by which the *vis essentialis* and the qualified vegetable matter produce living beings. And this requires demonstrating, as Wolff does in his treatise, the actions of the proximate cause by taking on, through the manipulation of instruments, the role of the cause. If Wolff is a rationalist, he is one who uses experiment to demonstrate empirically the action of the causes with which he is primarily concerned: proximate efficient causes.

Haller surely relies heavily upon the empirical. His anatomical starting point establishes this. But his appeal to God as the efficient cause of the generation of all organic forms just as surely establishes his reliance upon speculation reached through reason. Indeed, strictly speaking, his experiments on developing chickens can establish pre*formation* at best and not preexistence. Haller's appeal to God as efficient cause of generation also precludes the possibility that he, like Wolff, is searching for a natural proximate cause of generation. Some commentators have taken this turn to God as a breakdown of Haller's own strict adherence to experiment. John Neubauer, for example, believes that Haller's religious commitments, together with the especially hard case of fetal formation, force him in the end to abandon his own empirical commitments, thus exposing an irreparable rift between science and religion in his thought.[48] I am more swayed by Maria Teresa Monti's and Richard Toellner's approach, which is to argue that there is unity in Haller's thought.[49] This entails granting Haller his empiricist credentials and recognizing that it is true that observation and experiment are crucial to him. But his empirical study of the anatomical structure of fully formed organisms predates his investigations of developing chicken embryos – his "Dissertation" was written in 1752 but his first articulation of his mature endorsement of ovist preexistence did not appear until 1758. Moreover, anatomy is conceptually prior to physiology, for Haller, and the latter depends upon investigations pursued in the former (e.g., Haller 1747, 5; Haller 1777). Whatever else anatomy may include, for Haller, it certainly includes the study of organic structure and the functions of specific organic parts. And he consistently includes generation in his physiological treatise (Haller 1747; EP 8), thus implying that a study of generation is conceptually posterior to and parasitic upon empirical findings regarding

---

[48] John Neubauer, "Albrecht von Haller's Philosophy of Physiology," *Studies on Voltaire and the Eighteenth Century* 215 (1983): 321; cf. Neubauer, "La philosophie de la physiologie d'Albrecht von Haller," *Revue de Synthèse*, nos. 113–14, p. 135–42, and Hochdoerfer, *The Conflict between the Religious and the Scientific Views of Albrecht von Haller (1708–1777)*, 13–14.

[49] Monti, "Théologie physique et mécanisme dans la physiologie de Haller," *passim*, esp. 71; Toellner, *Albrecht von Haller*.

functioning structures.[50] An empirical study of anatomical structures and their functions, then, leads Haller to hypothesize God as the only possible efficient cause of organic formation. It is true that Haller's empirical findings force him to a rationalism beyond anything we find in Wolff. But Haller is an experimentalist in the end, too, for he resolutely pursues experiment after experiment to garner empirical support for the very early existence of hidden organic parts, in order to match and surpass Wolff's experiments and thereby pose a constant challenge to epigenesis (Haller [1754–1777] 1983, 418).

Haller and Wolff, then, clearly embrace distinct theories of generation. We may call these preexistence and epigenesis only if we acknowledge the strong strand of preexistence theory in Wolff's mature view and only if we acknowledge that there is tremendous development of the fetus (development brought on by natural means) on Haller's view. Still, the difference remains: for Haller, the essential structure must be formed by God, a belief Wolff rejects. One source of this difference is the nature of their interventions and experimental techniques throughout their investigations on living beings. Wolff intervenes to reproduce the causal action of the *vis essentialis* and thus claims to have truly explained generation. He need not, according to this conception of explanation, deal with the more abstract questions of the relation of structure and function in organic bodies. But Haller must deal with such questions because the primary object of study in his early experiments is the completely structured animal, and this conditions what he must observe when turning his attention to fetal formation.

---

[50] For an account of the disciplinary boundaries between anatomy and physiology before 1800, together with the shift in these boundaries after 1800, see Andrew Cunningham, "The Pen and the Sword: Recovering the Disciplinary Identity of Physiology and Anatomy before 1800," pt. 1, "Old Physiology – the Pen," *Studies in History and Philosophy of Biological and Biomedical Sciences* 33 (2002): 631–65; pt. 2, "Old Anatomy – the Sword," *Studies in History of Philosophy of Biological and Biomedical Sciences* 34 (2003): 51–76.

12

Soul Power

Georg Ernst Stahl and the Debate on Generation

FRANCESCO PAOLO DE CEGLIA

1. PREMISES FOR A PHANTOM EMBRYOLOGY

Before beginning the analysis, it should be specified that it offers hardly any advantage for strictly medical purposes, except for a series of connections, which, being more useful in terms of our present disquisition, we shall describe, avoiding any excessive prolixity.[1]

Questions connected to generation are on the whole extraneous to the genuine study of medicine. They should therefore be dealt with summarily, and only because of their repercussions in prevention or cure. Such is Georg Ernst Stahl's firm belief. If, however, the author of *Theoria medica vera* really did attribute such a limited importance to the question, why, in the first decade of the eighteenth century did a debate emerge on the explanatory validity and meaning of Stahl's embryology? What were the various significations imputed to the dictum of the "Beacon of Halle University"? This essay is an attempt to reconstruct Stahl's position with regard to the problems of generation and trace the history of its interpretation in the first phase of the debate, from 1708 (the year of the publication of *Theoria medica vera*) to 1734 (the year of Stahl's death).

Stahl believed that the medical discipline should exclude anything that did not help the physician attain his immediate professional goal: keeping people

English translation by Lisa Adams.

[1] Georg Ernst Stahl, *Theoria medica vera, physiologiam et pathologiam, tanquam doctrinae medicae partes vere contemplativas, e naturae et artis veris fundamentis intaminata ratione et inconcussa experientia sistens* (Halae Magdeburgicae: Impensis Orphanotrophei, 1708a). Quotations are from the second edition, ed. Johann Joachim Juncker (Halae Magdeburgicae: Impensis Orphanotrophei, 1737). The third and last Latin edition, was edited by Ludwig Choulant, 2 vols. (Leipzig: Sumptibus Vossii, 1831).

healthy or helping them regain their health. This is why, though a chemist of some note, he considered almost every notion provided by chemistry or anatomy as superfluous to medicine.[2] From this viewpoint, his position was not very far from that of an eminent tradition of medical practitioners, whatever their background or education. A knowledge of embryology did not have repercussions of an immediate therapeutic nature, any more than a general knowledge of anatomy or chemistry. It was therefore of no use, except insofar as it could throw light on the diet that pregnant women should follow in order to have healthy children, the transmission from parent to child of a "predisposition" to illness (and consequently what therapy to apply), and how to make childbirth easier. These aspects, which will not be discussed here, take up at least half of Stahl's exposition.

The suspicion with which Stahl regarded embryological studies, though retrograde and shortsighted, was not a mere idiosyncrasy but was shared by a considerable proportion of the medical profession.[3] Even Friedrich Hoffmann, his mechanist alter ego at the University of Halle, expressed himself in similar terms in the same period:

> A doctor may well ignore the origins of life . . . since his one true and genuine task is – in a more circumscribed manner – to conserve the vital and healthy state of an already formed and living body, as well as to ward off every type of affliction, whether in a preventative or curative manner. Medicine is like horticulture, in which one may well ignore the way in which a plant sprouts from a seed, though not the way in which it becomes fecund and reproduces, nor all those things necessary for its cultivation.[4]

Thus, on the usefulness to medicine of the new embryological discoveries, the conflict of opinion was not first and foremost between animists and mechanists but rather between the champions of a therapeutic vocation and the upholders of an enquiry which was animated by a renewed speculative curiosity, whether of a biological or, as we shall see, ethical-religious character.

Stahl was a prolific writer, and more so in the field of medicine than in that of chemistry.[5] The monumental *Theoria medica vera* and the many doctoral

---

[2] Georg Ernst Stahl, *Theoria medica vera*, 43–64.

[3] Jacques Roger, *Les sciences de la vie dans la pensée françaises au XVIIIe siècle* (Paris: A. Colin, 1963), 7–19.

[4] Friedrich Hoffmann, *Commentarius de differentia inter ejus doctrinam medico-mechanicum, et Georgii Ernesti Stahlii medico-organica* (1739), in Hoffmann, *Operum omnium physico-medicorum supplementum* (Geneva: Apud Fratres de Tournes, 1749), 1:449.

[5] Johan Christof Goetz, *Scripta D. Georgii Ernesti Stahlii*, 2d ed. (Nuremburg: Impensis B. W. M. Endteri filiae Mayerin, 1729).

dissertations written under his supervision embrace a kaleidoscopic variety of themes, all of which concern problems relating to excretion and secretion and their repercussions on the individual's state of health. One fact is significant: none of the hundreds of theses discussed under Stahl's guidance are dedicated to embryological studies in the strict sense. Even the *Dissertatio inauguralis medica de abortu et foetu mortuo*, though it does allude to embryology, is oriented toward the definition of diagnosis, prognosis, and treatment in the case of abortion.[6] *Theoria medica vera* does indeed contain a section on generation, immediately following the chapter on nutrition, but perhaps this was included simply for the sake of expository completeness.[7] It represents, therefore, the only text to which Stahl's disciples could refer when they sought to attribute a theory of embryology to their master – a theory which Stahl himself perhaps never deliberately set forth.

Stahl's medical system is based on the premise that "organic" matter – as he was wont to call it – is in itself passive and, due to its mucous-fatty composition, chemically unstable; it is thus prone to rapid putrefaction. A veritable physiological paradox (*paradoxon physiologicum*) nevertheless occurs in the human body: despite its propensity to alteration, it is not prone to corruption, apart from exceptional cases.[8]

The main role is to be attributed to the actions, not of matter (*materiarum*) nor in matter (*in materiis*), but rather on matter (*in materias*).[9] We must, therefore, postulate an agent which is other than the body and which deflects organic matter from its otherwise inexorable course of instability and decay.

---

[6] Stahl, *De abortu et foetu mortuo....* (Halle: Litteris C. Henckelii, 1708b).

[7] Stahl, *Theoria medica vera*, 370–94.

[8] Georg Ernst Stahl, *Oeuvres medico-philosophiques de G. E. Stahl*, French translation by Théodose Blondin, 6 vols. (Paris: Ballière, 1859–64), ccxxv–xlii; F. Bouillier, *Du principe vital et de l'âme pensate: Examen des diverses doctrines médicales et physiologiques sur les rapports de l'âme et de la vie.* (Paris: Baillière et fils, 1862), i–xxv; Albert Lemoine, *Le vitalisme et l'animisme de Stahl* (Paris: Germer Baillère, 1864), 166–204; Lelland J. Rather, "G. E. Stahl's Psychological Physiology," *Bulletin of the History of Medicine*, 35, 1961; A. Pichot, *Histoire de la notion de vie* (Paris: Gallimard, 1993), 453–502; François Duchesneau, *Les modèles du vivant de Descartes à Leibniz* (Paris: Vrin, 1998), 287–311; Johanna Geyer-Kordesch, "Die *Theoria medica vera* und Georg Ernst Stahls Verhältnis zur Aufklärung," in *Georg Ernst Stahl (1659–1734)*, ed. Wolfram Kaiser and Arina Völker (Halle: Wissenaschaftliche Beiträge der Martin-Luther-Universität Halle-Wittemberg, 1985); Johanna Geyer-Kordesch, *Pietismus, Medizin und Aufklärung in Preußen im 18. Jahrhundert. Das Leben und Werk Georg Ernst Stahls* (Tübingen: M. Niemeyer Verlag, 2000), 140–220; Francesco Paolo de Ceglia, *Introduzione alla fisiologia di Georg Ernst Stahl* (Lecce: Pensa, 2000), 13–85; Kevin Chang Ku-Ming, *The Matter of Life: Georg Ernst Stahl and the Reconceptualizations of Matter, Body, and Life in Early Modern Europe* (Ph.D. diss., University of Chicago, 2002), 124–40.

[9] Stahl, *Theoria medica vera*, 383.

This can only happen by identifying an entity which is able to confer motion on matter, the only safeguard against corruption. Such an agent is not to be found in the conceptual equipment of the mechanists, who could only offer incomplete explanations, maintaining as they did that motion is life, not one of its instruments. They would not even take the trouble to discover the origin of such a motion, or at best they would say that it resides in God, using terms which Stahl, as a layman, could only judge to be blasphemous and totally unscientific.[10]

And so Stahl selects the soul as the source of the body's life. He attributes to the soul, as an intelligent entity, the role of ideal container of the "information" necessary to carry out physiological functions and, in the case of unforeseen morbid superventions, the recovery of health. The soul becomes (or perhaps returns as) the *fons vitae* and the ideal seat of physiopathological algorithms.[11] The pages of the *Theoria medica vera* dedicated to generation can thus be read as the sometimes unwitting precipitate (to borrow a term from chemistry) of such theoretical premises.

2.   THE SOUL AS ARCHITECT OF THE BODY

For Stahl, it is the soul that shapes the unborn child's body: "The soul must likely carry out its actions by means of a certain moral evaluation; and this is because everything is built and formed for its use, its ends and its needs."[12] The order and measured punctuality with which the new organism is formed and grows requires the postulation of an intelligent and somehow conscious being that guides every formative action. Indeed, the construction of a new body is too complex a process to carry out mechanically, without the *schema* or "project" of an architect. Moreover, it is not legitimate to affirm, as the mechanists do, that the order is determined by the peculiar and perfect structure of the parts. First, it would be impossible to demonstrate that the correct form and disposition of the bodily channels can produce such a marvellous result and account for its temporal rhythms. Second, whoever did so

---

[10] Stahl, *De abortu et foetu mortuo*, 151–4, 383.

[11] Francesco Paolo de Ceglia, "Ipocondria ed isteria nel sistema medico di Georg Ernst Stahl," *Medicina and Storia*. 2, no. 4 (2002): 51–86. In Cartesian philosophy there is a schism between soul and life. The soul is no longer the bearer of life but is hosted in the machine-body as long as the latter is functional, that is, alive. From this perspective, Stahl attempts to mend the schism opened by Descartes, by attributing a life-giving function to the soul.

[12] Stahl, *Theoria medica vera*, 373.

would not be able to identify a cause which could explain such a disposition; and so they would simply shift the problem to a different conceptual level.[13]

It has often been maintained that Stahl's choice of the soul was conditioned by his religious beliefs.[14] Yet in order to explain his choice of a theoretical system, in his works Stahl does not usually have recourse to religious arguments, which are used not infrequently, on the other hand, by some of his mechanist adversaries (albeit with a merely rhetorical purpose at times) even though the latter are considered "laymen," and by many of his disciples as well. Indeed, authors such as Hoffman, Heister, and Leibniz – perhaps in order to obviate accusations that might be directed against their respective approaches to medicine and philosophy – do not hesitate to accuse Stahl of atheism or materialism.[15] So is it necessary to resort to Pietism or is it sufficient to identify a uniquely intellectual heuristics, whether adequate or not, that could lead to the same theoretical results?

Julius Caesar Scaliger writes in his commentary on Aristotle's *Historia animalium*,

> It is not absurd that the soul should build its own dwelling place. Rather, this comes from the wisdom attributed to it by God.... So one should say, albeit against the will of the philosophers, that the vegetative soul possesses the power to give the form of an animal's body to the seed, not through instruments but *out of* itself, as a divine thing.[16]

---

[13] Ibid., 250–2. The claims of authors such as Hoffmann and his acolytes that matter is active, the bearer of endogenous information, are also vain; according to Stahl, they are simply trying to eliminate the problem of the soul (cf. Friedrich Hoffmann, *Commentarius de differentia inter ejus doctrinam medico-mechanicum, et Georgii Ernesti Stahlii medico-organica*, in *Operum omnium physico-medicorum supplementum* [Geneva: Apud Frates de Tournes, 1739], 1: 439–51). They would not be able to explain the difference between organic and inorganic (Stahl, *Theoria medica vera*, 9–16). The various intermediate beings, such as *archaei*, ferments, operative ideas, informing powers, and natural appetites, of which ancient and modern tradition speaks are moreover fictitious and futile. If they existed, their function could be more economically attributed to the soul, which is not denied by anyone for metaphysical or religious reasons. So *nec fit per plura quod potest fieri per pauciora* (ibid., 380–1).

[14] Johanna Geyer-Kordesch, *Pietismus, Medizin und Aufklärung in Preußen im 18. Jahrhundert. Das Leben und Werk Georg Ernst Stahls* (Tübingen: M. Niemeyer Verlag, 2000).

[15] Hoffmann, *Commentarius, de*, 437–8; Lorenz Heister, *De medicina mechanica praestantia prae stahliana*, in *Compendium medicinae practicae* (Amsterdam: Apud Janssonio-Waesbergios, 1738); Georg Ernst Stahl–G. W. Leibniz, *Negotium otiosum seu Skiamachia adversus positiones fundamentales Theoriae medicae verae a viro quodam celeberrimo intentata, ced armis converses enervate* (Halle: Litteris Orphanotropheis, 1720), 182.

[16] Julius Caesar Scaliger, *Aristotelis historia de animalibus. Iulio Caesare Scaligero interprete....* (Toulouse: typis R. Colomerij, 1619), 595–6.

This work was published long before Stahl's *Theoria medica vera*.[17] Though there are differences between the two, that is not my point here.[18] Rather it is that, in the light of his illustrious predecessor, Stahl's position no longer seems an isolated one.

But is Stahl's animism a direct derivation from Scaliger's medical and philosophical teachings? Stahl was familiar with Scaliger, whose works were widespread in seventeenth- and eighteenth-century German culture,[19] though this does not necessarily mean that Stahl was influenced by them. Nevertheless, we cannot ignore the fact that, apart from the author of *Theoria medica vera*, others had elaborated a similar animistic theory, though in widely differing cultural contexts and even before the birth of Pietism.[20] As we shall see, Scaliger was not alone but was part of a long and heterogeneous tradition of animistic physiology.

Though we could discuss at length the philological accuracy of the Aristotelian exegesis put forward by the Italian medical doctor,[21] this would be beyond the scope of the present essay. It should be noted, however, that Scaliger's animism took shape within the variegated universe of Renaissance Aristotelianism, particularly in relation to commentaries on *Historia animalium* and especially the second book of *De generatione animalium*.[22] In Stahl, the soul's function is to confer on matter, itself passive, the information "in

---

[17] In the *Exotericae exercitationes*, Scaliger reflects more generally as follows: "This force, architect of such a noble temple, has been judged most wise by all philosophers. The temple will thus have been built by a substance. This is the form of the seed.... Now, in growth and the reparation of lost parts of the body, such a work is carried out by the soul itself.... Since it builds the heart, it knows what life is, since it prepares the parts, it knows their purpose and use" (*Exotericarum exercitationum liber ...*, [Lutetiae: exoffina M. Vascosani, 1557], 13r).

[18] One is the notion of the soul's unity. For Scaliger, three different souls guide the body in succession: "The soul, which at first lives in the fetus, is considered merely vegetative even by our theologians. Next comes the sensitive soul. Once these two have been annulled, ours is instilled" (ibid., 16r–v).

[19] Peter Petersen, *Geschichte der aristotelischen Philosophie im protestantischen Deutschland* (Leipzig: F. Meiner, 1921), 258.

[20] Bouillier, *Du principe vital et de l'âme pensate*, 114–15; Jacques Roger, *Les sciences de la vie*, 429.

[21] Guido Giglioni, "Girolamo Cardanoe e Giulio Cesare Scaligero: Il dibattito sul ruolo dell'anima vegetative," in *Girolamo Cardano: Le opera, le fonti, la vita*, ed. Marialuisa Baldi and Guido Canziani (Milan: Franco Angeli, 1999).

[22] Scaliger, *Exotericarum exercitationum liber*, 6r. The interpretation was hardly original. For instance, Themistius had already advanced an animist interpretation of Aristotle's theory of generation (Themistius, *In libros De anima paraphrases*, ed. R. Heinze (Berlin: G. Reimerum, 1899).) For Hartmann, Stahl's medical doctrine has its roots in a tradition which includes Empedocles, Hippocrates, Galen, Alexander of Aphrodisias, and Averoes (Georg Volcmar Hartmann, *Schediasma apologeticum quo sentential illustris Stahlii de natura humana ... ita confirmatur...*. [Erfordiae: Sumptibus C. F. Jugnicolii, 1735], 1–2). For Alberti, referring to the

view of which" an animal is able to act. In Aristotelian terms, the *anima* is the form. This undergoes a process of hypostatization that turns it into an unextended substance. Such a result had already been alluded to in Aristotle's own writings.[23]

Neither in this case do we intend to posit a direct derivation of Stahl's theory of generation from Aristotle's. Yet Stahl is Aristotelian in spite of himself, in the measure in which he seeks a single principle to explain the integrated complex of physiopathological functions and, in this case, those of ontogenetic propulsion. Galen's faculties, van Helmont's *archaeus*, and Willis's ferments all fragment the information present in the body, attributing it to relatively independent centers. Stahl, on the other hand, seeks a single "switchboard" for all vital algorithms.

While Aristotelian form is distinct from matter and yet requires it, Stahl's soul is completely "other" with respect to the body. In Aristotle the "blindness" of matter only makes sense within an ontological framework, since in experienced reality there is no matter without form. For Stahl, on the other hand, bodies are by nature "inorganic" and therefore blind. The soul is joined to them to make them "organic," but it is a surplus, a magnanimous gesture on nature's part. In Aristotle matter always finds a reservoir of information on which to draw, while in Stahl a dark mechanical fate awaits it from which it can only occasionally escape.

3. MASCULINE AND FEMININE

Stahl came relatively late to the study of the theory of generation; by this time his animism had taken on its definitive shape. The doctrine of the soul could at most give an account of fetal growth as a process analogous to nutrition. Yet it left open many of the questions about conception and heredity, questions on which Stahl never took a clear position.

Contrary to the views of some historians,[24] Stahl's approach was not speculative, nor did he have great theoretical interests. Furthermore, there is

---

interpretation of the *Coimbricenses*, together with Themistius, Plato, Simplicius, and Johannes Philoponus were all animists too (Michael Alberti, *De paedantismo medico*, 31).

[23] Polycarp Gottlieb Schacher, *De anima rationali, an sit corpori vitali principi* (Leipzig: Fleischer, 1715); Johann Daniel Longolius, *Systema stahlianum de vita et morte corporis humani ab incongruis medicorum mechanizantium opprobriis vindicatum....* (Budissae: Literis Richterianis, 1732), 3–4.

[24] Stahl, *Oeuvres medico-philosophiques*; Richard Koch, "War Georg Ernst Stahl ein selbständiger Denker?" *Sudhoffs Archiv für Geschichte der Medizin und der Naturwissenschaften* 18 (1926): 20–50.

considerable confusion in the pages of *Theoria medica vera* dedicated to generation. And this is not just a matter of stylistic sloppiness, as is the case elsewhere. The author, who is generally parsimonious with his citations, includes a number of references to the observations and opinions of others. Yet the very physician who had often underlined the nonhomogeneity of animal physiology in relation to that of humans, adducing as proof the peculiarly ratiocinative dimension of the human soul,[25] comments on and makes his own the results of observations conducted on animals. Moreover, he introduces arguments only to reject them immediately as *otiosae quaestiones*.[26] The impression one has is that he is unable to disentangle himself from the mass of opinions and observations, and so he airs the material and the shaky arguments he has managed to find, asking the reader to discover a way out on his own, as long as it leads to this one certainty: that the soul is the body's architect.

Stahl presupposes what should instead be demonstrated. A kind of "heuristics of the invisible" pervades the whole discussion. In embryology the visible is not in itself important, since whatever one sees – whether an indistinct substance or an *ingens numerus* of *animalculi* – will only be matter which the soul has organized as its host. Thus the conveyor, which does not hold any interest, will be visible, but not the conveyed.[27]

The author reveals a rather provincial knowledge of contemporary literature on embryology. Moreover, he reads "modern" texts in search of an answer to "ancient" questions. Central to his interests, for example, is the identification of the parent who possesses the active generating principle. Scripture, as well as Harvey's and Malpighi's observations, maintains that this principle is conveyed by the male and that the female simply provides nourishment.[28] Stahl agrees, betraying once again an ill-concealed Aristotelian influence. Nevertheless, he is cautious; in the *De abortu et foetu mortuo* he writes,

> In the whole issue there is nothing so certain and sure which, once solicited by opinions and observations to the contrary, cannot change or indeed be transformed into the diametrically opposite position.[29]

The origin of the fetus is unknown, too, as is the mechanism of conception:

> Let us avoid becoming entangled *a priori* in such sterile and truly otiose questions. It will be sufficient to investigate *a posteriori* the truth. Those who

---

[25] Georg Ernst Stahl, *De frequentia morborum in homine prae brutis* (Halle: Litteris C. Henckelii, 1705).

[26] Stahl, *Theoria medica vera*, 374.

[27] Ibid., 372–3, 384–6.

[28] Ibid., 377–8.

[29] Stahl, *De abortu et foetu mortuo*, 16–17.

have the time and the inclination may well write comments in which they say whether there is a transmission [*traductio*] of souls or a new individual creation [*individua creatio*].[30]

Unlike elsewhere, Stahl does not bother to reach sure and definitive conclusions. Hybridizations, for example, would demonstrate that the two sexes participate equally in generation. Nevertheless, he wonders,

> why a mare that has coupled with a donkey and given birth to a mule, the next year, when coupled with a horse, gives birth to a colt similar to a donkey. And again, one cannot be reticent over the *vulgaris observatio* according to which the child would inherit the physical features as well as the moral and intellectual propensities of the most energetic parent in the act of conception.[31]

These questions throw light on the author's method of argumentation; and this has little to do with the published findings of a scientist such as Malpighi, which already herald the methods of modern biology. Stahl does not deviate from the *modus* typical of a certain practical medicine and natural philosophy of the time: general assumptions and recourse to the *communis opinio* and *diuturna experientia*,[32] as well as to the narration of clinical cases which he either treated personally or read about in medical compilations.

4. BETWEEN EPIGENESIS AND PREFORMATION

Stahl's physiology could be defined as a teleologically inspired hydraulicism: physiological functions are carried out thanks to humoral motion and the related processes of secretion and excretion, regulated by the soul with a view to the maintenance or recovery of health. After the problems of conception have been put aside, so to speak, the growth of the unborn child's body may be interpreted from the following perspective: once conceived, the embryo is slowly nurtured by its own soul through the processes of "filtration" of matter which is provided by the mother, and these processes are analogous to the processes of nutrition.[33] In this sense Stahl is a classical epigenesist.

[30] Stahl, *Theoria medica vera*, 374.

[31] Ibid., 376–7.

[32] Gabriele Baroncini, *Forme di esperienza scientifica* (Florence: Leo S. Olschki Editore, 1992), 182–5.

[33] Stahl, *Theoria medica vera*, 378–9.

270

Malpighi had discerned the outline of a bodily structure in an *unincubated* chicken egg.[34] Stahl, together with some of his contemporaries,[35] misinterprets the anatomist's words, reading into them the fact that a bodily outline may be discerned in an *unfertilized* egg:[36] "Malpighi," he writes, "has confirmed the presence of this same bodily lineament or outline even in chicken's eggs conceived and laid without any participation from the cock."[37]

Contrary to Aristotle, in Stahl the *punctum saliens* is composed of the brain, medulla oblongata, spinal cord, and nerves. Stahl does not seem particularly bothered that the existence of an organization prior to fecundation – that is, to the participation of the male principle – could ruin his system. On the contrary, he embraces, at least at this point, a sort of uncertain ovist preformation theory which discerns in the corporeal outline the maximum form that the female principle – the mother's soul – can elaborate on its own.[38] The soul of the unborn child – a principle deriving from the male – is grafted onto this lineament, and on the basis of an epigenetic scheme, it builds the rest of the body. Stahl declares that he does not know whether the soul of the fetus is part of the parents' soul or a being created *ex novo*.[39] Thus, although Stahl's sympathies lie with a spermist model, what we are really faced with is "an ovi-spermatism linked to animism and independent of any traducianism."[40]

[34] Marcello Malpighi, *On the Formation of the Chick*, trans. Howard B. Adelmann, in Adelmann, *Marcello Malpighi and the Evolution of Embryology* (Ithaca: Cornell University Press, 1966), 2:224–6.

[35] Walter Bernardi, *Le metafisiche dell'embrione: Scienze della vita e filosofia da Malpighi a Spallanzani, 1672–1793* (Florence: Leo S. Olschki Editore, 1986), 51–2.

[36] Stahl, *Theoria medica vera*, 379, 384.

[37] Ibid., 384.

[38] Duchesneau, *Les modèles du vivant de Descartes à Leibniz*, 296–311.

[39] Stahl, *Theoria medica vera*, 374.

[40] Joseph Tissot, *La vie dans L'homme*, 2 vols. (Paris: Victor Masson et fils, 1861), 2:390. Just one part of the section on generation of the *Theoria medica vera* might well merit Leibniz's accusation (Georg Ernst Stahl–G. W. Leibniz, *Negotium otiosum*, 18): that Stahl conceived of the soul as divisible and so as matter. "If it is the soul that builds the body, how is the former communicated along with the sperm? . . . As for its division, one may have a general picture: the soul's essence . . . consists particularly in motor activity, and movement consists in perpetual numerical division. . . . Now nothing would prevent us from transferring this same consideration to the motor; in other words, it would not be futile to think that since movement is a divisible thing, so is the motor" (Stahl, *Theoria medica vera*, 374). This passage, however, expresses a kind of rhetorical doubt, or at any rate a "philosophical inkling," which Leibniz, who had not read all of the *Theoria medica vera*, does not take into account (Albert Lemoine, *Le vitalisme et l'animisme de Stahl* [Paris: Germer Baillère, 1864], 150–3).

Despite a certain incongruity between epigenesis and preformation theory, one aspect is worth noting. Unlike some of his contemporaries – Hoffman for instance[41] – Stahl the animist does not ascribe any importance to the nervous system in the regulation of physiological functions. Even in the pages dedicated to perception, he reduces its role to a minimum.[42] The action of the soul on the body is effected through tonic motion, which consists in an alternating activity of constriction and relaxation of the fleshy parts: a vital movement, the purpose of which is to direct the humors toward and especially through these parts.[43] The soul is thus the regulator of the body's humoral system. Since Stahl denies resolutely the existence of a nerve fluid, or any other *succus* present in the nervous system, the provisional collocation of such a "hydraulic" principle in the brain, medulla, and nerves seems incompatible with the role which he later attributes to it, though in conformity with the most recent observations in embryology of his day. Moreover, it is not clear to the reader how, even under the guidance of the soul, from the primitive outline one can arrive at the construction of the remaining parts, for instance, the blood. Stahl writes, "It does not seem to me so far from common sense . . . to believe that it is the same active principle which, arranging point by point the solid parts, carries out the 'mixture' of the fluid parts thanks to this same action."[44] Stahl's physiology is thus a physiology without anatomy: whatever the structure on which one operates, there will always be an immaterial principle with sufficient information to fulfil its own purpose.[45]

5. THE TRANSMISSION OF IDEAS

A desire or a fear conceived in the mind of a pregnant woman can provoke alterations in the unborn child's body. This is one of the questions with which Stahl deals at length in numerous publications:

> Although we cannot explain the way in which this fictitious image [*imago fictitia*] – which some call more precisely an *ens rationis ratiocinantis* – is communicated by the mother's soul to that of the fetus, nevertheless it is an incontrovertible truth that, regarding imagination and evaluation, there is

[41] Hoffmann, *Commentarius*, 452–3.
[42] Stahl, *Theoria medica vera*, 396–406.
[43] Georg Ernst Stahl, *De motu tonico vitali*. . . . (Halle: Litteris C. Henckelii, 1692), 29–30.
[44] Stahl, *Theoria medica vera*, 379.
[45] Stahl's physiology is without anatomy but not without a "histology" (i.e., the study of the mixture of tissues in the broad sense). Rather, we might say that "histological health" – the prevention of humoral corruption and tissue decay – was itself the end of physiology, the purpose of the vital motions (de Ceglia, *Introduzione alla fisiologia di Georg Ernst Stahl*, 40–6).

mutual consent and communication between thinking, imagining and evaluating beings.[46]

When ideas possessed by the mother pass to the soul of the fetus, they cause changes in the latter; and this is proved a posteriori by the existence of deformities. Two arguments may be adopted a priori. The first is that ideas as such may only be received by a suitable, or rather thinking, container, and this is the soul. The second is that, in order for them to produce effects, they must somehow be represented, compared, and morally evaluated by the host container; and only a soul can do this.[47] If it is true that some individuals bear the signs on their body of objects that were either loved or feared by their mother during pregnancy, then the ideas conveying the information could not simply have been transmitted from one soul to another. In that case they would have remained inert and would not have produced any effect. They must have been subjected to a fresh evaluation, however unconscious, by the soul of the fetus: activating the same desire or fear that pervaded the mother's soul, it somatizes these ideas. Not all ideas become motion; that is, not all ideas turn into a process which leads to somatic alteration, but only those which the unborn child's soul evaluates as a good to be pursued or an evil to be avoided.

Souls come into contact with one another, though Stahl admits that he does not know how. In this way they exchange "packages of information" – pathogens in the case of a fetal monstrosity. The soul is a thinking *ens*, whose contents can only be "intellectual": that is, the contents of the soul are ideas.[48] The generating nucleus of a monstrosity is therefore a thought, a manifestation of an unconscious deliberation. It corresponds to a schema, which in Stahl's language indicates a set of instructions needed to carry out a motion. The soul of the fetus, guided by the received idea, communicates anomalous motions to the organism, which in turn become abnormal formations in the tender little body.

As we have seen, the conveyed idea sometimes results in a particular physical conformation. Nevertheless, the schema of a monstrosity is no different from any other "genetic" information communicated by the mother to the fetus. And so the transmission of a *similtudo* between the features of the mother and those of the child is regulated by the same dynamics. Likewise for the communication of the mother's *impressiones*: the child may have

[46] Ibid., 373.

[47] Ibid., 375.

[48] Peter H. Niebyl, "Sennert, van Helmont, and the Medical Ontology," *Bulletin of the History of Medicine* 45 (1971): 115–37.

characteristics that recall, for better or worse, people, animals, or things which disturbed the mother's soul during pregnancy.

It is wrong, however, to think that the idea necessarily has a physical outlet. It can be hosted by the unborn child's soul more generally as an *adsuetudo*, a habit of carrying out certain motions, which are either *stricto sensu* "moral," as in the propensity to commit petty thefts,[49] or "physiological," as in a tendency to hemorrhoidal blood loss.[50] There is a strong resemblance between parents and children:

> Regarding the solid parts, structure and texture share this similarity, whilst regarding the fluids, it is their motion which is similar. The places [*loca*] and more especially the particular times [*tempora*] at which these motions occur, emulate those to which the parents were accustomed.[51]

Of these *adsuetudines*, the *haereditaria dispositio ad varios affectus* – the congenital tendency to contract illnesses – is significant for the purposes of medical praxis. A knowledge of this tendency allows the physician to conduct more accurate diagnoses and prognoses, as well as to prescribe the most appropriate therapy. As Stahl explains,

> On the basis of these presuppositions, we may say that hereditary disposi-tions are primarily, and fundamentally, based on an idea of action which has been communicated [*communicata idea agendi*]. We choose to call it a "habit" [*adsuetudinem adpellari judicamus*], through which the ideas of actions gen-erally carried out in the economy of the parents' bodies are communicated, whether such actions be positive – when they benefit the body in some way – or negative – when they consist in fear, aversion or trepidation.[52]

Although Stahl refers to "parents" in his analysis, he seems to concentrate more on the relationship between mother and child, attributing to the father the capacity to influence the soul and hence the body of the unborn child only at the moment of conception.

6. *RATIO* AND *RATIOCINATIO*

In order to obtain a theoretical model which is able to counter the objections of those who claim that vital actions – and thus the transmission of ideas – are not conscious and so cannot be attributed to the soul, Stahl introduces a

[49] Stahl, *Theoria medica vera*, 375.
[50] Georg Ernst Stahl, *De Haereditaria dispositione ad varios affectus* (Halle: Litteris C. Henckelii, 1706), 31–2.
[51] Ibid., 34.
[52] Ibid., 36.

distinction which, while it resolves some problems, introduces others. On the one hand, there is *logos*, or *ratio*, which represents immediate and simple intellection, without any real discursive apparatus, and this is what Stahl generally means when he speaks of the *anima*. On the other hand, there is *logismos* – or *ratiocinatio* or *animus* – which presupposes an awareness of sensible or imaginative data around which reasoning, or ratiocination in the strict sense, is articulated. Memory and imagination are only associated with the objects of ratiocination. It would thus be the *ratio* that controls bodily functions without the *ratiocinatio* being aware of this.[53]

The above distinction between *ratio* and *ratiocinatio* is not ontological but functional, though it is neither altogether clear nor perfectly intelligible. The terms *ratio* and *ratiocinatio*, like *anima* and *animus*, are often loosely interchangeable. The postulation of unconscious information residing in the soul had numerous precedents. To cite just one, which we have already mentioned, Scaliger had used terms which were, once again, notably similar to Stahl's: "Who can say that the faculty which moves the heart and transforms food into the humors acts without reason [*sine ratione*]? Without ratiocination [*sine ratiocinatione*] – that is, without the deduction of arguments – but not without reason."[54] Thomas Feyens, Johann Sperling, and Daniel Sennert, with whom some of Stahl's contemporaries associated him,[55] had expressed themselves in similar terms.[56] And the same solution was commonly adopted by those who, like these authors, believed that matter was merely passive and who sought an informing principle that could combine and integrate the physiological functions. Some placed the origins of the soul's knowledge in divine wisdom, whereas Stahl cut the umbilical cord, so to speak, between soul and God: the knowledge possessed by the vital principle was altogether "terrestrial."

As we have seen, the manner of transmitting ideas is unknowable. In the light of the distinction between *ratio* and *ratiocinatio*, however, not even the contents of such transmissions appear perfectly intelligible; when they give rise to monstrosities, for instance, ideas are called *entia rationis ratiocinantis*. So do they belong to the *ratio* or *ratiocinatio*? Schematically, it may be said

[53] Georg Ernst Stahl, *De differentia rationis et ratiocinationis* (Halle: Litteris C. Henckelii, 1701).

[54] Scaliger, *Exotericarum exercitationum liber*, 391v–92r.

[55] Peter Christoph Burgmann, *Succinctum hypotheseos stahlianae examen de anima rationali corpus humanum struente motusque vitales tam in statu sano quam morboso administrante* (Leipzig: Apud I. F. Gleditschii B. filium, 1731), 5–10; Hartmann, *Schediasma apologeticum quo sententia illustris Stahlii de natura humana*, 1–3, 10–11, 23–27.

[56] Thomas Feyens, *De viribus imaginationis tractatus* (Lugduni Batavorum: Elzevirus, 1635), 290; Johann Sperling, *Tractatus physicus de formatione foetus in utero* (Vitembergae: Apud Haeredes T. Meni, 1641), 33; Daniel Sennert, *Hypomnemata Physica* (Frankfurt: Schleichus, 1683), 131.

that they are "beings of *ratio*," since they are ideas that convey patterns of involuntary motion. The *ratio* is the most salutary dimension of the soul because it is the most "natural": indeed, animals, that possess only *ratio*, fall ill less frequently than human beings.[57] Yet man is also endowed with *ratiocinatio*, which disturbs and alters the *ratio* and distracts it from its purpose. It is in this sense that such ideas are *entia rationis ratiocinantis*: they are "beings of *ratio*" that err due to their commerce with ratiocination. They are ideas that correspond to the "soul's passions": "The soul's passions cannot therefore be different from certain untimely and premature decisions which the soul makes, whether through the perception of objects, or through a mere fictitious elaboration formed on the basis of elements provided by the memory."[58]

In other words, *ratio*, weighed down by the difficult task of preserving the body and besieged by perceptions and fantasies wrought by *ratiocinatio*, sometimes errs in carrying out those actions which fall within its province. The anomalous motions are then transferred, by means of ideas, from mother to child, where they act on the plasticity of the little body.

7. AFTER STAHL: BURGMANN'S CRITIQUE OF STAHLIAN MEDICINE

Succinctum hypotheseos stahlianae examen de anima rationali is the title of a lecture published in 1731 by Peter Christoph Burgmann.[59] The work aims to discuss and demolish Stahl's doctrine of the soul: as the body's architect, as a regulator of physiological functions, and as a healer in the case of illness: "It is not without reason that I have reviewed the paradoxical doctrine of the soul's power over the human body, as it is sustained by the famous Stahl, together with his disciples [*illustrissimus Stahlius una cum asseclis suis*]."[60] Burgmann does not make a clear distinction, however, between Stahl's writings and later commentaries on them. And this point is a crucial one for our historical reconstruction. From the beginning the distinction seems unclear between the

---

[57] Stahl, *De frequentia morborum in homine prae brutis.*

[58] Stahl, *Theoria medica vera*, 340–1.

[59] The work was preceded by two scathing attacks on Stahl's doctrine of the soul, both published in 1729 (Martin Nagy Borosnyai, *De potentia et impotentia animae humanae in corpus organicum....* [Halle: typis I. C. Hilligeri, 1729]; Christian Gottfried Stenzel, *Bini tractatus....* [Vintembergae: Stanno B. Gaeberdti, 1729]). But they contain practically no reference to the theory of generation.

[60] Peter Christoph Burgmann, *Succinctum hypotheseos stahlianae examen de anima rationali corpus humanum struente motusque vitales tam in statu sano quam morboso administrante* (Lipsiae: Apud I. F. Gleditschii B. Filium, 1731), iii r.

genuine dicta of the master and his disciples' comments. Stahlism becomes a sort of semantic "microcosm" that could be revised in a relatively independent manner by its various interpreters. Its basic premise was simple: that matter is passive and that the only source of motion is the soul. The rest was subject to free interpretation and could even be reformulated on the basis of the fresh empirical data that the sciences were gradually accumulating. This sort of flexibility was one of the major causes of the longevity of Stahlism,[61] which would otherwise be inexplicable.

Burgmann's discourse seems designed as a scene from the Judgement Day of Animism presided over by an intransigent mechanistic judge. Yet in the *Examen*, the object of the polemic is not animism in general, nor even Stahl perhaps, who is not often directly cited, but rather Michael Alberti, whose passages on the theory of generation are meticulously quoted, dismantled, and rejected.

Alberti was the first disciple of Stahl for whom he was able to find a secure academic post.[62] His work, which is in some ways original, was explicitly presented as a continuation of the master's, whose official spokesman he became. Indeed, in some controversies Alberti acted as a kind of verbal "body-guard."[63] In some of his writings, Alberti takes up – though almost always in an impromptu manner – Stahl's positions on embryology, extending and reorganizing what in *Theoria medica vera* had been expressed in a vague and incomplete way. It was his *Introductio in universam medicinam* that sparked off the controversy. While remaining basically faithful to the text of *Theoria medica vera*, it expounds its arguments in a more assertive manner. It is also more comprehensible linguistically, if only because it consists of a list of fundamental theses, which could thus be more easily dealt with in a debate. And so it was easier to dismantle piece by piece.[64]

Burgmann's criticisms are centered on the impossibility for the animists – Stahl, Alberti, Nenter, and Coschwitz[65] – of explaining the origins of the

---

[61] Bernnard Joseph Gottlieb, "Bedeutung und Auswirkungen des hallischen Professors und kgl. Preuß. Leibartzes Georg Ernst Stahl, auf den Vitalismus des XVIII Jahrhunderts, insobesondere auf die Schule von Montpellier," *Nova Acta Leopoldina: Neue Folge* 89, no. 12 (1943): 423–502.

[62] Wolfram Kaiser, "Der Lehrkörper der Medizinischen Fakultät in der halleschen Amtszeit von Georg Ernst Stahl," in *Georg Ernst Stahl (1659–1734): Hallesches Symposium 1984*, ed. A. Völker and W. Kaiser (Halle: Martin-Luther-Universität Halle-Wittemberg, 1984), 59–66.

[63] Michael Alberti, *De paedantismo medico*, 1708.

[64] Michael Alberti, *Introductio in universam medicinam....* (Halle: Litteris et Impensis Orphan-otrophei, 1718).

[65] Georg Phillip Nenter, *Theoria hominis sani sive physiologia medica* (Argentorati: typis et sumt. J. Beckii, 1714); Georg Daniel Coschwitz, *Organismus et mechanismus in homine vivo* (Leipzig: F. Lankisius, 1725).

information possessed by the soul. He does not attack the animist position *tout court* but rather the hard-line animism of Stahl and the Halle school. This is the reason why Burgmann shows a certain appreciation for Sennert, who, though he considered the soul as the body's architect, at least maintained that "the forms do not act because of some exceptional power of their own, but rather they are an instrument, and it is the hand of God, that most wise Creator and Artificer, that gives them their power and efficacy."[66]

Whence does the soul acquire its knowledge? From the father? And the latter from whom? We could go right back to Adam! The soul of the latter, however, never built its own body – which was shaped directly by God – and so it could not have communicated anything to later generations. Given the passivity of matter, only God can know how conception occurs:[67]

> All the parts of the fetus are present from the first instant [*a primo inde tempore iam adsint*] and they don't need any application or coordination.... It is enough for them to be nourished and to grow daily thanks to the addition of the nutritional fluid.[68]

Like Stahl, Burgmann seems to avoid the problem of conception and reduces generation merely to the growth of the unborn child's body. Once matter has been considered passive, the origins of life are excluded from research. Stahl places such origins in the soul, defined by the contradictory *asylum ignorantiae*, while Burgmann places them in divine wisdom, whose function is not altogether different from that of the soul. Both study the living, not life. Yet Stahl's position is not tenable for Burgmann, who recounts it thus: "Once the informing principle *has been communicated by the father* [italics added], conception immediately takes place."[69] This sentence, quoted in a lecture by Burgmann that should have been on Stahl, is as usual taken from Alberti. The master, as we have seen, had been much more cautious, and Burgmann knew it, though this was irrelevant for him: "In fact, the affirmations of the famous Stahl in Section Four of the *Physiology*, dedicated to generation, are not so intelligible. The doctrine is there expressed in words that aren't clear enough to allow for the easy comprehension of what he is trying to say."[70]

So in the "vulgate," Alberti's assertions replace the hesitant statements actually made by Stahl. The only point at which Burgmann quotes Stahl's exact words is when he discusses the presumed transmission of ideas from

---

[66] Peter Christoph Burgmann, *Succinctum hypotheseos stahlianae examen*, 185.

[67] Ibid., 39–44.

[68] Ibid., 36–37.

[69] Alberti, *Introductio in universam medicinam* thesis XXXIII.

[70] Peter Christoph Burgmann, *Succinctum hypotheseos stahlianae examen*, 12.

the mother's soul to that of the fetus, perhaps solely because of the considerable space dedicated to it in *Theoria medica vera*. The three arguments used are interesting, since they penetrate to the core of animism, using Stahl's own doctrine against him. They are not, however, formally perfect, nor "philologically" correct. Above all, for Burgmann the fetus's soul in the womb cannot form ideas of sensual (*sensuales*) things. Yet it is Stahl himself who maintains that a soul without a body cannot know the external world,[71] though he is referring – and Burgmann doesn't consider this – to a soul which not only does not have a body but cannot enter into contact with another soul which has. The second objection springs from the belief that in Stahlian medicine the fetus acts only through *ratio*. And yet in Stahl's psychology ideas connected with the passions of the soul, which are the most violent and commonly transmitted, have to be elaborated by *ratiocinatio*. Indeed, they are *imagines fictitiae* or *entia rationis ratiocinantis*.[72] Apart from the problematic nature of the expression *ens rationis ratiocinantis* (discussed earlier), Burgmann does not adequately evaluate the fact that the "activation" in the fetus of *ratio* alone is Alberti's theory, not Stahl's, who never pronounced on the subject.[73] Finally, Burgmann objects, communication is seen to derive from a harmony between souls, but how this happens is not explained. Once again, the reference is to Alberti's "bold" affirmations,[74] not to statements made by Stahl, who had admitted more modestly that "the way in which this fictitious image [*imago fictitia*] is communicated from the mother's soul to that of the fetus cannot be explained."[75] Burgmann is aware, therefore, that many of the positions attributed to Stahl are not actually his but rather those of his followers, in particular Alberti, but this does not interest him. From now on, when Stahlian medicine is discussed, the references will be increasingly to that of his commentators.

8. PIOUS REACTIONS AGAINST MECHANISM: LONGOLIUS AND HARTMANN

To recapitulate, Stahl dedicates a section of *Theoria medica vera* to a declaredly vague theory of generation. Alberti offers an initial interpretation of

---

[71] Stahl, *Theoria medica vera*, 29.

[72] Ibid., 250.

[73] Michael Alberti, *Medicinische und philosophische Schriften....* (Halle: J. C. Hendel, 1721), 149, 154.

[74] Ibid., 154, 613.

[75] Burgmann, *Succinctum hypotheseos stahlianae examen*, 50–8.

it, which, in his "Introductio," becomes a veritable reference manual. In reply, Burgmann elaborates a critique of the *systema stahlianum*, which is really a critical paraphrase of Alberti. And so a game of Chinese boxes begins, in which each writer declares that he is offering a faithful commentary on Stahl's words while he is actually falling into a kind of exegetical *emboîtement*.

Burgmann had dared to contradict the master, and the reactions of the animists were quick to arrive: for instance, Johann Daniel Longolius's *Systema stahlianum vindicatum* was published in 1732. From the beginning there were many commentaries on Stahl's writings, both from supporters and detractors. He was not, however, widely read, since his Latin was difficult and prolix, especially in *Theoria medica vera*. So his disciples' texts, which were taken to be "inspired" by the master, were used as a primary source. They were written in both Latin and German, the former being reserved for academic, especially medical, circles, the latter for a public with a lower level of formal education. As part of a project of cultural reform, the German texts, including those by Storch, Weisbach, and Richter, had an explicitly didactical purpose and were published as manuals of medico-moral pedagogy.[76] Thanks to a wider circulation than the Latin texts, those in *Muttersprache* spread to all levels, especially outside academic circles, the image of Stahl as a theorist of Pietist medicine based on the idea of a virtually omnipotent soul.[77]

Longolius, who was familiar with German medico-moral publications, is the first to reveal a "Germanic" stylistic influence in his Latin writings on the subject, in a debate which had previously been conducted in a more academic, "Latinate" manner. From the outset the author lines up the two opposing teams, revealing yet again that the confrontation was not between two individuals – Stahl and Burgmann – but between two opposing schools of thought: "What Stahl and his followers call 'tonic' and claim to be immaterial, due to its immaterial principle, the anti-Stahlians, who ignore the union between body and soul, prefer to call 'elastic' and they believe it to be matter."[78]

Longolius sees Burgmann's mechanistic philosophy as incomplete: "What are you waiting for in order to abandon your metaphysical mechanism?" he

---

[76] Johann Storch, *Leitung zur Historie des Höchsten Gottes.* . . . (Eisenach, 1752); Christian Weisbach, *Wahrhaffte und gründliche Cur aller dem menschlichen Leibe zustossenden Kranckheiten.* . . . (Strassburg: Dulsecker, 1712); Christian Friedrich Richter, *Die Höchst-Nöthige Erkenntnis des Menschen: Zum Drittenmal vermehret und verbessert heraus gegeben.* (Leipzig: bey J. F. Gleditsch und Sohn, 1710).

[77] Geyer-Kordesch, *Pietismus, Medizin und Aufklärung in Preußen im 18. Jahrhundert*, 118–39.

[78] Johann Daniel Longolius, *Systema stahlianum de vita et morte corporis humani ab incongruis medicorum mechanizantium opprobriis vindicatum.* . . . (Budissae: Literis Richterianis, 1732), xi.

asks. Stahl too is a mechanist in his own way, since he sees physiological dynamics as governed by mechanical laws, though he is aware that he cannot attribute to matter information which it does not possess. If one opts for a complete mechanism, like Stahl's, one must have recourse to an immaterial principle:

> Be careful, for when you are instructed by a true mechanist on true mechanistic philosophy and its proper application to the human body, you will have to admit, whether you like it or not, that an immaterial principle forms the human body. And this is the rational soul.[79]

Longolius's position does not represent an orthodox Stahlism, however. In his commentary on a biblical passage, he does what the author of *Theoria medica vera* had always refused to do: he places the soul under divine guidance:

> I must nevertheless admit that the soul as such, that is, the soul as giving life to its body, also gives form and produces under Divine guidance [*sub tutela divina etiam formantem et efformantem*]. Since a soul which doesn't animate [*anima non animans*] is a contradiction in terms.[80]

This is not the only time that Longolius takes such interpretative licence. Moreover, the object of his polemic is not always perfectly clear: he is opposed to Burgmann's theory of generation, yet his objections are leveled against a "mechanicism without a soul" rather than against preformation theory in itself. Indeed, after discussing certain biblical passages, he even admits the plausibility of preexistence, as long as it is inspired by animism:

> Thus all the parts of the fetus preexist in the ovule, as the soul preexists in the paternal loins. They are united in the act of conception. The union could otherwise only take place thanks to tonic motion, which awakens the delineated parts, giving them form and structure. This solution is no less rational than the formation of parts through time.[81]

The position is clearly a long way from Stahl's, though it shares with it an openness toward the various options concerning a material substratum and an intransigence regarding the soul's formative role.

While Burgmann, as a convinced mechanist, tends to deny that the information could be possessed by someone other than God, Longolius seems at times to extend this information to subjects which Stahl had never even mentioned. To counter Burgmann's objection that a fetus cannot have sensual

---

[79] Ibid., 7.
[80] Ibid., 11.
[81] Ibid., 27–28.

ideas, he claims that it behaves like an angel: it can receive these ideas without being able to formulate them independently. Moreover, he favors the attribution of a soul even to plants, and on the authority of the Bible he does not rule out that the stars may be moved by angelic intelligences.[82]

Longolius was not the only follower of Stahl to express his indignation. An occurrence in 1732 caused a stir in the German provinces: a woman, frightened by the sight of a monkey, had given birth to a child with anomalous features, similar to those of the animal. Georg Volcmar Hartmann considered the event as evidence in support of his old master Stahl's animist theory, and so the following year he published the *Epistola de bruto ex homine*.

Posse ex homine brutum nasci: a human can give birth to an animal was Hartmann's conclusion.[83] Despite its title, the short work is not a true disquisition on teratological embryology. The birth of a *monstrum* is little more than a pretext to discuss other issues. At the center of Hartmann's interests are the many problems related to preformation theory, and his declared opponent is Burgmann. So in this part of the debate, Stahl becomes the champion of an opposition to preformation, whereas he had actually seemed to accept it at times. Besides, just the year before, Longolius (in Stahl's name) had seemed rather open to the theory himself.

Problems related to monstrous births were one of the recurrent themes in texts hostile to preexistence and preformation.[84] In applying preformation theory, one would have to conclude that a simian seed was somehow present in the woman before she even saw the monkey. According to Burgmann all this would have occurred because of the Divine Will. For Hartmann, on the other hand, the explanation is blasphemous and merits the very accusation of *asylum ignorantiae* that Burgmann had leveled against Stahl himself. The *suppositum mechanicorum* is thus false: "So there is an intelligent cause that can make mistakes. This cause ... isn't God."[85]

9. CONCLUSION: WAS STAHL A PIETIST PHYSICIAN?

Most of the critical literature on Stahl depicts him as a Pietist physician. Stahl's adherence to Spener's spiritual movement is an indisputable fact, though one

[82] Ibid., 39–47.

[83] Georg Volcmar Hartmann, *Epistola de bruto ex homine*....(Erfordiae: Sumptibus C. F. Jungnicolii, 1733), 15.

[84] Roger, *Les sciences de la vie*, 397–418.

[85] Hartmann, *Schediasma apologeticum*, 18.

that still needs to be interpreted.[86] As for the influence of Pietism on Stahl's concept and elaboration of animism, the present writer is still doubtful, despite meticulous scholarship on the subject. The notion of the soul, which dominated Stahl's physiopathology, especially his presumed embryology, was appreciated and upheld by the Pietists. Spener's followers considered it the bulwark of *pietas* against all forms of atheism. They loaded it with meanings which Stahl had never given it. The most devout of his disciples increased the dose: they published commentaries, paraphrases, and works of Stahlian inspiration, full of religious considerations. Thus they transformed the master's doctrine into a *medicina theologica*.[87]

Perhaps, one might conjecture, Stahl simply let things be, realizing that a Pietist interpretation of his system allowed for its diffusion among an ever-wider range of society. With his transfer to Berlin, he stopped writing on medicine, confining himself to reediting his earlier works and having them translated. On the basis of available documents, it is therefore difficult to know how his ideas evolved. Those who speak of Stahl as the founder of a Pietist medical system or theory of generation quote not his works but those of his pupils. In the literature on the subject, passages from *Theoria medica vera* are, at best, quoted in the incomplete and corrupted versions by Ruf and Ideler, who often leave the reader to infer that the text, presented by them in an abbreviated form, contains further reflections on the soul in the sections not translated. Yet these "reflections" are nowhere to be found in the Latin edition. In this perspective, the *affaire Stahl* tells the story of a text, *Theoria medica vera*, which was badly written and translated worse. But we will not end here.

What makes the intelligence of Stahl's work difficult to grasp is above all the lack of an explanation of soul as he conceives it. This soul has little to do with the spiritual being of the theologian, for they share scarcely more than a name. Stahl's is a minor vegetative and animal soul which finds itself endowed with intellective capacities by accident. It is not the Cartesian *res cogitans*; if it occasionally seems so, then this is due more to the author's inability to develop his system coherently than to a deliberate choice on his part. Those who would like to associate it with Descartes's *homme* should look not to his *res cogitans* but to his spirits; it is their functions which are

---

[86] Jürgen Helm, "Das Medizinkonzept Georg Ernst Stahls und seine Rezeption im Halleschen Pietismus und in der Zeit der Romantik." *Berichte zur Wissenschaftsgeschichte* 23 (2000): 167–90.

[87] Michael Alberti, *Specimen medicinae theologicae* [...] (Halle: Hendel, 1726).

carried out by Stahl's soul. Pietism would thus explain the reception of Stahl's system, not its formation.

Stahl never elaborated a complete theory of generation. It was his disciples who conceived of one and attributed it to their master, often more for ethico-religious than for scientific reasons. This fact helped forge the image of a Pietist Stahl, *adversus atheos* theorist of the soul's power, that divine architect of the human body.

13

Charles Bonnet's Neo-Leibnizian Theory
of Organic Bodies

FRANÇOIS DUCHESNEAU

The Genevan naturalist Charles Bonnet (1720–93) occupies a central place in the history of theories of generation. In his *Considérations sur les corps organisés* (1762), as in his later works, he strove to establish a coherent, empirically grounded theory of generation capable of overturning the epigenetic hypotheses stemming from the natural philosophies of Pierre-Louis Moreau de Maupertuis (1698–1759), Georges-Louis Leclerc de Buffon (1709–88), and John Turberville Needham (1713–81). In order to achieve this task, he benefited from the experimental discoveries and analyses of his friend, the illustrious Swiss physiologist Albrecht von Haller (1708–77). After long doubt as to what position to adopt, Haller firmly committed himself to the hypothesis of the preexistence of the preformed organism in the egg, which was to account for all the phenomena involved in the embryo's fertilization and development.[1] He was particularly concerned with disproving epigenetic explanations that seemed to follow from empirical arguments, such as those set forth by Caspar Friedrich Wolff (1733–94) in his *Theoria generationis* (1759) and in his *Theorie von der Generation* (1764).[2] Still, as a partisan of Newtonian methodology, Haller hesitated to use his research to create a theoretical system that could guide the interpretation of specific processes taking place beyond the field of visible appearances. This was precisely the task which Bonnet set himself in a "philosophical spirit, [consisting] principally in the *analysis* of facts, in the judgment of these facts, in their comparison, in the art of determining their consequences, of linking them together, and of

---

[1] Albrecht von Haller, *Commentarius de formatione cordis in ovo incubato*, ed. M. T. Monti (Basel: Schwabe & Co. AG, 2000); Maria Theresa Monti, *Congettura ed esperienze nella fisiologia di Haller* (Florence: Leo S. Olschki Editore, 1990); Amor Cherni, *L'Épistémologie de la transparence: Sur l'embryologie de A. von Haller* (Paris: Vrin, 1998).

[2] Shirley Roe, *Matter, Life and Generation: Eighteenth-Century Embryology and the Haller-Wolff Debate* (Cambridge: Cambridge University Press, 1981).

revealing in this way the principles that naturally result from the best observed facts."[3]

One significant aspect of Bonnet's approach lies in his appeal to a model of organic bodies that he ties to Leibniz's conceptual inventions in the representation of the living. The most recent studies of Leibniz's natural philosophy, notably my own contributions to the Leibnizian theory of organisms, organic bodies, and generation,[4] have raised the issue of the influence of the Leibnizian methodology and of its analytic or synthetic schemes on the subsequent development of the life sciences. My objective here is to locate and describe the intellectual evolution that drove Bonnet to integrate Leibnizian explanatory concepts and models within a significantly different life science. In his work after the *Considérations*, Bonnet made increasingly frequent reference to the similarities and differences between Leibniz's views and his own. This is the case when he treats the serial representation of the chain of being in *La contemplation de la nature* (1765–66), or when he theorizes, in *La palingénésie philosophique* (1769–70), about the transformation of organisms from birth to death and the destiny of corpuscular beings at the end of the physiological process. Keeping in mind the principal elements of the Leibnizian model of organized bodies, I will attempt to reveal the concepts specific to Bonnet's theory of generation and of the metamorphosis of living things, first in the *Considérations*, then in *La palingénésie*. This theoretical construction offers a paradigmatic representation of the organic order which is worth piecing together and analyzing.

1. BONNET'S RECEPTION OF LEIBNIZ'S MODEL

We may ask what Charles Bonnet knew about, incorporated from, and rejected in this Leibnizian model of the organism.[5] In his day, Bonnet could only consult part of the Leibnizian corpus. As he himself admitted, his principal source of information on Leibniz's philosophy was *Essais de théodicée* (1710), where the theory of the organism is given only incidental treatment. Bonnet also refers to the copy of a letter addressed to Hermann and attributed

---

[3] Charles Bonnet, *La Palingénésie philosophique*, in *Oeuvres d' histoire naturelle et de philosophie*, 18 vols. (Neuchatel: Samuel Fauche, 1779–83), ii iv, O XV: 203.

[4] François Duchesneau, *La physiologie des Lumières: empirisme, modèles et théories* (The Hague: Martinus Nijhoff, 1982), 65–102; François Duchesneau, *Les modèles du vivant de Descartes à Leibniz* (Paris: Vrin, 1998); François Duchesneau, "Leibniz's Model for Analyzing Organic Phenomena," *Perspectives on Science* 11 (2003) 378–409.

[5] Olivier Rieppel, "The Reception of Leibniz's Philosophy in the Writings of Charles Bonnet (1720–1793)," *Journal of the History of Biology* 21 (1988): 119–45.

to Leibniz (and dated October 16, 1707), which Samuel König had produced in 1751, inciting a quarrel on the principle of least action at the Academy of Berlin. After the publication of *Considérations sur les corps organisés* (1762) and *La contemplation de la nature* (1765–66), it is clear that Bonnet deepened his knowledge of Leibniz's oeuvre. The year 1765 marked Rudolph Eric Raspe's publication of Leibniz's *Nouveaux essais sur l'entendement humain*, and Louis Dutens published the *Opera omnia* in Geneva in 1768. Bonnet may also have had access to some of Leibniz's other texts, including the *Système nouveau de la nature et de la communication des substances* (1695), the *Principes de la philosophie ou Monadologie* (1714) and the *Principes de la nature et de la grâce* (1714), but the traces of these works, especially those which involve the notion of the monad, if they exist, are certainly rare and insufficient to show that Bonnet initially had a profound understanding of Leibniz's theory of the organism. It is essentially with *La palingénésie* that Bonnet's Leibnizianism comes to the fore. But even then, Bonnet asserts the originality of his theses and their autonomy from those held by Leibniz and the neo-Leibnizians. He does so, for instance, during his polemic with Pierre Sigorgne after the anonymous publication of the latter's *Institutions leibnitiennes* in 1768,[6] which came as he was working on *La palingénésie*. His position is also captured in the epistolary exchange between Johann Kaspar Lavater (1741–1801) and Moses Mendelssohn (1729–86) which he followed closely. Everything supports Mendelssohn's ultimate judgment in March 1770:

> I would never want to challenge [the] merit of Mr. Bonnet's [originality]. I would simply like (and all intelligent readers will be convinced of the same connection) to make it clear to Mr. Lavater that the philosophical principles from which Mr. Bonnet begins are no longer new for the Germans; that, after Leibniz, all the monadists, and especially Hansch, Bilfinger, Cantz, Baumgarten have arrived by subtle speculations to the point where the "Palingenesist" has been conducted by means of observations. We cannot blame a man like Mr. Bonnet for never having read the German metaphysicians. Only Leibniz should be familiar to him, and the "Palingenesist" gives all possible credit to that man, who brings glory to Germany. (published in Raymond Savioz, [Paris: Vrin, 1948a])

The Leibniz-Bonnet relationship – which, stemming from a summary knowledge of Leibniz's writings, developed and grew over time – concerns certain themes of biological interest: the definition of the organic body, the hypothesis of the preexistence and preformation of germs, the theoretical sketch of the chain of being (*scala naturae*), and the conception of the mind-body

---

[6] Pierre Sigorgne, *Institutions leibnitiennes ou Précis de la monadologie* (Lyon: Frères Périsse, 1768).

relationship in mixed beings. To this network of ideas are tied speculations about what can be called a theory of vital organization *stricto sensu*. Here we are more interested in the positive aspects of this theory than in its speculative aspects, limiting our inquiry especially to the themes of individual organization and the genesis of organic bodies and considering, first, the state of these issues in 1762.

The *Considérations sur les corps organisés* attempts to present a "type of logic" which would analytically develop hypotheses inferred from the facts to act as explanatory principles.[7] The main issue concerns the generation, origin, and development of organized bodies. Unquestionably, the method of analysis is intended to be empiricist: in no place does it attempt to deal directly with concepts going beyond observational analogies, based on a network of rationally derived conditions of intelligibility. Such a framework may play a part, no doubt, in theory building, but in a rather implicit and partial way. However, some premises reveal a vein that one might be justified in considering "Leibnizian." In this respect one could mention the architectonic principle of continuity to which Bonnet often resorts in his later work. But there are also two other Leibnizian epistemological suppositions that one could consider primordial in this instance: one deals with the complex order of organisms, an order that cannot be reduced by any mechanistic analysis; the other involves the infinite, or at least indeterminate, combinatorial analysis of organicity in nature. The first premise brings about the return of the distinction between "machines of nature" and "machines of art" which underlay Leibniz's definition of the organism. The second involves the *emboîtement* of organic structures in one another in an analytic regression and accounts for the preexistence of germs and for the constitution of individual organisms: it develops in close affinity with Leibniz's definition of organic bodies and with the preformationist approach to the phenomena of generation.

But these three presuppositions which we find in Bonnet's analysis, notwithstanding their apparent similarities, reveal deep differences that suggest a selective and inventive, or deviant, assimilation of Leibnizian theory. Bonnet in no way grasped the whole theory, and he could only suspect its sophisticated methodology. Moreover, had Bonnet himself not claimed that, if reading Leibniz's *Essais de théodicée* in 1748 led him to outline the program for all of his later theoretical investigations, the monadological conception discouraged him so deeply that he did not integrate its elements with his philosophy until late in life (Savioz 1948b 100–1)? Still, these similarities and

---

[7] Charles Bonnet, preface to *Considérations sur les corps organisés* (Paris: Fayard, 1985), 19.

differences are very instructive and justify a closer look at the suppositions that tie the *Considérations* to the Leibnizian heritage.

Bonnet invokes the principle of continuity at length in a framework of reflection on the vegetative modes of reproduction that characterize polyps.[8] He insists on the nonanalogical characteristic of these modes, if one refers to the normative representation according to which all organic animal systems reproduce from a sufficiently integral and integrated state of the original organisms. Following Abraham Trembley's (1710–84) observations, the reproduction of the polyp from a very incomplete portion of its body suggested a system of order internal to the living animal situated at the frontier of a vegetable state.[9] From this came the demand for a gradualist extension of the notion of an organic system to represent forms which become more complex and differentiate themselves through continuous transitions. The general law of organic life, one might say, should take into account the continuous series of modes of reproduction that experience of metamorphosis and regeneration reveals. When formulating his principle "Nature does not move in leaps" (*Natura non facit saltus*), Leibniz anticipates just this idea of a serial sequence of living forms in hierarchical progression. Leibniz's presumed letter which König revealed to the world testifies to this.[10]

But how did Bonnet interpret that text? Essentially, he came away with the thesis of a continuous order of specific living forms placed in a hierarchy of complexity and perfection. We can conceive this order empirically insofar as we are able to list the diverse intermediate forms between the kingdoms. The polyps establish such a transition in the natural class of the zoophytes. But the problem resumes with the transition of mineral to vegetal forms. In this connection, Louis Bourguet's analysis of organized crystalline structures[11] to which Bonnet refers seems flawed, since elementary crystals fail

---

[8] Bonnet, *Considérations sur les corps organisés*, §209, 177–9.

[9] Abraham Trembley, *Mémoires pour servir à l'histoire d'un genre de polypes d'eau douces, en bras en forme de cornes* (Leiden: Jean & Herman Verbeek, 1744); Virginia Dawson, *Nature's Enigma: The Problem of the Polyp in the Letters of Bonnet, Trembley, and Réaumur* (Philadelphia: American Philosophical Society, 1987).

[10] Leonhard Euler, *Commentationes mechanicae*, in *Opera omnia*, II 5 (Lausannae: Auctoritate et impensis Societatis scientiarium naturalium helveticae, 1957), 265–6. Certain of the most authoritative specialists tend to consider this letter authentic, but many of Bonnet's contemporaries, including Leonhard Euler, thought it was a forgery (Pierre Costabel, "L'Affaire Maupertuis-Koenig et les 'questions de fait': Arithmos-Arrythmos," in *Skizzen aus der Wissenschaftsgeschichte*, ed. K. Figala and E. Berniger, [Munich: Minerva Publikation, 1979].)

[11] Louis Bourguet, *Lettres philosophiques sur la formation des sels et des crystaux et sur la generation et le méchanisme organique des plantes et des animaux.* (Amsterdam: François L'Honoré, 1729); Olivier Riepel, "Organization in the Lettres Philosophiques of Louis Bourguet Compared to the writings of Charles Bonnet," *Gesnerus* 44 (1987): 125–32; François Duchesneau, "Louis

to correspond to animal and vegetable germs: only these imply a preformed organism subject to development by intussusception.[12] Bonnet holds, however, that the organic forms which we classify as different species according to our limited comprehension of the connections between resembling individuals should be seen as composing a graduated table of living structures. This table is all the more accurate, since one can compare the organisms on the basis of the modes of composition and development of their germs. To sum up, the principle of continuity, according to Bonnet, should act as a principle of order for the analysis of perceptible appearances: it is not invoked to act as an instrument for passing to a rational symbolic order beyond the confused level of empirical concepts. Above all, it could not help to unveil the conditions of intelligibility for complex organizations which present apparent discontinuities at the level of empirical analysis. But this was exactly the role the principle played in Leibnizian science. In the famous controversial letter of 1707, Leibniz presented its usage in this way for the determination of natural classes:

I believe therefore that there are good reasons to believe that all the different classes of beings, which together form the universe, are in God's ideas, who clearly knows their essential gradations, but like so many coordinates of the same curve, whose union does not permit that others be inserted between any two of them, since this would mark disorder and imperfection. Men are related then to animals, animals to plants, and plants to fossils, which are linked in turn to bodies that the senses and the imagination cause us to see as entirely dead and unformed. But since the law of continuity demands that, *when the essential determinations of a being approach those of another, . . . then in consequence all the properties of the first should gradually approach those of the other*, it is necessary that all the orders of natural beings form a single chain, in which the different classes, like so many links, are so closely connected that it is impossible for the senses and the imagination to precisely determine the point where one begins, or ends: all species which border, or occupy what might be dubbed regions of inflection and return, must be equivocal and endowed with characteristics that can relate them to their neighboring species. . . . [From this comes the reversal] of common rules, built on the supposition of a perfect, absolute separation of the different orders of simultaneous beings which fill the universe.[13]

Bourguet et le modèle des corps organiques," in *L'edizione del testo scientifico d'età moderna*, ed. M. T. Monti (Florence: Leo S. Olschki Editore), 3–31.

[12] Bonnet, *Considérations sur les corps organisés*, §210, 181; Bernardino Fantini, "Le crystal comme métaphore de la vie," in *Charles Bonnet savant et philosophe (1720–1793)*, ed. M. Buscaglia et al. (Geneva: Éditions Passé Présent, 1994).

[13] Euler, *Commentationes mechanicae*, in 265–6.

According to Leibniz, the complete understanding of species, if it were possible, would be in terms of coordinates representing all the properties of each type of organism.[14] This conjunction of properties would be differentially distinct from the conjunction of properties defining the closest neighboring species. In a certain way, the second conjunction can be understood as the limit of what determines the first; at the same time, though, it constitutes a heterogenic determination in relation to the first. All effectively realized organic types would express the same basic similarity from the point of view of their living structures and functions, but this similarity involves degrees, and these degrees correspond to the characteristic, yet distinct, neighboring positions on the curve and therefore to forms of organization and modes of integration of properties differentially distinct from each other. Considering only phenomena limits us to the construction of hypothetical models: these are based on the choice of characteristics that permit the formation of a possible graduated series of structural and functional determinations. But the real differences between species can only depend on an integral concept implying an infinite number of such properties. This is why it is necessary to conceive the proper internal constitution of a type of organism as a sort of matrix permitting the combination of diverse series of structural and functional determinations. The chain of classes of beings appears then like a curve that integrates these combinatorial determinations forming matrices for the apparent development of organic structures and functions.

The second Leibnizian postulate in the *Considérations* involves the complex order of organisms, which cannot be dealt with by means of a strictly mechanical analysis. In many places in the text, Bonnet uses Leibniz's distinction between "machines of nature" and "machines of art" to emphasize that the organization of the primordial living beings escapes the possibility of any epigenetic conceptualization. Of course, one must be able to explain the processes of germ transformation mechanically, and even the ensemble of processes characterizing the development of organic structures, yet what escapes mechanical explanation altogether is vital organization itself, both for the harmonic, structural combination that it implies and for the diverse functions it is destined to perform. But Bonnet's thesis boils down to the assumption that an initial extra-mechanical action inscribes an immanent design of functionality within the hypersubtle organic machines. From this the need arises to represent minimal structures as an integrated and functional organization which must be reconciled with the analysis of the mechanisms through

---

[14] François Duchesneau, *Leibniz et la méthode de la science* (Paris: Presses Universitaires de France, 1993), 372–4.

which this representation unfolds and actualizes itself. The first chapter of the *Considérations* is devoted to "germs as principles of organized bodies" and poses the possibility of two preformationist hypotheses, that of *emboîtement* and that of dissemination, based on the premise that it appears impossible to "mechanically explain the formation of organized beings."[15] The only issue, then, is to presuppose the embryonic preexistence of a germ or an organized corpuscle. The description of germs, especially according to the hypothesis of *emboîtement*, seems conceptually dependent on the Leibnizian tradition in three respects: the analogical resort to infinitesimals, the appeal to a priori intellection beyond mere imagination, and the conception of a world of organized beings in serial *emboîtement* or envelopment in the least germs. The hypothesis of dissemination, whose modern inspiration derives from Claude Perrault (1608–80),[16] does not seem to involve specific Leibnizian traits except for the fact that Bonnet emphasizes the germ's all-embracing organic character and provides empirical reasons in favor of its indestructibility, since the germ's tiny size allows it to avoid the dissolution affecting mixed substances.

In general as well as in Bonnet's case, the principle of continuity implies the principle of sufficient reason: "Nature does not move in leaps. Everything has its sufficient reason or its proximate and immediate cause."[17] For Bonnet, the principle justifies a twofold analytic approach of apparently Leibnizian allegiance. First of all, this approach concerns development, that is to say, a process of growth affecting the minimal structure of the organism and producing observable transformations. Bonnet assumes that the mechanism of growth is inaccessible. It is useful to substitute for this unattainable causal explanation a description of phases of evolution that form the links of a continuous serial chain. The strategy of analysis involves establishing the link between the germs as outlines or schemata of the organized bodies and the empirical description of the resulting organisms. If we were to presume that the outline is very dissimilar and of a frankly inferior order to the organism that emerges, it would be necessary to resort to mechanistic models of partial epigenesis which are unable to account for the emergent organization. The mechanistic explanation turns out to be inadequate to account for the characteristics of organic integration and functionality. This leads to a solution that

---

[15] Bonnet, *Considérations sur les corps organisés*, §1, 21.

[16] Jacques Roger, *Les sciences de la vie dans la pensée française au XVIIIe siècle* (Paris: A. Colin, 1963 [2nd ed. 1971]), 339–44.

[17] Bonnet, *Considérations sur les corps organisés*, §6, 23.

consists in attributing to the germ the essential organization of the resulting organism:

> It is said that the germ is an outline or sketch of the organized body. This notion can be insufficiently precise. Either it is necessary to attempt to mechanically explain the formation of organs, which good philosophy recognizes to be beyond its scope; or it must be admitted that the germ actually contains a miniature version of all the essential parts of the plant or the animal it represents.[18]

Obviously, Leibniz held a stronger version of the methodological maxim: "Everything occurs mechanically in nature" (*Omnia fieri mechanice in natura*). For him, no limitation should affect the formulation of mechanistic explications of vital phenomena in principle. He presupposes, moreover, the integral correspondence of teleological reasons and causal explanations in the analysis of physiological processes. Of course, the method of functional description is more immediate and less profound, and it should only be provisionally substituted for the explanation of the enveloped micromechanisms. As the correspondence between Leibniz and Bourguet well illustrates, Leibniz tends also to discern a considerable disparity of order between the seminal organic bodies and the bodies which these seminal bodies become as a result of drastic transformation. Of course, the continuity of simple and complex forms, from the embryonic to the developed, is the absolute rule, but the curve which would symbolize the transformation of the germ into a complex organism could not result from a simple linear progression alone from the imperceptible to the perceptible; on the contrary, it is necessary to assume it implies major inflections, transitions at the limit, and other equally remarkable transformations. This opens up the possibility of admitting models of partial epigenesis into the framework of a theory that only presupposes a deployment of infinite organization in the production of animated bodies. According to Bonnet, at least in the period preceding *La palingénésie philosophique*, the continualist schema implies the extensive preexistence of the organism in the corpuscle; according to Leibniz, it was more a question of the dynamic preexistence of a structuring disposition, of a tendency toward organization and complexity, and so of the preexistence of a design reflected in the microsystem of points of force correlated with and interacting under the form of organic bodies.

The third Leibnizian premise in the *Considérations* involves the infinite combination of the conditions of organicity. By presenting the thesis of the

---

[18] Ibid., §35, 33.

emboîtement of germs, which he considers the most plausible explanation of generation, Bonnet sustains a notion of the infinite "envelopment" of organic structures:

> The different orders of the infinitely small merged within one another, which this hypothesis supposes, overwhelm the imagination without repulsing reason. Accustomed to distinguishing that which springs from the understanding from that which comes from the senses, reason envisages with pleasure the grain of a plant or the egg of an animal as a tiny world inhabited by a multitude of organized beings, succeeding one another through the centuries.[19]

Typically Leibnizian, this representation gains support from the double thesis that the systems of bodies are actually infinitely divided and that each animalcule belonging to such a system integrates a world of other bodies and therefore of still more elementary living beings: "An animal," Bonnet affirms, "is a world inhabited by other animals; these are in turn their own worlds, and we have no idea where this ends."[20] "It is with pleasure," he adds later, "that I contemplate this magnificent series of organized beings, enclosed like so many tiny worlds within each other."[21] But the hypothesis is immediately modified in a less Leibnizian sense: it is agreed that there occurs a gradual passage of the most embryonic forms of the individual organism to its developed forms, but it is affirmed that the transition could not reduce to a series of mechanical transformations. A more Leibnizian physiologist would have attempted to determine a law for this series of transformations, employing diverse techniques of approximation. Bonnet opts instead for the hypothesis of a preordained minimal structure which he calls the "secret mechanism of growth."[22] Influenced by the physiology of Haller, this hypothesis involves a collection of simple fibers capable of integrating the particles conveyed by the vital fluids through their special material disposition as "types of netlike frameworks" (*espèce[s] d'ouvrage[s] à réseau*).[23] The processes of assimilation and elimination in the fibril networks are invoked in this way to allow the growth, preservation, and diversified functioning of the plant or animal structure. According to Bonnet, the realities of the physical universe result from combinations of diverse elements endowed with essential or primordial properties (extension, solidity, inertial force) and with modal properties that

[19] Ibid., §3, 21.
[20] Ibid., §72, 54.
[21] Ibid., §128, 88.
[22] Ibid., §15, 25 n. 1.
[23] Ibid., §14, 25.

imply variations of shape, proportion, and quality.[24] The primary or inorganic elements of an atomic nature combine to form the secondary or organic elements that represent invariable and everlasting combinations: these include the organic corpuscles or germs through which development, itself impossible to analyze mechanically, produces the visible organisms. The organization of the germ as a basic framework of fibers determines the nutrition, growth, and reproduction that characterize all organic beings. From this perspective, there is no indefinite analytic regression of the conditions of organicity in the composition of an individual organism but rather a conceivable reduction of the complex organism to a primordial germ state that would represent the minimal organic combination. In these conditions, the individual machines of nature, due to the corpuscular combination from which they derive, can in no way display the infinite organization that the Leibnizian perspective requires. The notion of organized bodies which Bonnet adopts entails that they are finite in their respective compositions, and this contradicts Leibniz's model, even if in other places the author of the *Considérations* seems to come closer to it in the way in which he conceives the serial *emboîtement* of germs or organic corpuscles. But the apparently infinite combinatorial analysis of the generative processes rests in fact on an organization of individual living beings which ought to be analyzable into fibril structures, themselves dependent on networks of corpuscles. We find here a convincing indication of the fact that Bonnet places his neo-Leibnizian analytic principles in a general empirical framework conforming to a Newtonian-inspired physics and, in this way, significantly modifies the Leibnizian concepts of organism and organic body. These modifications reach their full expression with *La palingénésie philosophique*, in which Bonnet develops *in extenso* his "Leibnizian" model of the living being.

2. ORGANIC ORDER IN *LA PALINGÉNÉSIE*

The principal thesis of *La palingénésie* concerns the belief that the sufficient reason for vital phenomena does not end with the rough organic body constituting the animal. Of course, this coarse body which observation reveals to us, along with its structures and functional properties, depends on the transformations which the germ constituting its organic preformation undergoes. A tiny indestructible organic body corresponds to the germ, acting as the seat of a soul capable of sensitive and volitional operations of different degrees and

---

[24] Ibid., §§130–2, 61–3.

providing a foundation for the animal's later development. The substantial mix formed by the soul and the tiny organic body constitutes the animal's "self" (*personne*). This self is capable of subsisting beyond the present state and conserving the organic and psychic properties which pertain to it. Bonnet himself goes so far as to presume a possible heightening of these properties in a future world state resulting from the resurrection of animated organisms. In all, these tiny indestructible bodies are conceived as "organic points," playing the role of "germs of restitution" (*germes de restitution*). These produce organic combinations endowed with sense and relatively more perfect limbs in the future state,[25] but after metamorphoses which preserve and prolong the essential organism underlying the actual state. This perfectibility is therefore inscribed *ab origine* in the substantial compound of the soul and the tiny indestructible body, which is revealed in a transitory mode while the germ evolves into the developed organism and which could be revealed in a more permanent mode in a future world. In the present state, the organic body is capable of reproduction insofar as the purpose of the germ is the conservation of the species. It is possible to imagine a heightened state of perfection where the organisms bearing individual selves would no longer be subject to the challenges of reproduction and would persist beyond the actual limits of generation and death.

Bonnet certainly invokes Leibnizian preestablished harmony to support this hypothesis beyond the teachings that follow from theological revelation. From this premise he infers a cosmogony which supposes a serial linking of global transformations or revolutions affecting the composition of our world in its successive states. The idea suggested is that the law of preestablished harmony implies the most integral correspondence of the concomitant and successive stages of beings which compose the created world:

> The existence and particular determinations of each being are always in relation to the existence and determination of corresponding or neighboring beings. The present has been determined by the past, the subsequent by the antecedent. The present determines the future. The *universal* harmony is in this way the *result* of all the *particular* harmonies of *coexisting* and *successive* beings.[26]

The particular harmonies insert themselves in a double scale of perfection, bodily and intellectual, and they come in a variety of forms involving degrees and nuances that spread out to infinity. If we suppose that revolutions affect

---

25 Bonnet, *La palingénésie philosophique*, I iv, O XV: 186.
26 Ibid., VI ii, O XV: 262.

the earth's economy, we must imagine that the organic kingdom has undergone changes corresponding to its *Umwelt*. Hence no doubt organized beings underwent and will undergo changes in form and structure that make them diverge from their first state; but whatever they are, the revolutions of the organic world will not break its unity and harmonic combination of states. "The organic kingdom could therefore . . . have constantly conserved this sort of *unity* that makes each species a unique and always surviving whole which takes from time to time new forms and new modalities."[27] From this harmonic hypothesis which applies as much to substantial mixes as to the organic bodies that express them in the order of natural phenomena, Bonnet infers a genuine architectonic of primordial germs called to unfurl in actualized germs, which themselves will develop into accomplished organisms. "Who could deny," Bonnet asks, "that the Absolute Power could contain in the first germ of each organized being that rest of the germs corresponding to the diverse revolutions that our planet would undergo?"[28] This, however, is a contingent vision of the serial expression of germs in their successive development. More profoundly, a combination of all the existing germs capable of actualizing themselves in given circumstances is posed *ab origine* following the scheme of harmony. "I therefore believe," Bonnet affirms, "that the *germs* of all organized beings have been originally constructed and calculated in terms of *determined relationships* to the diverse *revolutions* that our planet must undergo."[29]

In the spirit of *La contemplation de la nature*, Bonnet insists on the scalar differentiation of species and on the relative perfection of the organic system which characterizes individuals belonging to each of these species. The complexity of organization and cerebral functions is without doubt the most convincing evidence of the rank in the natural order that the diverse substantial compounds of soul and organic unit occupy. But it is still necessary to hold that a complex organic integration underlies the different intellectual properties of the types of animals: every animal is in effect "a particular system, in which all its parts are related or in harmony."[30] Under these conditions, the tiny body seems to hold "an abridgment of a compound organic system,"[31] analogous to that of the developed organic body in its actual state. Regarding the chain of beings, Bonnet reminds us that the determination of species is

---

[27] Ibid., VI v, O XV: 277–8.
[28] Ibid., 283.
[29] Ibid., 272–3.
[30] Ibid., VI IV iv, O XV: 209.
[31] Ibid., III i, O XV: 213.

based on identifying the characteristics that define "nominal essences," following Locke's terminology. Fixing species in this way is a first step toward conceiving taxonomies with greater generality than the simple hierarchy of species. Confronting the results of this intellectual exercise with the more fluent divisions of nature allows us to conceive the possibility of adjoining species (*espèces mitoyennes*), which renders our methodological classifications more relative. The architectonic principle of continuity suggests the hypothesis of an order of graduated differentiation of possible organic systems. Through an a posteriori approach based on the observation of facts and empirical inference, the naturalist reinforces this metaphysical representation of nature's order situated at the heart of Leibnizian analysis. If we conceive through metaphysical extrapolation that, in some later phase of our world, the tiny animated bodies can occasion, through the modulated expression of undying germs, the development of organic beings with enhanced properties, the particular and more perfect organic systems should reflect the same law of continuous gradation, term by term: "so that all the degrees of the ladder are continually variable in a determined and constant proportion: that is to say that the mutability of each degree will always have its reason in the degree which has immediately preceded it."[32] In the same vein of ideas, Bonnet takes into consideration the germs which, in the present world, remain unemployed and do not undergo full and complete organic expression but whose tiny indestructible bodies subsist indefinitely and contribute in their subsequent expression to a general organic system analogous to the actual one, though, he presumes, with greater richness: "Nothing is lost in the immense stores of Nature; everything has its use, its end, and the best possible end."[33]

In the theoretical framework provided by the hypothesis of the chain of beings, the analogy between animals and plants should be analyzed by taking into account the generic differences which experience reveals. These differences testify to the complexity of facts at the same time that they reveal their insertion in the same grid of analysis. Bonnet in this way gives credence to the possibility that plants possess sensibility in addition to the structural and functional similarities that link them to organic systems reflecting animality. But sensibility depends not only on a material organization but also on the presence of a soul. Plants, in their essence, can only be seen as mixed beings, just like animals. Still, this analogical link, based on the similarities of

[32] Ibid., III iii, O XV: 220.
[33] Ibid., III iv, O XV: 223.

functional properties, should not hide the differences between the organic systems involved. Among these differences is the fact that the animal is defined as an individual whole, while the plant more easily appears as a sum of individual wholes, as an "organic society,"[34] This feature signifies a particular constitution of germs representing the capacity of sets of individuals and therefore of the underlying substantial mixes for reproduction. Under these conditions, the unity of the general system comprising such diversified particular systems can only rest on the identity of a developmental law: this law is a law of "evolution" and entails the unfolding of all preformed structures so that they can conform to a germ pattern capable of supporting the implied analogy of the diverse organic systems. Bonnet employs the class of zoophytes to this end. The paradigmatic cases in this instance are the polyps and the "animalcules of infusions." For these animals at the frontier of animality, development seems to derive from a specific vegetative force working in and on relatively unstructured organic matter, and this could have lent support to the hypotheses of Buffon and Needham, whom Bonnet designates as "epigenesists." The analysis proposed to counter these hypotheses relies on an interpretation of experiments by Trembley and his followers identifying germs not as nested forms of structurally and functionally complex animals but as simple "organic preformations": "I took the word *germ* in a much looser sense, for all *organic preformation* from which a *polyp* can result as from its *immediate principle*."[35] This structurally simplified meaning of "germ," in contrast to the habitual norms of animal complexity, joins up with the assignation of a functional property of irritability that may explain the effects of the vegetative forces in operation. Beyond these structural and functional features of the germ that warrant the integration of the zoophytes to the hierarchy of beings, the metaphysical model of the tiny indestructible bodies applies to "insects" like the polyp. Following Lazzaro Spallanzani's (1729–99) experiments on the animalcules of infusions, the idea of a mode of organization and of reproduction analogous to that of the polyps applies, at an inferior level of the hierarchy, to the micro-organisms which the supporters of epigenesis, like Needham, believed were formed through a process of vegetation from the simple organic corpuscles.[36]

---

[34] Ibid., IV iii, O XV: 241.

[35] Ibid., V ii, O XV: 249.

[36] Lazzaro Spallanzani, *Nouvelles recherches sur les découvertes microscopiques et la generation des corps organisés. . . . avec des notes. . . . par M. de Needham* (London and Paris: Lacombe, 1769); Marta Stefani, *Corruzione e generazione: John T. Needham e l'origine del vivente* (Florence: Leo S. Olschki Editore, 2002).

François Duchesneau

3. THE MACHINES OF NATURE RESHAPED

Bonnet constructs his theory of organized bodies following a model not too far removed from Leibniz's, but he integrates considerations based on recent observations and experiments of contemporary naturalists. He also integrates theoretical propositions, especially from Haller's *Elementa physiologiae corporis humani* (1759–66). For him, it was a matter of assuring the compatibility and coherence of these diverse sources. Speculating on the enhanced knowledge which a transcendental intelligence could claim, Bonnet resorted to a comparison which identifies *a contrario* the two most essential sources for the science of vital phenomena that, according to him, finite understandings could construct: "While a Leibniz tries to reveal *universal* harmony or a Haller attempts to penetrate the mysteries of organization, these [other supramundane] intelligences smile, and only see in these great philosophers talented Hottentots who attempt to discover the secret of a watch."[37] This metaphor suggests that the theory of organized bodies to which Bonnet, as a finite mind, aspires would combine elements of Hallerian physiology with Leibnizian principles.

Bonnet holds that living beings are "organic machines," which he represents by employing the contrast which Leibniz had made between "machines of nature" and "machines of art."[38] The organized bodies are, in effect, machines formed by a prodigious number of smaller machines (*machinules*) whose actions "conspire," the model of these tiny machines being provided by the fiber, which is a "very compound small organic whole" (*petit tout organique très composé*).[39] The fiber itself is composed of infinitely small machines called fibrils, endowed with their own functions.[40] A distinctive trait of these organic machines is their capacity to compensate for the wear stemming from movement through the integration of new molecules. But this trait is associated with another even more discriminating one, the capacity for development:

> *Development* presupposes a secret and most sophisticated mechanism in the organic whole. As it gradually spreads in all directions, each piece essentially retains the form it had on a very small scale. It is therefore necessary that its *integrative* parts be fashioned and arranged in respect to each other with such an art that they constantly conserve the same relationships, the same proportions,

[37] Bonnet, *La Palingénésie philosophique*, XII viii, O XVI: 29.
[38] Ibid., IX ii, O XV: 351–2.
[39] Ibid., 354–5.
[40] Ibid., 350.

300

the same agency, at the same time that the new integrative particles combine with the old ones.[41]

Bonnet refers to contemporary physiology, dominated by Haller's work, in order to explain the composition of integrated organic machines capable of assimilation and growth. He spots in the structure of secretory organs a classical example of machines organized for assimilation and development in an organism functioning for the preservation of the whole. The emphasis is, however, on the fibers endowed with the functional property of irritability as well as the fibers which exert sensibility, since irritability and sensibility are the two fundamental properties of organic elements or systems in Haller's physiology. These properties, which occur in the tiny machines inside the organic machine, cannot be reduced to properties of inorganic compounds. In Hallerian physiology, they represent functional effects strictly linked to combinations of living microparts.[42]

Along with these essential properties of the living fiber which provide the capacity for dynamic preservation and development we find the organic machines' aptitude for reproduction, perpetuating their specific model of configuration.[43] Rediscovering Louis Bourguet's Leibnizian concept of "organic mechanism,"[44] Bonnet seems for an instant to transcend the limits of preformationism in determining the type of replication specific to individual organic systems:

> Whatever the manner in which this reproduction of living beings operates, whichever system we embrace in order to explain it, it will not appear less admirable to those who at least perceive the prodigious art required in its organization and in the diverse means by which it is executed in the *plant* and the *animal*, and in each of the different species of the two kingdoms. In this way, whether this reproduction depends on preexistent germs or, if one prefers, forms daily in the individual *procreating* similar entities, the conservation of the species in either hypothesis will be nonetheless one of the most beautiful aspects of the *organic machines'* perfection. And if it were possible that the sole laws of this mechanism sufficed to form new individual wholes, things would appear even more admirable.[45]

---

[41] Ibid., 352.

[42] Duchesneau, *La physiologie des Lumières*, 141–70.

[43] Bonnet, *La Palingénésie philosophique*, IX iii, O XV: 56.

[44] Bourguet, *Lettres philosophiques sur la formation des sels et des crystaux et sur la generation et le méchanisme organique des plantes et des animaux*; Duchesneau, "Louis Bourguet et le modèle des corps organiques."

[45] Bonnet, *La Palingénésie philosophique*, IX iii, O XV: 356–7.

On this point, it is useful to remember Bonnet's firm adherence to the thesis of the preformation of organisms in the form of germs and the preexistence of germs from the initial creation. Between the two conceivable modalities of preexistence – dissemination and *emboîtement* – which appear in the first chapters of the *Considérations*, Bonnet prefers the second modality, particularly for the analysis of generation in complex organisms formed by multiple dissimilar parts. Apparently, the consideration of zoophytes, whose modes of reproduction approach those of the vegetable kingdom (as indicated by Trembley's observations of polyps and Spallanzani's observations of spermatozoa and of the "animalcules of infusions"), encourages him to conceive another type of preformation, one which implies the autonomization and development of new organisms from similar parts integrated in the organic systems which, following the example of the polyps, appear to be "all ovaries."[46]

As all living beings are organic machines, they require preformation, since, following the thesis presented in the *Considérations*, no conceivable mode of mechanical construction can explain the genesis of such harmonic and functionally integrated complex systems: "in the actual order of our physical knowledge, we find no reasonable means of mechanically explaining the formation of an animal, nor of the least organ."[47] From this comes the crucial declaration: "all organic wholes are originally preformed, and those of the same species have been contained in one another, and therefore develop from one another; the small from the large; the invisible from the visible."[48] It is far from the case, however, that this principle applies in a univocal manner: "I have in no way assumed," Bonnet affirms, "that this *preformation* was *identical* in all species."[49] Bonnet rejoins Leibniz, who, in his correspondence with Bourguet, specified that chaos – that is to say, an unorganized state of material realities – cannot bring about the formation of any organism whatsoever. It is always necessary to presuppose something already organized which can, in developing and in transforming, induce a structural and functional delineation of the resulting organism in the form of a germ: "Thus, this word [germ] will not only designate an organized body reduced to a small scale; it will also designate all species of original preformation from which an organic whole can result as from its principle."[50] This reiterated formula is capable of including many modes of reproduction under the idea of emergence through the simple development of an organic system. In fact, according to Bonnet,

---

[46] Bonnet, *Considérations sur les corps organisés*, §319, 347.
[47] Bonnet, *La Palingénésie philosophique*, X i, O XV: 386.
[48] Ibid.
[49] Ibid.
[50] Ibid., X ii, O XV: 393.

it may account for the types of regeneration in the formed organism which modulate the principle of preformation. Of relevance here are Spallanzani's observations on the reproduction of the snail's amputated head or the salamander's severed feet. Hence the postulation of "germs which one can call *repairer germs* and which only contain precisely what needs to be replaced."[51] Hence, as well, the hypothesis of a dissemination of these latent germs in the substance of the organs following a preordained pattern fitted to respond to accidents that may affect the organism in question. These repairer germs are endowed with less capacity than the germs of whole organisms. This gives us a hierarchy of germs which, duly analyzed, displays the range of modalities and possible varieties gathered under the general law of development or "evolution": this last term is used in its eighteenth-century sense to signify an unfurling of structures and functional properties encapsulated in the original organic whole.

For a partisan of Hallerian physiology, the most natural starting point for establishing this hierarchy rests in the fiber. Bonnet thus examines the regeneration of the fiber, which he considers "an organic whole, which nourishes, grows, vegetates"[52] through "a very organized small whole much more compound than we imagine."[53] If we amputate the extremity of the fiber, we bring about a derivation of nutritive sap toward the organic points that form its elements. These organic points can be identified as fibrils, secondary elements formed from a multitude of molecules or organically linked first elements. In virtue of their assimilative power, these elements configured in a network undergo an evolution that restores the complete form of the fiber's organic whole without it being necessary to incorporate the development of a germ, properly speaking. The growth of fibrils destined to restore the fiber's integral structure "appears to depend at the last resort on the nature, the number and the respective arrangement of elements, and on a secret relationship of all this to the force which presses the nutritive sap into the knots of the fiber and spreads out its elements."[54] According to Bonnet, this force at the level of the fiber can be assimilated to Hallerian irritability, at least for those organs where it is found; we can imagine that at other levels or in other organic systems it will be conceived along the analogy of irritability but be identified and defined in diverse ways according to the functional effects that express its action. "It has been proved that *irritability* is the *vital principle* in the animal. Irritability

---

[51] Ibid., 392.
[52] Ibid., X iii, O XV: 392.
[53] Ibid., 395.
[54] Ibid., 394–5.

is the true cause of the heart's movements. We are still ignorant of the vital principle of the plant: it may possess many such principles subordinate to one another."[55]

Correlatively, a system is put in place in which organic wholes of diverse orders integrate with one another to create more and more complex networks that reproduce in response to a plurality of physical factors, following the design of their immanent "organic preordination."[56] At least, this vision of things prevails in the case of reproduction of all similar organic wholes: Bonnet maintains in effect that the reproduction of dissimilar organic wholes cannot be conceived without the intervention of "properly specified germs" (*germes proprement dits*),[57] involving the embryonic configuration of the entire resulting organisms.

Faithful to continuous graduation and progressive differentiation, Bonnet distinguishes four principal types of organic preformation. The first involves the regeneration of similar compounds through a structuration which takes hold of the "organic points" of fibrillar structures and acts to assure their expansion. Bonnet considers it likely that certain organs of the lesser classes of animals, including the class of the polyps, can form in this way, without true preexistence, except of "organic elements from which they were due to emerge:"[58] these elements, the foundation of a true organic construction, would still be preordained to produce the emergence of specific properties that fiber compounds elicit.

The second type of organic preformation appears in the regeneration of integral parts (heads, tails, legs, etc.) that involve the combining of dissimilar parts. It presupposes the latent preexistence of complex parts capable of replacing amputated parts through development. It is inconceivable that the processes involved can be reduced just to the mechanical apposition of corpuscles following the analogy of crystallization.

The third type of preformation corresponds to the simultaneous reproduction of complex ensembles of organic parts. Bonnet illustrates this type with the production of vegetable parts that include, in the state of envelopment, an autonomous plant's ensemble of parts. Analogous cases in the animals are found, for example, in the severed earth worms whose anterior and posterior segments develop the lost organic parts through microdispositives

[55] Ibid., X vii, O XV: 408–9; Bonnet, *Considérations sur les corps organisés*, §168, 136–7.

[56] Bonnet, *La Palingénésie philosophique*, X iii, O XV: 396.

[57] Ibid., X iv, O XV: 397.

[58] Ibid., X iv, O XV: 399.

endowed with the capacity to restore specific elements of the complete developed structure.

The fourth type of preformation is that of the germ to which "the entire organized body owes its origin"[59] and is more directly related to the conservation of the species. The dominant model is the generation from eggs and grains, but this model is shown to be flawed in numerous cases where the offspring emerges more directly from the parent, as with the polyp. Other exceptions to this model are the animalcules of infusions and no doubt a number of other microscopic beings whose laws of development elude us. "I make a point," Bonnet states, "of ignoring the laws that determine the evolution of this crowd of *microscopic* beings about which the best lenses teach us nothing more than their existence, and which belong to another world I would name the world of the *invisible*."[60] These mysterious modes of reproduction can, it seems to him, be analogically referred to the inferior types of preformation. An indirect corroboration of this hypothesis is inferred, on the one hand, from the fact that the diverse modes of development seem to operate at the same time in the most complex organisms and, on the other, from the fact that the appeal to germs requires widening their definition in order to permit their application to organisms of the lesser classes: a widening based on the subordinate modalities of preformation. Hence, through analytic reduction comes the inference of fibrillar and quasi-molecular modes of reproductive generation: these modes lie at the limit of a strict preformationism which itself forms a more essential requisite when the focus is on the generation of the most complex organisms.

4. RECASTING LEIBNIZIAN PRINCIPLES

If Bonnet employs a Leibnizian framework of analysis and imports some of Leibniz's metaphysical theses and architectonic principles, he attempts to specify his points of disagreement with what he considers the closest system to his own philosophy of nature. In *La palingénésie*, the difference becomes especially evident with regard to the theses which Leibniz had set out in articles 89 and 90 of *Essais de théodicée* and in certain passages of *Nouveaux essais sur l'entendement humain* dealing with the essential relation of the soul to the organic body, the preformation of the living in the state of germs, the envelopment of the organism, and the survival of animals' psychic states after death.

[59] Ibid., X vii, O XV: 404.
[60] Ibid., 407.

Bonnet credits Leibniz with having held, on one hand, that all souls or substantial forms, in short, all the monads (a term which he does not hesitate to appropriate in his later presentations), were indestructible, and, on the other hand, that every formal principle or soul was perpetually united to an organic body. It is this double thesis that he notes in the parts of Leibniz's works with which he is familiar and which he considers a permanent accomplishment in Leibnizian philosophy. At the same time, he still finds an inadequate representation in Leibniz of the persistence of animal organisms after the dissolution of their macroscopic body. This representation was inadequate because it implied a purely analogical and therefore fictional parallelism between the emergence of organized bodies by evolution or development and their decay into an ultimate stage of envelopment. Bonnet rejects this hypothesis of envelopment consecutive to vital development as if the two were strict counterparts.

Leibniz could only consider the soul of beasts and other entelechies to be indestructible in so far as they were immaterial and indivisible, but he reserved, at the same time, immortality proper for spirits, that is to say, souls endowed with "consciousness or reflexive perception" of what they are, which makes them immortal moral subjects capable of reward or punishment. On this point, Bonnet believes he may attribute to the souls of beasts a form of self or personhood (*personnalité*) based on the empirical consecution of sensations which Leibniz considered at best an inferior degree of apperception, since it is not reflexive and thus is incapable of reaching an understanding of abstract relations and of the reasons underlying empirical consecutions. The Leibnizian argument restricted itself to positing that souls are entirely simple substances that cannot begin except through creation and end through annihilation: hence the necessity of conceiving them as imperishable. At the same time, the "organism" of animated bodies can only arise from "an already organic preformation:"[61] all generation of these bodies requires only the transformation and growth of the minimal structures which constitute their germs, and the souls which must be assigned to these will found the sequences of perceptions that correlate with their organic development. This development therefore implies the serial mutations of the primitive organism until its end, its apparent death, that is to say, the dissolution of the integral structure of the large machine which leaves nothing in place except the minimal organic body corresponding to the extreme withdrawal of the living structure. The Leibnizian conception of organic bodies implied that the totality of the phenomena affecting the organism in its phases of development

---

[61] Leibniz, *Essais de théodicée*, §90, GP vi: 152.

and envelopment can be mechanically explained, but the mechanism in question was subordinated to a transcendental principle which implied that the functional integration of its parts correlates with the sequence of states of the soul through the phases of transformation of the substantial compound. This is well expressed in the preface to *Essais de théodicée*, cited by Bonnet: "Mechanisms suffice to produce the organic bodies, . . . provided that we add to them an already entirely organic *preformation* in the seed of the bodies in which they are born, up to the primeval seeds; this could only come from the Author of things who, doing everything from the start with order, had preestablished all order and all future artifice."[62]

Already in the first sections of *Considérations sur les corps organisés*, Bonnet had attempted to produce as rigorous a physical representation as possible of the development and envelopment of the organic structures. This representation rested on the preexistence of organic bodies in the nested germs, on a conception of the body conforming to Haller's model of fiber properties, on a theory of assimilation and elimination of nutritive and stimulating elements, including those of the seed, and on a mechanist interpretation of apparent death. According to this transposition, Leibnizian envelopment did not appear to be anything more than a reduction of the global structure of bodies as transformed and altered by the successive phases of organic life. As Bonnet points out in *La palingénésie*,

> Leibniz thought that all souls had always preexisted in some kind of organized bodies; and his great principle of sufficient reason persuaded him that they would remain united after death to an organic whole: it in no way appears, he said, that there are souls entirely separated from all bodies in the order of nature. But he did not explain the nature of future bodies, their place, their relationship with previous bodies, etc. One even sees . . . that he seemed to believe that this would be the same body, only concentrated or enveloped. What we call generation, he had said, is nothing except an augmentation; the apparent death is nothing except envelopment"[63]

Bonnet refers to the preface of *Nouveaux essais sur l'entendement humain* in order to emphasize Leibniz's doctrine of the union of souls to organized bodies, a union which he interprets in the light of the metaphysical principle of preestablished harmony. He particularly insists on recourse to the law of continuity in order to guarantee the linkage of series of states related to the soul and the organic body, the two series being strictly required to express

[62] Ibid., GP: 40.
[63] Bonnet, *La Palingénésie philosophique*, VII v, O XV: 31.

each other. He finds confirmation that Leibniz held that "the animal reduced to a small scale" is conserved after death, and he cites two crucial passages that corroborate his interpretation. Thus Leibniz held that "in general no disturbance of the visible organs can bring things to a complete confusion in the animal, nor destroy all organs, and deprive the soul of all its organic body and the indelible remains of all precedent traces."[64] At the same time, opposing any conception of the separated soul as being subject to metempsychosis, the author of the *Nouveaux essais* claimed that the soul

> always keeps, even after death, an organic body, part of the previous one, even if what it keeps is always subject to imperceptible dispersing and mending and even suffers great changes for a time. In this way, in place of a transmigration of the soul, there is a transformation, envelopment, or development and drastic change of the body of this soul.[65]

But this reduction would not result in germs of restitution capable of regenerating the former organism – indeed transforming it into a more perfect living entity according to the natural order. Hence the acknowledgement of failure that Bonnet associates with the Leibnizian doctrine of envelopment, which he contrasts with the main tenets of his own doctrine:

> These expressions "reduced to a small scale" are not equivocal anymore, and I have carefully analyzed the author's "envelopment." He in no way imagined an *indestructible germ* lodged in the visible brain from the beginning; he did not consider this germ as the true seat of the soul; he failed to make it the residence of the self.[66]

For Bonnet, the Leibnizian envelopment of organic bodies after death is tantamount to admitting their irredeemable dissolution. Besides, the assumption that Leibniz had recognized the personhood of the animal is in no way obvious. For him, the personhood, linked to self-consciousness, required access to an apperception of abstract notions and principles, which required in turn a true reflexive capacity: unlike human beings, animals could not be considered persons, since their intellect is at best limited to the empirical consecutions of sensible representations, without being capable even of partially raising itself to the stage of rational argument. Bonnet detects a double danger in restricting personal immortality to minds, that is to say, to a special type of monad associated with corresponding organic bodies. There is a risk in reducing the immortality of beasts to mere persistence of minimal

---

[64] Leibniz, A VI vi: 58.
[65] Leibniz, *Nouveaux Essais* II xxvii §6, A VI vi: 233.
[66] Bonnet, *La Palingénésie philosophique*, VII vii, O XV: 326.

organic compounds: this position, which verges on materialism, would cast doubt on the postulation of a nonmaterial origin for organisms and on the postulation of psychic processes at the onset of organic functioning, providing a global sufficient reason for the integrated effects that we associate with organic mechanism. Similarly, there is a risk in considering the rational soul and the corresponding self as contingent enhancements of the psychic principles within the organic compound resulting from a particular divine volition. This status would imply a precariousness of duration that threatens the persistence of the self after death. Once again, this danger implies insufficient discrimination between the Leibnizian position and an interpretation of organic mechanisms that approaches materialism. This could not be the fundamental orientation of a truly Leibnizian metaphysics. It is therefore preferable to hold that Leibniz had not sufficiently considered the implications of his theory of envelopment and that he would have corrected it in light of an analysis better grounded in empirical evidence about preformation, like that provided by naturalists who have studied the metamorphosis of insects.

As a consequence, Bonnet feels he has no choice but to revise the Leibnizian notion of envelopment. He assumes that the soul of sensible beings is always linked with a tiny indestructible body with which it forms a substantial compound. This in some way constitutes the essential germ which will produce the macroscopic body and its characteristic sequences of states. Furthermore, the intellectual operations of sensible souls imply a continuum of phenomenal expressions and consequently a gradation of forms of personality which even encompasses the inferior stages of the animal kingdom. The intellectual operations and acts of self-consciousness that underlie the attribution of personhood constantly require the presence of an integrative and functional network of nervous fibers, which can be identified as the "brain." The preexistence of such an enveloped structure constitutes an essential condition for the emergence of living organisms from their germinal delineation, but, beyond this, this structure finds its sufficient reason in the timeless substrate formed by the mix of animal soul and tiny indestructible body. The immortality of living beings relies on this substantial mix, which traverses sequential metamorphoses and underlies the organism's diverse stages of phenomenal expression, even beyond the death and massive dissolution of the macroscopic body.

At the end of the seventh part of *La palingénésie*, Bonnet reacts to the attribution of his own ideas to Leibniz, an attribution which Abbot Sigorgne had made in his *Institutions leibnitiennes ou Précis de la monadologie* (1768), depriving the author of *La contemplation de la nature* of proper recognition for his special blend of natural philosophy.

Among the important differences modifying Bonnet's adhesion to the Leibnizian model of the organism, one should note the presupposition according to which there exists a final analysis of any material organization. Of course, Bonnet emphasizes the perfection of means of analysis, which, from his epoch and even more in the future, should provide access to the currently unknown microstructural and microfunctional dimensions of living beings, whether these means would come from mechanics, chemistry, or microscopy.[67] The decomposition of material bodies uncovers a plurality of levels of composition and, in the case of organized bodies, a plurality of stages of integration. Leibniz assumed the infinite progression of such an analytic decomposition of material parts and the necessity of postulating an extraneous sufficient reason for such compounds in the form of monads. By according a real status in some way to the infinitesimal quantities that Leibniz had only considered as simple limits and therefore as *entia rationis*, Bonnet assumes that if the actual progression involves an attainable end, this end would be made up of "elements" representing "absolutely simple substances."[68] There is no reference here to atoms, since these would involve some material extension, which in turn would require a sufficient reason for its composition. Bonnet relies on Leibniz to dissolve the continuity paradox which seems to affect the ultimate reduction of material parts: "If we make the effort," he affirms, "to deepen these general principles, we will recognize with the inventor of the famous *monads* that material extension is only a pure phenomenon, a simple appearance relative to our way of perceiving."[69] The monad thus becomes the true element in the order of decomposition, beyond all appearances that analysis would use to empirically characterize the ultimate ingredients of organic compounds. Bonnet is sufficiently Leibnizian to emphasize that the world of phenomena, beyond the relativity of appearances, refers to the constitutive laws which provide access to the underlying order of the phenomena themselves. In this way, nature is a "regular system of appearances; because these appearances are determined by the most sensible laws, and these are the only laws which we are capable of discovering, from which we derive our most beautiful *theories*, and which constitute the most precious foundation of our *natural* knowledge."[70] But, for Bonnet, this knowledge is a posteriori. It is only through knowledge of empirical effects that finite understandings can reach some expressions of the laws of nature, and these will

---

[67] Ibid., XIII i, O XVI: 31–2.
[68] Ibid., XIII ii, O XVI: 34.
[69] Ibid., 35.
[70] Ibid.

only provide probable explanations for the observed phenomenal relations.[71] At the same time, according to Bonnet, the nature of the active principles underlying the properties to which we link the causal production of observable effects remains unknowable beyond the formulation of hypotheses. On this point, the methodological model that was inherited from Newton and that implies the recourse to explanatory unknowns, such as force of attraction or irritability, to symbolize the latent causes of phenomena moves us away from the Leibnizian model of constructing hypotheses and providing mechanical schematizations under the aegis of architectonic principles.

Thus, the analogy of observable phenomena warrants the representation of an indefinitely graduated ladder of living species without the postulation of underlying mechanisms capable of explaining such a scalar disposition of forms. The analogy also guides us toward the representation of a mode of composition of organized bodies that would reflect their functional integration and intimate relationship to the phenomena of sensibility. In this way we also obtain laws of development and reproduction of organized bodies that follow from the observation of a primordial organizational design weaving the continuity of successive generations. In this precise spirit, the "preordained foundation of organization"[72] whose essential characteristics progressively reveal themselves through experience, without our being able to discover the very mechanism of its production, guarantees the serial linkage of the particular organic systems formed by species as well as individuals. If a future configuration of the organisms should see the light, it is already implied in this initial foundation of organization: "The actual constitution of the animal – I refer to its *organic* and *psychological* constitution – therefore contains secret particularities which are the foundation of the connection between this constitution and the one that follows it."[73] Articulated by the principle of continuity, this proposition attests to a link which unites the preformed structure of the germ to the developed structure of the organism, but in Bonnet's mind it assumes a more metaphysical significance: beyond the "envelop" or "mask" or "skin" (*écorce*) which the phenomenal germ and its organic expansion form it posits the presumed existence of an "undying germ" which, with the soul, forms the "personhood of the animal."[74] Beyond sensible appearances, this personality would be grounded in the continuity between the germinal microorganism and its expression under the form of a larger organized body: it

[71] Ibid., XIII viii, O XVI: 59–60.
[72] Ibid., XIV iii, O XVI: 70.
[73] Ibid., XIV i, O XVI: 65.
[74] Ibid., XII XII i, O XVI: 9.

would involve an analogous continuity between the phases of transformation characterizing the basic constitution or the substantial nucleus of the living organism.

With this assimilation of the monad to an immortal germ, which would be the source of the living organism's personhood responsible for the ultimate restitution of the organic body to a state of perfection, Bonnet effaces Leibniz's crucial distinction between the phenomenal order where the mechanisms of organized bodies unfurl and the substantial order of the monads as formal integrative principles for the functional processes proper to machines of nature. Leibnizian science did not attempt to outpace the analysis of organic mechanisms, nor replace it by a speculative representation of tiny inalterable and undying bodies associated with so many soul-like principles. In his representation of the living, Bonnet ended up fusing monadological and physical considerations. In his conception of physiological science, Leibniz carefully distinguished mechanical considerations linked to organized bodies and functional considerations linked to monads as formal principles, while presuming that the harmonic correspondence between the two series would be corroborated by the analysis of "well-founded phenomena." Bonnet, on the other hand, places these two types of considerations on the same level. He merges them analytically, for instance, when he describes the unitary organization involving the imperishable germ and its essential corpuscular body, or when he envisions all organic phenomena as diverse modalities of the same everlasting primordial organization.

5. CONCLUSION

The continuities with, and divergences from, Leibniz's model of organisms which we have shown first in Bonnet's *Considérations sur les corps organisés* determine, to a large extent, the pursuit of a tangential intellectual trajectory relative to that aspect of the Leibnizian philosophy of nature. The continuities and divergences bear on the use of the principle of continuity in order to understand the link between the structural and functional states of the individual organism, which determine at a more global level the graduated transition of species: they also bear on the complex order of organized bodies which accounts for the fact that they constitute machines of nature irreducible, in their initial formation, to mechanical analysis and on the representation of the *emboîtement* of germs underlying the emergence of organized bodies. Later, the assimilation and differentiation of Leibnizian theses become even more pronounced. The serial representation of the chain of beings in *La contemplation de la nature* presents an indication of the disagreement with the theses which Bonnet found in Leibniz's letter copied by König and with the analyses that Leibniz had developed in his *Nouveaux essais sur l'entendement humain*. Despite Sigorgne's insinuations of plagiarism and Mendelssohn's reference to similar conclusions formulated by German Leibniz-inspired philosophers, the difference from and continuity with the Leibnizian model are still more distinctly marked by the developments in *La palingénésie philosophique* relating to the state of survival of animal organisms and the destiny of mixed corpuscular beings after physiological death. If Bonnet employs preestablished harmony to represent the predelineation of all organized beings in the original germs, he imagines, at the foundation of these germs, souls endowed with personhood and tiny immortal organic bodies, of which the bodies subjected to the vicissitudes of generation and death are only contingent products. In this instance, he moves away from Leibniz's relationship of dominant monads to phenomenal bodies. Furthermore, inspired not only by Trembley's experiments on the regeneration and reproduction of polyps but also by Spallanzani's on the animalcules of infusion, Bonnet ends up relativizing the concept of germ and extending its meaning to represent the preformed element of organic structures underlying the ontogenesis of organisms essentially constituted of similar parts. He thus parts with the notion of integrated, infinite organic structures, a notion that lay at the heart of the Leibnizian presentation of machines of nature. The model which Bonnet favors is closer to Haller's conception of fiber architectures and their emergent functional properties, such as irritability and sensibility. Reproaching Leibniz for his too relativistic conception of the persistence of animal organisms after death, Bonnet substitutes for the Leibnizian notion of envelopment his own vision of the dissolution of macroscopic organic bodies. In this dissolution, the substantial mixes which constitute the tiny animated immortal bodies remain: these are the true "germs of restitution" capable of producing organisms that will eventually compose a system of nature more perfect and more complete than our own. From an epistemological point of view, Bonnet reduces the infinite analysis of conditions, which, according to Leibniz, determines the composition and order of organic processes. He assumes an end of analysis applied to the material organization of the living: in this way analysis would attain "the preordained foundation of organization." He situates the monads and the last elements of organized bodies, which make up the primordial germs, on the same level. It is this original reinterpretation that is lucidly reflected, even beyond the developments of *La palingénésie*, in several writings found in the last tome, published in 1783, of *Oeuvres d'histoire naturelle et de philosophie* (1779–83). At the same time that Bonnet stresses the metamorphosis of the framework of reference that he inherited from Leibniz's philosophy of the living, he keeps deepening his knowledge of Leibnizian and neo-Leibnizian texts. With considerable reservation, linked to the crystallization of his own philosophical views, he participates in the rediscovery of those Leibnizian models that could contribute to fostering a renewed comprehension of organisms in their individuality and complex interrelations. Bonnet testifies in this way to a Leibnizian "renaissance" which, during the second half of the eighteenth century, radically influenced epistemological reflection on the living. This reflection took very diverse forms: it implied apparently antinomical choices of Leibnizian theses and annexed them to theories which tended to diverge from the central tenets of Leibniz's theory of the organism. From Maupertuis to Needham, from Caspar Friedrich Wolff to Blumenbach and Kant, from Bonnet to the first theoreticians of biology, remarkable models developed along partly convergent and partly divergent paths. These fruitful variations and adjustments no doubt expressed the richness and polymorphism of the original model without exhausting its potential influence on emerging theories.

VI

Kant and His Contemporaries on Development and the Problem of Organized Matter

14

Kant's Early Views on Epigenesis

The Role of Maupertuis

JOHN ZAMMITO

Epigenesis was an important idea for Immanuel Kant from at least 1787, with his famous analogy in the B-version of the first *Critique*, and it played a central role in the "Critique of Teleological Judgment" of the third *Critique* (1790), where he identified his views with those of the great German physiologist Johann Friedrich Blumenbach.[1] Yet neither the general concept of epigenesis in the eighteenth century nor its place in Kant's thinking has ever been stabilized in the scholarship.[2] According to the historian of biology C. U. M. Smith, epigenesis is the idea in embryology that "organs ... are progressively formed from, or emerge from, an originally undifferentiated, homogeneous [material]."[3] I suspect that Kant was *never* comfortable with epigenesis, that it was a strain for his critical philosophy even when he explicitly invoked it, and that his attachment to Blumenbach was a case, as Robert Richards aptly put it, of mutual *mis*understanding.[4] That makes his involvement with epigenesis all the more interesting. To get at what troubled Kant, early and late, with the concept of epigenesis, I think it fruitful to go back to his earliest

---

[1] Immanuel Kant, *Critique of Pure Reason*, B167; *Critique of Judgment* §§80–1, AA 5: 417–24.

[2] See my "'This Inscrutable *Principle* of an Original *Organization*': Epigenesis and 'Looseness of Fit' in Kant's Philosophy of Science," *Studies in History and Philosophy of Science* 34 (2003): 73–109, for a discussion of the literature.

[3] C. U. M. Smith, *The Problem of Life: An Essay in the Origins of Biological Thought* (New York: Wiley, 1976), 264.

[4] Richards puts it succinctly: "The impact of Kant's *Kritik der Urteilskraft* on the disciplines of biology has, I believe, been radically misunderstood by many contemporary historians.... Those biologists who found something congenial in Kant's third *Critique* either misunderstood his project (Blumenbach and Goethe) or reconstructed certain ideas to have very different consequences from those Kant originally intended (Kielmeyer and Schelling)" (Robert Richards, *The Romantic Conception of Life: Science and Philosophy in the Age of Goethe* [Chicago: University of Chicago Press, 2002], 229). See Robert Richards, "Kant and Blumenbach on the *Bildungstrieb*: A Historical Misunderstanding," *Studies in the History and Philosophy of Biology and the Biomedical Sciences* 31 (2000): 11–32.

discussion, his response to the version propagated by a figure whom at the time he regarded extremely highly – Pierre Louis Moreau de Maupertuis, the president of the Prussian Academy in Berlin from 1746 to his death in 1759. The key textual locus for consideration is a passage in Kant's work of 1762–3, *The Only Possible Argument in Support of a Demonstration of the Existence of God*.[5]

Methodologically, my endeavor picks up two important motifs that seem to characterize current Kant studies: first, the effort to discern continuities from the "precritical" to the critical period, overriding Kant's own disposition to suppress his earlier thought; and, second, in terms of that bridging enterprise, to assay whether the continuities nonetheless entail *developmental shifts* – notably, but not exclusively, the "critical" epistemology itself – or, by contrast, whether these continuities simply reflect Kant's *intermittent engagement* with motifs or arguments which *remain similar*. Paul Guyer, for example, has pressed hard the notion that this latter mode of continuity characterizes Kant's aesthetic thought from the precritical period up through the *Critique of Judgment*.[6] That raises many interesting questions. My inquiry into epigenesis asks about a similar continuity in terms of Kant's thought on the life sciences.

The strategy here is first to establish the importance of Maupertuis for Kant in the 1750s and 1760s. I am inspired in this endeavor, as in so much of my work on Kant, by suggestions from Giorgio Tonelli. Then I expand upon two specific issues in Maupertuis himself: first, his revisonist physicotheology, and, second, his place in the life sciences of the eighteenth century. That will bring me to Kant's *Only Possible Argument*, and to contesting readings of it. Finally, I will compare the result with Kant's famous exposition of epigenesis in §81 of the *Critique of Judgment* of 1790.

1. THE IMPORTANCE OF MAUPERTUIS FOR KANT

If we turn to the most recent biography of Kant, we find that Manfred Kuehn mentions Maupertuis only once, noting that *The Only Possible Argument*

---

[5] Immanuel Kant, *Der einzig mögliche Beweisgrund zu einer Demonstration des Daseins Gottes*, AA2:113–15 (*The Only Possible Argument in Support of a Demonstration of the Existence of God*, in Kant, *Theoretical Philosophy 1755–1770*, trans. and ed. David Walford and Ralf Meerbote [Cambridge: Cambridge University Press, 1992], 155–7).

[6] Paul Guyer, editor's introduction to Immanuel Kant, *Critique of the Power of Judgment* (Cambridge: Cambridge University Press, 2000), xiii-xxxiv, and editorial notes, pp. 351–98.

"was influenced by the *Essai de cosmologie* and the *Examen philosophique* [*de la preuve de l'existence de Dieu employée dans l'Essai de Cosmologie*] (1758) of Maupertuis."[7] Kuehn offers no elaboration. In *Kant's Philosophy of the Exact Sciences*, Michael Friedman refers to Maupertuis's connection to Kant also only once, in a footnote, where he observes that the "basic idea of [Kant's *Universal Natural History and*] *Theory of the Heavens* corresponds to Maupertuis's polemic against Wolffian physicotheology of 1748."[8] Friedman's work concentrates on the critical period, and he acknowledges his student Alison Laywine as a more thorough interpreter of the precritical period.[9] However, the name of Maupertuis does not present itself at all in the index of her book, *Kant's Early Metaphysics and the Origins of the Critical Philosophy*.[10] Nor does his name appear in the index of the recent, outstanding collection of essays *Kant and the Sciences*, edited by Eric Watkins.[11] Martin Schönfeld's careful archaeology of the "precritical project" mentions Maupertuis a couple of times, but with no direct connection to Kant.[12]

From this quick dip into recent English-language literature, it would appear that Maupertuis was of very little significance for Kant. Only in a lengthy monograph by the German scholar Hans-Joachim Waschkies, *Physik und Physikotheologie des jungen Kant*, does the question of Maupertuis's importance for Kant get a careful discussion.[13] Waschkies is the only scholar mentioned who acknowledges that a case had been made for that importance by Giorgio Tonelli.[14] What was the case Tonelli made, and does it deserve the obscurity into which it has fallen? The epitome of Tonelli's view is that,

---

[7] Manfred Kuehn, *Kant: A Biography* (Cambridge: Cambridge University Press, 2001), 141. The earlier standard biography in English (translation), Ernst Cassirer's *Kant's Life and Thought* (New Haven: Yale University Press, 1981), only mentions Maupertuis once as well (p. 51), and at even greater remoteness from Kant.

[8] Michael Friedman, *Kant and the Exact Sciences* (Cambridge, MA: Harvard University Press, 1992), 11n.

[9] Ibid., xvii, on the "precritical" period. See also Laywine, *Kant's Early Metaphysics and the Origins of the Critical Philosophy*, North American Kant Society Studies in Philosophy, vol. 3 (Atascadero, CA: Ridgeview, 1993).

[10] Laywine, *Kant's Early Metaphysics and the Origins of the Critical Philosophy*.

[11] Eric Watkins, ed., *Kant and the Sciences* (Oxford: Oxford University Press, 2001).

[12] Martin Schönfeld, *The Philosophy of the Young Kant: The Precritical Project* (Oxford: Oxford University Press, 2000), 28, 116, 123, 162–3, 210–11.

[13] Hans-Joachim Waschkies, *Physik und Physikotheologie des jungen Kant* (Amsterdam: B. R. Grüner, 1987), esp. 562–8, 573–8.

[14] Ibid., 48, 50, 562–8.

with Christian August Crusius, Maupertuis was one of the two essential *Gewährsmänner* (intellectual mentors) for the early Kant.[15] It was Tonelli who claimed that "one of the basic motives [of *Universal Natural History*] is that of the polemic against physicotheology, in line with Maupertuis's principles."[16] I wish to take up his persistent claim that Maupertuis was a decisive influence on the early Kant.

Waschkies praises Tonelli for having been the only scholar seriously to endeavor to contextualize Kant's work in the natural-scientific milieu of contemporary Prussia, but he rejects Tonelli's thesis about Maupertuis and Kant's *Universal Natural History*.[17] As Waschkies sees it, Maupertuis had condemned Newton's physicotheology, yet Kant, "despite the reservations Maupertuis had expressed," attempted in *Universal Natural History* "to do a better job of it than Newton."[18] Still, Waschkies affirms that Kant had read (the German translation of) *Essai de cosmologie* by the time he composed *Universal Natural History* and that it served as one of three crucial textual stimuli for Kant's own meditation – along with the extended summary of Thomas Wright's cosmology in a Hamburg magazine, to which Kant made explicit reference, and the translation of the first volume of Buffon's *Histoire naturelle*, which appears to have inspired the first part of Kant's title.[19]

A critical issue in dispute between Waschkies and Tonelli is the relative contributions of Martin Knutzen and Maupertuis to Kant's initiation into Newtonianism. According to Waschkies, Tonelli dismissed Knutzen's grasp of Newtonianism as "timid" and generally ascribed little influence to Knutzen.[20] Waschkies goes to great lengths to demonstrate the contrary. Ironically, after all his effort, Manfred Kuehn cites Waschkies as his source for the suspicion that Knutzen was *not* important for Kant, especially not Kant's Newtonianism![21] It is noteworthy that Martin Schönfeld has argued strongly

---

[15] Giorgio Tonelli, "Der Streit über die mathematische Methode in der Philosophie in der ersten Hälfte des 18. Jahrhunderts und die Entstehung von Kants Schrift über die 'Deutlichkeit,'" *Archiv für Philosophie* 9 (1959): 57–66, citing 66.

[16] Giorgio Tonelli, "Conditions in Königsberg and the Making of Kant's Philosophy," in *Bewußt sein: Gerhard Funke zu eigen*, ed. Alexius Bucher, Hermann Drüe, and Thomas Seebohm (Bonn: Bouvier, 1975), 126–44, citing 140.

[17] Waschkies' praise for Tonelli is in *Physik und Phyikotheologie*, 561; his rejection of the hypothesis comes at 573–5. Clearly, this is in direct conflict with the claims of Michael Friedman noted earlier, who seems to have followed Tonelli.

[18] Ibid., 576.

[19] Ibid.

[20] Ibid., 564.

[21] Kuehn, *Kant*, 84, and n. 108.

that there is little evidence of Kant's sophistication concerning Newton in his early work on *Living Force* and that the real turn to Newtonianism came in the 1750s – which would be far too late for Knutzen (d. 1751) to have been instrumental.[22] All this should occasion renewed interest in the Tonelli hypothesis. Maupertuis may very well have been important for Kant in ways that have not yet been taken fully into account, especially in terms of his critical initiation into Newtonian mechanics and cosmology.[23] But that is not my concern here. Rather, I wish to pursue the obvious importance of Maupertuis's physicotheology for Kant's *Only Possible Argument*, to set the terms of my problem of epigenesis in Kant.

The French scholar Jean Ferrari has established that, among scholars cited favorably by Kant, "Maupertuis occupies a more than honorable place."[24] Ferrari shows that Kant defended Maupertuis, with Buffon, as scientific pioneers persecuted by superficially brilliant practitioners of *belles lettres*.[25] Kant mentioned Voltaire's wicked *Akakia* pamphlet deriding Maupertuis more frequently than any other text by Voltaire, but always with distaste.[26] Identification with stern science over against irresponsible wit would have lifelong significance for Kant.[27] Most importantly, Ferrari recognizes that the early Kant found in Maupertuis metaphysical impulses paralleling, if not even inspiring, his own.[28] Kant regarded Maupertuis as an ally in the *metaphysical* endeavor of reconciling science and religion, a central preoccupation of German philosophy in the eighteenth century. Maupertuis decisively reoriented physicotheology, revamping in a distinctly original manner the link between physics and metaphysics.

---

[22] Schönfeld, *Philosophy of the Young Kant*, 69–70, 73–95.

[23] An important starting point for investigating Maupertuis's transmission of Newtonianism to Kant would be the publication in 1742 of the second edition of the *Discours sur les différentes figures des astres* and of the *Lettre sur la comète*, in which, according to David Beeson, "Maupertuis presents the analysis from the *Principia* clearly and concisely, in terms wholly accessible to the intelligent layman" (*Maupertuis: An Intellectual Biography* [Oxford: Voltaire Foundation, 1992], 148–52, esp. 151).

[24] Jean Ferrari, "Kant, Maupertuis et le principe de moindre action," in *Pierre Louis Moreau de Maupertuis: Eine Bilanz nach 300 Jahren*, ed. Hartmut Hecht (Berlin: Arno Spitz, 1999), 225–34, citing 225.

[25] Ibid., 227–30. Ferrari points to two *Reflexionen* by Kant that make this point: A A 15: 389 and 587.

[26] Ibid., 232.

[27] See my "'Method' vs. 'Manner'? Kant's Critique of Herder's *Ideen* in the Light of the Epoch of Science, 1790–1820," *Herder Jahrbuch / Herder Yearbook 1998*, ed. Hans Adler and Wulf Koepke (Stuttgart: Metzler, 1998), 1–26.

[28] Ferrari, "Kant, Maupertuis," 231–3.

John Zammito

2. MAUPERTUIS: LEAST ACTION, PHYSICOTHEOLOGY, AND EPIGENESIS

Jean D'Alembert paid Maupertuis the sound compliment that he was the first Frenchman to take up the cause of Newtonianism.[29] Maupertuis was not a mathematical physicist of the order of D'Alembert or Leonhard Euler, and the further he moved from his studies with Jean (I) Bernoulli, the more his metaphysical impulses overshadowed his mathematical endeavors.[30] Yet he remained one of the most cited and respected Newtonian physicists of his day, and D'Alembert and Euler both assiduously sought to find in his work elements they could affirm.[31] When Maupertuis, then, came to believe that he had found a grand synthetic principle for rational mechanics in the "principle of least action," his friends in the first circle of mathematical physics struggled both to give as much credit to this achievement as they could and to defend Maupertuis from Samuel König's charges of plagiarism from Leibniz and especially the ensuing scurrilities of Voltaire.[32] Maupertuis announced his new "principle" first to the Paris Academy of Science in 1744.[33] He reiterated it as his inaugural presentation to the Prussian Academy in Berlin in 1746.

His assumption of the presidency of the Berlin Academy brought him into an already charged force field of tension between Newtonian and Leibnizian factions.[34] Maupertuis inherited a situation in which Euler and the academy secretary, Samuel Formey, had already contrived hard lines of confrontation.[35] To be sure, Maupertuis sympathized with Euler, but he tried to play a role

---

[29] Jean D'Alembert, *Preliminary Discourse to the Encyclopedia*, trans. Richard Schwab (Chicago: University of Chicago Press, 1995).

[30] Mary Terrall, *The Man Who Flattened the Earth: Maupertuis and the Sciences in the Enlightenment* (Chicago: University of Chicago Press, 2002), 173.

[31] Ibid., 290, 307; Beeson, *Maupertuis*, 225.

[32] On December 10, 1745, Euler wrote to Maupertuis: "I am convinced that nature acts everywhere according to some principle of maximum or minimum.... It seems to me also that it is here that we must seek the true principles of metaphysics" (cited in Terrall, *The Man Who Flattened the Earth*, 230n).

[33] Maupertuis wrote, "I presented the principle upon which the following work [his Berlin Academy address] was based on April 15, 1744 at a public assembly of the Royal Academy of Sciences in Paris, as the proceedings of that Academy will attest" (cited in Patrick Tort, "Bio-bibliographie" to his edition of the *Vénus physique* [Paris: Aubier-Montaigne, 1980], 65).

[34] Terrall, *The Man Who Flattened the Earth*, 257 ff. Compare the earlier studies: Ronald Calinger, "Frederick the Great and the Berlin Academy of Sciences (1740–1766)," *Annals of Science* 24 (1968): 239–49; Harcourt Brown, "Maupertuis *Philosophe*: Enlightenment and the Berlin Academy," *Studies on Voltaire* 24 (1963): 255–69; Hans Aarsleff, "The Berlin Academy and Frederick the Great," *History of the Human Sciences* 2 (1989): 193–206.

[35] I believe Terrall (*The Man Who Flattened the Earth*, 257) has a better case than Beeson (*Maupertuis*, 109).

above the fray befitting the president of the academy. That was not simply an *institutional* commitment, however.[36] We have come to realize that there was a good deal of Leibniz in Maupertuis himself.[37] For too long he has been seen merely as the Newtonian archenemy of Wolff at Berlin.[38] That role really belongs to Euler.[39] The metaphysics which became the central preoccupation of Maupertuis by the end of the 1740s, culminating in the *Essai de cosmologie* and the *Système de la nature*, have an unmistakably Leibnizian cast.[40] For too long, as well, the story of Maupertuis in the 1750s has been consumed with the König affair.[41] Instead, I want to consider the substantial experimental and theoretical work Maupertuis devoted to life science.

As Mary Terrall rightly maintains, "Maupertuis was working on quite other problems . . . questions of the generation of organisms, the properties of matter, and the possibility of a science of life."[42] Maupertuis was a close friend of the Comte de Buffon starting in the 1740s.[43] He was, sometimes uncomfortably, the patron of his countryman from St. Malo, Julien Offray de La Mettrie, first in Paris and then in Berlin.[44] La Mettrie paid Maupertuis the "compliment" of dedicating *Natural History of the Soul* (1745) to him.[45] Finally, Maupertuis

---

[36] See Claire Salomon-Bayet, "Maupertuis et l'institution," in *Actes de la journée Maupertuis (Créteil, 1er décembre 1973)*, issued by the C. N. R. S. (Paris: Vrin, 1975), 183–202.

[37] Hartmut Hecht, "Pierre Louis Moreau de Maupertuis oder die Schwierigkeit der Unterscheidung von Newtonianern und Leibnitianern," in *Leibniz und Europa: Vorträge des VI. Internationalen Leibniz-Kongresses, Hannover, 18. bis 23. Juli 1994* (Hannover: Leibniz Gesellschaft, 1994), 1:331–8.

[38] That was certainly Wolff's suspicion. In a letter of 1748 he wrote of the Berlin Academy as the combination of "so-called Newtonian philosophy with the French world of flattery" (cited in Terrrall, *The Man Who Flattened the Earth*, 251n). Terrall elaborates: "The Wolffians looked at Berlin and saw foreign mathematicians overreaching the bounds of their competence to dictate to German metaphysicians" (ibid., 265).

[39] And even Euler was no pure Newtonian – whatever that might mean in the eighteenth century. See *Leonhard Euler 1707–1783: Beiträge zu Leben und Werk*, Gedenkband des Kantons Basel-Stadt (Basel: Birkhäuser, 1983); Rüdiger Thiele, *Leonhard Euler* (Leipzig: Teubner, 1982).

[40] Ernst Cassirer, *Philosophy of the Enlightenment* (Princeton: Princeton University Press, 1951), 86. But Terrall cautions, "It is not clear whether Maupertuis appreciated how close his position was to that of Leibniz" (*The Man Who Flattened the Earth*, 284n).

[41] Even the new collection, ostensibly a balance of three hundred years of historical consideration, still weighs this controversy very heavily in its treatment; see Hartmut Hecht, ed., *Pierre Louis Moreau de Maupertuis: Eine Bilanz nach 300 Jahren* (Berlin: Arno Spitz, 1999).

[42] Terrall, *The Man Who Flattened the Earth*, 310.

[43] This very important relationship deserves further study, and without the presupposition that Buffon was the senior partner, as in Jacques Roger, *Les sciences de la vie dans la pensée française du XVIIIe siècle* (Paris: Armin Colin, 1963), 475–6.

[44] Beeson gives an extended account of the relationship between Maupertuis and La Mettrie, ascribing great influence to the latter in the development of Maupertuis's thought in life science (see Beeson, *Maupertuis*, 191, 206–14, 241).

[45] Ibid., 191.

engaged in a crucial exchange with Denis Diderot on the life sciences and the "new Spinozism."[46] That is, Maupertuis has a place among the most important proponents of "vital materialism" in French thought at midcentury.[47]

In the years after 1746, Maupertuis engaged intensely in empirical study of the problem of heredity, foremost through his famous analysis of polydactyly in the Ruhe family of Berlin, whereby Maupertuis not only established decisive empirical evidence for biparental contributions in generation but went beyond this to a theory of hereditary variation which took into account both the spontaneity of change and the question of its inheritance across numerous generations.[48] He supplemented this investigation with his own controlled

---

[46] See Aram Vartanian, "Diderot and Maupertuis," *Revue Internationale de Philosophie* 38 (1984): 46–66. I do not agree with much in this essay, but it sets the issues in place; see also, for a more persuasive account, André Robinet, "Place de la polémique Maupertuis-Diderot dans l'oeuvre de Dom Deschamps," in *Actes de la journée Maupertuis*, 33–46, and the discussions in Beeson and Terrall.

[47] On the question of vital materialism, see Peter Hanns Reill, *Vitalizing Nature in the Enlightenment* (Berlekey: University of California Press, forthcoming), and the many earlier articles of which this monograph is the culmination: "Science and the Science of History in the *Spätaufklärung*," in *Aufklärung und Geschichte*, ed. H. E. Boedeker, George Iggers, Jonathan Knudsen, and Peter H. Reill (Göttingen: Vandenhoeck and Ruprecht, 1986); "Anti-Mechanism, Vitalism and Their Political Implications in Late Enlightened Scientific Thought," *Francia* 16, no. 2 (1989): 195–212; "Science and the Construction of the Cultural Sciences in Late Enlightenment Germany: The Case of Wilhelm von Humboldt," *History and Theory* 33, no. 3 (1994): 345–66; "Anthropology, Nature and History in the Late Enlightenment: The Case of Friedrich Schiller," in *Schiller als Historiker*, ed. Otto Dann, Norbert Oellers, and Ernst Osterkamp (Stuttgart: Metzler, 1995), 243–65; "Between Mechanism and Hermeticism: Nature and Science in the Late Enlightenment," in *Frühe Neuzeit – Frühe Moderne?* ed. Rudolf Vierhaus (Göttingen: Vandenhoeck and Ruprecht, 1992), 393–421; "Analogy, Comparison, and Active Living Forces: Late Enlightenment Responses to the Skeptical Critique of Causal Analysis," in *The Skeptical Tradition around 1800*, ed. Johann van der Zande and Richard Popkin (Dordrecht: Kluwer, 1998), 203–11. See also Theodore Brown, "From Mechanism to Vitalism in Eighteenth-Century Physiology," *Journal of the History of Biology* 7 (1974), 179–216, and my chapter "Kant versus Eighteenth Century Hylozoism," in *The Genesis of Kant's Critique of Judgment* (Chicago: University of Chicago Press, 1992), 189–99, together with the literature there cited.

[48] See above all Bentley Glass: "[V]irtually every idea of the Mendelian mechanism of heredity and the classical Darwinian reasoning from natural selection... is here combined, together with De Vries' theory of mutations as the origins of species" ("Maupertuis, Pioneer of Genetics and Evolution," in Bentley Glass et al., *Forerunners of Darwin: 1745–1859* [Baltimore: Johns Hopkins University Press, 1968], 51–83, citing 60). The specific claims about Mendel are demolished neatly by Iris Sandler, "Pierre Louis Moreau de Maupertuis – A Precursor of Mendel?" *Journal of the History of Biology* 16 (1983): 101–36. The claims for Maupertuis regarding Darwin and evolution, or "transformism," as it has been called, were stated enthusiastically by Paul Ostoya, "Maupertuis et la biologie," *Revue d'Histoire des Sciences et de Leurs Applications* 7 (1954): 60–78, and A. C. Crombie, "P. L. Moreau de Maupertuis, F. R. S. (1698–1759): Précurseur du transformisme," *Revue de Synthèse* (3e série) 58 (1957): 35–56, and demolished with equal enthusiasm by Anne Fagot, "Le 'transformisme' de Maupertuis," in *Actes de la journée Maupertuis* (Paris: Vrin, 1975), 163–78. As David Beeson has it, "the

breeding experiments with dogs and other animals in the personal menagerie he kept in his Berlin home.[49]

Another key feature of these Berlin years was Maupertuis's attention to the section for speculative philosophy of the Berlin Academy. This was a section without parallel in the academies of science with which Maupertuis had been familiar in the West, and, in addition, it was a section beset with mediocrity of accomplishment and dogmatic provincialism. He set about systematically trying to improve its tone, not only by actively recruiting talent throughout Europe to bolster its ranks (with modest success in the persons of Jean Baptiste Mérian and Johann Georg Sulzer), but also, and more importantly for our considerations, by working extensively in this area himself.[50] The themes of his philosophical endeavors and those of his life science, at least one commentator has realized, had systematic connections motivated by this institutional commitment.[51]

Linking life science with physicotheology at midcentury is key. The importance of physicotheology for natural science in the first half of the eighteenth century has not been sufficiently recognized.[52] Not only was the explosion of physicotheological writing one of the most striking features of the period, but it both expressed and elicited central issues in the practice and philosophy

---

very idea of a scientific precursor is dangerous" (*Maupertuis*, 180). Still, one must establish fairly the status of Maupertuis in the history of life science, and Glass is not wrong to esteem it highly. I hold with François Russo that "it is hard for one to underscore too heavily the novelty, the epistemological audacity represented in [Maupertuis's] conceptions. Of course, these hypotheses were very crude and they were [at the time] incapable of experimental confirmation. Nevertheless..." ("Théologie naturelle et sécularisation de la science au XVIIIème siècle," *Recherches de Science Religieuse* 66 (1978): 27–62, citing 51n). In the most balanced assessment of Maupertuis and life science, Michael Hoffheimer concludes that the work, for all its empirical problems, "nonetheless marks an important turning point in the history of biology" ("Maupertuis and the Eighteenth-Century Critique of Preexistence," *Journal of the History of Biology* 15 (1982): 119–44, citing 137).

[49] Samuel Formey's *Souvenirs* contain an account of visiting Maupertuis's home and making his way through all the animals, as Terrall notes (*The Man Who Flattened the Earth*, 337–8n).

[50] Brown, "Mauperptuis, *Philosophe*," 256.

[51] Salomon-Bayet, "Maupertuis et l'institution."

[52] Waschkies makes this his key theme. For other discussions, see esp. François Russo, "Théologie naturelle"; Udo Krolzik, "Das physikotheologische Naturverständnis und sein Einfluß auf das naturwissenschaftliche Denken im 18. Jahrhundert," *Medizinhistorisches Journal* 15 (1980): 90–102; Wolfgang Philipp, "Physicotheology in the Age of Enlightenment: Appearance and History," *Studies on Voltaire and the Eighteenth Century* 57 (1967): 1233–67; Richard Toellner, "Die Bedeutung des physico-theologischen Gottesbeweises für die nachcartesianische Physiologie im 18. Jahrhundert," *Berichte zur Wissenschaftsgeschichte* 5 (1982): 75–82; Manfred Büttner, "Zum Übergang von der teleologischen zur kausalmechanischen Betrachtung der geographisch-kosmologischen Fakten: Ein Beitrag zur Geschichte der Geographie von Wolff bis Kant," *Studia Leibnitiana* 5 (1973): 177–95.

of science during this epoch. The inspiration of this burst of physicotheological writing can be traced directly to Isaac Newton and the Boyle Lectures, together with the ensuing controversy between Leibniz and Clarke.[53] The driving impetus was the need to demonstrate to the orthodox that the pursuit of rational-mechanistic natural science was not a threat to religion. Accordingly, physicotheology attacked not only atheism but pantheism with ever more "proofs" of the existence of a traditional theistic God from the order and beauty of nature. One striking feature of this physicotheological literature was its anthropocentric teleology.[54] God's existence and providence seemed most demonstrable in features of the world construed as beneficial for man. The tradition of physicotheology suffered some intrinsic tensions, however.[55] One strand of the tradition presumed not only the transcendent importance of man in nature but also a complete and determinate structure of nature from the moment of creation. Another strand viewed man as only a part of nature, and nature itself as incomplete, allowing man a role in its ongoing perfection.

These tensions were exacerbated by the interjection of the modern rational-mechanist model of natural science. In this light, three considerations became important. First, the complexity of phenomenal nature led to an emphasis on function, as against structure, to secure systematic unity.[56] Second, the

---

[53] Isaac Newton, "General Scholium," in *Mathematical Principles of Natural Philosophy*, 1714 edition; Newton, Query 31, *Opticks* (1706 and subsequent editions); G. W. Leibniz and Samuel Clarke, *Correspondence*, ed. Roger Ariew (Indianapolis: Hackett, 2000). See Margaret Jacob, *The Newtonians and the English Revolution, 1689–1720* (Ithaca: Cornell University Press, 1976), on the Boyle Lectures. On the Leibniz-Clarke correspondence, see Alexandre Koyré, "The Work-Day God and the God of the Sabbath," *From the Closed World to the Infinite Universe* (Baltimore: Johns Hopkins University Press, 1957), 235–72; F. E. L. Priestley, "The Clarke-Leibniz Controversy," in *The Methodological Heritage of Newton*, ed. Robert Butts and John Davis (Toronto: University of Toronto Press, 1970), 34–56; G. H. R. Parkinson, "Science and Metaphysics in the Leibniz-Newton Controversy," *Studia Leibnitiana Supplementa* 2 (1969): 79–112; Margula Perl, "Physics and Metaphysics in Newton, Leibniz, and Clarke," *Journal of the History of Ideas* 30 (1969): 507–26; Carolyn Iltis, "The Leibnizian-Newtonian Debates: Natural Philosophy and Social Psychology," *British Journal for the History of Science* 6 (1973): 343–77; and, from a political perspective, Steven Shapin, "Of Gods and Kings: Natural Philosophy and Politics in the Leibniz-Clarke Disputes," *Isis* 72 (1981): 187–215. The metaphysical issues of the eighteenth century were all spelled out in these crucial formulations, making the essential debate one between Newtonians and Leibnizians, with Cartesians playing a fading role, especially by midcentury. Kant's metaphysical horizon was fixed by these controversies.

[54] Philipp, "Physicotheology," 1245; Krolzik, "Das physikotheologische Naturverständnis," 94.

[55] Krolzik, "Das physikotheologische Naturverständnis," 92–6.

[56] Toellner, "Die Bedeutung des physico-theologischen Gottesbeweis," 77; Krolzik, "Das physikotheologische Naturverständis," 98. A key figure in this turn to function was the German theorist Fabricius, who was known to Kant.

coherence and simplicity of order that (at least after Newton) seemed possible for the inorganic world just could not be found in the organic world; hence the latter demanded a far more immediate intervention of divinity, triggering a flood of specifically biological physicotheological works.[57] Finally, as anomalies proliferated, the notion that nature was not constant in its order but rather developmental stimulated the *historical-genealogical* study of nature and consequently forced a critical appraisal of the methodological resources science might possess for such study.[58]

The most obvious locus of this shift to a literal "history of nature" was in the study of the earth.[59] But there were indeed *four* fundamental problems of origin driving natural science in the eighteenth century: (1) the problem of the origin of the regularities observed in galaxies, solar systems, and planets; (2) the problem of the history of the earth; (3) the problem of the emergence of life and its organization in species; and (4) the generation of individual living beings.[60] If, in all these, not only physicotheology but natural science in the eighteenth century postulated the intervention of God at some point, the decisive questions were when and how? These were problems not merely of ontology (how did these respective things come into being?) but equally of epistemology (how could humans attain knowledge of this?)[61] And in terms of a domain for natural science itself, how much would be left to nature once this divine role had been taken into account? The roughly "deist" resolve of much of the eighteenth-century scientific community was to postulate a God who created "primary matter" and endowed it with fundamental *laws*, whereupon nature could run on its own.[62] The danger, spotted already in the seventeenth century by Blaise Pascal, was that this made place for a creator God, but thereafter divinity became irrelevant.[63] This threat inspired Newton to the argument of the "General Scholium" and informed the Boyle Lectures

---

[57] Russo, "Théologie naturelle," 39; Toellner, "Die Bedeutung des physico-theologischen Gottesbeweis," 79.

[58] Russo, "Théologie naturelle," 43–50. A crucial role was played here by *comets*, starting with Halley and Whiston and carrying through the first half of the eighteenth century. See Sara Schechner-Genuth, *Comets, Popular Culture, and the Birth of Cosmology* (Princeton: Princeton University Press, 1997). Maupertuis and Kant were both intrigued by this element.

[59] See Roy Porter, *The Making of Geology: Earth Science in Britain, 1660–1815* (Cambridge: Cambridge University Press, 1977); Rachel Laudan, *From Mineralogy to Geology: The Foundations of a Science, 1650–1830* (Chicago: University of Chicago Press, 1987). Kant took a strong interest in what he called "physical geography," which had a substantial geological dimension, from the 1750s onward.

[60] Russo, "Théologie naturelle," 43.

[61] Ibid., 49–50.

[62] Ibid., 45.

[63] Blaise Pascal, *Pensées*, trans. A. J. Kraisheimer (Harmondsworth, England: Penguin, 1966).

and the Leibniz-Clarke debate. Its general name from the late seventeenth century onward was "Spinozism."[64]

By the mid-eighteenth century, not only was the mechanist paradigm hitting anomalies in virtually every domain, physicotheology had concurrently carried its search for marvels to ridiculous – and ridiculously anthropocentric– extremes.[65] A feature of both was that God appeared increasingly a stop-gap "solution" to any problem in human understanding of the natural world, a haven for the lazy and the sloppy.[66] And the more particular the marvel appearing so inexplicable without recourse to divinity, the more shaky the *proof structure* of the recourse became. Ignorance was being offered as the ground for a crucial certainty.[67] The result was a crisis in physicotheology directly related to methodological impasses in natural science itself. There were two paths that might be taken out of the imbroglio. The first was to isolate natural science from all such metaphysical adventures. This was the (proto-positivist) path advocated by D'Alembert, especially in his important essay "Cosmologie" in the *Encyclopédie* (1754).[68] The other was to *reform* metaphysics to accommodate the realities of natural-scientific practice. This was the path eventually taken by Immanuel Kant.

His decisive predecessor on this path was Maupertuis.[69] Maupertuis sought to *save* physicotheology by shifting from an emphasis on particular wonders to an emphasis on universal laws of nature, especially mathematical ones.[70] The decisive strategy was to make *theological* room for natural necessity without making divinity redundant and to make *scientific* room for contingency in the

---

[64] See Diderot's famous article "Spinosiste," in the *Encyclopédie* (reprinted in Diderot, *Oeuvres complètes.* ed. John Lough and Jacques Proust (Paris: Herman, 1975), 8:328–9); the early works by Paul Vernière, *Spinoza et la pensée Française avant la Révolution* (Paris: PUF, 1954), and Emile Callot, *La philosophie de la vie au XVIIIe siècle* (Paris: Rivière, 1965); and the recent work of Jonathan Israel, *Radical Enlightenment: Philosophy and the Making of Modernity 1650–1750* (Oxford: Oxford University Press, 2001), 6–7, 704–13.

[65] On the exhaustion of the mechanist paradigm, see Smith, *The Problem of Life*, 268, and François Jacob, *The Logic of Life: A History of Heredity* (New York: Pantheon, 1973), 36; on the ridiculous excesses of physicotheology, see the writings of Maupertuis and Kant themselves.

[66] Kant, *Der einzig mögliche Beweisgrund* A A 2:119; trans. 161.

[67] Of course, the greatest critic here proved to be David Hume. On Maupertuis and Hume, see Lionel Gossman, "Berkeley, Hume and Maupertuis," *French Studies* 14 (1960): 304–24, and Roger Oake, "Did Maupertuis Read Hume's *Treatise of Human Nature*?" *Revue de Littérature Comparée* 20 (1940): 81–7.

[68] Jean D'Alembert, "Cosmologie," in *Encyclopédie, ou Dictionaire raisonée des sciences, des arts et des métiers*, ed. Denis Diderot and Jean D'Alembert, (Paris, 1754), 4:294–7.

[69] This claim for the *philosophical* importance of Maupertuis should boost the endeavor to establish his rightful place in the intellectual history of the eighteenth century.

[70] Maupertuis, *Essai de cosmologie*, in Maupertuis, *Essai de cosmologie: Système de la Nature: Réponse aux Objections de M. Diderot*, ed. François Azouvi (Paris: Vrin, 1984).

natural world, to account (objectively) for its dynamism and (subjectively) for the limits of human comprehension.[71] The epistemological and ontological resolution Maupertuis endeavored to achieve wove chance and lawfulness together into a theory of process and emergence.[72] It was a distinctly *naturalist* and *historicist* philosophy of science. This was the burden – and the commonality – of his two most important works, the *Essai de cosmologie* (1751) and the *Système de la nature* (1756). In the next decade, these would be the works upon which Kant meditated as he formulated his own position in *The Only Possible Argument*.

"A crowd of natural scientists, after Newton, have found God in the stars, in insects, in plants, in water," Maupertuis pronounced acerbically in the *Essai de cosmologie*.[73] As an illustration, he described one text which made an argument to God from the hide of a rhinoceros. Such proofs verged on the "indecent," and Maupertuis urged, "[L]et us leave such trivialities [*bagatelles*] to those who do not sense their frivolity."[74] He chose to address their great forebear himself, critiquing Newton's invocation not only of organic form but of the order of the solar system to claim that only *choice* by an intelligent artificer could explain what nature presented to observation.[75] Maupertuis noted that Newton believed that "it was impossible for blind fate to make [the planets] move in the same direction and in almost concentric orbits."[76] But "the alternative between choice and extreme chance was founded on nothing more than Newton's incapacity to provide a physical cause for this uniformity."[77] The very possibility of a physical cause, however, sufficed in Maupertuis's view to discredit Newton's move. Maupertuis conceded that the argument from organic form appeared "more solid," but he suggested that such proofs were in fact epistemologically quite problematic: "The bodies of animals and plants are machines of such complexity that their ultimate parts utterly escape our senses, and we have too little knowledge of their utility and end for us to be able to judge of the wisdom and power necessary to construct them."[78]

---

[71] For a good recognition of the complexity of this balancing of chance and necessity, see Russo, "Théologie naturelle," 51n, 59.

[72] Annie Ibrahim, "Matière inerte et matière vivante: La théorie de la perception chez Maupertuis," *Dix-Huitième Siècle* 24 (1992): 95–103.

[73] Maupertuis, *Essai de cosmologie*, 7.

[74] Ibid., 12.

[75] Newton was most explicit in Query 31 to the *Opticks* from the 1706 edition forward.

[76] Maupertuis, *Essai de cosmologie*, 8.

[77] Ibid., 9.

[78] Ibid., 14.

He proposed a new principle for natural theology: "it is not at all in the tiny details, in those parts of the universe whose affinities we know too little, that one should be seeking the Supreme Being; it is in phenomena whose universality allows no exception and whose simplicity is exposed entirely to our scrutiny."[79] To be sure, Maupertuis went on, "the Supreme Being is everywhere, but he is not equally visible everywhere. We will see him best in the most simple objects; let us seek him in the first laws which he imposed upon nature, in those universal rules according to which motion is conserved, transmitted or destroyed."[80] He urged that physicotheology work from mathematical laws and the experimental-observational evidence that confirmed them. "This proves again the perfection of the Supreme Being: that all things are thus ordered that a blind and necessary mathematics carries out what the most enlightened intelligence and the freest will prescribed."[81]

During and after his last years in Paris, another domain became increasingly prominent in Maupertuis's thought: the life sciences. One of the glaring sore points both for natural science and for physicotheology by the mid-eighteenth century was preformation in the theory of generation, whether "ovist" or "animalculist" in orientation.[82] All the methodological issues of experimental life science, on the one hand, and all the metaphysical issues of a credible natural theology, on the other, assumed maximum acuity just there.

The thrust of Maupertuis's first intervention in this arena, *Vénus physique* (1745), was, as Michael Hoffheimer has contended, primarily critical, not constructive.[83] Its main aim was to demonstrate that "neither ovism nor animalculism could reasonably be accepted."[84] Maupertuis felt that the evidence from heredity and from what he took as the most rigorous embryological observation (by William Harvey, already more than a century old and without the benefit of microscopy) compelled return to the ancient view that the fetus arose from the "mixture of the seminal fluids given off by both sexes."[85] That meant that the idea of *emboîtement* – the complete preexistence of germs from the moment of divine creation of the world – had to be replaced by the idea that Harvey had enunciated: *epigenesis*.

[79] Ibid., 21.

[80] Ibid., 23.

[81] Ibid., 25.

[82] On the proliferation of ideas under these various headings and the problems with preformation, see Roger, *Les sciences de la vie*, 255–453.

[83] Hoffheimer, "Maupertuis," 124.

[84] Maupertuis, *Vénus physique*, in *Vénus physique suivi de la Lettre sur le progrès des sciences*, ed. Patrick Tort (Paris: Aubier Montaigne, 1980), 117; translated in Maupertuis, *The Earthly Venus*, trans. Simone Boas, *Sources of Science* no. 29 (New York: Johnson Reprint), 51.

[85] Ibid., 81, trans. 7.

The two claims Maupertuis believed observation and experience warranted – biparental contribution to and epigenetic development of the embryo – left a great explanatory void: *by what mechanism* did this take place? Maupertuis ventured a theory, suggesting that in the parental seminal fluids there were elementary particles which were keyed to particular parts of the parental organisms and which, in the mixture of their seminal fluids, were "attracted" to one another in some naturally lawful manner to form the same parts in the offspring.[86] That implied a theory of *active* matter, of "vital materialism." It was *materialism*, in that it allowed to nature the power of self-formation.[87] But it was *vital* in that it recognized that the mechanical principles of physics did not suffice to account for the phenomena.[88] The project was to move cautiously from established physical forces to whatever new ones proved necessary to save the phenomena. "Might they not depend on the same mechanisms and on similar laws? Could the ordinary laws of motion suffice, or should we call upon new forces for help?"[89] Maupertuis noted that already in chemistry the phenomena demanded a theory that surpassed the impact transmission of force. Geoffroy proposed *affinity* (*rapport*) as the essential principle required, and Maupertuis accepted this.[90] Indeed, he insinuated that in physics already departure from strict mechanism had become necessary: the recognition of *attraction* as a fundamental force beyond impact.[91] "Why should not a cohesive force, if it exists in Nature, have a role in the formation of animal bodies? ... [I]f these particular particles had a special attraction for those which are to be their immediate neighbors in the animal body, this would lead to the formation of the fetus."[92] Such a theory, Maupertuis argued, could handle all the anomalies that had accumulated so fatally in the theory

---

[86] Beeson and others have stressed that in the original (1745) edition, this caution extended to the crucial question of the source of the particles in the seminal fluids of parental organisms (Beeson, *Maupertuis*, 176; Hoffheimer, "Maupertuis," 125; Glass, "Maupertuis," 67).

[87] "These active properties allow for many possible outcomes, within certain parameters, and they locate the capability to produce order in matter itself" (Mary Terrall, "Salon, Academy, and Boudoir: Generation and Desire in Maupertuis's Science of Life," *Isis* 87 [1996]: 217–29, citing 223).

[88] That is, it embraced "the idea that matter contained a plastic, vital, even divine principle continuously at work" (Smith, *The Problem of Life*, 268).

[89] Maupertuis, *Vénus physique*, 120, trans. 55.

[90] Ibid., 120–1, trans. 55–6.

[91] Ibid., 121, trans. 55.

[92] Ibid., 121, trans. 56. Sandler can object, with two hundred years of hindsight, that Maupertuis here blurs phenotype with genotype, failing to observe the cardinal principle from Weisman governing modern reproduction theory. All true, but useful only for her project of debunking the "precursor of Mendel" argument, not for assessing Maupertuis's place in eighteenth-century life science or the history of biology.

of preformation. It could account for inheritance of traits from both parents, for hybrid animals, for the emergence of "monsters," etc.[93]

Maupertuis moved from the level of individual generation to the question of heredity across generations and therewith to the question of variation in species. It was clear from human breeding of animals that "[n]ature holds the source of all these varieties, but chance and art sets them going."[94] Breeders produced plants and animals that "did not exist in nature. At first they were individual freaks, but art and repeated generations turned them into new species."[95] To explain this, Maupertuis suggested that a plethora of elementary particles floated about in the seminal fluids, most of them derived from the immediate parents, but others of more remote heritage or altered by circumstance. "Chance or a shortage of family traits will at times cause variant combinations," he wrote.[96] Moreover, "I do not exclude the possible influence of climate and food ... I simply do not know how far this kind of influence of climate and food may go after many centuries."[97] What he did affirm strongly was that variants of this sort, left to nature, tended to degenerate and vanish; nature returned to her original patterns: "after a few generations, or even in the next generation, the original species will regain its strength."[98]

Such were the conjectures Maupertuis felt prepared to offer in 1745. He admitted he was not entirely satisfied with himself. He had managed only to "put forth doubts and conjectures."[99] That was too modest. As Mary Terrall argues, Maupertuis was engaged in "synthesizing evidence from anatomy, natural history, animal breeding, and travel literature" to propose a "theoretical model of active matter," ascribing to it "properties responsible for organization and heredity" and even affirming the "possibility of change in

---

[93] Ibid., 123, trans. 58. As Hoffheimer writes, the "theoretical limitations of preexistence were most trenchantly fixed by familiar macroscopic phenomena: hereditary resemblances, hybrids, and monsters" ("Maupertuis," 123).

[94] Maupertuis, *Vénus physique*, 134, trans. 71.

[95] Ibid., 134, trans. 72. Hoffheimer observes that "Maupertuis has a wholly elastic notion of species" ("Maupertuis," 127), and Sandler notes that "Maupertuis had a confused idea about what constituted a species" ("Maupertuis," 114n). Small wonder, given not only the eighteenth-century debates about that but even the current ones! There is a want of historical sense in these observations, true as they are.

[96] Maupertuis, *Vénus physique*, 140, trans. 79.

[97] Ibid., 141, trans. 80. "The final outcome of any mating is never completely determinate, since factors such as climate, diet, and even chance combinations of elements, contribute to individual variations" (Terrall, *The Man Who Flattened the Earth*, 217).

[98] Maupertuis, *Vénus physique*, 140, trans. 79, and again 141, trans. 80.

[99] Ibid., 144, trans. 85.

organic forms."[100] We can scarcely conceive a more revolutionary agenda in life science at the middle of the eighteenth century.[101]

The original edition of *Vénus physique* went through three printings by 1750.[102] It exercised a major influence on the work of the Comte de Buffon. It was part of a dramatic paradigm shift in the life sciences confirmed by contemporary writings of La Mettrie, Buffon, Diderot, and the school of Montpellier.[103] In the first volumes of his *Histoire naturelle* (1749), Buffon vigorously affirmed Maupertuis's achievement: "The *Vénus physique* ... although very brief, gathers together more philosophic ideas than there are in many large volumes on generation." Its author was "a man of spirit who seemed to me to have reasoned better than all those who have written before him on this matter."[104]

Once Buffon had taken up his ideas and confirmed them against his massive experimental and historical knowledge, Maupertuis set about elaborating a general theory of active matter, a plausible theory of *hylozoism*.[105] That was the endeavor of the *Système de la nature*. It sought, in the words of David Beeson, "to bring the whole natural world within a single system."[106] The thematic with which Maupertuis opened his *Système de la nature* picks up directly on all the considerations of the *Essai de cosmologie*. Maupertuis suggested that the recourse to divinity was a direct consequence of the failures of rational-mechanical thought to account for such things as chemical affinity and organic life, but such was bad science and demeaning theology. Recourse to the doctrine of preexistence constituted a cardinal instance. Whether one placed the divine miracle of generation at the origin of creation or at each conception, the miracle remained, and time was surely immaterial to an eternal being.[107]

---

[100] Mary Terrall, *The Man Who Flattened the Earth*, 202.

[101] As Bentley Glass characterizes the context, "By 1740 there remained very few epigenesists indeed, and the encasement theory of preformation prevailed almost universally" ("Maupertuis," 62).

[102] Terrall, *The Man Who Flattened the Earth*, 209n.

[103] For the crucial role of the medical school of Montpellier, see Elizabeth Williams, *The Physical and the Moral: Anthropology, Physiology, and Philosophical Medicine in France, 1750–1850* (Cambridge: Cambridge University Press, 1994).

[104] Buffon, *Histoire naturelle*, vol. 2 (1749), cited in Hoffheimer, "Maupertuis," 129.

[105] Hoffheimer, "Maupertuis," 126. Hoffheimer observes, very strikingly, that this "hylozoism converges with various forms of the Leibniz-Wolffian philosophy" then current in Germany (136). Terrall makes the important point that Maupertuis endeavored to make this theologically and scientifically palatable (*The Man Who Flattened the Earth*, 329).

[106] Beeson, *Maupertuis*, 211.

[107] Maupertuis made this argument in all three of his major publications in life science. Jacques Roger scoffed at it, but Immanuel Kant employed it both in 1763 and in 1790.

"Original production, in all systems, is a miracle."[108] The issue accordingly shifted: "[T]he universe once formed, by what laws is it preserved? What are the means whereby the Creator provided for the reproduction of individuals that perish? Here we have a free field [for inquiry]."[109]

Maupertuis asked, *by what specific mechanism* did certain parts become eyes and others ears?[110] "One will never explain the formation of a single organized body strictly by the physical properties of matter"[111] A single blind force of attraction in the universe could not suffice to explain the distinctive character of organized life-forms. Clearly the phenomena of the physical world required more of an explanation than Descartes could achieve with his dualism of spirit and matter; that had led to the conundrum of *forces* and especially to the problems of *attraction* in cosmology and *affinity* in chemistry. But if even these were insufficient to account for life-forms, "it seems appropriate to admit more new ones, or better to explore the properties [organic forms] possess."[112] "If one wishes to offer a conception about this, though it works only by analogy, it would seem necessary to have recourse to some principle of intelligence, to something similar to what we call *desire, aversion, memory*."[113] Thus, in embryology, each particle in the seminal fluid of father and mother retained "a kind of memory of its original placement [in the adult organism] and it will seek to take it up again as often as it can to form the same part in the fetus."[114] This language of desire, aversion, and memory was intentionally analogical and metaphorical.[115] Maupertuis offered an alternative language of "instinct," should his readers prefer.[116] He was even prepared to surrender all the metaphors, so long as the phenomena he had identified received the scientific attention they merited, which meant that a mechanistic physics could not suffice.[117]

But Maupertuis went decisively further, suggesting that generation should be construed as part of a universal dynamism of nature, advancing to higher and higher degrees of organization from inanimate matter to living organisms

---

[108] Maupertuis, *Système de la nature*, in Maupertuis, *Essai de cosmologie: Système de la nature: Réponse aux Objections de M. Diderot*, 157.

[109] Ibid., 155.

[110] Ibid., 146–7.

[111] Ibid., 155–6. Thus, Glass is wrong to suggest that Maupertuis found vitalism "abhorrent," that he was "a consistent mechanist" ("Maupertuis," 62).

[112] Maupertuis, *Système de la nature*, 154.

[113] Ibid., 147.

[114] Ibid., 158–9.

[115] Terrall (*The Man Who Flattened the Earth*, 328) makes this point clearly.

[116] Maupertuis, *Système de la nature*, 179.

[117] Ibid., 180.

to man.[118] He found evidence of this self-formative capacity in crystallization, as in the "tree of Diana."[119] Maupertuis concluded that plants and even crystals exhibited some elements of the self-organization that he discerned definitively in animal sexual reproduction, and accordingly he offered the prospect of a unified theory of natural process.[120] He conjectured that in the early phases of the formation of the earth, the globe had a fluid surface in which all elements floated freely. The least active elements eventually gave form to metal and stone, while the most active emerged as animals and man.[121] This plasticity persisted. Reflecting on catastrophe theories in geology, Maupertuis speculated that upon a new geological catastrophe, "new unions of elements, new animals, new plants, or things utterly unprecedented" might well ensue.[122] Early in the *Essai de cosmologie*, Maupertuis had written a passage as crucial for his life science as for his physicotheology:

> Could one not say that in the chance combination of natural products, given all that present themselves possess a certain aptness [*convenance*] enabling them to survive [*subsister*], it is no marvel that such aptness is to be found in all the species that actually exist? Chance, one might say, had produced a numberless multitude of individuals; [only] a small number found themselves constructed in a manner that the parts of the animal could satisfy its needs; in another, infinitely larger number, there was neither aptness nor order: all these last have perished; animals without mouths cannot live, others which lacked reproductive organs could not perpetuate themselves: the only ones that remain are those where order and aptness happened to arise, and these species which we see around us today are but the smallest part of that which a blind destiny produced.[123]

In *Système de la nature*, Maupertuis picked up the thread of this argument from chance for the variety of species and situated it in this general theory of dynamism in nature:

> Might one not explain in this manner how from a mere pair of individuals the multiplication of the most diverse species might have ensued? They will not have owed their initial formation to anything more than random productions, in which the elementary particles did not maintain the order that prevailed in the

---

[118] Ibid., 166.

[119] Ibid., 167. Already in *Vénus physique* Maupertuis developed this idea of self-formation in inorganic forms and offered the example of the "tree of Diana" from crystallography (119; trans. 54).

[120] Maupertuis, *Système de la nature*, 166–7.

[121] Ibid., 169.

[122] Ibid., 170.

[123] Ibid., 11–12.

male and female parents; each degree of error will have created a new species, and as a result of repeated deviations there will have arisen the infinite diversity of animals which we see today, which may grow further with the passing of time, but to which the passage of centuries may [just as well] supply only the most imperceptible augmentations.[124]

It is passages like these that lend some plausibility to arguments for Maupertuis as a "forerunner" of later biological theories.

But let us stay with the eighteenth-century context. David Beeson can discern little difference between Maupertuis's position and that of La Mettrie. That is, "both La Mettie and Maupertuis seem to suggest that materialism is the inevitable conclusion of empiricism."[125] Diderot made that the thrust of his famous "accusation" that Maupertuis was virtually a "Spinozist."[126] Indeed, Maupertuis had been accused of all this already upon the publication of *Vénus physique*, in the review of that work in the *Bibliothèque raisonée* (1745).[127] The connection with Spinozism in 1753 really came as no surprise to Maupertuis, since his friend La Condamine had raised it in their correspondence, and Maupertuis himself speculated, before his contact with Diderot, that it was this Spinozist tinge that may have put off Buffon from responding to the new work, which Maupertuis had sent him.[128] Indeed, his "response" to Diderot was only a half-hearted repudiation, and as Mary Terrall concludes, "Ultimately, Maupertuis's defense amounted to showing how close his position was to that of his interlocutor."[129]

What was Maupertuis's ultimate position? In his life science, "his challenge to preexistence theories went beyond anatomy and physiology to a theory of organization (on a submicroscopic scale) *and* a theory of heredity (on the macroscopic scale)."[130] On the first score, "Maupertuis's dynamic organicism made activity fundamental to matter."[131] On the second, "the theory historicized the problem of organization."[132] In a word, Maupertuis's life science culminated in *vital materialism*, in *hylozoism*. That was crucial for its reception by Immanuel Kant.

[124] Ibid., 164–5.

[125] Beeson, *Maupertuis*, 214.

[126] Diderot, *Pensées sur l'interprétation de la nature*, in *Oeuvres complètes*, ed. Jean Verloot, (Paris: Hermann, 1981), vol. 9.

[127] Terrall, *The Man Who Flattened the Earth*, 224.

[128] Ibid., 323.

[129] Ibid., 346. Vartanian is unpersuasive in trying to dismiss the "Spinozist" connection entirely in his "Diderot and Maupertuis."

[130] Terrall, "Salon, Academy, and Boudoir," 225.

[131] Ibid., 224.

[132] Terrall, *The Man Who Flattened the Earth*, 340.

3. KANT'S *ONLY POSSIBLE ARGUMENT*:
    EPIGENESIS AND HYLOZOISM

One of the merits of Martin Schönfeld's study is to highlight and problematize the presence of physicotheology in Kant's *Only Possible Argument*. "Why does a book that purports to present *one* possible argument contain two?" Schönfeld queries.[133] Scholars of Kant have often found it fruitful to use such oddities of textual organization as clues to philosophical complexities in Kant's position or intention. Schönfeld believes this one reveals the continuity of the text with Kant's "precritical project" of the 1750s, namely, "the unification of natural science and metaphysics into a philosophical model of nature."[134] As Kant himself put it, "[M]y intention ... has been focused on the method of using natural science to attain cognition of God."[135]

Maupertuis was a direct inspiration. Kant's whole treatment of physicotheology in *The Only Possible Argument* is derived from Maupertuis. His distinction between a flawed traditional form of physicotheology and a revised form which could have greater warrant picked up and elaborated all the key points, down to the very examples in Newton which Maupertuis had adduced. Kant was explicit about this connection. Maupertuis's name appeared frequently in the text, and uniformly favorably, with the one possible exception – which is of course my primary interest – regarding epigenesis. Crucially, Kant accepted without reservation Maupertuis's "important discovery" regarding the principle of least action. From this principle in physics, Maupertuis had drawn the essential metaphysical implication: "This acute and learned man immediately sensed that ... such a universal cohesiveness [*Zusammenhang*] in the simplest natures of things afforded a far more fitting foundation [for theology] than any perceptions of various contingent and variable arrangements instituted in accordance with particular laws."[136]

The challenge faced by this revisionism in physicotheology was to overcome the scruple among the orthodox that *any* such register of natural necessity might constitute a threat to religious conviction regarding the existence of God.[137] Like Maupertuis, Kant wished to "remove the baseless suspicion ... that explaining any of the major arrangements in the world by appealing to the universal laws of nature opens a breach which enables the wicked

[133] Schönfeld, *The Philosophy of the Young Kant*, 192.
[134] Ibid., 8.
[135] Kant, *Der einzig mögliche Beweisgrund*, AA 2:68, trans. 114.
[136] Ibid., AA 2:99, trans. 142.
[137] Ibid., AA 2:118, trans. 160.

enemies of religion to penetrate its bulwarks."[138] That is, many felt that "if one admitted that the operations of nature could produce such results, the admission would be tantamount to ascribing the perfection of the universe to blind chance."[139] Kant, like Maupertuis, contended that "in spite of its *prima facie* similarity ... the atomistic system of ... *Epicurus*" was quite distinct from what he was proposing.[140] Kant insisted that the ancient hypothesis of Epicurus and Lucretius of "blind chance" in the "swerve of atoms" to account for motion was an "absurdity and deliberate blindness."[141] But it was just as important to deny the modern reassertion of such ideas associated with "Spinozism." Spinoza's God was tantamount to atheism: "Possessing neither cognition nor choice, it would be a blindly necessary ground of other things and even of other minds, and it would differ from the eternal fate postulated by some ancient philosophers in nothing except that it had been more intelligently described."[142] Kant wanted nothing to do with pantheism: "The world is not an accident of God."[143] For Kant, "there is a distinction to be drawn between that which is the effect of constant and necessary laws and that which is the product of blind chance."[144] Thus, "the necessity perceived in the relation of things in regular combinations, and the connection of useful laws with a necessary unity, afford proof of a Wise Author, just as well as the most accidental and artificially devised provision."[145] This was the core of the revisionist physicotheology which Kant shared with Maupertuis. Whether either could ultimately succeed in upholding this distinction, it was the centerpiece of a common metaphysical endeavor to reconcile natural science with religion, for them a most urgent issue for philosophy at the middle of the eighteenth century.

Kant postulated an "order of nature" which "embraced a complex harmony in a necessary unity."[146] Now, Kant recognized that "all natural things are contingent in their existence," and thus, too, the natural laws relating them.[147] Still he contended that "there nonetheless remains a kind of necessity which is very remarkable. There are, namely, many laws of nature, of which the

---

[138] Ibid., AA 2:148, trans. 188.

[139] Ibid., AA 2:119, trans. 161.

[140] Ibid. Maupertuis endeavored to draw the same distinction in *Système de la nature*, 182.

[141] Ibid., AA 2:125, trans. 164.

[142] Ibid., 2: 89, trans. 133.

[143] Ibid., AA 2:90, trans. 134.

[144] Ibid., AA 2:126, trans. 167.

[145] Ibid., AA 2:123, trans. 164.

[146] Ibid., AA 2:107, trans. 150.

[147] Ibid., AA 2:106, trans. 148.

unity is necessary."[148] This principle of a necessary order of nature expressed in the unity of its laws proved fundamental to Kant's philosophy of science throughout.[149] Here, Kant elaborated: "[T]he necessity of these laws is such that they can be derived from the universal and essential constitution of all matter."[150] A central heuristic maxim for natural science followed: one must "not immediately suppose the existence of new and diverse operative causes to explain different effects because of some seemingly important dissimilarity between them."[151] Thus, "it is the mark of excessive haste to ascribe an arrangement immediately to the act of creation."[152]

Following the bold example of Maupertuis, Kant chastised Newton for just this failing. By name, he made the case that Newton had no hesitation in ascribing the sphericity of the earth and its polar flattening to physical-mechanical laws but that he fell back directly to divine intervention to account for the common direction and plane of the orbits of planets in the solar system.[153] Later, without explicitly mentioning Newton's name, Kant suggested that Newton

> would not normally be willing to admit such an account in natural sciences. He can name no purpose to explain why it should be better for the planets to move in one direction rather than in a number of different directions, nor why they should revolve around the sun in orbits approximating to a single common plane of reference.[154]

Maupertuis had been willing to leave the argument against Newton at that point, holding out simply the *possibility* of a natural explanation.[155] Kant went further, urging his own hypothesis from *Universal Natural History* as an available, theoretically consistent natural explanation.[156]

---

[148] Ibid.

[149] On Kant's philosophy of science, see Gerd Buchdahl, *Metaphysics and the Philosophy of Science: The Classical Origins: Descartes to Kant* (Cambridge, MA: MIT Press, 1969); Peter Plaass, *Kant's Theory of Natural Science* (Dordrecht: Kluwer, 1994); Robert Butts, ed., *Kant's Philosophy of Physical Science* (Dordrecht: Reidel, 1986); Friedman, *Kant and the Exact Sciences*, n. 8; and Watkins, *Kant and the Sciences*, n. 11.

[150] Kant, *Der einzig mögliche Beweisgrund*, AA 2: 99, trans. 142.

[151] Ibid., AA 2:113, trans. 155.

[152] Ibid., AA 2:135, trans. 175.

[153] Ibid., AA 2:121, trans. 162.

[154] Ibid., AA 2:142, trans. 182.

[155] Maupertuis, *Essai de cosmologie*, 8.

[156] Kant adverted to his *Universal Natural History* first in the preface of *Der einzig mögliche Beweisgrund*, in a note regarding the parallel hypothesis advanced by Johann Lambert (AA 2:69; trans., 114), then summarized the argument of his earlier study in the seventh reflection of Part II (ibid., 137–51, trans. 177–91).

More significantly still, he made this the token for a crucial distinction in philosophical rigor: "All explanations of provisions in the world, not just those relating to the animal- and plant-kingdoms, which are presented in terms of laws which have been artificially instituted with a view to realizing some specific objective," failed to achieve full philosophical rigor.[157] This lapse of rigor arose, Kant suggested, "when something which could perhaps be accounted for in terms of ordinary mechanical forces, is explained in terms of the plant- and animal-kingdoms, simply because order and beauty are prominent there."[158] Kant adduced the example of snowflakes: "crystals which are so regular, so delicate, so far removed from all the clumsiness which blind chance could bring about" that even "art has nothing at all to offer which displays greater precision [*Richtigkeit*]."[159] He went on: "And yet it has occurred to no one to explain their origin in terms of a special snow-seed [*Schneesamen*], or to imagine an artificially instituted arrangement of nature to account for them."[160] In a footnote he suggested that mildew might well be a chemical phenomenon, and, in contrast with Maupertuis, he definitely chose to view the crystal "tree of Diana" as "an effect of the universal laws of sublimation" rather than as something quasi-organic.[161]

"What is objectionable," he claimed, "is ... that the order of nature which produced these useful consequences is construed as being artificially and deliberately connected with the other orders of nature, whereas, in fact, it may perhaps be necessarily connected with them."[162] "Something is attributed to an artificially devised order of nature before it has been properly established that nature is incapable of producing that phenomenon in accordance with her universal laws."[163] In a footnote, Kant articulated the key point: "It is, however, to be remarked that any law which is instituted for the sake of some special use is artificially devised, for *it is then no longer connected with the other laws of nature with necessary unity*"(emphasis added).[164]

Yet just this problem for the unity of the order of nature is what the entire world of living things presents. A fundamental line of division in physicotheological argument arose between the organic and the inorganic. Whatever naturalism might obtain in the interpretation of the inorganic world, living

---

[157] Ibid., AA 2:135–6, trans. 176.
[158] Ibid., AA 2:135, trans. 176.
[159] Ibid., AA 2:113–14, trans. 155.
[160] Ibid., AA 2:114, trans. 156.
[161] Ibid., AA 2:114n, trans. 156n.
[162] Ibid., AA 2:136, trans. 177.
[163] Ibid., AA 2:135, trans. 176.
[164] Ibid., AA 2:136n, trans. 176n.

forms required direct divine intervention. Kant unequivocally affirmed living forms as prime examples of "an artificially devised order of nature."[165] "The structure of plants and animals . . . cannot be explained by appeal to the universal and necessary laws of nature. . . . [I]t would be absurd to regard the initial generation of a plant or animal as a mechanical effect incidentally arising from the universal laws of nature."[166] Life had to be recognized as "a contingent, purposeful phenomenon," that is, one which was the direct "product of choice."[167] As for the necessary unity he had adduced for the order of nature, Kant could only rationalize that "the creatures of the plant- and animal-kingdoms everywhere offer the most admirable examples of a unity which is at once contingent and yet in harmony with great wisdom."[168] Great wisdom, perhaps: but the necessary unity of an order of nature? Living forms, because radically contingent, stood in a "connection" that was, Kant admitted, "quite alien to the nature of things themselves."[169] In those terms, this radical contingency of the entire organic world represented a formidable stumbling block for the unity of the order of nature.

Kant recognized the methodological inference that should follow: "One is inclined to suppose that perhaps even when, in organic nature, many perfections may seem to be the product of provisions which have been especially made, they may, notwithstanding, be the necessary effect of a single ground."[170] That is, "even in the case of the structure of an animal, it can be assumed that there is a single disposition, which has the fruitful adaptedness to produce many different advantageous consequences [*daß eine einzige Anlage eine fruchtbare Tauglichkeit zu viel vortheilshaften Folgen haben wurde*]."[171] How, though, should this be conceived?

We have reached the crucial discussion of preformation and epigenesis.[172] Kant wrote,

> Is each individual member of the plant- and animal-kingdom directly formed by God, and thus of supernatural origin, with only propagation, that is to say, only the periodic transmission for the purpose of development, being entrusted to a natural law?[173]

---

[165] Ibid., AA 2:136n, trans. 176n.
[166] Ibid., AA 2:114, trans. 156.
[167] Ibid., AA 2:96, trans. 140.
[168] Ibid., AA 2:107, trans. 149–50.
[169] Ibid., AA 2:96, trans. 140.
[170] Ibid., AA 2:107, trans. 150.
[171] Ibid., AA 2:126, trans. 167.
[172] Without the mention of either technical term, it should be noted.
[173] Kant, *Der einzig mögliche Beweisgrund*, AA 2:114, trans. 156.

Such was his description of preformation. The questions that arise are these. First, *whose* specific formulation of preformation is Kant invoking here? Second, how exactly is he parsing two key issues, the moment of divine intervention and the residual role of natural laws? It would appear that he had in mind the form of preformation elaborated by Malebranche, linked to his occasionalist doctrine, which spreads divine intervention across phenomenal time.[174] It is not immediately clear what Kant meant in claiming this theory allowed natural law a role only in "periodic transmission for the purpose of development [*Fortpflanzung, das ist, der Übergang von Zeit zu Zeit zur Auswickelung*]." This is a very obscure formulation, which only gets clarified in subsequent passages in the text. Kant continued:

> Or do some individual members of the plant- and animal-kingdom, although immediately formed by God and thus of divine origin, possess the capacity, which we cannot understand, actually to generate their own kind in accordance with a regular law of nature, and not merely to unfold them?[175]

This is Kant's formulation of epigenesis. Three features deserve note. First, like Maupertuis, Kant ascribed the *original* emergence of all species to direct divine intervention. Second, he ascribed to all subsequent reproduction an immanent capacity, "in accordance with a regular law of nature," for *Erzeugung*, not simply *Auswicklung*, as in preformation. But, finally, crucially, he claimed that the capacity for such *Erzeugung* was *nicht begreiflich* – literally, *inconceivable*: very strong language betokening unintelligibility.

Kant followed his presentation of these two alternative theories with this comment: "There are difficulties on both sides, and it is perhaps impossible to make out which difficulty is the greatest."[176] That would seem to suggest that Kant took a thoroughly neutral stand as between the two theories, objecting to both. But his elaboration of their difficulties complicates that assessment. For Kant, we should note at the outset, the issue between the two theories was *metaphysical*, not scientific.[177] Presumably with reference to preformation, he objected that "it is utterly unintelligible to us that a tree should be able, in virtue of an internal mechanical constitution, to form and process its sap in such a way that there should arise in the bud or the seed something containing a tree like itself in miniature, or something from which such a tree could

---

[174] On the prominence of Malebranche's formulation for the first half of the eighteenth century, see Smith, *The Problem of Life*, 266.

[175] Kant, *Der einzig mögliche Beweisgrund*, AA 2:114, trans. 156.

[176] Ibid.

[177] Ibid.

develop [*werden*]."[178] There are several problems with this formulation, if it is to be grasped as a consideration of the preformation theory. All mention of divine intervention, at the dawn of the Creation or over time, has disappeared in this passage. What is left is the suggestion that by purely mechanical means the generation of an identical offspring could be achieved. That smacks far more of a *third* option altogether – what the eighteenth century regarded as *spontaneous generation*, a phenomenon of materialist *chance* – than of preformation.[179] This third option will occupy us further below.

Kant accorded no metaphysical significance to the timing of divine intervention: "[W]hether the supernatural generation occurs at the moment of creation, or whether it takes place gradually, at different times, the degree of the supernatural is no greater in the second case than in the first."[180] This is a strong echo of Maupertuis, and to the same end: to urge that moving the miracle back to the moment of the Creation is only meaningful if it allows to natural laws a more substantial part in the subsequent process of reproduction. Kant did not find it comforting that in preformation theory "natural philosophers have been left with something when they are permitted to toy with the problem of the manner of gradual propagation."[181] He elaborated: "As for the natural order of unfolding . . . it is not a rule of the fruitfulness of nature, but a futile method of evading the issue."[182] We must wonder that Kant found this aspect of preformation theory so unsatisfactory, for it won considerable praise from the greatest physiologist of the day in Germany, Albrecht von Haller, and it is not all that remote from his own theory of 1775.[183]

In any event, Kant was no kinder to the recently revived theories of epigenesis: "the internal forms proposed by Buffon, and the elements of organic matter which, in the opinion of Maupertuis, join together as their memories dictate and in accordance with the laws of desire and aversion, are either as incomprehensible as the thing itself, or they are entirely arbitrary inventions [*sind entweder eben so unverständlich als die Sache selbst, oder ganz willkürlich erdacht*]."[184] It is important to note that Kant read Maupertuis's

---

[178] Ibid., AA 2:114–15, trans. 156–7.

[179] John Farley, *The Spontaneous Generation Controversy from Descartes to Oparin* (Baltimore: Johns Hopkins University Press, 1977).

[180] Kant, *Der einzig mögliche Beweisgrund*, AA 2:114–15, trans. 156–7.

[181] Ibid.

[182] Ibid.

[183] Shirley Roe, *Matter, Life and Generation: Eighteenth-Century Embryology and the Haller-Wolff Debate* (New York: Cambridge University Press, 1981); François Duchesneau, "Haller et les théories de Buffon et C. F. Wolff sur l'epigenèse," *History and Philosophy of the Life Sciences* 1 (1979): 65–100.

[184] Kant, *Der einzig mögliche Beweisgrund*, AA 2:115, trans. 157.

metaphors quite literally. Of the whole approach of epigenesis, he made two *dismissive* criticisms: First, these theories achieved no explanatory purchase on the phenomenon itself, and second, they were literally fantastical.

The choice boiled down, as he saw it, to attributing generation "immediately to a divine action... at every mating" or granting "to the initial divine organisation of plants and animals a capacity, not merely to develop [*entwickeln*] their kind thereafter in accordance with a natural law, but truly to generate [*erzeugen*] their kind."[185] Given this extended discussion, it is not hard to grasp that interpreters like Philippe Hunemann and Susan Shell have inferred that Kant was not really neutral at all but leaned toward the epigenetic option – *erzeugen* rather than *entwickeln*.[186] It behooves us, however, to consider, before reaching a conclusion, a second discussion of the problem, later in the text. That discussion takes place in the context of Kant's ranking of approaches to physicotheology in terms of their philosophical rigor, a discussion we have already stressed at a more universal level of this argument. Kant appears to give convincing evidence for the interpretation of Hunemann and Shell when he writes as follows:

> The philosophical character of the mode of thought which maintains that each individual animal or plant is immediately subsumed under a special act of creation is then even less than that of the view which maintains that, with the exception of a few directly created organisms, all other creatures are subsumed under them in accordance with a law which governs the ability to generate (not merely one which governs the capacity to unfold). This latter type of theory is more philosophical because it explains a greater number of phenomena in terms of the order of nature.[187]

There follows, however, a qualifying statement: "Its philosophical superiority could only be challenged if it could be clearly demonstrated that the order of nature was incapable of explaining the phenomenon under examination."[188]

What is to be made of that? Hunemann argues that Kant could not accept epigenesis any more than preformation on the terms he had laid out and that both forms had to be found wanting in philosophical rigor, despite the

---

[185] Ibid.

[186] Susan Shell, *The Embodiment of Reason: Kant on Spirit, Generation, and Community* (Chicago: University of Chicago Press, 1996), 384 n. 5; Philippe Hunemann, *Métaphysique et biologie: Kant et la constitution du concept d'organisme* (Villeneuve: Presses Universitaires du Septentrion, 2002), 169–200.

[187] Kant, *Der einzig mögliche Beweisgrund*, AA 2:135–6, trans. 176.

[188] Ibid., AA 2:136, trans. 176.

preferability of the *possibility* of epigenesis.[189] Hunemann goes on to suggest that when Caspar Friedrich Wolff developed a more scientific mechanism for the explanation of epigenesis, Kant could and did endorse it.[190] I disagree. Far from finding Wolff's epigenesis congenial, Kant seems clearly to have accepted Haller's side of the dispute with Wolff through the 1770s.[191] Kant followed Haller's view very closely in his first essay on race, adopting eagerly the latter's modified preformation theory both because it seemed more methodologically viable and also – perhaps even more – because it vigorously asserted the metaphysical objections against hylozoism.[192]

Epigenesis as an empirical scientific theory had *no* prospect of realization for Kant, because he held firm to the conviction that "one is incapable of rendering distinct the natural causes which bring the humblest plant into existence."[193] Thus, for Kant, the hypotheses of Buffon and of Maupertuis were *not* scientific but only fanciful or metaphysical, that is, *ganz willkürlich erdacht*. Kant allowed no prospect, notwithstanding the purported superiority of the scientific *motivation* of their enterprise, of any real scientific *method* or evidentially warranted *explanation*.[194] What was it that made these hypotheses appear irredeemably fanciful to Kant? The answer is *hylozoism*. I suggest that Maupertuis was the early Kant's paradigmatic instance of a modern hylozoist.[195] Kant's *Dreams of a Spirit-Seer* treats him in exactly that context:

> *Hylozoism* invests everything with life, while *materialism*, when carefully considered, deprives everything of life. Maupertuis ascribes the lowest degree of life to the organic particles of nourishment consumed by animals; other philosophers regard such particles as nothing but dead masses, merely serving to magnify the power of the levers of animal machines.[196]

Kant could only view the assertion of epigenesis as hylozoism, and, early or late, there was nothing toward which he felt a stronger metaphysical animus.

---

[189] Hunemann, *Métaphysique et biologie*, 181–3.

[190] Ibid., 306.

[191] This is the position from which he composed the first essay on race: Kant, "Von den verschiedenen Racen der Menschen," AA 2:427–44.

[192] Phillip Sloan, "Preforming the Categories," *Journal of the History of Philosophy* 40 (2002): 229–53, has made this perfectly clear.

[193] Kant, *Der einzig mögliche Beweisgrund*, AA 2:138, trans. 179.

[194] Modern critics advance the same scruple against these eighteenth-century theories, but their objections do not sanction Kant's, for his objection was not "prescience" but metaphysical fiat.

[195] His former student, Johann Gottfried Herder, became the later Kant's exemplar of hylozoism.

[196] Kant, *Träume eines Geistersehers*, AA 2:330; trans. "Dreams of a Spirit-Seer," in Kant, *Theoretical Philosophy 1755–1770*, 317–18.

In his ultimate consideration of the issue in the *Critique of Judgment* (1790), Kant wrote,

> We perhaps approach nearer to this inscrutable property if we describe it as an *analogue of life*, but then we must either endow matter, as mere matter, with a property that contradicts its very being (hylozoism) or associate it with a foreign principle *standing in communion* with it (a soul).[197]

Kant could accept neither. That was what fatally compromised the epigenetic hypothesis in his eyes in 1763, as it later would in 1790. So, in keeping with my original suggestion about trends in recent Kant studies to relate ideas of the precritical period to those of the critical period, I would like to conclude by juxtaposing Kant's comments on Maupertuis in *The Only Possible Argument* with his treatment of epigenesis (without mention of Maupertuis) in the *Critique of Judgment*.

4.   KANT'S LATER THOUGHT ON EPIGENESIS

In his dispute with Johann Gottfried Herder and Georg Forster from 1784 through 1788 – that is, up through the time of his revision of the first *Critique* – Kant rejected as an insupportable hylozoism what was simply epigenesis as Herder derived it from the writings of Caspar Friedrich Wolff and probably Johann Friedrich Blumenbach.[198] Indeed, one can find no unequivocal

---

[197] Kant, *Kritik der Urteilskraft*, AA 5:374–5.

[198] In the second installment of his review, Kant wrote, "As the reviewer understands it, the sense in which the author uses this expression [i.e., *genetische Kraft*] is as follows. He wishes to reject the system of evolution on the one hand, but also the purely mechanical influence of external causes on the other, as worthless explanations. He assumes that the cause of such differences is the vital principle [*Lebensprinzip*] which modifies *itself* from within in accordance with variations in external circumstances, and in a manner appropriate to these. The reviewer is fully in agreement with him here, but with this reservation: if the cause which organizes *from within* were limited by its nature to only a certain number and degree of differences in the development of the creature which it organises (so that, once these differences were exhausted, it would no longer be free to work from another archetype [*Typus*] under altered circumstances), one could well describe this natural development of formative nature in terms of germs [*Keime*] or original dispositions [*Anlagen*], without thereby regarding the differences in question as originally implanted and only occasionally activated mechanisms or buds [*Knospen*] (as in the system of evolution); on the contrary, such differences should be regarded simply as limitations imposed on a self-determining power, limitations which are inexplicable as the power itself is incapable of being explained or rendered comprehensible" (Kant, "Recensionen von J. G. Herders *Ideen zur Philosophie der Geschichte der Menschheit*," Theil 1.-2. AA 8:43–66, citing 62–3).

affirmation of epigenesis in any of Kant's writings before 1787.[199] In his *Metaphysics Lectures*, Kant quite explicitly rejected it: "The system of epigenesis does not explain the origin of the human body, but says far more that we don't know a thing about it."[200]

What could Kant possibly have been thinking, then, in his famous analogy at B167 of the first *Critique*, evoking "as it were, ... the *epigenesis* of pure reason?"[201] The fundamental analogy structure at B167 invoked the disjunction: *either* spontaneous generation *or* epigenesis. Preformation was introduced only negatively, in connection with the misguided endeavor to insert a third, intermediate position. What did "epigenesis" signify here? What did the immediately ensuing phrase "*self-thought* first principles *a priori*" betoken?[202] Was there something that Kant now saw in the idea of epigenesis that could help him elucidate the peculiar and essential *spontaneity* of the understanding in his transcendental deduction?

What drew Kant to epigenesis at all? The beginning of an answer is that Kant appropriated the notion from Herder.[203] He found it in a form that was too radical for his taste, yet it charmed him to seize it from Herder and make it stand precisely for his own position. All that was required was a two-step process. First, Kant had to insist that even epigenesis implied preformation: at the origin there had to be some "inscrutable" (transcendent) endowment – and with it, crucially, inalterable fixity in species. Granted that, the organized matter within the natural world could proceed on adaptive, even mechanistic lines. This made epigenesis over into Kant's variant of preformation. Even so, this seemed to postulate the objective *actuality* of these forces for natural science, something Kant found inconsistent with his "Newtonian" prescriptions. Hence Kant faced the ultimate need for a second step: to transpose the whole matter from the constitutive to the regulative order.

Kant employed the contrast of *educt* and *product* to discriminate preformation from epigenesis.[204] In an educt, all the relevant material preexists, and only its aggregation is shuffled, whereas in a product, altogether new

---

[199] Günter Zöller, "Kant on the Generation of Metaphysical Knowledge," in *Kant: Analysen – Probleme – Kritik*, ed. H. Oberer and G. Seel (Würzburg: Königshausen and Neumann, 1988), 71–90, citing 80–4, discusses uses of epigenesis in Kant's lectures and *Reflexionen*, but there is no reason to suspect any of these date significantly before 1786.

[200] Kant, *Vorlesungen über Metaphysik*, A A 29:761.

[201] Kant, *Critique of Pure Reason*, B 166–7 (1787).

[202] Ibid.

[203] Zöller, "Kant on the Generation of Metaphysical Knowledge," 81. It is even possible that Kant adopted the very *term* from Herder.

[204] It reappears in a crucial context: Kant, *Kritik der Urteilskraft*, AA 5:423.

things emerge.[205] Kant believed this distinction already established in chemistry, and theories of generation in the life sciences seemed variants of the same principle. Thus there were, for him, only *two* theoretical possibilities for generation, namely, preformation (the educt theory) and epigenesis (the product theory). In terms of the educt/product distinction, we gain a clearer sense of what Kant found essential in the idea of epigenesis. But we also learn that it was problematically spontaneous, from Kant's vantage point, ascribing too much power to material substances. That is, the metaphysical issue with epigenesis was still hylozoism.

Crucial for Kant's more positive disposition toward epigenesis was the revolution in thinking about this phenomenon inaugurated by Johann Friedrich Blumenbach in 1781.[206] Blumenbach repudiated hylozoism, too: "No one could be more totally convinced by something than I am of the mighty abyss which nature has fixed [*befestigt*] between the living and the lifeless creation, between the organized and the unorganized creatures."[207] This was what Kant found most gratifying in Blumenbach's new book on the *Bildungstrieb*, as he reported in his letter of acknowledgment to Blumenbach.[208] What bound the two thinkers most together was their commitment to the fixity of species.

But how could the transformist implications of epigenesis be contained within the limits of the fixity of species?[209] This was the essential question that Kant posed in his second essay on race in 1785, and the stakes of the question were not small. Without some regulation in the history of generation, the prospect of the scientific reconstruction of the connection between current and originating species (*Naturgeschichte*, in Kant's new sense, or the "archaeology of nature," as he would call it in the third *Critique*) appeared altogether hopeless.[210] Yet it was not simply a *methodological* issue, however dire. There was also an essential *metaphysical* component. Consider Kant's highly charged language in the 1785 essay on race:

[I]f some magical power of imagination...were capable of modifying...the reproductive faculty itself, of transforming Nature's original model or of making

[205] Kant, *Vorlesungen über Metaphysik*, AA 28:689, AA 29:760–1.
[206] Timothy Lenoir, "Kant, Blumenbach and Vital Materialism in German Biology," *Isis* 71 (1980): 77–108; Richards, "Kant and Blumenbach on the *Bildungstrieb*," n. 4; Peter McLaughlin, "Blumenbach und der Bildungstrieb: Zum Verhältnis von epigenetischer Embryologie und typologischem Artbegriff," *Medizinhistorisches Journal* 17 (1982): 357–72; and my own essay "'This Inscrutable *Principle* of an Original *Organization*,'" n. 2.
[207] Johann Friedrich Blumenbach, *Über den Bildungstrieb* (Göttingen: Dieterich, 1789), 71.
[208] Kant to Blumenbach, August 5,1790, AA 11:176–7.
[209] See especially McLaughlin, "Blumenbach und der Bildungstrieb."
[210] Kant, "Bestimmung des Begriffs einer Menschenrace," AA 8:89–106, citing 101–3.

additions to it,... we should no longer know from what original Nature had begun, nor how far the alteration of that original may proceed, nor... into what grotesqueries of form species might eventually be transmogrified.... I for my part adopt it as a fundamental principle to recognize no power... to meddle with the reproductive work of Nature... [to] effect changes in the ancient original of a species in any such way as to implant those changes in the reproductive process and make them hereditary.[211]

We cannot but discern that again it is the idea of *hylozoism* – of any radical spontaneity in matter itself – that Kant could not abide. All organic form had to be fundamentally distinguished from mere matter. "Organization" demanded *separate creation*. Eternal inscrutability was preferable to any "speculative" science. In the third *Critique*, Kant would twice insist that no human could ever achieve a *mechanist* (he meant, as well, a *materialist*) account of so much as a "blade of grass."[212] Kant remained adamant that the *ultimate* origin of "organization" required a *metaphysical*, not a physical, account: "How this [original] stock arose, is an assignment which lies entirely beyond the borders of humanly possible *natural philosophy*, within which I believe I must contain myself," Kant wrote in 1788.[213] He invoked Blumenbach for support in these *metaphysical* reservations.[214]

By the time Kant came to write the *Critique of Judgment*, he was fully conversant with Blumenbach's sophisticated theory.[215] But what progress had Kant made on the philosophical conundrum of preformation versus epigenesis? It is important to distinguish two quite distinct sets of discriminations in the *Critique of Judgment* that both point back to B167, but with different implications. The first discriminations come in a footnote to §80; the second come in the main text of §81. The footnote to §80 evokes the familiar term *generatio aequivoca* in order, as before, to disparage it. The contrast, however, is not to epigenesis or to preformation but rather to *generatio univoca*, which Kant further subdivides into *generatio homonyma* and *generatio heteronyma*. While he once again dismissed spontaneous generation as contradictory, Kant asserted that transmutation of species (*generatio heteronyma*) was not contradictory, only unfound in experience. Thus, the issue at stake in this discrimination was the principle of the fixity of species. In §81, however, we come upon a different schematization. Here, Kant postulated

---

[211] Ibid., 97.

[212] Kant, *Kritik der Urteilskraft*, ΛΛ 5:400, 409.

[213] Kant, "Über den Gebrauch teleologischer Principien," AA 8:179.

[214] Ibid., 180n.

[215] One indication, as Phillip Sloan has noted, was that Kant suppressed any mention of *Keime* in that work, though it still thronged with the term *Anlage* ("Preforming the Categories," 247–8).

that we must think of organisms on the analogy of an intelligent causality and that when we do so we face alternatives that can best be grasped *in terms drawn from metaphysics* (i.e., the obverse of the analogy at B167). The categories Kant offered were *occasionalism* and *prestabilism*. He dismissed occasionalism as curtly as he had dismissed spontaneous generation (though, of course, for different reasons), and in turning to "prestabilism" he distinguished two subsets: *individual preformation*, which he identified with the "theory of evolution" (i.e., *emboîtement*) and termed an "educt," and *generic preformation*, which he suggested was the proper sense of *epigenesis*. That is, though a "product," epigenesis "still performed in accordance with the internally purposive predispositions that were imparted to its [original] stock."[216] Kant expressed a clear preference for epigenesis over individual preformation, since it entailed "the least possible application of the supernatural" in scientific theory.[217]

Note that Kant configured his whole conceptualization *under the aegis of preformation*. Even as he was prepared to affirm epigenesis, Kant set strict limits upon it: ultimately this was merely "generic preformation," that is, it, too, was restricted by the original intervention of a *transcendent* causality.[218] There is no strict parallelism between the distinctions of §80 and of §81. The distinction between *generatio homonyma* and *generatio heteronyma* does not map neatly onto that between individual and generic preformation. That suggests that a different point is being made in the latter distinction, and indeed this point has to do with the character of the *causality* that must be employed in conceptualizing organic forms altogether, namely, the inadequacy not merely of *mechanism* but above all of *materialism*.[219] Yet there is some spillage between the two patterns of discrimination. Kant found the idea of the transmutation of species – *generatio heteronyma* – to induce the very sorts of loose thinking in science that might read epigenesis as hylozoism, as a *vital* materialism. It was this above all that he wished to circumvent, both with his ontological argument that even epigenesis depended upon an original creation which instilled organization into inert matter and with his epistemological argument that an empirical science of life-forms could only work with maxims of reflective judgment imputing purposiveness and thus that the very idea of a natural purpose was merely a heuristic fiction suited to

---

[216] Kant, *Kritik der Urteilskraft*, AA 5:423.
[217] Ibid., 424.
[218] Ibid., 423.
[219] A. Genova,"Kant's Epigenesis of Pure Reason," *Kant-Studien* 65 (1974): 259–73; Zöller, "Kant on the Generation of Metaphysical Knowledge," 90.

our limited reason. It was just these elements in Blumenbach's new work on the *Bildungstrieb* which Kant found so gratifying.

> Blumenbach... rightly declares it to be contrary to reason that raw matter should originally have formed itself in accordance with mechanical laws, that life should have arisen from the nature of the lifeless, and that matter should have been able to assemble itself into the form of a self-preserving purposiveness by itself; at the same time, however, he leaves natural mechanism an indeterminable but at the same time also unmistakable role under this inscrutable *principle* of an original *organization*, on account of which he calls the faculty in the matter in an organized body (in distinction from the merely mechanical *formative power* [*Bildungskraft*] that is present in all matter) a *formative drive* [*Bildungstrieb*] (standing, as it were, under the guidance and direction of that former principle).[220]

The leading life scientist of the day seemed to be affirming just the same metaphysical and methodological discriminations he himself demanded.

I think it is essential to dwell for a moment on Kant's suggestion that there is a *radical incongruity* between his notion of organic form as "intrinsic purposiveness" and the conventions of natural science: "[I]ts form is not possible according to mere natural laws, i.e., those laws which can be cognized by us through the understanding alone when applied to objects of sense."[221] In Gerd Buchdahl's terms, is Kant here addressing "transcendental lawlikeness" or only "empirical lawlikeness"?[222] Given that it is "understanding alone ... applied to objects of sense," one might infer the most extreme construction, that *organisms are incoherent according to the transcendental possibility of objective experience.*[223] But let us settle for the weaker claim, that organisms are not amenable to empirical laws after the fashion of mechanism: "It is indeed quite certain that we cannot even become sufficiently knowledgeable of, much less provide an explanation of organized beings and their internal possibility according to mere mechanical principles of nature."[224]

---

[220] Kant, *Kritik der Urteilskraft*, AA 5:424.

[221] Ibid., 370.

[222] Gerd Buchdahl, "The Conception of Lawlikeness in Kant's Philosophy of Science," *Synthese* 23 (1971): 24–46.

[223] Timothy Lenoir recognizes this radical possibility ("The Göttingen School and the Development of Transcendental *Naturphilosophie* in the Romantic Era," *Studies in History of Biology* 5 [1981]: 111–205, citing 149).

[224] Kant, *Kritik der Urteilskraft*, AA 5:400.

Of course, Kant's escape was to suggest an epistemological evasion of this unpalatable ontological prospect. He argued for the *"irreducibility* of biology to physics" but not because ontological reductionism was unacceptable. Indeed, it was *possible* – noncontradictory, though incomprehensible for finite human reason – that there could be a physicochemical basis for organic forms.[225] Kant went even further and supported the methodological program to *seek* reduction to mechanical explanation.[226] However, he argued that just here the methodological program would come up against an insuperable *epistemological* stumbling block – grounded in the limitations of human reasoning, not in the "order of nature" itself.[227] This is the famous argument of Kant's Dialectic of Teleological Judgment, and his resolution was that in order to make organic forms intelligible at all we must have recourse to the analogy of purpose or design.[228] Clark Zumbach phrases it suitably: Kant "is actually claiming that living processes must be viewed in terms of the *idea* of a free cause."[229] Kant transposed his metaphysical problem into a methodological one, his ontological need into an epistemological constraint: "[N]ature [i.e., the "order of nature" as a system] can only be understood as meaningful if we take it at large to be designed."[230] That is, "we need to be able to comprehend all of nature, not as a living being, but as a rational analog of a living being."[231] That is an *epistemic* strategy, a heuristic, not a fact. In Kantian terms, there is a subjective necessity – a "need of reason" – for this move, but no objective necessity, no natural law evident in the matter at hand (the "order of nature").[232]

Kant purchased the determinacy of his metaphysical principles of nature at a significant cost. The binding constraint of the concept of matter he adopted in the *Metaphysical Foundations of Natural Science* was that the laws he

---

[225] Ibid., 388.

[226] Ibid., 417–8.

[227] Ibid., 382.

[228] Kant, *Kritik der Urteilskraft*, AA 5:405–10.

[229] Clark Zumbach, *The Transcendent Science: Kant's Conception of Biological Methodology* (The Hague: Nijhoff. 1984), 99.

[230] Robert Butts, "Teleology and Scientific Method in Kant's *Critique of Judgment*," *Nous* 24 (1990): 1–16, citing 5.

[231] Ibid., 7; Peter McLaughlin, *Kant's Critique of Teleology in Biological Explanation: Antinomy and Teleology* (Lewiston, NY: Edwin Mellen Press, 1990).

[232] "[T]his claim has a decidedly negative import; it is essentially just an affirmation that the mechanical conception of nature and its conception of causality fails to provide a complete characterization of living systems. . . . Thus, the claim that there are free causes in living systems has no ontological force. It is rather a transcendental claim, i.e., one concerning the possibility of our judgments" (Zumbach, *The Transcendent Science*, 107).

generated could apply *only* to outer sense. This was a dramatic *restriction* in scope relative to the transcendental principles, which held for all aspects of possible experience, including *inner sense*. It "has the effect of restricting our attention to nonliving material substances," as Michael Friedman notes.[233] "Thus, the metaphysical principles of pure natural science apply only to the activities and powers of nonliving, nonthinking beings: beings represented solely through predicates of outer sense."[234] By defining matter as essentially *lifeless* in order to construe Newton's inertia, Kant excluded all aspects of life *in principle* from conformity to the metaphysical foundations of natural science and hence precluded *by definition* any "bottom-up" integration of empirical concepts and laws in these domains with the "top-down" foundation of his physical science.[235] The constitution of his Newtonian metaphysics of nature obstructed *in principle* what Friedman terms the "necessary convergence of constitutive and regulative procedures ... absolutely essential to Kant's entire project."[236] Indeed, this is the point toward which my whole exposition has been aiming, for it brings into glaring salience the problem of reconciling biology *at all* with Kant's prescriptions for science.

With epigenesis, the "order of nature" proved greater than the order of Kant's version of Newtonian physics, and the program for any life science necessarily exceeded the "Newtonian" constraints Kant wished to impose upon it.[237] His demand that the life sciences submit to the methodological principles of his "Newtonianism" (as in his critique of Herder and his dispute with Forster and above all in his preface to the *Metaphysical Foundations of Natural Science*) was misguided.[238] Epigenesis incited a fundamental erosion of Kant's boundary between the constitutive and the regulative, between the transcendental and the empirical: a naturalism beyond anything Kant could

---

[233] Michael Friedman, "Causal Laws and the Foundations of Natural Science," in *The Cambridge Companion to Kant*, ed. Paul Guyer (Cambridge: Cambridge University Press, 1992), 161–99, citing 185.

[234] Ibid., 182.

[235] "The inertia of matter is and signifies nothing but its lifelessness, as matter in itself. Life means the capacity of a substance to determine itself to act from an internal principle, of a finite substance to determine itself to change, and of a material substance to determine itself to motion or rest as change of its state. Now, we know of no other internal principle of a substance to change its state but desire and no other internal activity whatever but thought" (Kant, *Metaphysische Anfangsgründe der Naturwissenschaft*, AA 4:465–566, citing 544).

[236] Friedman, "Regulative and Constitutive," *Southern Journal of Philosophy* 30 suppl. (1991): "System and Teleology in Kant's *Critique of Judgment*," 73–102, citing 95.

[237] And thus his effort to "police" the practices of the experimental physics of his day was unavailing. See my "'Method' versus 'Manner'?"

[238] Kant, *Metaphysische Anfangsgründe der Naturwissenschaft*, AA 4:467–9 and passim.

countenance, though his own thought carried him there, just as empirical evidence carried the life scientists of the late eighteenth century to that same realization. To be consistent, Kant should have *qualified* his conception of Newtonian science in order to make room for the ontological actuality of life.[239] Some scholars take this to have been the guiding impulse in his *Opus Postumum* and in subsequent German *Naturphilosophie*.[240]

---

[239] "Whilst the extensionalist mathematical Newtonian approach offers the potential for (mathematical) a priori processing of physical nature, the price which this pays is that since forces do not have in this scheme any basic or 'essential' place, they have (because of the conceptual doubt attaching to them) to be introduced *ad hoc* (from 'without'), by way of hypothesis only. The objection to this, of course, ... [is] that such a basic and powerful notion as force (let alone the force of attraction) ought not to be surrounded with the suspicion which – particularly during the seventeenth and eighteenth centuries – surrounded anything 'hypothetical' in science" (Gerd Buchdahl, "Kant's 'Special Metaphysics' and the *Metaphysical Foundations of Natural Science*," in *Kant's Philosophy of Physical Science*, ed. R. E. Butts [Dordrecht: Reidel, 1986], 121–61, citing 150–1).

[240] For the beginnings of a revision of the relation between Kant and *Naturphilosophie*, see Frederick Beiser, *German Idealism* (Cambridge, MA: Harvard University Press, 2002), and Robert Richards, *The Romantic Conception of Life*, n. 4.

15

Blumenbach and Kant on Mechanism and Teleology in Nature

The Case of the Formative Drive

BRANDON C. LOOK

1

In the middle of his detailed and subtle discussion of teleology in the *Critique of Judgment*, Kant makes the following remarkable claim:

> [I]t is quite certain that we can never adequately come to know the organized beings and their internal possibility in accordance with merely mechanical principles of nature, let alone explain them; and indeed this is so certain that we can boldly say that it would be absurd for humans even to make such an attempt or to hope that there may yet arise a Newton who could make comprehensible even the generation of a blade of grass according to natural laws that no intention has ordered; rather, we must absolutely deny this insight to human beings.[1]

Kant's relatively well known denial of the possibility of a "Newton for a blade of grass" might lead one to think that he was universally skeptical of the work of those engaged in what we would now call "biology." Yet Kant took work in natural history very seriously, as John Zammito shows in chapter 14 of this volume, and the *Critique of Judgment* actually represents one of the

Research on this paper was made possible by a grant from the Alexander von Humboldt Foundation.

[1] AA 5:400. Citations to the works of Kant are to the Akademie edition (= AA), that is, *Kants gesammelte Schriften*, edited by the Königlich Preußische Akademie der Wissenschaften, Berlin, 1903/11, followed by volume number and page number. Translations are from the *Cambridge Edition of the Works of Immanuel Kant*. More specifically: *Critique of Pure Reason*, ed. P. Guyer and A. Wood (here references will be, as is customary, to the first [A] and second [B] editions); *Theoretical Philosophy after 1781*, eds. H. Allison and P. Heath, 2002 (includes the *Metaphysical Foundations of Natural Science*); *Critique of the Power of Judgment*, ed. P. Guyer, 2000 (despite Guyer's more accurate title, I shall refer to this work in the text as the *Critique of Judgment*); *Lectures on Metaphysics*, ed. K. Ameriks and S. Naragon, 1997; *Correspondence*, ed. A. Zweig, 1999. All other translations are the author's.

most important moments in philosophical thinking regarding the science of life. Moreover, in the appendix of the third *Critique*, Kant reveals himself to be a strong supporter of a very specific position in the modern debate regarding animal generation: the epigenesis of his younger contemporary Johann Friedrich Blumenbach (1752–1840), who had advocated the existence of a fundamental force – the *Bildungstrieb*, or "formative drive" – in matter that explains reproduction, nutrition, and regeneration.

The purpose of this essay is to analyze Blumenbach's views regarding the *Bildungstrieb* in general and the generation of life in particular and Kant's appropriation or misappropriation of these views in his writings on the nature of matter and the living world. This analysis will also address a more substantial philosophical issue, namely, the manner in which Blumenbach and Kant provide an explanation of the union of the mechanistic and teleological ways of explaining organized nature.[2]

2

Johann Friedrich Blumenbach was one of the most important scientists of his day. A professor of medicine and natural sciences at the University of Göttingen, he produced work of such variety and significance that he has come to be called "the father of anthropology," "the founder of race theory," "the path breaker for ethnology," and even "the first Egyptologist."[3] My concern, however, is his work in physiology, in particular, *Über den Bildungstrieb und das Zeugungsgeschäfte (On the Formative Drive and Matters of Reproduction)* (1781) and, to a lesser extent, his *Institutiones Physiologicae* (1786),[4] work that garnered him great acclaim in Europe generally and

---

[2] This subject has also been addressed by the following: Timothy Lenoir, "Kant, Blumenbach, and Vital Materialism in German Biology," *Isis* 71 (1980): 77–108; Lenoir, "The Göttingen School and the Development of Transcendental Naturphilosophie in the Romantic Era," *Studies in History of Biology* 5 (1981): 111–205; Lenoir, *The Strategy of Life: Teleology and Mechanics in Nineteenth Century German Biology* (Dordrecht: Reidel, 1982); James L. Larson, *Interpreting Nature: The Science of Living Form from Linnaeus to Kant* (Baltimore: Johns Hopkins University Press, 1994); Robert J. Richards, "Kant and Blumenbach on the *Bildungstrieb*: A Historical Misunderstanding," *Studies in the History and Philosophy of Biology and Biomedical Sciences* 31, no. 1 (2000): 11–32; John H. Zammito, "'This Inscrutable *Principle* of an Original *Organization*': Epigenesis and 'Looseness of Fit' in Kant's Philosophy of Science," *Studies in History and Philosophy of Science* 34 (2003): 73–109.

[3] See Johann Friedrich Blumenbach, *Über den Bildungstrieb und das Zeugungsgeschäfte*, with a foreword by L. v. Károly (Stuttgart: Gustav Fischer, 1971), which contains a facsimile edition of the original edition (Göttingen: Johann Christian Dietrich, 1781), v.

[4] Göttingen, 1786. Translated by Charles Caldwell (Philadelphia: Thomas Dobson, 1795).

in Germany especially and of which Kant apparently thought highly. It is in the former book especially that Blumenbach argues forcefully against the doctrine of preformationism and in favor of epigenesis, offering his account of a unique force present in the beings of the organic world. As other contributions to this volume have laid out the intellectual landscape of seventeenth- and eighteenth-century thought concerning the question of generation and reproduction, let me go straight to a sketch of Blumenbach's observations in *Über den Bildungstrieb.*

We are confronted with many examples of organic life and its processes in this work, but Blumenbach focuses on two separate instances before presenting his grand theory. First, he recounts how, while in the country on a summer holiday, he found a many-armed green polyp in a mill pond. He decided to conduct the well-known experiment of vivisection on the polyp's members and observed the growth of new, albeit shorter and thinner, arms in place of those that had been amputated. Later, back in Göttingen, Blumenbach visited a patient who had suffered from an abscessed wound stemming from tuberculosis in his knee.[5] He observed that now, after the healing process, there was a scar and a slightly depressed region of flesh: "thus," he reasoned, "*mutatis mutandis* the same case as with my green polyps from the mill pond."[6] In other words, in both cases there was the same natural tendency for the body to return to its original form, while, at the same time, the regenerated flesh was somewhat dwarfed. These two instances allow Blumenbach to make an even stronger claim:

> That in all living creatures, from man down to maggot and from the cedar down to mold, there is a peculiar, innate, lifelong active and effective drive [*Trieb*] to take a particular form in the beginning, then to maintain it, and, even when it has been destroyed, to repair it whenever possible. A drive (or tendency or endeavor, however one wants to call it) that is completely different both from the common properties of the body in general as well as from the other characteristic powers [*Kräfte*] of the organized body in particular; that seems to be one of the first causes of all generation, nutrition, and reproduction; and for which I, in order to anticipate any misinterpretation and to distinguish it from other natural powers, reserve the name "formative drive" [*Bildungs-Trieb*] (*Nisus formativus*).[7]

Blumenbach is adamant, however, that this force, tendency, or drive is not to be confused with the *vis plastica* of John Turberville Needham (1713–81),

---

[5] That is, *spina ventosa*, the now rare case in which the tuberculosis bacteria afflict the bones (rather than the lungs).

[6] Blumenbach, *Über den Bildungstrieb und das Zeugungsgeschäfte*, 11.

[7] Ibid., 12–13.

which, according to Blumenbach, amounts essentially to an occult quality, or the *vis essentialis* of Caspar Friedrich Wolff (1738–94) or any other purely mechanical forces.[8]

Given his evidence, it might seem surprising that Blumenbach should make the bold claim that reproduction, nutrition, and generation are all reducible to or dependent upon this *same* force. But, in his view, "a truth that one ought not to lose sight of and whose neglect may have often already hampered its happy progress [*Fortgang*] is that reproduction, nutrition, and regeneration [*Wiederersetzung*] are essentially just modifications of one and the same force, which in the first case builds, in the next provides [*unterhält*], and in the third repairs!"[9] Blumenbach does not present an explicit argument that shows that these three distinct activities are modifications of the same force; he asserts it as a fundamental fact of his science.[10] One might assume that Blumenbach is reasoning in the following way: The abilities to reproduce, gain nourishment from the surroundings, and heal oneself wholly or in part are what distinguish the living from the dead, the animate from the inanimate; these powers are the fundamental features of life in general; and, therefore, insofar as we can speak of a single force of life, these activities of reproduction, nutrition, and regeneration must be mere modifications of this same life force (i.e., they are merely conceptually distinct). Yet such an argument borders on animism, something that Blumenbach opposed, for one could very well suppose that this single force, which can manifest itself in a variety of ways, is simply due to a soul that exists along with or within a given body. At the same time, Blumenbach believes that the formative drive is linked to the original matter, indeed, that a feature of matter itself is that it has this force. After all, he argues that the regenerated parts of the polyp and the tissue of a human wound are smaller precisely because there is less matter and the regenerative power is consequently diminished. The project for Blumenbach is, then, to explain

---

[8] Ibid., 14. For more on the forces of Needham and Wolff, see Shirley Roe, *Matter, Life and Generation: Eighteenth-Century Embryology and the Haller-Wolff Debate* (Cambridge: Cambridge University Press, 1981), 18–20, 112–20. See also François Duchesneau, "Vitalism in Late Eighteenth-Century Physiology: The Cases of Barthez, Blumenbach, and John Hunter," in *William Hunter and the Eighteenth-Century Medical World*, ed. W. F. Bynum and Roy Porter (Cambridge: Cambridge University Press, 1985), 269 ff.

[9] Blumenbach, *Über den Bildungstrieb und das Zeugungsgeschäfte*, 18–19.

[10] If one thinks of the fundamental forces in physics – gravity, electricity, magnetism, and the nuclear forces – one can begin to see some problem with Blumenbach's presentation. While contemporary physicists dream of a unified theory (one that would show how all these forces are "modifications" of the same force), they recognize that this unified theory will require evidence and argument. This does not seem to be the case with Blumenbach, even though Needham, for example, had argued that there are different kinds of forces active in an organism.

how the formative drive can be a primitive feature of matter itself without, at the same time, reducing all biological phenomena to purely mechanistic modes of explanation. If the *Bildungstrieb* is an inherent property of matter, however, then it is easy to see how reproduction, nutrition, and generation *must* be modifications of it.

The remainder of *Über den Bildungstrieb* is devoted to a development of this notion of a formative drive, especially as it concerns reproduction, and to a critique of rival theories of generation, especially the pure mechanist or preformationist programs. Blumenbach claims that a variety of difficulties can be removed if one adopts the doctrine of a formative drive and epigenesis. For example, when bedeguar develops on wild roses, naturalists seemed to be confronted with two unhappy choices: either this is a case of *generatio aequivoca* (spontaneous generation) or the preformationist must assert that there were all along seeds of the gall on the branches and leaves of all rose bushes. Blumenbach attributes this phenomenon simply to the contact of gall wasps.[11] Further, against the preformationist (ovist) view, he argues that one can find no evidence in fertilized chicken eggs of blood or blood vessels, which ought to be observable *ab initio*, and that therefore it is wrong to hold that the animals were present in some form prior to fertilization.[12] Concerning the spermist view, Blumenbach is absolutely contemptuous, arguing that the forms of the little spermatic worms (*Saamenwürmen*) had been depicted by different microscopists in widely disparate ways that nevertheless still failed to bear any relation to the future adult animal.[13] In contrast to the evolutionists and preformationists, Blumenbach argues that the first signs of the organized body appear a short time after fertilization, and he gives the following picture of reproduction: "[T]he paternal and maternal reproductive fluids [*Säfte*] – the raw material of the new creature-to-be – need a certain time to prepare their mixture and inner combination and other necessary changes before the *Bildungstrieb* is excited in them and the formation of the as yet unformed material can begin."[14] After fertilization and the excitation of the *Bildungstrieb*, however, the development of the organism proceeds very rapidly; in fact, according to Blumenbach, the new organism has its greatest formative drive at its conception, and the *Bildungstrieb* trails off as it ages.[15]

---

[11] Blumenbach, *Über den Bildungstrieb*, 23–6.

[12] Ibid., 28–30.

[13] Ibid., 32–6.

[14] Ibid, 42.

[15] Blumenbach here seems to contradict himself, endorsing the view, on the one hand, that the formative drive is dependent upon the quantity of matter (which explains why the polyp's new

Again, it will be this *Bildungstrieb*, in its different modifications or manifestations, that will account for the nourishment and regeneration of the
organism throughout its life and for its further reproduction. But this is perhaps
still a vague picture, and the vagueness has contributed to a dispute between
scholars writing on the nature of the *Bildungstrieb* and Kant's relation to
Blumenbach.[16] According to Timothy Lenoir, for example, "the *Bildungstrieb*
did not exist apart from its material constituents, but it could not be explained
in terms of those elements. It was an emergent property having a completely different character from its constituents."[17] On the other hand, Robert
Richards claims that, in the 1781 edition of *Über den Bildungstrieb* at
least,

> Blumenbach considered the drive to be an independent vital agency. It caused
> the formation of the embryo out of homogeneous seminal material and contin-
> ued to operate in maintaining the vitality of the organism and in repairing its
> injuries. In this respect the *Bildungstrieb*, despite Blumenbach's asseverations
> to the contrary, did appear rather like Wolff's *vis essentialis*, that "orders ever
> thing in vegetative bodies on account of which we ascribe life to them."[18]

Moreover, according to Richards, "for Blumenbach, the *Bildungstrieb*
endowed the homogeneous, formless mixture of male and female semen with
its most essential character – form, organization – and set the various parts
so articulated into mutually harmonious operation. This was a teleological
cause fully resident in nature."[19]

There are two issues here that need to be distinguished: first, whether the
Bildungstrieb can be considered to be independent of the matter; second,
what the causal role is of the *Bildungstrieb*. The first issue points to a problem in the life sciences in general and can be highlighted by asking, What
would it mean for the *Bildungstrieb* to be independent of matter? According to the first edition of Blumenbach's work, the *Bildungstrieb* is *not* to
be considered something that could in fact or in principle exist apart from
matter; it is neither a soul nor a general "life-force" instantiated in various
organic beings. In this regard, Richards's claim seems misleading, and Lenoir
is most likely correct: in Blumenbach's view, the *Bildungstrieb* cannot exist

---

limbs are smaller than the original limbs) and, on the other hand, that the formative drive is
strongest from conception on (when, presumably, there is the *least* amount of matter).

[16] I address Kant's understanding of Blumenbach in the next section.

[17] Lenoir, *The Strategy of Life*, 21.

[18] Richards, "Kant and Blumenbach on the *Bildungstrieb*," 19, containing a passage from Wolff,
Theorie von der Generation, 160.

[19] Richards, "Kant and Blumenbach on the *Bildungstrieb*," 20.

without matter; it is, as it were, an "emergent property."[20] Yet, if we say that the *Bildungstrieb* is an emergent property, then it would seem that its qualities derive in some respect from the matter. But the *Bildungstrieb* acts to organize and increase that matter. What kind of picture do we have here, then, of causality? Is it possible that the *Bildungstrieb* is both an emergent property of matter and a causal principle active in the development and organization of matter? Blumenbach apparently thinks so. In short, Richards's interpretation of Blumenbach's view, as expressed in the first edition of *Über den Bildungstrieb*, is consistent but not true to Blumenbach, whose view is, if not inconsistent, at least deeply problematic. Blumenbach, as I read him, really does believe that the *Bildungstrieb* is a force that (1) cannot have existence apart from matter, (2) is "emergent" in the sense described by Lenoir, but (3) does exert causal powers on the organization of the organism whose matter is its "source."

Yet this suggests a tension in his view that will cause Blumenbach to modify or at least clarify his conception of the *Bildungstrieb*. In one of his other works in the life sciences, the *Institutiones Physiologicae*, Blumenbach makes clear that, in the debate between mechanists and vitalists, he belongs to the latter camp. He writes the following in the opening of this work:

> In the living human body, the healthy functions of which constitute the exclusive object of the science of physiology, there are three things worthy of our immediate attention and regard; namely, The solids, or parts containing; The fluids, or parts contained within the solids; And lastly, the Vital Energies, which in the consideration of the defense of physiology, constitute the most interesting and important objects of our regard. It is in consequence of these energies that the solids are rendered alive to the impulse of the fluids, endowed with a power to propel the same, and also to perform a variety of other motions.[21]

And Blumenbach states later that the *Bildungstrieb* (or *nisus formativus*) is among the most important vital energies, serving as "an efficient cause of the whole process of generation."[22] Moreover, in the second edition of *Über den Bildungstrieb* (1789), Blumenbach consciously appeals to Newtonian ideas and language:

> I hope it will be superfluous to remind most readers that the word *Bildungstrieb*, like the words attraction, gravity, etc. should serve, no more and no less, to

---

[20] Lenoir, "The Göttingen School and the Development of Transcendental Naturphilosophie in the Romantic Era," 155, where Lenoir calls Blumenbach's view "emergent vitalism."

[21] Blumenbach, *Über den Bildungstrieb*, 1–2.

[22] Ibid., 34–5.

signify a power whose constant effect is recognized from experience and whose cause, like the causes of the aforementioned and the commonly recognized natural powers, is for us a *qualitas occulta*.[23]

In other words, the *Bildungstrieb* arises because of some unknown primary cause and, in turn, acts as a secondary cause of the structure and development of organic beings. Whether this will ultimately be a satisfactory account of the nature of reproduction, nourishment, and regeneration is another matter. But, certainly, Blumenbach's account in the 1780s of the *Bildungstrieb* represented one of the final blows to all theories of preformationism and evolution and, at the same time, explained the fundamental features of the living world in a way that echoed Newton's account of the physical world. Insofar as his view gives an account of the tendency within organisms to reproduce, nourish, and heal themselves and to develop their organization, and at the same time describes this force or tendency as an original property of matter, it deserves the phrase "vital materialism."[24]

3

While Blumenbach attempts to wrap himself in the mantle of Newtonianism, Kant, as we saw earlier, claims in §75 of the *Critique of Judgment* that there can be no Newton for a blade of grass. At the same time, Kant offered both public and private praise for Blumenbach's views on animal generation and the *Bildungstrieb*, and there can be little doubt where he stood on the matter (at least by 1790). Consider the following, from §81 of the *Critique of Judgment*:

> No one has done more for the proof of this theory of epigenesis as well as the establishment of the proper principles of its application, partly by limiting an excessively presumptuous use of it, than Privy Councillor **Blumenbach**. He begins all physical explanation of these formations with organized matter. For he rightly declares it to be contrary to reason that raw matter should originally have formed itself in accordance with mechanical laws, that life should have

---

[23] Blumenbach, *Über den Bildungstrieb* (1789), 25–6. Translation from Richards, "Kant and Blumenbach on *Bildungstreib*," 24. To the best of my knowledge, Kant had only the first edition of *Über den Bildungstrieb* in his library. Yet, when he writes to Blumenbach in 1790, Kant thanks him for a copy of his book that Blumenbach sent the previous year. I shall assume, then, that Kant did have access to the second edition (as well as the *Institutiones*) when writing the *Critique of Judgment*.

[24] Cf. the title to Lenoir, "Kant, Blumenbach, and Vital Materialism in German Biology." Lenoir's other phrase, "teleomechanism" (Lenoir, *The Strategy of Life*, 24.) is somewhat problematic because the notion of "mechanism" involved is unclear.

arisen from the nature of the lifeless, and that matter should have been able to assemble itself into the form of a self-preserving purposiveness by itself; at the same time, however, he leaves natural mechanism an indeterminable but at the same time also unmistakable role under this inscrutable **principle** of an original **organization**, on account of which he calls the faculty in the matter in an organized body (in distinction from the merely mechanical **formative power** [*Bildungskraft*] that is present in all matter) a **formative drive** [*Bildungstrieb*] (standing, as it were, under the guidance and direction of that former principle).[25]

And, in a letter to Blumenbach, Kant lauds his "excellent essay" (*Über den Bildungstrieb*) and continues,

> I have found much instruction in your writings, but the latest of them has a close relationship to the ideas that preoccupy me: the union of the two principles that people have believed to be irreconcilable, namely the physical-mechanistic and the merely teleological way of explaining organized nature.[26]

Moreover, in his lectures on metaphysics, the *Metaphysik Vigilantius*, Kant states explicitly that the "system of epigenesis is now generally assumed," due in no small part, of course, to the work of Blumenbach.[27]

But did Kant really understand Blumenbach and his project? How does Kant, in his mature writings, address the issues of teleology and mechanism in the life sciences and the problem of animal generation? In what follows, I wish to explain the relation between Kant and Blumenbach in greater detail, hoping to put Kant's glowing remarks on Blumenbach into proper perspective.[28]

---

[25] AA 5:424.

[26] AA 11:185

[27] AA 29:1031.

[28] In particular, I hope to show that there is something correct about Robert Richards's claim that Kant fundamentally misunderstood Blumenbach. However, it is unclear to me how "creative" their misunderstanding might be. See Richards, "Kant and Blumenbach on the *Bildungstrieb*." My exegesis and interpretation have been influenced not only by Richards's article but also by the following: Henry Allison, "Kant's Antinomy of Teleological Judgment," *Southern Journal of Philosophy* 30 (1991): 25–42; Wolfgang Bartuschat, *Zum systematischen Ort von Kants Kritik der Urteilskraft* (Frankfurt: Klostermann, 1972); Christel Fricke, "Explaining the Inexplicable: The Hypothesis of the Faculty of Reflective Judgement in Kant's Third Critique," *Noûs* 24 (1990): 45–62; Hannah Ginsborg, "Kant on Understanding Organisms as Natural Purposes," in *Kant and the Sciences*, ed. E. Watkins (Oxford: Oxford University Press, 2001), 231–58; Paul Guyer, "Organisms and the Unity of Science," in *Kant and the Sciences*, 259–81; Konrad Marc-Wogau, *Vier Studien zu Kants Kritik der Urteilskraft* (Leipzig: Otto Harrassowitz, 1938); Peter McLaughlin, *Kant's Critique of Teleology in Biological Explanation: Antinomy and Teleology* (Lewiston, NY: Edwin Mellen Press, 1990); Clark Zumbach, *The Transcendent Science: Kant's Conception of Biological Methodology* (The Hague: Martinus Nijhoff, 1984).

With respect to the role of teleology in philosophy and the explanation of animal generation, I think that one can show a natural trajectory from Kant's views as expressed in the first edition of the *Critique of Pure Reason* (1781/87) through his *Metaphysical Foundations of Natural Science* (1786) and the oft-overlooked essay *On the Use of Teleological Principles in Philosophy* (1788) to the *Critique of Judgment* (1790). The *Critique of Pure Reason* and the *Metaphysical Foundations of Natural Science*, despite their brilliance, might be thought to lead Kant into a difficult position with respect to the science of the living world. Indeed, it is clear that both works were written first and foremost with Newtonian physics in mind. The "Transcendental Analytic" of the *Critique of Pure Reason* attempts, of course, to establish the fundamental categories of "substance," "cause," and "community," among the other "dynamic" categories, for all objects of possible experience; meanwhile, the "Transcendental Dialectic" seeks to show how reason, left unchecked, invariably leads us into error, especially when concerned with matters relating to the nature of the soul, world concepts, and the nature of God. Of the many things that follow from Kant's *Critique*, that which is important to our study is the priority given to efficient causation and the consequent limitation of accounts of purposiveness and final causality. The *Metaphysical Foundations of Natural Science* carries the Kantian project further by providing a "metaphysics of nature" that grounds pure natural science, which ultimately entails an explication of the doctrine of body and that which is knowable of matter as such. Moreover, according to Kant in this work, "what can be called *proper* science is only that whose certainty is apodictic; cognition that can contain mere empirical certainty is only *knowledge* improperly so-called."[29] Thus, not only is the mere possibility of a science of the living world called into question with this definition, but, given the conceptual apparatus available through the *Critique of Pure Reason*, it is also difficult to understand how we can understand the causal mechanisms at work in biological phenomena.

Moreover, in the appendix of the "Transcendental Dialectic" of the *Critique of Pure Reason*, Kant claims that reason establishes for us the goal of reducing all of nature to a single principle: "If we survey the cognitions of our understanding in their entire range, then we find that what reason quite uniquely prescribes and seeks to bring about concerning it is the **systematic** in cognition, i.e., its interconnection based on one principle."[30] Ultimately, however, the idea of a systematic unity of nature under one principle or causal law is what Kant terms a "regulative" idea, "bringing unity into particular

[29] AA 4:468.
[30] A645/B673.

cognitions as far as possible and thereby **approximating** the rule to universality."[31] While we might not be able to find this systematic unity in nature,
Kant's view seems to be that there is nothing in nature that clearly disproves
this fundamental regulative idea.[32]

In the *Metaphysical Foundations of Natural Science*, Kant stresses a particular view of the nature of matter, arguing that matter must be considered
essentially lifeless. And one might naturally conclude that this view of matter
would provide the basis of a fundamental or universal rule in our investigations of the natural world. The third chapter of the *Metaphysical Foundations
of Natural Science*, for example, is devoted to mechanics, or the relational
nature of matter and its original moving force.[33] It is here that Kant reworks
Newton's first and third laws of motion[34] and goes on to provide his own second law of mechanics – that every change of matter has an external cause. His
argument is the following: "[M]atter, as mere object of the outer senses, has
no other determinations except those of external relations in space, and therefore undergoes no change except by motion"; the causes of change in matter
cannot be internal, "for matter has no essentially internal determinations";
therefore, "every change in a matter is based upon external causes."[35] Kant
continues with the following argument against hylozoism:

> The inertia of matter is, and means, nothing else than its *lifelessness*, as matter
> in itself [*Materie an sich*]. *Life* is the faculty of a *substance* to determine itself to
> act from an *internal principle*, of a *finite substance* to change, and of a *material
> substance* [to determine itself] to motion or rest, as change of its state. Now
> we know of no other internal principle in a substance for changing its state
> except *desiring*, and no other internal activity at all except *thinking*, together
> with that which depends upon it, the *feeling* of pleasure or displeasure, and
> *desire* or willing. But these actions and grounds of determination in no way
> belong to representations of the outer senses, and so neither [do they belong]
> to the determinations of matter as matter. Hence all matter, as such, [*Materie
> als solche*] is *lifeless*.[36]

According to Kant, there is a great deal at stake here: the possibility of
a natural science rests on the law of inertia, and, indeed, the "opposite of

---

[31] A647/B675.

[32] Cf. Guyer, "Organisms and the Unity of Science," 260.

[33] AA 5:477.

[34] Cf. Isaac Newton, *The Principia: Mathematical Principles of Natural Philosophy*, trans. I.
Bernard Cohen and Anne Whitman (Berkeley: University of California Press, 1999), 416–17.

[35] AA 4:543.

[36] AA 4:544.

this, and thus the death of all natural philosophy, would be hylozoism."[37] The tension that Kant sees here is between the following:

1. Hylozoism: All actions of a material substance result from the inner determinations of the substance.
2. Law of Inertia: All actions of a material substance result from the determinations of things external to the substance.

In other words, in order for the law of inertia to be true, hylozoism must be shown to be false, that is, matter must be shown to be lifeless. Now, according to Kant, "life is the faculty of a substance to determine itself to act from an internal principle," and we know of no *inner* principles of substances other than desire and no *internal* activities but thought. Further, these determining grounds are not accessible to us via our *external* senses. Therefore, Kant wishes us to arrive at the following two-part conclusion: (1) these determining grounds do not belong to matter as matter, and (2) all matter as such is lifeless. The curious and suspect feature of Kant's argument here, however, is that it is cast in terms of constitutive principles. Kant does not make the modest proposal that, when we do natural science, we ought to consider matter as lifeless; he seems to be saying that it must be considered so, that it is so.[38]

Having given primacy to efficient causation and a view of matter as lifeless in the first *Critique* and the *Metaphysical Foundations of Natural Science*, Kant turns his attention to the science of life and teleology in his short essay *On the Use of Teleological Principles in Philosophy*. He writes,

> The concept of an organized being already conveys that it is matter, in which everything stands in a reciprocal relation of ends and means. And this can even be thought of simply as a system of final causes, whose possibility leaves to human reason only a teleological and in no way physical-mechanical mode of explanation. Therefore, one cannot investigate in physics where all organization comes from. The answer to this question, if it were accessible to us at all, would clearly be outside of natural science and in metaphysics. For my part, I derive all organization of organic beings (through reproduction) and later forms (of these kinds of natural things) from laws of gradual development from original structures [*Anlagen*] (like those often found in the transplantation of plants), which were found in the organization of its type [*Stamm*].[39]

---

[37] AA 4:544.

[38] More exactly, Kant's point seems to be here that matter must be cognized a priori as lifeless. To anticipate: it will be in the *Critique of Judgment* that Kant backs down from this claim by distinguishing between determinate and reflective judgments and thereby allowing our concept of matter to accommodate the purposiveness attributable to organisms.

[39] AA 8:179.

For our purposes it is important to highlight a few points in this passage. First, an organized being is said to be *ipso facto* a material being, but also something more than this: a being that can be thought of simply as a system of final causes. Consequently, according to Kant, we cannot investigate the origin of all organization within physics. If matter is essentially lifeless, as we were told in the *Metaphysical Foundations of Natural Science*, then we might think that we should be able to approach the study of organized (i.e., material) beings from the standpoint of pure natural science. In other words, we should be able to study it purely mechanically. But, of course, the crucial fact here is that any such organized being is, for Kant, a "system of final causes," and therefore its study must be outside the realm of physics *narrowly conceived* and in the realm of metaphysics.

The *Critique of Judgment*, of course, represents the crucial moment in Kant's thinking regarding teleology and the problems associated with the life sciences. Unfortunately, the "Critique of Teleological Judgment" and, in particular, the "Dialectic of the Teleological Power of Judgment," which contain Kant's reflections on biology, are notoriously difficult to interpret. Thus, while we saw that Kant's endorsement and praise of Blumenbach were clear, we might find that the argument leading up to this point is more obscure.

One problem seems to be that, while Kant argued in the *Critique of Pure Reason* that we should attempt to bring all our knowledge under one unifying principle (presumably a principle of mechanism and efficient causality), he suggests in the *Critique of Judgment* that this cannot be done. Kant's reason is quite simply that it is impossible to cognize organisms in the terms of mechanism.[40] Thus, a great deal depends upon our understanding of mechanism and the nature of matter. In §65 of the "Dialectic of Teleological Judgment," Kant writes the following:

> An organized being is thus not a mere machine, for that has only a **motive** power, while the organized being possesses in itself a **formative** power [*bildende Kraft*], and indeed one that it communicates to the matter, which does not have it (it organizes the latter): thus it has a self-propagating formative power, which cannot be explained through the capacity for movement alone (that is, mechanism).... Strictly speaking, the organization of nature is therefore not analogous with any causality that we know.[41]

Seemingly following Blumenbach, Kant here clearly rejects the idea of a mechanical explanation of the living and of its characteristic feature: its ability

---

[40] This is, of course, also the point that comes out in his essay *On the Use of Teleological Principles in Philosophy* considered just previously.

[41] AA 5:374–5.

to reproduce. But Kant's point is not just that the mode of explanation must be different – appealing to a form of causation other than mechanical or efficient causation – but that the nature of matter must be differently conceived. For matter must be capable of being affected by the "*bildende Kraft*," which is not simply a special effect of motion.

This point leads us to that section of the "Dialectic of the Teleological Power of Judgment" which Kant calls "Antinomy of Teleological Judgment." Here we might find echoes of the distinction between hylozoism and the principle of inertia present in the *Metaphysical Foundations of Natural Science* that we considered earlier; only this time Kant concerns himself with the generation of material things. He writes in §70,

> The **first maxim** of the power of judgment is the **thesis**: All generation of material things and their forms must be judged as possible in accordance with merely mechanical laws.
>
> The **second maxim** is the **antithesis**: Some products of material nature cannot be judged as possible according to merely mechanical laws (judging them requires an entirely different law of causality, namely that of final causes).
>
> Now if one were to transform these regulative principles for research into constitutive principles of the possibility of the objects themselves, they would run:
>
> **Thesis**: All generation of material things is possible in accordance with merely mechanical laws.
>
> **Antithesis**: Some generation of such things is not possible in accordance with merely mechanical laws.[42]

In what follows, Kant appeals to a different distinction, a distinction between determining powers of judgments and reflecting powers of judgments: the former "determines" "an underlying concept through a given empirical representation"; the latter "reflects" "on a given representation, in accordance with a certain principle, for the sake of a concept that is thereby made possible."[43] And Kant claims that, when we appeal to the laws of mechanism in our study of the natural world, we "do not thereby say that they are possible only in accordance with such laws," only that we "should always reflect" on the natural world "in accordance with the principle of the mere mechanism

---

[42] AA 5:387.

[43] AA 20:211. These definitions are from the unpublished "First Introduction" to the *Critique of Judgment*.

of nature."[44] In other words, Kant seems to conclude that we can resolve any contradiction or antinomy if we reinterpret the conflict between the constitutive principles of determining judgment in terms of regulative principles of reflecting judgment. As he writes in the next section, "All appearance of an antinomy between the maxims of that kind of explanation which is genuinely physical (mechanical) and that which is teleological (technical) therefore rests on confusing a fundamental principle of the reflecting with that of the determining power of judgment."[45]

Yet is this really the way that Kant intends to resolve the antinomy? Does he mean to relegate the causal principle of the *Critique of Pure Reason* to a regulative principle in the *Critique of Judgment*? Does an understanding of the fundamental features of a mechanism carry with it no constitutive claims? Most likely not. Indeed, there is good reason to look (briefly) for another way of solving the conflict set up in §70.[46] In order to establish a different resolution to the antinomy of teleological judgment, that is, in order to explain how we can resolve the apparent conflict between mechanical and teleological modes of explanation, we need to consider in greater detail the notion of "mechanism" at work in Kant's *Critique of Judgment*. In the most general sense, mechanism is any kind of causal explanation that does not appeal to purposiveness.[47] In another sense – the sense most familiar to philosophers from the time of Descartes onwards – mechanism is the scientific account of the world in terms of matter in motion. But Kant also suggests another kind of relation when he speaks of mechanism: the explanation of wholes in terms of their component parts. Consider the following passage from §77 of the *Critique of Judgment*, in which Kant is concluding his account of the antinomy:

> Now if we consider a material whole, as far as its form is concerned, as a product of the parts and of their forces and their capacity to combine by themselves (including as parts other materials that they add to themselves), we represent a mechanical kind of generation. But from this there arises no concept of a whole as an end, whose internal possibility presupposes throughout the idea of a whole on which even the constitution and mode of action of the parts depends, which is just how we must represent an organized body. But from this, as has just

---

[44] AA 5:387.

[45] AA 5:389.

[46] In the next several paragraphs my discussion follows Marc-Wogau, *Vier Studien zu Kants Kritik der Urteilskraft*, 214–45; Allison, "Kant's Antinomy of Teleological Judgement"; and McLaughlin, *Kant's Critique of Teleology in Biological Explanation*, chap. 3.

[47] Cf. AA 5:406, 20:219.

been shown, it does not follow that the mechanical generation of such a body is impossible; for that would be to say the same as that it is impossible (i.e., self-contradictory) to represent such a unity in the connection of the manifold for every understanding without the idea of that connection being at the same time its generating cause, i.e., without internal production.[48]

If the whole is determined by its parts, then it can be explained in the typical mechanistic sense, the sense that underlies the causal principle of the "Second Analogy of Experience." If, however, the whole is to be considered an end, if the whole contains the grounds for the possibility of the parts, and if the whole orders its own parts, then we must appeal to another notion of mechanism. And this notion of mechanism or mechanistic explanation need not conflict with a teleological mode of explanation.

Ultimately, however, Kant's resolution of the antinomy of teleological judgment depends upon additional argumentative moves. In order for both mechanistic and teleological modes of explanation to be unifiable in judging nature, Kant claims, there must be a principle that is outside of both and "hence outside of the possible empirical representation of nature" but that nevertheless "contains the ground of both, i.e., in the supersensible." Now, according to Kant, "we can have no concept" of this supersensible ground "except the undetermined concept of a ground that makes the judging of nature in accordance with empirical laws possible."[49] While this is quite vague, we can see that Kant's general point is that, with the assumption of a supersensible ground, both modes of explanation are possible. Further, as becomes evident in the concluding sections of the "Dialectic" (§§77–8), Kant makes an important distinction between *our* "discursive" understanding and an "intuitive" understanding. It is a contingent fact regarding our understanding that we can only cognize the natural world in terms of certain concepts or categories, and these require that we should explain the world mechanistically – in terms of part-whole relations in which the parts determine the wholes but not in which the whole is conceivable as a cause of its parts. On the other hand, a purely intuitive understanding – for example, that of God – could in fact cognize a whole as a cause of its parts. Furthermore, it need not be the case that the world is constituted simply by matter in motion. The principles associated with mechanism, therefore, will be regulative, while we can still have a constitutive sense of causality consistent with the "Second Analogy" of the *Critique of Pure Reason*. Or, in other words, a reductionist mechanist view underlies our discursive understanding, but it is not constitutive of objects of experience,

[48] AA 5:408.
[49] AA 5:412.

as it seemed to be in the *Metaphysical Foundations of Natural Science* and in the presentation of the antinomy of teleological judgment. According to Kant, then, when a mechanistic account of phenomena is possible for us, it is true, but when it is not possible, we may attempt to give a teleological account of the phenomena, recognizing purposiveness in nature. As Kant writes in the final paragraph of the "Dialectic of Teleological Judgment,"

> Now on this is grounded the authorization and, on account of the importance that the study of nature in accordance with the principle of mechanism has for our theoretical use of reason, also the obligation to give a mechanical explanation of all products and events in nature, even the most purposive, as far as it is in our capacity to do so (the limits of which within this sort of investigation we cannot determine), but at the same time never to lose sight of the fact that those which, given the essential constitution of our reason, we can, in spite of those mechanical causes, subject to investigation only under the concept of an end of reason, must in the end be subordinated to causality in accordance with ends.[50]

What, ultimately, are we to make of Kant's evaluation of Blumenbach's argument? It certainly is the case, as Kant claims in his letter, that both men are concerned with the unification of the diverse modes of explanation in the study of life. And Blumenbach, like Kant, rejects preformationism and a simplistic form of mechanism in biology. Further, it would seem that Kant praises Blumenbach in part simply because the latter recognizes that living beings manifest the properties of both formed and forming beings, designed and designer, *gebildete* and *bildende*.[51] But Kant's argument is significantly more nuanced than Blumenbach's. Or, put the other way around, Blumenbach's view is so philosophically simplistic in comparison to Kant's that we have good reasons to think that the sage of Königsberg should not have been genuinely impressed. First, Blumenbach's understanding of the mechanism and causality at work in an organism fails to meet the high standards that Kant established in the "Critique of Teleological Judgment." After all, Blumenbach had argued in the *Institutiones* that the *Bildungstrieb* be considered a vital energy that acts as an *efficient* cause of reproduction. How exactly this should work remains, not surprisingly, a mystery. Second, as should be clear from his commitment to the idea of the *Bildungstrieb* as a form of energy, Blumenbach remains a vitalist, and vitalism is a position that Kant rejects. Although the *Bildungstrieb*, as suggested in the previous section, should not be considered an independent life force present in matter, Blumenbach nevertheless

---

[50] AA 5:415.
[51] This point is made by Zumbach, *The Transcendent Science*, 85.

represents the *Bildungstrieb* as constitutive (in Kantian parlance) of the matter of organized beings. If this formative drive is constitutive of matter, then it must be a feature of *all* matter.[52] But if this were the case, then we would have no means to distinguish between "animate" matter and "inanimate" matter or between "living" matter (organisms) and "nonliving" matter, which is not an attractive position. Third, although Blumenbach compares the *Bildungstrieb* to an occult quality in the second edition of his treatise, this should not lead us to think that he is making anything like Kant's appeal to a "supersensible ground" for both mechanical and teleological modes of explanation. Rather Blumenbach's position seems to be that the formative drive is a "something-I-know-not-what" active in the world, not a "something-unknowable-to-me" that makes possible the ascription of purposiveness in the world. In other words, Blumenbach's views would lead us ultimately into precisely the antinomy of teleological judgment that Kant sought to avoid.

None of the foregoing should be seen as disparaging Blumenbach's achievement, for his views regarding the issue of animal generation did establish a flourishing research program for much of eighteenth- and nineteenth-century biology.[53] Yet, from Kant's perspective, Blumenbach could not be the "Newton for a blade of grass" – for there still can be no such figure. Ultimately, Kant's praise of Blumenbach ought to be seen as self-praise: it is Kant, not Blumenbach, who genuinely unifies teleology and mechanism, and he does so without a *Bildungstrieb*.

---

[52] It is perhaps for this reason that Salomon Maimon juxtaposes Blumenbach's idea of a *Bildungstrieb* with the idea of a world-soul (see AA 11:174).

[53] Cf. Lenoir, *The Startegy of Life*.

VII

Kant and the Beginnings of Evolution

16

Kant and the Speculative Sciences of Origins

CATHERINE WILSON

1. INTRODUCTION

Kant pretended to expertise in many nonphilosophical subjects, including the history of civilizations and infant care. He offered up his thoughts on human evolution, skin coloration, formative forces, and competitive social behavior in numerous essays and reviews, and he also composed an essay on animal form, the *Critique of Judgment*, Part II. One might suppose him to be a key contributor to the flowering of the human and social sciences in the eighteenth century, not their unhappy observer. Yet, as John Zammito has noted in his insightful study of the *Critique of Judgment*, there is something remarkable about Kant's attitude. His desire to affirm the utter mystery of life and the inexplicability of its origins and the distinction between man and the rest of nature left him, in Zammito's terms, "sharply estranged from the [eighteenth century's] most effective currents."[1] "A large part of [Kant's] critical philosophy," he suggests, "can be interpreted as an effort to balance his recognition of the limitations of speculative rationalism or 'dogmatic metaphysics' with his recognition of the essential human interest in metaphysics, the unavoidable problems of God, freedom and immortality."[2] Zammito argues that "Kant's personal commitment to a 'theistic' if not outright Christian posture strongly colored the ultimate shape of his work."

An example of this pattern of partial engagement with and partial estrangement from the life sciences can be found in Kant's writings on generation. Kant wrestled throughout his life with the problem of the origins of life and the moral significance of the solution to this problem. His various

---

[1] John A. Zammito, *The Genesis of Kant's Critique of Judgment* (Chicago: University of Chicago Press, 1992), 190.
[2] Ibid., 176.

pronouncements on creation, development, and speciation reveal the predicament he faced. Some basis for a belief in free will, some hope of immortality, and some sense of the divine origins of the world had to be established, he thought, as a defense against civil disorder and personal libertinage:

> [T]hat the world has a beginning: that my thinking self is of a simple and hence incorruptible nature; that this self is also, in its voluntary actions, free and raised above the constraint of nature; and that finally, the entire order of the things that make up the world stems from an original being, from which everything takes its unity and purposive connection – these are so many foundation stones of morals and religion. The antithesis robs us of all these supports, or at least seems to do so.[3]

Yet Kant knew, not only that these foundations were speculative and indemonstrable, but that quite opposed accounts of the origins of the world, the nature of the thinking subject, and the reasons for the appearance of order and harmony were available. Kant's solution to the dilemma was to attack materialism as speculative metaphysics, thereby confounding skeptics and philosophical libertines. His discussion of the origins of life reflects this strategy, even while he struggled to acknowledge the possibility of evolution and transformation mediated by natural powers alone.

Kant could not avoid considering these possibilities, for the researches and ponderings of eighteenth-century investigators pointed to the conclusion that the earth was a very ancient and mutable planet, that many forms of life had arisen earlier and vanished from its surface, that sensibility was a property of suitably organized matter, and that humans were animals who possessed especially nimble fingers and a facility for language and whose religions and moralities were as varied as their dress. Bernard de Maillet in geology; Peter Camper and Alexander Munro in comparative anatomy; Herman Boerhaave, Albrecht Haller, and Theophile Bordeu in physiology; Étienne de Condillac and Denis Diderot in philosophy; and the Comte de Buffon in natural history were inquirers whose results were as significant in their own way as the seventeenth century's innovations in mathematical physics, astronomy, and microscopy. These results had to be integrated with the traditional concerns of metaphysics, even if in Kant's system integration involved drawing certain boundaries, between science and metaphysics, in favour of science, but also between science and morality, in favour of morality.

---

[3] Kant, *Critique of Pure Reason*, trans. Werner S. Pluhar (Indianapolis: Hackett, 1996), A466, B494. (Hereafter CPR.)

2. THE SCIENCES OF ORIGINS

The learned periodical literature of the 1770s and 1780s, especially the *Teutsche Merkur*, the *Goettingischen Gelehrten Anzeigen*, and the *Brittisches Museum fuer die Deutschen*, is an excellent guide to contemporary theorizing on such topics as the natural history of mankind, the difference between humans and animals, the nature of instinct and reason, and the origins of language and morals. These discussions signaled an emerging historicism about the earth and its creatures that antedated Darwin by more than a century.

The speculative science of origins, as it might be called, begins in the modern period with the revival and partial Christianization of Epicureanism in works such as Descartes's *Principles of Philosophy* (1644), which recapitulated some features of his unpublished *Treatise on the World*. According to the Epicurean tradition, set out in Titus Carus Lucretius's great Latin poem, *De rerum natura*, which was recovered in manuscript in 1417 and which went through many printed editions and translations beginning in 1486, multiple worlds come into existence from randomly combining atoms without intelligent guidance or divine fiat:

> For certainly it was no design of the *primordia rerum* that led them to place themselves each in its own order with keen intelligence, nor assuredly did they make any bargain what motions each should produce; but because many *primordia rerum* in many ways, struck with blows and carried along by their own weight from infinite time up to the present, have been accustomed to move and to meet in all manner of ways, and to try all combinations, whatsoever they could produce by coming together.[4]

These multiple worlds contain living creatures, originally brought forth by our mother earth under the influence of the sun and rain. While the animals we observe maintain their kinds from generation to generation, no object and no world endures forever. Each dissolves into its constituent atoms, and long before this happens, many types of animal have died out, including those that happened to be born without hands, or mouths, or moveable limbs. "So with the rest of like monsters and portents that [the earth] made, it was all in vain; since nature banned their growth, and they could not attain the desired flower of age nor find food nor join by the ways of Venus.[5] Only the

[4] Lucretius, *De rerum natura*, trans. W. H. D. Rouse, rev. Martin Ferguson Smith (Cambridge, MA: Harvard University Press, 1992), V:416–32.
[5] Ibid., V:837–54.

better equipped kinds have survived:

> [M]any species of animals must have perished at that [early] time, unable by procreation to forge out the chain of posterity: for whatever you see feeding on the breath of life, either cunning or courage or at least quickness must have guarded and kept that kind from its earliest existence.[6]

Descartes referred to his own combinatorial account of origins as a "fable" and did not try to make precise his conception of the difference between a scientific hypothesis and an entertaining myth. Nor did his successors insist overmuch on the literal truth of their accounts.[7] Nevertheless, studies of erosion and stratification and of the ever-increasing supply of fossils, now agreed to be remains of actual plants and animals, supported the conclusion that the earth was very old, that it had suffered major geological catastrophes, that many types of animal had perished, as Lucretius suggested, or had been transformed into other types of animal. Many "systems of nature" were reformulated from Descartes's time onwards.[8] The hypothesis of the emergence of a settled order that only appears to be the product of a benevolent intelligence is revived by Hume in the *Dialogues Concerning Natural Religion* (1755). Philo describes the Epicurean-Lucretian hypothesis of the spontaneous self-production of the world as "commonly and I believe justly esteemed the most absurd system, that has yet been proposed." But the unstated conclusion of the *Dialogues* is that it is not a bad hypothesis, certainly no worse than its creationist competitor.

One of the most impressive contributions to the speculative science of origins, popular, outrageous, and of great interest to Buffon and Cuvier,[9] was the *Telliamed* of Bernard de Maillet, published posthumously in 1748. Purporting to be a letter from a French missionary recounting his interview with an Indian philosopher, Telliamed's Third Conversation offers a cosmogony in which our planetary system is one of hundreds of millions, each controlled by its own sun. A series of revolutions turns flaming stars into planets and planets into stars as their suns burn out and the planets abandon one vortex

---

[6] Ibid. V:855–61.

[7] Thomasso Campanella and Johannes Kepler published fictions about life and culture on the sun and moon. They were followed by such works as Cyrano de Bergerac's *Other Worlds* (1657), trans. Geoffrey Strachan (London: Oxford University Press, 1965), and Bernard Fontenelle's *Conversations sur la pluralité des mondes* of 1686.

[8] E.g., Pierre Louis Moreau de Maupertuis, *Éssai de cosmologie* (Leyden, 1751). A German translation appeared in the same year.

[9] Bernard de Maillet, *Telliamed, ou Entretiens d'un philosophe indien avec un missionaire francois sur la diminution de la mer, la formation de la terre, l'origine de l'homme, etc.* (Amsterdam, 1748 [trans. Albert V. Carozzi (Urbana: University of Illinois Press, 1968), 4]).

for another. Beneath the surface of the earth, according to Maillet, are super-posed worlds with "entire cities, durable monuments . . . the bones of men and animals, some petrified, others not, stones and marbles."[10] The "fate of the earth" is uncertain. It may one day be completely burnt up or it may leave our vortex for a neighboring one.[11] Maillet calculated a surprisingly accurate age of four billion years for the earth, with human history going back, he thought, an impressive 500,000. The Comte de Buffon veiled his true intentions and adjusted his language in his *Histoire naturelle* (1749–68), but he gave a history of the cosmos and of the evolution of animals that was not orthodox. He posited a single unit of vegetable-animal life, the organic molecule. He noted the underlying similarity of internal structure in man, quadruped, cetacean, bird, reptile, and fish, suggesting that the Supreme Being, "in creating animals, employed only one idea, and at the same time diversified it in every possible manner."[12] He suggested that the variety of external forms might be produced by "mixture and successive variation and generation of the primary species"[13] and that it was even possible that "all animated beings have proceeded from a single species."[14] The discovery of vestigial limbs, such as the complete leg and foot in the skeleton of the seal, and the interesting similarly of the whale's ear to the human ear noted by Munro,[15] furthered the suspicion of a single mammalian *Bauplan* – perhaps indeed a common ancestor for all living things.

The gulf traditionally supposed to exist between sensibility and reason, animal and human, had been narrowed by Locke and by Hume, and the notion that the intellect resided in a separate immaterial soul was attacked by La Mettrie, claiming Descartes as his inspiration. Condillac, the destroyer of metaphysical systems, tried to show how a statue given increments of sensory capability could metamorphose into a fully-functioning human being.[16] The anatomical similarity between humans and the apes brought back from the Dutch colonies of Java and Sumatra in the first years of the eighteenth century – especially the orangutan – suggested either that the great apes had degenerated from men or else that men had "improved" themselves.

---

[10] Ibid., 174.

[11] Ibid., 170.

[12] G. L. L. Buffon, "Ass," in *Natural History, General and Particular*, 3rd English ed., trans. William Smellie, 9 vols. (London: Strahan and Cadell, 1791), 3:401–2

[13] Ibid.

[14] Ibid., 403.

[15] Alexander Munro, *The Structure and Physiology of Fishes Explained and Compared with Those of Man and Other Animals* (Edinburgh, 1785).

[16] Etienne de Condillac, *Traité des sensations* (Paris, 1754).

The young Kant was interested in questions concerning origins that he later deemed to be unanswerable by human inquiry. His first question was how the cosmos might have emerged from a disorganized chaos. Later, he wondered how plants and animals had come to be and what role formative forces in nature might play. Especially towards the end of his life he pondered the meaning of the appearance of rational, moral, and aesthetically sensitive humans. The first question seemed the easiest, though Kant did not dare to venture an answer except anonymously, and seemingly inspired as well by Albrecht Haller's cosmic poetry,[17] he drew on extant cosmogonies, including those of Descartes, Maupertuis, Pierre Esteve, Thomas Wright, and Buffon in his *Universal Natural History and Theory of the Heavens* of 1755.[18] Following the usual scheme, which he allows to be Epicurean,[19] Kant "set[s] the world in the simplest chaos" and then applies two opposing forces from the Newtonian system, attraction and repulsion, "two forces which are both equally certain, equally simple, and equally primary and universal."[20]

> I confidently apply this [reasoning] to my present undertaking. I assume the matter of the entire world to be universally scattered and I make a perfect chaos out of it. I see, in accordance with the established laws of attraction, the stuff forming itself and through repulsion modify its motion. I enjoy the satisfaction of seeing, without the aid of arbitrary notions, a well-ordered whole arise under the direction of established laws, a whole so similar to that world system which we have before our very eyes that I cannot prevent myself from holding it to be the same.[21]

The universe is in constant transition from one state to another, Kant thinks. Our world exists somewhere "between the ruins of a collapsed and the chaos of an undeveloped nature."[22] Planets and comets will crash into their suns, the suns will expand with new nourishment, explode and redisperse their masses. Because "the futility which is grafted onto their finite natures" destroys every world system, eternity will "usher in finally the moment of their doom through

[17] According to his translator, Kant quotes from Haller's "Unvollkommene Ode ueber die Ewigkeit," published in his *Versuch schweizerische Gedichte* (Danzig, 1743).

[18] Kant, *Allgemeine Naturgeschichte und Theorie des Himmels*, translated and edited by Stanley L. Jaki as *Universal Natural History and Theory of the Heavens* (Edinburgh: Scottish Academic Press, 1981). There is a detailed discussion of Kant's sources and his influence in the translator's introduction.

[19] Ibid., 88, A I:226 ("A" references here and below are to Kant's *Gesammelte Schriften*, ed. Akademie der Wissenschaften [Berlin: Reimer, later de Gruyter, 1910–]).

[20] Ibid., 91, A 1:234.

[21] Ibid., 84–5, A 1:225.

[22] Ibid., 159, A 1:319.

a universal decay."[23] The annihilation of every temporal object is compensated for by nature's "inexhaustible capacity of generating":

> Uncounted animals and plants become destroyed daily and are a sacrifice to transitoriness; but nature, through an inexhaustible capability of generating, brings forth no fewer [animals and plants] in other places and fills out the emptiness. Considerable pieces of the earth's surface, which we inhabit, become again buried in the sea, from which a favorable period had brought them forth; but in other places nature completes the defect and brings forth other regions, which were hidden in the depths of water, in order to expand over these [regions] new riches of fruitfulness. In such a way, worlds and world-orders fade away and are devoured by the abyss of eternity; however, creation is always busy in setting up new formations in other celestial regions and in repairing the loss with the gain.[24]

Insects, flowers, men, cities, and nations are meanwhile destroyed by cold, earthquakes, and floods. Even "man who seems to be the masterpiece of creation is himself not exempted" from the law of universal decay. However, Kant appears to allow that some humans will be able to contemplate all this wreckage and destruction as sublime. "Oh, happy is that soul when under the tumult of elements and the dreams [rubbles] of nature he is always set at a height from where he can see the ravages, which the frailty of the things of the world causes, roar by under his feet" (*sic*).[25] Eventually, the chains of matter will fall away from us, we will experience communion with the Infinite Being, and we will abandon the "changing scenes of nature" for rest and bliss.[26]

Kant notes that it is difficult to give a plausible and rigorous account of the origins and destiny of the universe. "To discover the systematic factor which ties together the great members of the created realm in the whole extent of infinity" seems to "surpass very far the forces of human reason."[27] There is a problem of theological and moral import as well in the appeals to chance, to blind mechanism, and to combinatorial processes. It is solved through recourse to a seventeenth-century cliché. The emergence of order from forces described by simple, general, uniform laws is not inconsistent with but, on the contrary, expresses the wisdom and power of the Creator.[28] There must

---

[23] Ibid., 157, A 1:317.
[24] Ibid.
[25] Ibid., 161, A 1:322.
[26] Ibid.
[27] Ibid., 81, A 1:221.
[28] Ibid., 82 ff., A 1:222.

be a God, Kant says, "precisely because nature can proceed even in chaos in no other way than regularly and [in] orderly [fashion]."[29]

Kant was not alone in worrying about the warrant for his theory. The speculative sciences of origins were recognized by their own authors as methodologically problematic, and autocritique and autoresponse were often incorporated into the exposition. Telliamed's interlocutor, for example, complains that Telliamed's views are "an intellectual game based on hypotheses, phenomena, reasonings, and deductions which are very remote from the proofs concerning the works of the sea and its diminution."[30] They should be considered "pleasant dreams like those we find even in the authors of our religion."[31] And the *Universal Natural History* incorporates similar epistemological considerations: Kant notes the satisfaction for man of "launching with his imagination across the frontier of completed creation." However, he says, "I am rightly concerned that hypotheses of this sort are usually held in no higher regard than are philosophical dreams."[32] He reviews the arguments for his scheme: There must be a center, else there would be no stability; the center must contain the densest matter; hence the creation of world must take place over time. But he claims to know a good deal more than this, as is especially apparent in his discussion of the intellectual and moral characteristics of the inhabitants of the inner and outer planets, deduced from their distance from the sun.[33]

In the *Dreams of a Spirit-Seer* of 1765, knowledge acquired in a state of reverie is treated by Kant with disdain. His role as speculative cosmologist was set aside up to the writing of what became the *Opus Postumum* following his death in 1804.[34] The "critical turn," supposed to have occurred around 1770, ushered in a new mood of caution about speculative hypotheses concerning the beginning and end of the world and fixed Kant's ethical programme. For, in the meantime, Kant had been worried about the foundations of ethics, seeking a nonsentimental basis for moral motivation, but reluctant to, as he expresses it in the *Dreams*, start up the big engines of the other world.

Kant's critical idealism is popularly said to show a third way between empiricism and rationalism. The *Critique of Pure Reason* has been ascribed various targets – dogmatic theologians who base their systems on the

---

[29] Ibid., 86, A 1:223.

[30] Maillet, *Telliamed*, 235–6.

[31] Ibid.

[32] Kant, *Universal Natural History*, 91, A I:234.

[33] "Von den Bewohnern der Gestirne," *Universal Natural History*, 167 ff., A 1:353 ff.

[34] See John Zammito on this phase and its sequel in *Kant, Herder, and the Birth of Anthropology* (Chicago: University of Chicago Press, 2002), esp. chaps. 3 and 5.

Revelation and appeal to biblical authority; visionary dreamers or enthusiasts who intuit theology and accumulate fanatical followers; Wolffian metaphysicians who claim to deduce the truths about God, the soul, and the world from the principles of non-contradiction and sufficient reason. Kant's "critique" is opposed to dogmatism, to reveries, and to rationalism. But Wolffian metaphysics *was no longer a powerful intellectual force by the 1770s*, and the intellectual anxiety prompting the writing of the three Critiques is inadequately explained by reference to school metaphysicians and enthusiasts, none of whom advocated doctrines that were *directly* threatening to morality and civil order, for God and a future state are presupposed by theologians, visionaries, and metaphysicians alike. The danger Kant saw in traditional and enthusiastic theology and metaphysics lay not in the undermining power of their doctrines but in their intellectual weakness against undermining doctrines.

The new proponents of the speculative sciences of origins, the inheritors of the Epicurean tradition – Buffon, Maupertuis, Hume, Holbach, and Herder – were directly threatening. Kant's announcement in the preface to the second edition of the first *Critique* leaves no doubt that it is these authors who concern him. It is the duty of the Schools, Kant says there,

> to investigate thoroughly the rights of speculative reason, in order to forestall once and for all the scandal that sooner or later must become apparent, even to the people, from all the controversies – the controversies in which, when there is no critique, metaphysicians . . . inevitably become entangled, and which thereafter even corrupt their teaching. Solely by means of critique can we cut off, at the very root, *materialism, fatalism, atheism*, free-thinking *lack of faith, fanaticism*, and *superstition*, which can become harmful universally; and, finally, also *idealism* and *scepticism*.[35]

Kant's aim, as he states here, was to stifle the growth of all inflammatory and socially disruptive "-isms" and to reassert the authority of academic philosophy over popular philosophy and *belles lettres*, and he compares his role to that of the police whose function in civil society is to prevent violence and permit the citizenry to go about their daily affairs. The critical philosophy was intended to provide solid foundations for a formally secularized (if hardly de-theologized) non-hedonic ethics. This would be accomplished through an attack on invitations to venture outside the experienced life-world – the specially legitimated extensions of mathematics and Newtonian physics constituting an exception – to acquire knowledge. His correspondence with

---

[35] Kant, preface to the second edition, CPR Bxxxiv.

Lambert in the 1760s meanwhile had suggested to him the possibility of a renewal of metaphysics not on pseudo-mathematical but on genuinely "constructive" grounds, on analogy with Euclidean geometry. Kant's noncommittal brand of teleology – teleology that is "thought" and that serves as a heuristic for the naturalist but is not asserted dogmatically – is a third way between physicotheology and materialism. It not only permits but also encourages "thought," though not knowledge, about purposes and God, while not laying itself open to skeptical assaults and ridicule.

The deployment of the critical philosophy against materialism is extensively discussed in the sections of the *Critique of Pure Reason* concerned with what is perhaps best translated as the theory of transcendental method (*Transzendentale Methodenlehre*). Kant's aim in these sections is to deliver on his promise to limit the claims of knowledge to make room for faith, that is, for commitment to non-natural ethical ideals. He employs an array of transempirical hypotheses in an effort to show up the hypotheses of materialism and mortalism as arbitrary and unfounded. The critical metaphysician is permitted to articulate transcendental hypotheses for dialectical purposes – such as the hypothesis that our earthly life is merely a dream – though if he were to advance them as theories, they would have to be rejected.[36] Kant's named target is Hume – more precisely, the "cold-blooded Hume," not the skeptic about the ontology of causal relations but, as far as one can determine from the context, the author of the posthumous *Dialogues*, which undermined so thoroughly Kant's favoured proof for the existence of God, the argument from design.[37] In these passages, Kant faces the question whether religion is an illusion, whether, as Epicurus had argued somewhat ambiguously, the gods exist only insofar as men believe them to exist. He is also concerned with the apparent absurdity and contingency of procreation, the sense of futility that he thought attended any reflective, nondebauched person's thinking of himself as a chance product of nature, doomed like everything else to dissolution and extinction, or worse, as a chance product of purposeless congress driven by sexual need. As he says in the *Critique of Pure Reason*,

> Generation, in man as in non-rational creatures, is dependent upon opportunity, indeed often upon sufficiency of food, upon the moods and caprices of rulers, nay, even upon vice. And this makes it very difficult to suppose that a creature

[36] Kant, CPR A782, B810.

[37] See D. J. Loewisch, "Kants *Kritik der reinen Vernunft* und Hume's *Dialogues Concerning Natural Religion*," *Kant Studien* 56 (1965): 1170–207; Catherine Wilson," Interaction with the Reader in Kant's Transcendental Theory of Method," *History of Philosophy Quarterly* 10 (1993): 87–100.

whose life has its first beginning in circumstances so trivial and so entirely dependent upon our own choice, should have an existence that extends to all eternity. As regards the continuance (here on earth) of the species as a whole, this difficulty is negligible, since accident in the individual case is still subject to a general law, but as regards each individual it certainly seems highly questionable to expect so potent an effect from causes so insignificant.[38]

How can we believe that our coming into existence, depending as it does on moods and whims, opportunity, and vice, entitles us to eternal life or at least demands meritorious conduct of the sort historically supposed to confer it?[39]

3. CRITICAL IDEALISM AS A RESPONSE TO MATERIALISM

The title *Système de la nature*, along with its variants, was one that eighteenth-century authors often reached for. La Mettrie wrote one, Maupertuis wrote one, Buffon wrote one, Diderot wrote one. In his own *Système de la nature*, Holbach wrote,

> Man seeks to range out of his sphere: notwithstanding the reiterated checks his ambitious folly experiences, he still attempts the impossible; strives to carry out his researches beyond the visible world.... He quits the contemplation of realities to meditate on chimeras, he neglects experience to feed on conjecture, to indulge in hypothesis. He pretends to know his fate in the indistinct abodes of another life.... Man is the work of Nature: he exists in Nature: he is submitted to her laws: he cannot deliver himself from them; nor can he step beyond them even in thought.... The beings which he pictures to himself as above nature, or distinguished from her, are always chimeras formed after that which he has already seen.... There is not, there can be nothing out of that Nature which includes all beings.... Let him [man] with a murmur yield to the decrees of a universal necessity.[40]

To see in the laws of the senses objective grounds and to yield without a murmur to the decrees of a universal necessity is to be, in Kant's emerging moral ontology, an animal, and, as Edmund Burke would later say about women, probably not the best kind of animal; the suggestion of erotic swooning in

---

[38] Kant, CPR A779, B807.

[39] Ibid. See Bernard Edelman, *La maison de Kant*, translated by Graeme Hunter as *The House That Kant Built* (Toronto: University of Toronto Press, 1997). Also Robin May Schott, *Cognition and Eros* (Boston: Beacon, 1983).

[40] Paul-Henri Dietrich, Baron d'Holbach, *System of Nature, or Laws of the Physical and Moral World*, trans. H. D. Robinson (New York: Burt Franklin, 1868 [repr.1970, originally published 1770]), viii.

Holbach is probably calculated. How was the mockery of the French materialists, who held the concepts of the *mundus intelligibilis*, the pure intellect, the
pure understanding, and even pure philosophy for useless, to be dealt with?
Kant's critical idea was to turn the accusatory rhetoric of the materialists
against materialism itself. Holbach's confidence in atheism and determinism
would be shown up as another form of speculative enthusiasm. Kant would
take the position that it was not the theist but rather the materialist who "ranges
out of his sphere," "meditates on chimeras," and "pretends to know his fate
in the indistinct abodes of another life." Lambert, it might be observed, had
found Kant's theory of universal decay and cosmic renewals too harsh and
Lucretian, too insensitive to the continuity of species and the needs of the
creatures for whom God constructed the world. He restricted development
and decay to "things whose birth and demise take place under our eyes," and
this form of empiricism, displayed in his *Cosmologische Briefe ueber die
Einrichtung des Weltbaus* of 1761,[41] may have had a sobering effect on Kant.

In fact, this strategy of accusing the materialist of dreaming mad dreams
was not new. Accusations of fictitiousness were hurled by both skeptics and
the orthodox against one another.[42] Diderot refers to the "wings of theology,
pretty little bat wings with which one circles in darkness."[43] The physciotheol-
ogist H. S. Reimarus, author of *Die vornehmsten Wahrheiten der natuerlichen
Religion* (1754), addressed to Lucretius, Buffon, Maupertuis, Rousseau, La
Mettrie, "and other ancient and modern followers of Epicurus," attacks their
"fictitious hypotheses." His opponents "talk in the strain of poets; who create, destroy, and recreate at pleasure.... Their world is a fairy-land."[44] The
pantheists are victims of "reveries."[45] Spinoza is said to construct a world in
his imagination, "the reverse of what it really is," made up from an "arbitrary
conjunction of ideas."[46]

Kant perceived that the theologians, despite their control of the universities,
had little power to suppress unorthodoxy and libertinage in a literate culture.
The Schools, in his view, had failed to inculcate the foundations of morality
because they relied on a method of proof that had been exposed by the critics

---

[41] Translated and edited by Stanley L. Jaki as *Cosmological Letters on the Arrangement of the
World Edifice* (Edinburgh: Scottish Academic Press, 1976); for this aspect of Lambert's critique
of Kant, see 58–9, 63.

[42] Father Lactantius had described atomism as a "dream." It was common knowledge that Lucretius
had written *De Rerum natura* in a period of dementia brought on by a philtre given him by his
mistress.

[43] Diderot, *Oeuvres completes*, ed. J. Assezat, 20 vols. (Paris, 1875–7), 2:298.

[44] H. S. Reimarus, *The Principle Truths of Natural Religion*, trans. R. Wynne (London, 1766), 61.

[45] Ibid., 91.

[46] Ibid., 132.

of Descartes, by Locke and Hume, and in the futilities of the Leibniz-Clarke debate as hollow. The simplicity and immortality of the soul, the freedom of the will, and the existence of God were simply not demonstrable. Belief in these articles had therefore to be assured by something other than the old methods of demonstration.

The first *Critique* approaches the problem by arguing that even if the search for explanations in the special department of natural science requires us to posit a world in which Newtonian mechanics holds sway, materialism and determinism can never be *known* to be true for all phenomena. While they are just as thinkable as theism and the metaphysics of the soul, the position of critical idealism is that things-in-themselves cannot be construed either as material atoms or Leibnizian monads since they are not accessible by observation or by reason. They constitute an unknowable substratum. For nonscientific purposes, we must, and therefore may, posit the traditional subjects of metaphysics: God, the soul, and free will.

The main harmonization strategy at Kant's disposal for reconciling the need for the assumptions of determinism and materialism in the physical sciences with morality was the deployment of the phenomena-noumena distinction (though not as an item of positive knowledge) and the relativization of space, time, and causality to human minds. According to the doctrine of critical idealism, the spatio-temporal and causal structuring of experience and events follows from our necessary epistemological capacities rather than being intrinsic in external nature. Despite the evident power of the Newtonian system of mechanical causality, no one, on Kant's view, could claim to know that we were formed by nature and circumscribed by her laws, since we are, in a sense, the authors of nature and its laws ourselves. We conscious creatures make mechanism true by perceiving material objects and by producing judgments of cause and effect regarding them. With the concept of the rational self he has provided, Kant says, we are freed from

> any fear that if we removed matter then all thinking and even the existence of thinking beings would thereby be annulled.... [On the contrary] if I removed the thinking subject then the whole corporeal world would have to go away, since the world is nothing but the appearance in the sensibility of, and a kind of presentations of, ourselves as subject.[47]

The hypothesis that the material world is only a dream is of course a fantasy that has only a dialectical use – namely, for challenging the materialist to prove it false. But the positively asserted doctrines in the body of the first

[47] Kant, CPR A383.

Critique deliberately invoke this dogmatic idealism to motivate acceptance of the critical variant. The first edition is more transparent in this respect than the more cautiously and cumbersomely worded second edition:

> The concept of appearances...provides us...with the objective reality of noumena and justifies the division of objects into phenomena and noumena; and hence also the division of the world into a world of sense and a world of understanding.... For if the senses represent to us something merely *as it appears*, then surely this something must also in itself be a thing, and an object of a nonsensible intuition, i.e., an object of understanding. That is, a cognition must be possible in which no sensibility is to be found and which alone has reality that is objective absolutely.[48]

In fact, noumena are not objects of cognition. There is no nonsensible world for us to intuit, no world that "could engage our pure understanding...much more nobly."[49] Yet this residual thought-object continues to trail its elevated mystical-Platonic associations, even as it is deemed admissible "only in the *negative* signification," that is, as real but as neither perceptible nor cognizable as a determinate thing.[50]

In the A version of the paralogisms, there is another virtually positive idealist moment: "[W]e have proved undeniably that bodies are mere appearances of our outer sense and not things in themselves. In accordance with this, we may rightly say that our thinking subject is not corporeal."[51] What we call "matter" is said there to be

> nothing but a mere form, or a certain way of presenting an object, with which we are unacquainted, by means of that intuition which is called outer sense. Thus there may well be something outside us to which this appearance that we call matter corresponds; but in the same quality that it has as appearance it is not outside us, but merely within us as a thought.[52]

Having so dispensed with matter, and having shown that its ground is not cognizable and that it is not known to be a different kind of thing than the soul, Kant is in a position to designate certain kinds of materialist theorizing about the origins of the world as speculative and transcendental. Taking a stance on the question whether the world had a beginning in time is purposeless, for its end points are not the subjects of any possible experiences.[53] With his

[48] Kant, CPR A249.
[49] Kant, CPR A250.
[50] Kant, CPR B306–8.
[51] Kant, CPR A357 ff.
[52] Kant, CPR A370.
[53] Kant, CPR B353; cf. B454 ff.

delicensing of claims about the beginning of the world, Kant licenses other speculative hypotheses – the doctrines of free will, unlimited moral potency and immortality – on the grounds that, unlike hedonism, atheism, and mortalism, they serve human interests. These speculations are not merely available for dialectical use, like the fanciful hypothesis that life is a dream; they are actually mandatory. Rhetorically, this strengthening of their status occurs through the pressing of an analogy. As certain ways of perceiving – of objects, in space and time, etc. – are mandatory for us, so certain ways of conceiving ourselves as free, responsible, and accountable agents are unavoidable.

The overall aim of Kant's *Critique of Pure Reason*, then, was not merely to come to terms with the particular views of empiricist and rationalist predecessors concerning the roles of experience and a priori intuition in the constitution of knowledge and to offer necessary correctives. Least of all was the intention to encourage young atheists, as Kant's shocked reaction to some of the responses to the first edition demonstrates. Kant was not a cool skeptic, a German Hume disdainful of the metaphysical excesses of the Leibniz-Wolffian philosophy, and able to recommend suspension of belief in the face of antinomies. The task of the *Critique* was to fabricate metaphysical foundations for natural science that would preclude the possibility of the encroachment of nature and nature's laws into the moral realm.

4. KANT'S VIEWS ON THE ORIGINS OF LIFE AND MAN

Atomism and mechanism, submission to the laws of nature, were, then, sufficiently dealt with in the first *Critique*, but the specific question of the relationship between animals and humans and of the roles played by instinct, desire, and sensibility in human existence was not touched on there. In his *System of Nature* of 1735, Linnaeus had notoriously classified "man" in the genus *Anthropomorpha* with the apes, monkeys, sloths, and the "tailed men" reported by travelers.[54] This classification had caused him some trouble. Reproached, he responded,

> I demand of you, and of the whole world, that you show me a generic character . . . by which to distinguish between Man and Ape. I myself assuredly know of none. I wish somebody would indicate one to me. But, if I had called man an ape, or vice-versa, I should have fallen under the ban of all the ecclesiastics. It may be that as a naturalist I ought to have done so.[55]

---

[54] Linnaeus, *Systema Naturae* (Leyden 1735), quoted in John Greene, *The Death of Adam: Evolution and Its Impact on Western Thought* (Ames: University of Iowa Press, 1959), 176.

[55] Linnaeus, Letter to J. G. Gmelin, 1747, quoted in Greene, *Death of Adam*, 184.

Already in 1698 Edward Tyson had dissected a small chimpanzee, which, he claimed, constituted a link in the chain of being between monkey and man and resembled man more than the monkey, though it was difficult to explain why it did not speak.[56] Noting that the Dutch had found orangutans in 1702 and had shown one at a fair in 1720, Maillet asked through his interlocutor in the *Telliamed* whether "[i]f males and females of these Oran-soutans had been captured," and then had offspring, "it might have been impossible, through a couple of generations, to make them capable of using a true language and reaching a more perfect shape than they originally had?"[57] The playfulness, cunning, sociability, and parental care of the larger apes were remarked on, and though it was clear that they were a distinct species, as Linnaeus put it, not "of common descent or blood with us," their absence of material culture and language did less to reinforce the notion that humans were ontologically distinct in having a rational soul than to suggest that the manifestations of reason were related to the possession of speech. When the Dutch anatomist Peter Camper declared in 1779 that apes could not speak because of the structure of their vocal organs, this suspicion deepened.

Kant was somewhat favorably inclined, on anatomical evidence, to believe that humans had developed from something not quite human. His willingness to entertain the animal-to-man hypothesis is established by a passage cited by Lovejoy from Kant's 1771 review of a book by Dr. Moscati on "Upright Posture."[58] There, Kant seems to agree with Moscati that man was once a quadruped, whose "germ of reason" blossomed into culture when he attained bipedality. Yet this admission threw the notion of a species into confusion. In his 1775 essay on the "races," Kant seems to regard a species as a group constituted by descent from common ancestors and, at the same time, consisting of individuals capable of interbreeding.[59] In this connection, Kant regards it as probable that an original white race gave rise to black, brown, and yellow races, just as, according to Buffon, the wolf, fox, jackal, and hyena may have sprung from a common ancestor.[60] As Lovejoy points out, the "argument by

---

[56] Ibid., 176–7. Tyson's memoir published in London in 1699 was entitled *Orang-outang, or . . . the Anatomy of a Pygmie compared with That of a Monkey, an Ape, and a Man.*

[57] Maillet, *Telliamed*, 202.

[58] Quoted by A. O. Lovejoy, "Kant and Evolution," in *Forerunners of Darwin: 1745–1859*, ed. Bentley Glass, Owsei Temkin, and William L. Straus, Jr. (Baltimore, Johns Hopkins, 1959), 173–206, citing 177.

[59] On the background to and the importance and influence of this "historical" conception, see Philip R. Sloan, "The Historical Interpretation of Biological Species," *British Journal for the History of Science* 12 (1979) 109–53.

[60] Kant, *Von den verschiedenen Racen der Menschen* (1775), A II:430.

which Kant reasons that all men are of one *Stamm* [lineage] directly implies that men and other animals are not of one *Stamm*, i.e. not related through any lines of natural descent."[61] But did we humans have a nonhuman ancestor? Kant is unclear on this point. If we did, then, contrary to his claim, species can give rise to entirely different species, with which they presumably cannot then interbreed, or the notion of a species must be rather generous – to incorporate quadrupedal, irrational, early humans – and either the interbreeding criterion does not apply or we are more flexible than some of us might imagine. This puzzling dilemma may have given Kant one more reason to refuse to pronounce definitively on the question of ultimate origins and to declare the entire subject out of bounds.

It was one thing for natural philosophers to recognize the formal inclusion of men in the class of "animals" – sensitive, mobile beings – distinguished from the other animals by their rationality, as they always had. It was another matter entirely to regard men as essentially or importantly animals, or to regard them as cousins of the apes; Buffon described this as "une verité peut-être humiliante." The question of man's animal nature became pressing for Kant after the appearance of Herder's great evolutionary work, the *Ideen zur Philosophie der Geschichte der Menschheit* in 1784.[62] Citing Munro, Camper, Buffon, and Tyson, as well as the travelers Philibert Commerson and Georg Forster, Herder posited a unity in life-forms derived from an underlying animal prototype; cultural forms in turn were the various elaborations produced by the human animal. "Der Menschen aeltere Brueder sind die Tiere," said Herder.[63] Departing, perhaps, from Buffon's comment that "their instinct may appear more sure than his reason, their industry than his arts," Herder said not only that culture was zoologically and geographically determined but that men had learned almost everything worth knowing from animals, from those "living sparks of divine understanding."[64] In his view, "[s]ensibility [in] our species alters with education and environment; but it is everywhere the case that the use of the senses is what develops our humanity."[65] He also stated that "[r]eason, too carelessly, too uselessly diffused, may well weaken desires, instincts and vital activity – in fact has already done so."[66]

---

[61] Lovejoy, "Kant and Evolution," 181.

[62] Kant-Herder relations are given detailed treatment in Zammito, *Kant, Herder and the Birth of Anthropology*, 171 ff., 309 ff.

[63] Herder, *Ideen zur Philosophie der Geschichte der Menschheit*, ed. Martin Bollacher (Frankfurt: Verlag Deutscher Klassiker, 1989), I: 2: 3, 67.

[64] Herder, *Ideen*, I: 2: 3, 69.

[65] The title of *Ideen* I: 8: 1, 291.

[66] Herder, *Ideen*, 199. *Ideen*, II: 8: 5, 199.

The above theses – above all, the derogation of reason and the elevation of sensibility – were anathema to Kant. His views on the unity of life and the differentiation of species are hard to untangle, but we can reasonably take the third *Critique* as throwing obstacles in the way of the "uncritical" acceptance of the unseen formative and transformative forces that supported Herder's animalism.

In his 1785 review of the *Ideen*, Kant says that the idea of a relationship between species "such that one species should originate from another and all from one original species, or that all should spring from the teeming womb of a universal Mother," would lead to "ideas so monstrous that the reason shrinks from before them with a shudder."[67] A 1788 review of Georg Forster's account of his voyages refers in a similarly unappetizing vein to the earth's production of "animal and plants from her pregnant womb, fertilized by the sea slime."[68] Kant repeats his formulation from his review of Herder; such ideas cause the investigator to "shrink back from before them with a shudder of horror."[69] He reasserts the doctrine of the fixity of the species, denying in the first instance precisely what Darwin will lay great stress on, namely, that human art can produce monsters:

> I, for my part, adopt it as a fundamental principle to recognize no power in the imagination to meddle with the reproductive work of Nature, and no possibility that men, through external modifications, should effect changes in the ancient original of a species in any such way as to implant those changes in the reproductive process and make them hereditary.[70]

Otherwise, he says, any sort of grotesquerie might be produced. To be sure, this claim leaves open the possibility that nature, as opposed to human art, can modify animal forms, as well as the possibility that humans can modify forms to the extent permitted by the underlying dispositions of the species: Kant could not have failed to be aware that domestic animals, along with ornamental and edible plants, were modified by cultivation. On the basis of his racial doctrines, it seems that Kant allows for the possibility of modifications induced in one way or another, but restricts them to the elicitation of latent potentialities within a single species.

Why the references to "monsters" and the grotesque, one might wonder? The beauty of flowers, tropical birds, and seashells comes up for examination

---

[67] Kant, VIII:107–23, trans. Lovejoy, "Kant and Evolution," 195.
[68] Ibid., 196.
[69] Ibid., 197.
[70] Ibid., 184.

in the first part of the third *Critique*.[71] Our spontaneous appreciation of living forms – "free natural beauties" – indicates the presence of a preestablished harmony between nature and our faculty of taste. Nature can be studied and contemplated under her "good" aspect. But elsewhere, especially with respect to mammals, Kant expresses a kind of biophobic horror. Plants and animals seem to belong to a "separate world-system."[72] "Were experience not to furnish examples of [them]," he says in the *Opus Postumum*, "the possibility of living bodies would be dismissed by everyone as fantasies of the Prince of Palagonia."[73] Kant refers here to the Sicilian Ferdinando Francesco Gravina Agliata, who, according to travelers' reports, filled his palace with a tasteless and barbaric assortment of statues of animal-humans – "human torsos fitted with the wings of birds and fishtails, with the limbs of quadrupedal animals, the trunks of elephants, the tusks of boars, the claws of vultures, and the tail of a monkey or a fox."[74]

Creationism and the doctrine of preformation, it might be noted in this connection, secured the fidelity of the copy and sustained a discourse of aesthetic perfection. The new discourse and iconography of epigenesis and trans-species evolution that was replacing it was replete with disturbing Ovidian metamorphoses and monstrosities. Maillet tells us how fish grew wings and became birds when the seas receded and how their fins dried, cracked, and split, their beaks and necks lengthened and shortened, and colored down appeared on their bodies.[75] The early human fetus exposed to the eye of the unprejudiced anatomist is not a perfect miniature baby. It has gills like a fish and a tail like a salamander. In assembling a case against preformation, J. F. Blumenbach, in his treatise on generation, introduces us to teratomas; cysts filled with teeth and hair; two-headed creatures sharing a body; a polyp which, when slit open and laid flat, develops a new belly inside itself; nails growing from the stumps of severed fingers; and membranes and neovascularization surrounding tumors.[76]

Despite Kant's repulsion from the idea of entirely plastic organisms, as opposed to organisms that remain within their interbreeding categories and

---

[71] Kant, *Critique of Judgment*, trans. Werner S. Pluhar (Indianapolis: Hackett, 1987), A V:229 f. (Hereafter CJ.)

[72] Kant, *Opus Postumum*, trans. Eckhart Foerster and Michael Rosen (Cambridge: Cambridge University Press, 1993), A XXII:210. (Hereafter OP.)

[73] Ibid., A XXII:383.

[74] Jean Houel, *Voyage pittoresque des isles de la Sicile, de Malte, et de Lipari* (Paris, 1782), 41–50.

[75] Maillet, *Telliamed*, 187.

[76] J. F. Blumenbach, *Üeber den Bildungtrieb und das Zeugungsgechäfte* (Göttingen, 1781), translated by A. Crichton as *An Essay on Generation* (London, 1792), 72 ff.

that can exhibit only a limited and predetermined range of variations, there is a pattern of partial accommodation to transformism. In the 1788 essay on teleological principles, which Kant employed to explain racial diversification, he suggests that, though the production of new races of humans is not occurring, new varieties do appear, seeming to indicate "a nature inexhaustibly productive of new characters, both inner and outer."[77] In the *Anthropology*, he suggests that in some "future revolution of nature" a chimpanzee or an orangutan might "perfect the organs which serve for walking, touching, speaking, into the articulated structure of a human being, with a central organ for use of the Understanding, and would gradually develop itself through social culture."[78] Even Lovejoy, the implacable foe of Kant as evolutionary theorist, concedes that Kant's position in the third *Critique* appears to have softened in the two years following the Forster review. He no longer condemns transformism on a priori grounds. "Many genera of animals," he says, echoing Maillet, "share a certain common schema," on which not only their bone structure but also the arrangement of their other parts seems to be based: the basic outline is admirably simple yet was able to produce the great diversity of species, by shortening some parts and lengthening others, by the involution of some and the evolution of others. Despite all the variety among these forms, they seem to have been produced according to a common archetype, and this analogy reinforces our suspicion that they are actually akin, produced by a common original mother.[79]

"A hypothesis like this," he says, "may be called a daring adventure of reason, and one that has probably entered, on occasion, even the minds of virtually all the most acute natural scientists."[80] It is not absurd because it does not posit that organized beings have arisen from "crude, unorganized" matter; there is even reason to think that aquatic animals developed into marsh animals, then land animals.[81] Though experience gives us no examples of *generatio heteronyma*, Kant says, reason cannot show that it is impossible.[82]

Lovejoy notes ruefully that it is difficult to extract Kant's true position from a scheme of "elaborate self-contradictions."[83] His mature view can

---

[77] Lovejoy, "Kant and Evolution" 188.

[78] Kant, *Anthropology*, A VII, 327–8. This suggestion, Lovejoy says, cannot be taken as implying that in the past apes might actually have done so, but as evoking Charles Bonnet's claim that a series of "revolutions" will occur in which organisms are destroyed, encapsulated, and reborn into higher forms (Lovejoy, "Kant and Evolution," 203–4).

[79] Kant, CJ A V:419.

[80] Ibid., A V:419. n. 5.

[81] Ibid.

[82] Kant, CJ A V:420 n. 5.

[83] Ibid.

perhaps be summarized as follows: Physicotheology does not provide any proofs of a divine creator, and hexameralism is "poetic raving."[84] At the same time, formative forces that could work on dead matter to convert it into a living creature are unimaginable,[85] and hylozoism, the doctrine that all matter is living, would be "the death of all natural philosophy."[86] There are striking analogies between animal forms that have a powerful tendency to cause us to think that all life emerged from one, or some small number, of original living forms through the evocation of potentialities. However, we have never observed one species being changed into another. Further still, to obtain insight into a product of nature is to understand how it comes to be according to mechanical principles.[87] We are permitted to refer to an epigenetical force that works on matter already organized in some fashion, of the sort described by J. F. Blumenbach, "a power whose constant operation is known to us from experience, but whose cause, like the causes of most of the qualities of matter is a quality occult to us."[88] But the existence of the world system of mutually adapted plants and animals opens up a gap in our understanding: that they exist at all is a mystery, and we cannot imagine that forces resident in isolated particles of matter should have brought about the whole system which appears to us. Nature might have structured herself in numerous ways according to the laws of mechanics without producing living creatures, and we have no idea how mechanical forces could do so.[89] And there is a difficulty – the purposiveness of living creatures – that neither mechanism nor the theory of epigenetic forces solves and that theism does solve.[90] Organic bodies are those in which each of the parts is there for the sake of the rest and in which the operation of the whole determines the form of the parts. We can effectively distinguish these bodies from inorganic bodies, say, ice crystals, which, though they have complex forms, do not have parts which seem to serve the whole.

The conclusion is simple and inescapable: human beings cannot understand the origins of plants and animals. When we try, we go around in a circle; the weaknesses of theism lead us to mechanism or epigenetic forces, the

---

[84] Kant, CJ A V:410.

[85] Kant, CJ A V:425.

[86] Kant, *Metaphysical Foundations of Natural Science*, A IV:544.

[87] Kant, CJ A V:410. Unlike everyone from Bourguet to Holbach, as Zammito points out, Kant "chose to adhere to a relentless sense of mechanical causality" (*Genesis*, 190). He continues to try to revive the agenda, one might say, of 1715.

[88] J. F. Blumenbach, *An Essay on Generation*, trans. A Crichton (London: Cadell, Faulder Murray and Creech, 1792).

[89] Kant, CJ A V:360.

[90] Kant, CJ A V:257.

impotence of mechanism and the impossibility of systematizing epigenetic forces lead us back to theism, and there are no other philosophically coherent possibilities:

> It is quite certain that in terms of merely mechanical principles of nature we cannot even adequately become familiar with, much less explain organized beings and how they are internally possible. So certain is this that we may boldly state that it is absurd for human beings even to attempt it, or to hope that perhaps some day another Newton might arise who would explain to us, in terms of natural laws unordered by any intention, how even a mere blade of grass is produced.[91]

The generation of life presents a scientific mystery, as Buffon had already said, more elegantly and succinctly, at the start of the *Histoire naturelle*.[92] Kant had already said as much himself in the *Universal Natural History*: "The formation of all celestial bodies ... the origin of the whole present arrangement of the world-edifice will sooner be understood than the production of a single herb or of a caterpillar."[93] This time, however, Kant's assertion of ignorance had a dialectical use. It was intended to drive the reader into the acknowledgment of a complex set of antinomies in which the competing hypotheses of theism and Epicureanism both failed the test of being either knowable or intelligible, leaving a space for human interests to posit a morally satisfying answer to the question of origins. It is accordingly difficult to see the *Critique of Judgment* as making a definite contribution to the methodology or the theory of the life sciences.[94] Kant suggests that insofar as the objects of nature are appearances, a supersensible ground might exist for them that would be linked with higher purposes.[95] This supersensible ground would render our human existence more than gratuitous and arbitrary and would redeem our mode of reproduction from absurdity. However, Kant could not quite make out a parallelism between the unknowable but necessary immaterial thing-in-itself, the unknown but necessary free will of man, and an unknown but necessary building force of nature and thus be able to unify his system.[96]

---

[91] Kant, CJ A V:400.

[92] Buffon, *Natural History*, II:3.

[93] Kant, *Universal Natural History*, 88, A I:230.

[94] However, for the view that Kant, by contrast with the *Naturphilosophen*, provided a coherent, successful, influential groundwork for the life sciences, see Timothy Lenoir, *The Strategy of Life: Teleology and Mechanics in Nineteenth Century German Biology* (Dordrecht and Boston: D. Reidel, 1982), esp. the introduction and chap. 1.

[95] See Peter Mclaughlin, *Kant's Critique of Teleology in Biological Explanation* (Lewiston, NY: Edwin Mellen, 1990), 177.

[96] As McLaughlin points out, we have direct access to our freedom by introspection, but not to the supersensible element in nature (ibid., 178).

The *Critique of Judgment* makes a considerably less tidy impression than the *Critique of Pure Reason*. Yet Kant's anti-naturalist program in the second part of the *Critique of Judgment* is helped out by results obtained in the first part, "On the Judgment of Taste."

That we spontaneously judge the products of nature to be beautiful and that we require universal agreement in aesthetic judgment shows that one need not know oneself or natural things such as flowers and shells to be products of special creation in order to make objective normative judgments. Our aesthetic responsiveness helps to establish the existence of moral duties even if we cannot know that God created us and created the world for us. While the beauty of natural objects in our eyes does not license the inference to a divine creator, or show that the world was constructed especially for us, the harmony of our (emphatically non-hedonic) faculty of taste with the forms probably produced by the building forces of nature shows that we are *preadapted* to the world not only as system-building scientists but also as aesthetic observers who do not judge according to mere liking or interest. This aesthetic harmonization or preadaptation points to a disposition for morality, according to the *Critique of Judgment*.[97]

Two features strike Kant's reader forcibly in a survey of his oeuvre: the recurrence of optimistic ideas about progress, development, and the production of systems, forms, and cultures from inchoate beginnings, and his brooding pessimism, expressing itself in references to evil, sin, and the need for conquest and subordination. For Kant in his confident moments, moral and theological ideas were revelatory of freedom and its connection to the *summum bonum*, and in those moods he saw in the evolution of the orderly cosmos from its initial chaos and in cultural evolution from more primitive states prognostic signs of a good outcome for rational beings. He looked forward to the day when the heavenly city, the community of good beings – not necessarily men, perhaps rather the superior race which might follow men – living under universal regulations and engaged in a universal *commercium*, would be manifest on earth. This creation of moral-political order from the present semi-chaos would be the completion by reason of the work begun by nature in the formation of the cosmos and living creatures. That was the sanguine Kant. In the intervals, the choleric and melancholy Kant brooded on the evils of men, their possible extinction, and the indifference of nature.

The young Kant of the *Universal Natural History*, who confessed openly to his Epicurean leanings, looked unblinkingly on transitoriness and "waste,"

---

[97] Kant, CJ A 5:353.

but the older Kant could not. "Nature deals despotically with man. Men destroy one another like wolves," he mused. "Plants and animals overgrow and stifle one another. Nature does not observe the care and provision which they require. Wars destroy what long artifice has established and cared for."[98] Kant could not comprehend how it was possible to maintain the attitude of a Spinoza, who, because he cannot credit any pretended demonstration of the existence of a personal God, has resolved to be a moral and upright person without the hope of a future life:

> Deceit, violence, and envy will always be rife around him.... Moreover, as concerns the other people he meets: no matter how worthy of happiness they may be, nature, which pays no attention to that, will still subject them to the evils of deprivation, disease and untimely death, just like all the other animals on earth. And they will stay subjected to these evils always, until one vast tomb engulfs them one and all... and hurls them, who managed to believe that they were the final purpose of creation, back into the abyss of the purposeless chaos of matter from which they were taken.[99]

Kant denies that one can accept this picture and at the same time submit to the rigours of the moral law. The historical Spinoza, having lived before the era of the flourishing of the life and social sciences, did not face the test that Kant thinks no human being can pass.

To summarize, Kant's attitude to the flowering of anthropology and the speculative sciences of origins was ultimately one of partial resistance. His principle that "we have insight into nothing except what we can make ourselves"[100] implies not only that knowledge of nature is our production but that *nature* is in some sense our production, in keeping with Kant's carefully non-asserted but always dominant idealism. This metaphysical distancing coexisted with an omnivorous exploitation of factual material drawn from astronomy and cosmology, medicine and physiology, and travel literature. Yet Kant's political project – the extirpation of atheism, fatalism, and materialism – set him at odds with the most innovative and promising intellectual currents of the eighteenth century. His response to the discovery of the superabundance of life and the ubiquity of perishing – the confirmation of Lucretius's poetic speculations – did not lead him to an appreciation of the depth and individuality of the living body and of the losses and gains of civilization.

---

[98] Kant, OP A XXI:13.

[99] Kant, CJ A V: 452.

[100] Kant, OP A XXII:533; cf. CJ A V:384; Foerster gives additional references to Kant's Letter to Beck, 1794, A XI:515, and the *Reflexionen* R6353 and 6358, A XVIII:679, 683–4.

5. THE POSTCRITICAL KANT AND THE SPECULATIVE
SCIENCES OF ORIGINS

Did the elderly Kant reject critical idealism and embrace a new form of transcendental *Naturphilosophie*? Did he return to the speculative sciences of
origins and the hyperphysical leaps of imagination of the *Universal Natural
History*? This possibility seems increasingly likely. In the *Opus Postumum*,
Kant fills all of cosmic space with caloric, "not a mere thought object but mov-
able and in motion, ... everywhere homogeneous and unique [of] its kind."[101]
Caloric is the fundamental condition of experience, "[f]or without this motion,
that is, without the stimulation of the sense organs, which is its effect, no perception of any object of the senses, and hence no experience, takes place."[102]
Outer perceptions are "nothing other than the effect on the perceiving *subject*
of the agitating forces of matter, which are given *a priori*."[103]

Kant also appears more favourably inclined towards the hypothesis of a
single common ancestor so long as history can be understood to have an
intrinsic direction determined by an ethical principle:

> The organized creatures form on earth a whole according to purposes which
> [can be thought] a priori, as sprung from a single seed (like an incubated egg),
> with mutual need for one another, preserving its species and the species that are
> born from it. ... Also, revolutions of nature which brought forth new species
> (of which man is one).[104]

Kant's thinking culminates in a vision of the entire realm of living nature as
produced from mutual and compounded need. Species are, as it were, made
for one another, "indeed perhaps according to different primordial forms,
now vanished (but, among them, not men) – for the upheavals in the bosom
of the earth and its alluvial mountains give no evidence of such, according to
Camper."[105] The building forces of nature, he comes to insist, are progressive;
they bring into existence a complex "system" in which the lower parts exist
for the sake of the higher.

---

[101] Kant, OP A XXII:610–11.

[102] Kant, OP A XXII:551.

[103] Kant, OP A XXII:522.

[104] Kant, OP A XXII:241. Erasmus Darwin's hylozoistic *Zoonomia*, mentioned in passing by Kant,
appeared in English in 1794; a German translation, according to Foerster, was made by J. D.
Brandis and appeared in 1795–7. The elder Darwin's *Temple of Nature* describes the origins of
organic life in the ocean mud, but it appeared only in 1803, too late to have been intelligible to
the ailing Kant.

[105] Peter Camper refers to mammoths, rhinoceros, gazelles, dragons or pseudo-bears, lions, etc.
but claims never to have seen a fossilized human bone. Ibid., 67 n. 43.

Not only does the vegetable kingdom exist for the sake of the animal kingdom . . . , but men, as rational beings, exist for the sake of others of a different
species (race). The latter stand at a higher level of humanity either simultaneously (as, for instance, Americans and Europeans) or sequentially. For instance,
if our globe (having once been dissolved into chaos, but now reorganized and
regenerating) were to bring forth by revolutions of the earth, differently organized creatures, which in turn gave place to others after their destruction, organic
nature could be conceived in terms of a sequence of different world-epochs,
reproducing themselves in different forms, and our earth as an organically
formed body – not one formed merely mechanically.

How many such revolutions (including, certainly, many ancient organic beings, no longer alive on the surface of the earth) preceded the existence of man,
and how many (accompanying perhaps, a more perfect organization) are still
in prospect, is hidden from our inquiring gaze.[106]

"Nature organizes matter in manifold fashion – not just by kind, but also
by stages," says a footnote.

Not to be comprehended: That there are to be discovered in the strata of the
earth and in mountains, examples of the former kinds of animals and plants
(now extinct) – proofs of previous (now alien) products of our living, fertile
globe. That its organizing force has so arranged for one another the totality of
plants and animals, that they, together, as members of a chain, not merely in
respect of their nominal character (similarity) but their real character (causality)
which points in the direction of a world organization (to unknown ends) of the
galaxy itself.[107]

Vital forces, he now thinks, unlike the merely vegetative forces that operate in the organic machine, imply the existence of an immaterial principle
"possessing an indivisible unity in its power of representation." This vital
existence is distinct from the power of generation.[108] After the publication of
Schelling's *Von der Weltseele* in 1798, the organizing force is described in the
Opus Postumum as "outside space" but active in matter, both internally active
and externally, as an "*archeus*."[109] This principle, which "extends through all
parts of the world (transforming bodies and replacing dead ones with new
formations in their place)," is the *anima mundi*.[110] It is "beyond the limits of
our insight" whether it possesses understanding "or merely a capacity which

---

[106] Kant, OP A XXI:214–5.
[107] Ibid., A XXII:549.
[108] Ibid., A XXII:547.
[109] Ibid., A XXII:421.
[110] Ibid., A XXII:504.

is analogous to the understanding in its effects."[111] The world spirit not only builds bodies but also builds the world as an integrated system.

An immaterial moving principle in an organic body is its soul, and if one wishes to think of the latter as a world soul, one can assume of it that it builds its own body and even that body's dwelling-place (the world).[112] And in a deleted passage Kant contrasts the demiurge, which acts without any moral end, with God.[113] This is the terminus of his reflections: either there is an intelligent world soul that builds animal bodies and builds the integrated system of plant and animal bodies or else there are forces which, though unintelligent, we must conceive on analogy with intelligent agency. In either case, Kant concludes, our conception of ourselves as its most recent and significant products, laden with special responsibilities, is warranted.

---

[111] Ibid., A XXII:548.
[112] Ibid., A XXII: 97.
[113] Ibid., A XXI:214. Noted by Foerster and Rosen (276 n. 98), Kant also said in the *Opus Postumum* that he *was* God.

17

Kant and Evolution

MICHAEL RUSE

Where did Immanuel Kant stand on the question of organic evolution? This has been a topic of much debate, and it seems that it is still ongoing. Some think that he was close if not committed to evolution (see chapter 16 of this volume). This would fit with a general dynamic view of nature and harmonizes nicely with Kant's hypothesis about a natural origin of the universe – his formulation of what is known as the "nebular hypothesis." Others think that he was not in fact an evolutionist but that this was a contingent matter. Kant could have been an evolutionist; it was just that he did not think the facts were favorable to the idea (Richards 2003). And yet others – and I am one – think that Kant was not an evolutionist and that his opposition was deep and theoretical. Given his philosophy of nature, he simply could not have been an evolutionist.[1]

In major part, I think the confusion about Kant's position represents confusion in Kant's own thinking. To say something that a historian of philosophy can say but that a historian of science would never say, Kant simply was not dealing with a full deck and so he could not get things right. He did not know about the mechanism of natural selection, the mechanism that gives an adequate lawbound answer to the issue of organic origins. Hence, Kant was groping with the problems without the tools to give the full answers. However, I come now to praise Kant, not to bury him. Apart from the intrinsic interest of whatever Kant has to say on a topic – and although I think his position incomplete and inadequate, I think there are major insights of relevance to our understanding of organisms today – there is also some historical interest in

---

[1] Arthur O. Lovejoy, "Kant and Evolution," in *Forerunners of Darwin*, ed. B. Glass, O. Temkin, and W. L. Strauss Jr. (Baltimore: Johns Hopkins University Press, 1959); J. F. Cornell, "Newton of the Grassblade: Darwin and the Problem of Organic Teleology," *Isis* 77 (1986): 405–21.

trying to disentangle Kant's thinking. Kant's thinking was influential on others, specifically the German-trained French comparative anatomist Georges Cuvier, the greatest of all opponents of evolutionary thought. If the interpretation of Kant that I endorse is right, then this goes a long way to explaining the opposition of Cuvier – which was also more theoretical than empirical. And since Charles Darwin took Cuvier's concerns very seriously, accepting Cuvier's problem situation if denying Cuvier's nihilism, Kant's thought can then plausibly be said to be a factor in the formulation of the causal theory of evolution that almost all biologists endorse today. One can go one step earlier than the French historian and philosopher of science Michel Foucault, who wrote, "Seen in its archeological depth, and not at the more visible level of discoveries, discussion, theories, or philosophical options, Cuvier's work dominates from afar what was to be the future of biology."[2] More fully: seen in its archeological depth, and not at the more visible level of discoveries, discussion, theories, or philosophical options, Kant's work dominates from afar what was to be the future of biology.

1. KANT ON ORGANISMS

In the tradition that goes back to Aristotle, Kant saw the significant feature of organisms as being that they are organized – they and their parts seem as if fashioned for the ends of survival and reproduction. In other words, organisms demand explanation in terms of final causes as well as efficient and other causes. Yet, Kant saw that final causes are problematic. From the seventeenth century onwards many scientists – particularly scientists in the physical sciences – had tried to expel final causes as unneeded, confusing, and unacceptably theological. Francis Bacon, for instance, had wittily likened them to vestal virgins – beautiful but sterile. In line with this kind of thinking, Kant was uncomfortable with final-cause talk, because it does seem to imply design, and this is simply not acceptable in science. We are only allowed to talk in terms of material or mechanical causes. "Hence if we supplement natural science by introducing the conception of God into its context for the purpose of rendering the finality of nature explicable, and if, having done so, we turn round and use this finality for the purpose of proving that there is a God, then both natural science and theology are deprived of all intrinsic substantiality." Kant was unbending on this. "This deceptive crossing and

---

[2] Michel Foucault, *The Order of Things: An Archaeology of the Human Sciences* (New York: Pantheon, 1970), 274.

recrossing from one side to the other involves both in uncertainty, because their boundaries are thus allowed to overlap."[3]

But Kant recognized that we simply cannot do without final-cause thinking. Heuristically, in biology teleology is absolutely essential. We need the maxim, "An organized natural product is one in which every part is reciprocally both end and means." We simply cannot do biology without assuming final causes. "It is common knowledge that scientists who dissect plants and animals, seeking to investigate their structure and to see into the reasons why and the end for which they are provided with such and such parts, why the parts have such and such a position and interconnexion, and why the internal form is precisely what it is, adopt the above maxim as absolutely necessary." Scientists cannot do biology in any other way. Teleological thinking is not a luxury; it is a necessity. Life scientists

> say that nothing in such forms of life is in vain, and they put the maxim on the same footing of validity as the fundamental principle of all natural science, that nothing happens by chance. They are, in fact, quite as unable to free themselves from this teleological principle as from that of general physical science. For just as the abandonment of the latter would leave them without any experience at all, so the abandonment of the former would leave them with no clue to assist their observation of a type of natural things that have once come to be thought under the conception of physical ends.[4]

So how are we to solve the problem – what Kant called an "antinomy" – of needing to use final-cause talk and yet recognizing that only material-cause talk is acceptable in physical science or any science which claims to be talking of objective reality? Here Kantian metaphysics comes into play – phenomenally we can see no design in nature, but noumenally it is possible that there is design. God may be standing behind everything, but this is for things in themselves and not for the phenomenal world as we know it.

> It is at least possible to regard the material world as a mere phenomenon, and to think something which is not a phenomenon, namely a thing-in-itself, as its substrate. And this we may rest upon a corresponding intellectual intuition, albeit it is not the intuition we possess. In this way, a supersensible real ground, although for us unknowable, would be procured for nature, and for the nature of which we ourselves form part. Everything, therefore, which is necessary in this nature as an object of sense we should estimate according to mechanical laws.

[3] Immanuel Kant, "Von den verschiedenen Racen der Menschen" (1790), AA 2:427–44; Kant, *The Critique of Teleological Judgement*, trans. J. C. Meredith (Oxford: Oxford University Press), 31.
[4] Immanuel Kant, "Von den verschiedenen Racen der Menschen," 54.

But the accord and unity of the particular laws and of their resulting subordinate forms, which we must deem contingent in respect of mechanical laws – these things which exist in nature as an object of reason, and, indeed, nature in its entirety as a system, we should also consider in the light of teleological laws. Thus we should estimate nature on two kinds of principles. The mechanical mode of explanation would not be excluded by the teleological as if the two principles contradicted one another.[5]

We may (must) suppose God, but we cannot prove it. "All that is permissible for us men is the narrow formula: We cannot conceive or render intelligible to ourselves the finality that must be introduced as the basis even of our knowledge of the intrinsic possibility of many natural things, except by representing it, and, in general, the world, as the product of an intelligent cause – in short, of a God."[6]

For Kant then, teleological thinking is a regulative principle; it is a necessary heuristic. It is not a condition of rational thinking in the way that the mechanical philosophy is. We cannot think of the world except causally, for instance. We can certainly look at organisms without thinking of final causes. But as soon as we start to study them, to understand them, final-cause thinking comes into play – has to come into play. Final causes are part of the filter, the lens, through which we study the world. They are our doing: similar to things like causality in that we impute them to the world, but less strong than causality because we can think without them even though we cannot work without them. They are regulative. "Strictly speaking, we do not observe the *ends* in nature as designed. We only read this conception *into* the facts as a guide to judgement in its reflection upon the products of nature. Hence these ends are not given to us by the Object."[7]

2. KANT ON EVOLUTION

Now how does all of this relate to organic evolution? As I said earlier, I argue that Kant was not just an opponent of such evolution but was theoretically deeply against the idea. This follows precisely because of his commitment to adaptation and final causes. As I noted, you might be surprised at this. Particularly given his interest in the nebular hypothesis, you might think that Kant would have been at least empathetic to evolutionary speculations. However, far from Kant having seen evolution as being in some sense the

---

[5] Ibid., 65–6.
[6] Ibid., 53.
[7] Ibid.

answer to the nature of organisms, perhaps throwing light on final causes, he rather saw (organic) evolution as an impossibility. Specifically, he saw final causes – that is to say, situations where one feels obligated to invoke an end-directed kind of understanding – as being a barrier to true evolutionary thought.

> It is, I mean, quite certain that we can never get a sufficient knowledge of organized beings and their inner possibility, much less get an explanation of them, by looking merely to mechanical principles of nature. Indeed, so certain is it, that we may confidently assert that it is absurd for men even to entertain any thought of so doing or to hope that maybe another Newton may some day arise, to make intelligible to us even the genesis of but a blade of grass from natural laws that no design has ordered. Such insight we must absolutely deny to mankind.[8]

This does not mean that there cannot objectively be final cause without design (without a Designer, that is), for this would presume to know what we cannot know. What it does mean is that we cannot ourselves, at the phenomenal level, expect to get a mechanical explanation.

Kant did not think that the idea of organic evolution is silly. Indeed, like some of the evolutionists of his day – for instance, Erasmus Darwin (whose specific thinking Kant seems not to have known) – Kant thought that the isomorphisms between the parts of different organisms rather point in the direction of evolution. "This analogy of forms, which in all their differences seem to be produced in accordance with a common type, strengthens the suspicion that they have an actual kinship due to descent from a common parent." Kant even indeed went on to spell things out, speaking of our ability to "trace in the gradual approximation of one animal species to another, from that in which the principle of ends seems best authenticated, namely from man, back to the polyp, and from this back even to mosses and lichens, and finally to the least perceivable stage of nature."[9] But ultimately it appears that these connections are all ideal, connections in theory and not in actuality. There is no common descent. Evolution is untrue.

Why? In one way, let me agree that Kant did rather suggest that it is a contingent matter that evolution is false. "An hypothesis of this kind may be called a daring venture on the part of reason; and there are probably few even among the most acute scientists to whose minds it has not sometimes occurred. For it cannot be said to be absurd, like the *generatio aequivoca*,

[8] Ibid., 54.
[9] Ibid., 78–9.

which means the generation of an organized being from inorganic matter." In Kant's own language, there is nothing a priori self-contradictory about the idea of organic evolution. It is certainly logically possible that animals move from the water to the marshes and thence to the land, changing and adapting as they go along. Such a notion of evolution is not like a round square which never could exist, even in principle. Nevertheless, the facts of nature go against evolution:

> Experience offers no example of it. On the contrary, as far as experience goes, all generation known to us is *generatio homonyma*. It is not merely *univoca* in contradistinction to generation from an unorganized substance, but it brings forth a product which in its very organization is of like kind with that which produced it, and a *generatio heteronyma* is not met with anywhere within the range of our experience.[10]

But then Kant made it clear that the trouble with evolution is more than something just contingent. Evolution may not be a logical impossibility, but it goes against the final-cause thinking we use in biology. Organisms are organized in the way that they are, and moving across from one species to another would disrupt this organization in a way fatal to the intermediaries. "For in the complete inner finality of an organized being, the generation of its like is intimately associated with the condition that nothing shall be taken up into the generative force which does not also belong, in such a system of ends, to one of its undeveloped native capacities." Breaking from this inner finality would be too disruptive. Assuredly, "the principle of teleology, that nothing in an organized being which is preserved in the propagation of the species should be estimated as devoid of finality, would be made very unreliable and could only hold good for the parent stock, to which our knowledge does not go back."[11] Kant was not the clearest of writers, and these passages are no exception. But the basic idea is clear. It is evolution or final causation, and final causation wins. Contrivance is a fact of nature requiring final-cause understanding – the complexity of nature demands that we think in terms of ends appropriate to an intelligence. Evolution is a blind-law explanation, precluding final-cause understanding – no blind-law explanation can yield phenomena requiring understanding in terms of ends appropriate to an intelligence. Hence, evolution cannot in principle explain contrivance and so must be false. (Note that the phrase "ends appropriate to an intelligence" is not meant to imply that there is an intelligence. This is the whole point of Kant's intellectual wriggling.)

[10] Ibid., 79n–80n.
[11] Ibid., 80n.

3. OPUS POSTUMUM

What about the last, unfinished work the *Opus Postumum*? It is here that Catherine Wilson makes her strongest case for Kant's evolutionism, and let me say that it is a good case. For a start, we find here that Kant knew of Erasmus Darwin's evolutionary work *Zoonomia*. What is equally or more significant, we have several passages that look very close to endorsing some form of transmutationism. Just to requote passages that Wilson highlights:

> The organized creatures form on earth a whole according to purposes which [can be thought] *a priori*, as sprung from a single seed (like an incubated egg), with mutual need for one another, preserving its species and the species that are born from it. Also, revolutions of nature which brought forth new species (of which man is one).[12]

And again:

> Not only does the vegetable kingdom exist for the sake of the animal kingdom (and its increase and diversification), but men, as rational beings, exist for the sake of others of a different species (race). The latter stand at a higher level of humanity, either simultaneously (as, for instance, Americans and Europeans) or sequentially. For instance, if our globe (having once been dissolved into chaos, but now being organized and regenerating) were to bring forth, by revolutions of the earth, differently organized creatures, which, in turn, gave place to others after their destruction, organic nature could be conceived in terms of a sequence of different world-epochs, reproducing themselves in different forms, and our earth as an organically formed body – not one formed merely mechanically. How many such revolutions (including, certainly, many ancient organic beings, no longer alive on the surface of the earth) preceded the existence of man, and how many (accompanying, perhaps, a more perfect organization) are still in prospect, is hidden from our inquiring gaze – for, according to Camper, not a single example of a human being is to be found in the depth of the earth.[13]

"Nature organizes matter in manifold fashion – not just by kind, but also by stages," says a footnote. "Not to be comprehended: That there are to be discovered in the strata of the earth and in mountains, examples of the former kinds of animals and plants (now extinct) – proofs of previous (now alien) products of our living, fertile globe. That its organizing force has so

[12] Immanuel Kant, *Opus Postumum*, ed. Eckart Förster (Cambridge: Cambridge University Press, 1993), 57.

[13] Ibid., 66–7.

arranged for one another the totality of plants and animals, that they, together, as members of a chain, not merely in respect of their nominal character (similarity) but their real character (causality) which points in the direction of a world organization (to unknown ends) of the galaxy itself."[14]

Let me make three comments about these and related passages. First, they do not as such affect the different weights that Wilson and I would put on the biological thinking of the *Critique of Judgment*. The *Opus Postumum* comes in the decade after this work. Kant could have changed his mind, for better or for worse. He would not have been the first philosopher to have done this. Second, even if they represent a complete about-face for Kant, they do not affect the subsequent course of history. The *Opus Postumum* went unpublished for many decades, and only in the twentieth century did good editions start to appear (in English but ten years ago). Historically, the work and its contents are irrelevant.

Third, I am not at all convinced that they do represent an about-face. I am certainly not convinced that Kant now becomes an evolutionist. A number of people read *Zoonomia* almost as soon as it was translated into German (1795–9), but the mechanical-materialism of that work simply did not convince. No one was picking up on and endorsing the evolutionism. (Richards makes this point about Schelling.)[15] Kant's reference to *Zoonomia* seems to be about the teleological nature of organisms, not their evolution. Apparently the Englishman stimulates Kant to think of "organic bodies, every part of which is there for the sake of the other, and whose existence can *only* be thought in a system of purposes (which must have an immaterial cause)."[16] And the passages quoted above, although perhaps implying and endorsing an upward progress to life through time, does not make this an actual evolution. Others at the time – Schelling, Goethe, Hegel (soon after) – viewed life as having an upward thrust, with ever greater manifestations of basic forms (*Baupläne* or archetypes), but it is more an ideal pattern (perhaps seeds springing forth) than an actual physical evolution.

> Nature is to be regarded as a *system of stages*, one arising necessarily from the other and being the proximate truth of the stage from which it results: but it is not generated *naturally* out of the other but only in the inner Idea which constitutes the ground of Nature. *Metamorphosis* pertains

[14] Ibid., 85–6n.
[15] Robert Richards, *The Romantic Conception of Life: Science and Philosophy in the Age of Goethe* (Chicago: University of Chicago Press, 2003).
[16] Kant, *Opus Postumum*, 122.

only to the Notion as such, since only its alteration is development. But in Nature, the Notion is partly only something inward, partly existent only as a living individual: *existent* metamorphosis, therefore, is limited to this individual alone.[17]

Of course, you might say, none of this is very satisfactory. That is precisely my point. It is not very satisfactory. Kant and his fellows did not have natural selection. The issue of importance here is that one cannot consider even the aged Kant to have been an evolutionist, yet – even though his thinking was incomplete because of his ignorance of the needed science – this is not to deny the importance of Kant historically or the value of his philosophy of biology, especially as expressed in the *Critique of Judgment*.

4. GEORGES CUVIER

Why do I claim that there is much for us today to learn from Kant? Particularly important is the emphasis on final-cause thinking – its necessity whatever its status – and Kant's penetrating analysis of its nature. In fact, he anticipates (an anticipation that I have never seen acknowledged) one of the most popular contemporary analyses of such teleological thinking. The philosopher Larry Wright has argued that "when we say that Z is the function of X" – in traditional language, when we say that Z is the final cause of X – "we are not only saying that X is there because it does Z, we are also saying that Z is (or happens as) a result or consequence of X's being there." So, in terms of familiar examples: "Not only is chlorophyll in plants *because* it allows them to perform photosynthesis, photosynthesis is a *consequence* of the chlorophyll's being there. Not only is the valve-adjusting screw there *because* it allows the clearance to be easily adjusted, the possibility of easy adjustment is a *consequence* of the screw's being there."[18] Notice that all of this has to be understood in a generic sense, otherwise we run into missing-goal-object problems and the like. X does not always have to do Z in every case – it may be that X does Z on only a few occasions – but it must do it sometimes, and these must matter.

I need hardly say that this is pure and undiluted Kant, for like the great German philosopher, Wright is offering a proposal that involves a two-way causal connection. "A does B. A exists because it does B." In Kant's language: "I would say that a thing exists as physical end *if it is* (though in a double

---

[17] Georg Wilhelm Friedrich Hegel, *The Philosophy of Nature*, trans. A. V. Miller (Oxford: Oxford University Press, 1970), 21.

[18] Larry Wright, "Functions," *Philosophical Review* 82 (1973): 160.

sense) *both cause and effect of itself.*"[19] In Wright's language: "When we give a functional explanation of X by appeal to Z ('X does Z'), Z is always a consequence or result of X's being there (in the sense of 'is there' sketched above). So when we say that Z is the function of X, we are not only saying that X is there because it does Z, we are also saying that Z is (or happens as) a result or consequence of X's being there." Of course, what Wright knows and Kant did not know is that natural selection shows how something can be both cause and effect of itself – the success of the adaptation leads to its continuance, which in turn leads to the adaptation being used again. The hand and the eye lead to survival and reproduction, which lead to more hands and eyes. But it is Kant who has captured the logic of the situation – a situation that we do not find in the physical sciences. And of course, as Kant saw, the reason why we have this kind of peculiar situation in the life sciences is because organisms seem as if designed – they lead to their replication because of their distinctive natures. So even though we may judge Kant's analysis incomplete (as I would), we can nevertheless use his thinking for our own understanding of the logic of biological explanation (as I would).

But as I said above, more immediately pressing is the influence that Kant may have had on others – others who played a key role in the history of the idea of evolution. The opposition that Kant has to evolution – you cannot square it with teleology – is precisely the opposition that is expressed by Georges Cuvier. For the German-trained Cuvier, the key to understanding the organism lies in the fact that it is not simply subject to the physical laws of nature but is organized, with the parts directed to the end of the functioning whole – each individual feature playing its role in the overall end-directed scheme of things. This Cuvier referred to as the "conditions of existence," writing as follows:

> Natural history nevertheless has a rational principle that is exclusive to it and which it employs with great advantage on many occasions; it is the *conditions of existence* or, popularly, *final causes.* As nothing may exist which does not include the conditions which made its existence possible, the different parts of each creature must be coordinated in such a way as to make possible the whole organism, not only in itself but in its relationship to those which surround it, and the analysis of these conditions often leads to general laws as well founded as those of calculation or experiment.[20]

[19] Kant, "Von den verschiedenen Racen der Menschen," 18.

[20] G. Cuvier, *Le règne animal distribué d'après son organisation, pour servir de base à l'histoire naturelle des animaux et d'introduction à l'anatomie comparée* (Paris, 1817), 1:6; quoted in W. Coleman, *Georges Cuvier, Zoologist: A Study in the History of Evolution Theory* (Cambridge, MA: Harvard University Press, 1964), 42.

But how now is one to translate the teleology into a practical working science? Here we move to Cuvier's field of scientific interest and expertise, (animal) morphology or anatomy. He made detailed studies of animal after animal, thinking that the conditions of existence yielded a working guide for the investigator. This corollary, as we might call it, was referred to by Cuvier as the "correlation of parts." He argued that in order to be an integrated functioning being, every part of an organism had to be slotted in harmoniously with every other part. "It is in this mutual dependence of the functions and the aid which they reciprocally lend one another that are founded the laws which determine the relations of their organs and which possess a necessity equal to that of metaphysical or mathematical laws." And then, linking this back to the conditions of existence (of which the correlation of parts is really just a physical manifestation), "it is evident that the seemly harmony between organs which interact is a necessary condition of existence of the creature to which they belong and that if one of these functions were modified in a manner incompatible with the modifications of the others the creature could no longer continue to exist."[21]

The point is that the correlation of parts supposedly gave the anatomist a tool to make predictions. Pretend, for instance, you have only the tooth of an animal. From its design you can tell that it is for meat eating rather than for chewing vegetable matter. Then, you work outwards. If the owner is a carnivore, there is little point in its having hooves like a deer or a stomach like a cow or the armor of a tortoise or any of the other attributes that one associates with vegetarian prey. Rather, one needs claws and agility and intelligence and so forth. Thus you can infer what the whole animal must look like. And, one should say, Cuvier felt that this line of argumentation showed its worth again and again, for on several occasions having been given but a fossil fragment of an unknown beast, he was able to infer the whole form – a prediction triumphantly confirmed when later the whole animal was discovered.

The conditions of existence yield a second corollary, the "subordination of characters," and it was through this that Cuvier thought he was able to bring order to the animal world, dividing it into four basic groups or *embranchements*.

For a good classification . . . we employ an assiduous comparison of creatures directed by the principle of the subordination of characters, which itself

---

[21] Cuvier, quoted in Coleman, *Georges Cuvier, Zoologist*, 67–8.

derives from the conditions of existence. The parts of an animal possessing a mutual fitness, there are some traits of them which exclude others and there are some which require others; when we know such and such traits of an animal we may calculate those which are coexistent with them and those which are incompatible; the parts, properties, or consistent traits which have the greatest number of these incompatible or coexistent relations with other animals, in other words, which exercise the most marked influence on the creature, we call *caractères importants, caractères dominateurs*; the others are the *caractères subordonés*, and there are thus different degrees of them.[22]

Suppose you have in place a backbone. Then you know that many features to be found in the animal world no longer are possible for this particular animal. It must now have characters and only those characters which are part and parcel of being a vertebrate. Suppose now you have in place the backbone of a whale. It is no longer possible to have the limbs of a land predator like a tiger, nor the teeth of such an animal, nor its stomach or its brain, or many other things. Once you have gone the route of sea mammal, then you are constrained in many ways except for one outlet only. This gives rise to the proper way to classify – being ever more restricted in the choices open to the production of a functioning organism. And so we get the four basic groups (*embranchements*) of animals: the vertebrates, the molluscs, the articulates, and the radiates. "I have found that there exist four principal forms, four general plans, upon which all of the animals seem to have been modelled and whose lesser division, no matter what names naturalists have dignified them with, are only modifications superficially founded on development or on the addition of certain parts, but which in no way change the essence of the plan."[23]

5. AGAINST EVOLUTION

This then is Cuvier's brilliant teleological program for the animal world. It is very much in the spirit of Kant. But there is more than this, for Cuvier was right in line with Kant's denial that final causation and evolution can be held consistently by one and the same person. Like the philosopher, at one level Cuvier was happy to appeal to the empirical evidence. Napoleon, taking his savants with him, had gone campaigning in Egypt, and mummified humans and animals had been returned to France. But although these were

[22] Ibid., 77.
[23] Ibid., 92.

very old, there was no trace of evolutionary change. The Egyptians "have not only left us representations of animals, but even their identical bodies embalmed and preserved in the catacombs."[24] If not evolutionary, then what is the nature of organic origins? Cuvier admitted he did not have much positive idea. Beyond allowing that there seems to be a roughly progressive history to life (as revealed in the fossil record, which he himself was doing much to open up), Cuvier had little to say. "I do not pretend that a new creation was required for calling our present races of animals into existence. I only urge that they did not occupy the same places, and that they must have come from some other part of the globe."[25]

Ultimately, however, following in the steps of Kant, Cuvier believed it was final causation which spelt the doom of evolution. With the German philosopher, Cuvier just could not see how you could get change across a species barrier without disruption of the organism too great to be borne. The intricacies of adaptation exist specifically to aid the organism in the life and role that it occupies. Hybrids are neither fish nor fowl – adapted neither for water nor for air – and as such cannot possibly survive and reproduce. To use a mathematical analogy, the internal angles of an n-sided polygon number $2n - 4$ right angles. To take the example of the move from triangle to quadrilateral, you go from 2 right angles to 4 right angles. You simply cannot have a plane-sided figure whose internal angles add up to 3 right angles. That is simply an impossibility. And so also is an organism between two species, and especially an organism between two *embranchements*.

Speaking parenthetically, the mathematical analogy ought to have appealed to Kant, although actually the impossibility of a polygon with internal angles summing to $2n - 2$ right angles would seem to be synthetic a priori, as are the other truths of mathematics. Kant, as we have seen, does not want to give this kind of necessity to final-cause thinking. This surely shows that Kant, and probably Cuvier also, was caught trying to define a notion of necessity for teleological thought that does not sit altogether comfortably with our thinking on necessity in other instances. Speaking now both as a historian and as a Darwinian, this does not surprise me. Sophisticated though it may be, the Kant-Cuvier position cannot ultimately be correct. Conversely, however, this does not mean that it is philosophically worthless or scientifically without influence.

---

[24] Georges Cuvier, *Theory of the Earth*, 4th ed., trans. and ed. Robert Jameson (Edinburgh: William Blackwood, 1813 [1822]), 123.

[25] Ibid., 125–6.

6. CONCLUSION

Cuvier was wrong about evolution. It does happen. But, following Kant, he was right to stress the design-like nature of the world – even to the point where it led him to deny the possibility of organic change. Following Kant, he was right to stress that design is not some contingent thing that nature might show irrelevant but rather something deeply part of our understanding of the organic world and a real barrier to theories of change. Foucault was right to seize on this point and to argue that it is this that sets the agenda for the evolutionists – first Charles Darwin[26] and then his successors down to the present day. It was Darwin's genius to see that Cuvier (together with others, such as the British natural theologians, including Archdeacon William Paley[27]) had set the questions, posed the problems that must be answered, as it was also Darwin's genius to find the solution, natural selection. Final causation is a barrier to theories of change, and Darwin broke down the barrier. And this takes us back to Immanuel Kant. Like Moses, he was never to enter the promised land – Israel for the one, evolution for the other – but he did lead us to the borders.

---

[26] Charles Darwin, *On the Origin of Species* (London: John Murray, 1859).
[27] William Paley, *Natural Theology*, vol. 4 of the *Collected Works*, (London: Rivington, 1802 [1819]).

Bibliography

PRIMARY SOURCES

Alberti, Michael. 1708. *De paedantismo medico.* (No further references available.)

Alberti, Michael. 1718. *Introductio in universam medicinam....* Halae: Litteris et impensis Orphanotrophei.

Alberti, Michael. 1721. *Medicinische und philosophische Schriften....* Halle: J. C. Hendel.

Alberti, Michael. 1726. *Specimen medicinae theologicae....* Halae: Hendel.

Aldrovandus, Ulysses. 1642. *Monstrorum historia.* Bononiae: Marcus Antonius Bernia.

Aldrovandus, Ulysses. 1634. *Ornithologiae.* Vol. II. Bononiae. Nicolaum Tebaldinum.

Aquinas, Thomas. 1967. *Summa Theologiae.* Vol. 10, *Cosmogony.* Trans. William A. Wallace, O.P. New York: Blackfriars and McGraw-Hill.

Arantius, C. 1579. *De humano foetu libellus.* Basileae: Laurentius Scholzius.

Aristotle. 1963. *On the Generation of Animals.* Trans. A. L. Peck. Cambridge, MA: Harvard University Press. (Cited as GA.)

Aristotle. *Historia Animalium.* Vol. I (1965), trans. A. L Peck; vol. II (1970), trans. A. L. Peck; Vol. III (1991), trans. D. M. Balme and prepared by Allan Gotthelf. Cambridge, MA: Harvard University Press. (Cited as HA.)

Aristotle. 1992. *De partibus animalium I and De generatione animalium I (with passages from II. 1–3).* Ed. D. M. Balme. Oxford: Clarendon Press.

Avicenna. 1959. *Avicenna's De Anima, being the psychological part of Kitab al-Shifa.* London: Oxford University Press.

Baer, Karl Ernst von. 1827. *De ovi mammalium et hominis genesi.* Translated as *Within the Ovum* by C. D. O'Malley. *Isis* 48 (1956): 148.

Basson, Sébastien. 1621. *Philosophiae naturalis adversus Aristotelem libri XII.* Geneva: Pierre de la Rouiere.

Bauhin, Casper. 1614. *De hermaphroditorum monstrosorumque partuum natura et theologia, jurisconsultorum, medicorum, philosophorum et rabbinorum sententia libri II.* Basel.

Bayle, Pierre. 1982. *Oeuvres diverses.* Ed. Elisabeth Labrousse. Hildesheim: Olms.

Benito P. 1591. *De communibus omnium rerum naturalium principiis.* Venice.

Bibliography

Blumenbach, Johann Friedrich. 1792. *Üeber den Bildungtrieb und das Zeugungsgechäfte*. Göttingen. Translated as *An Essay on Generation* by A. Crichton. London.

Boaistuau, Pierre. 1561. *Histoires prodigieuses*. Paris.

Bonnet, Charles. 1779–1783. *Oeuvres d'histoire naturelle et de philosophie*. 18 vols. Neufchâtel: Samuel Fauche.

Bonnet, Charles. 1985. *Considérations sur les corps organisés*. Paris: Fayard.

Bourguet, Louis. 1729. *Lettres philosophiques sur la formation des sels et des crystaux et sur la génération et le méchanisme organique des plantes et des animaux*. Amsterdam: François L'Honoré.

Boyle, Robert. 1991. *Origins of Forms and Qualities*. In *Selected Philosophical Papers of Robert Boyle*. Ed. M. A. Stewart. Indianapolis: Hackett Publishing Company.

Boyle, Robert. 1772. *Works*. 6 vols. Ed. T. Birch. London.

Buffon, Georges-Louis Leclerc de. 1749. *Histoire naturelle*. Paris.

Buffon, Georges-Louis Leclerc de. 1749 [1780]. *Natural History*. Trans. William Smellie. Edinburgh.

Buffon, Georges-Louis Leclerc de. 1791. *Natural History, General and Particular*. 3rd English edition. 9 vols. Trans. William Smellie. London: Strahan and Cadell.

Burgmann, Peter Christoph. 1731. *Succinctum hypotheseos stahlianae examen de anima rationali corpus humanum struente motusque vitales tam in statu sano quam morboso administrante*. Lipsiae: Apud I. F. Gleditschii B. filium.

Charleton, Walter. 1652. *The Darknes of Atheism Expelled by the Light of Reason*. London: William Lee.

Charleton, Walter. 1654. *Physiologia Epicuro-Gassendo-Charletoniana: A Fabrick of Science Natural Upon the Hypothesis of Atoms*. London: Thomas Heath.

Charleton, Walter. 1659. *Natural History of Nutrition, Life, and Voluntary Motion*. London: Henry Herringman.

Charleton, Walter. 1659. *Oeconomia animalis, novis in medicina hypothesibus superstructa & mechanicè explicata*. London: Daniel & Redmann.

Charleton, Walter. 1659. *The Immortality of the Human Soul, Demonstrated by the Light of Nature*. London: Henry Herringman.

Charleton, Walter. 1666. *Dissertatio epistolica, de ortu animae humanae*. In Walter Charleton, *Oeconomia animalis, novis in medicina hypothesibus superstructa & mechanicè explicata*. 3rd ed. London: Roger Daniel.

Charleton, Walter. 1674. *Natural History of the Passions*. London: James Magnes.

Coimbra [Collegium Conimbricensis]. 1594 [1984]. *Commentarii ... in octo libros Physicorum Aristotelis Stagiritae*. Lugduni: Ioannes Baptista Buysson. Repr., Hildesheim: Olms.

Coiter, Volcher. 1572. *De ovis et pullis gallinaceis*. In *Externarum et internarum principialium humani corporis partium tabulae*. Noribergae: Theodorici Gerlatzeni.

Coiter, Volcher. 1572. *Externarum et internarum principialium humani corporis partium tabulae*. Noribergae: Theodorici Gerlatzeni.

Coiter, Volcher. 1933. De ovorum gallinaceorum generationis. Ed. Howard B. Adelmann. *Annals of Medical History* 5: 444–57.

Columbus, Realdus. 1559. *De re anatomica*. Venice: Bevilacqua.

Condillac, Etienne de. 1754. *Traité des sensations*. Paris.

Conway, Anne. 1982. *The Principles of the Most Ancient and Modern Philosophy*. Edited and with an introduction by Peter Loptson. The Hague: Martinus Nijhoff.

Conway, Anne. 1996. *The Principles of the Most Ancient and Modern Philosophy*. Ed. Allison P. Coudert and Taylor Corse. Cambridge: Cambridge University Press.

Coschwitz, Georg Daniel. 1725. *Organismus et mechanismus in homine vivo*. Lipsiae: F. Lankisius.

Cudworth, Ralph. 1678 [1845]. *The True Intellectual System of the Universe*. 3 vols. London: Tegg.

Cureau de la Chambre, Marin. 1647 [1989]. *Traité de la connoissance des animaux, où tout ce qui a esté dit Pour, & Contre le raisonnement des bestes est examiné*. Paris: Fayard.

Cuvier, Georges. 1813 [1822]. *Theory of the Earth*. 4th ed. Trans. and ed. Robert Jameson. Edinburgh: William Blackwood.

Cuvier, Georges. 1817. *Le règne animal distribué d'aprés son organisation, pour servir de base à l'histoire naturelle des animaux et d'introduction à l'anatomie comparée*. Paris.

D'Alembert, Jean. 1754. Cosmologie. In *Encyclopédie, ou Dictionnaire raisonée des sciences, des arts et des métiers*. Vol. 4. Ed. Denis Diderot and Jean D'Alembert. Paris.

D'Alembert, Jean. 1995. *Preliminary Discourse to the Encyclopedia*. Trans. Richard Schwab. Chicago and London: University of Chicago Press.

Daniel, Gabriel. 1702. *Voyage du monde de Descartes*. 2nd ed. Paris: V. Pepie.

Darwin, Charles. 1859. *On the Origin of Species*. London: John Murray.

Descartes, René. 1964–76. *Oeuvres de Descartes*. 11 vols. Ed. C. Adam and P. Tannery. Paris: J. Vrin. (Cited as AT, followed by volume and page number.)

Descartes, René. 1985. *The Philosophical Writings of Descartes*. Vol. 3. *The Correspondence*. Trans. John Cottingham, Robert Stoothoff, Dugald Murdoch, and Anthony Kenny. Cambridge: Cambridge University Press. (Cited as CSMK, followed by volume and page number.)

Descartes, René. 1996. *Le monde, l'homme*. Ed. Annie Bitbol-Hespériès and Jean-Pierre Verdet. Paris: Seuil.

Descartes, René. 2000. *Ecrits physiologiques et médicaux*. Ed. Vincent Aucante. Paris: Presses Universitaires de France.

Diderot, Denis. 1875–7. *Oeuvres completes*. Ed. J. Assezat. 20 vols. Paris.

Diderot, Denis. 1964. *Le Rêve de D'Alembert*. In *Oeuvres philosophiques de Diderot*. Ed. Paul Vernière. Paris: Editions Garnier.

Diderot, Denis. 1981. *Pensées sur l'interprétation de la nature*. In *Oeuvres complètes*. Vol. 9. Ed. Jean Verloot. Paris: Hermann.

Diderot, Denis. 1976. Spinosiste. In *Encyclopédie*. Repr. in Diderot, *Oeuvres philosophiques*, vol. 8, ed. John Lough and Jacques Proust, 328–9.

Digby, Sir Kenelm. 1644. *Two Treatises*. Paris: Blaizot.

Euler, Leonhard. 1957. *Commentationes mechanicae*. In *Opera omnia*, II 5. Lausanne: Auctoritate et impensis Societatis scientiarum naturalium helveticae.

Euler, Leonhard. 1983. *Leonhard Euler 1707–1783: Beiträge zu Leben und Werk*. Gedenkband des Kantons Basel-Stadt. Basel: Birkhäuser.

Fabricius d'Acquapendente, Hieronymus. 1604. *De formato fœtu*. Venice.

Fabricius d'Acquapendente, Hieronymus. 1621. *De formatione ovi et pulli*. Patavii: Aloysii Bencii Bibliopolae.

Fernel, J. 1655. *Physiologie*. Paris: Jean Guignard le Jeune.

Bibliography

Feyens, Thomas. 1620. *De formatrice fœtus liber*. Antwerp: Willem van Tongheren.

Feyens, Thomas. 1635. *De viribus imaginationis tractatus*. Lugduni Batavorum: Elzevirus.

Galen. 1625. *Opera*. Venetiis: Juntas.

Galen. 1854–6. *Oeuvres anatomiques, physiologiques, et médicales de Galien*. Trans. Charles Daremberg. 2 vols. Paris: Baillière.

Galen. 1963. *Galen on the Natural Faculties*. Trans. Arthur John Brock. Cambridge, MA: Harvard University Press.

Galen. 1985. *Three Treatises on the Nature of Science*. Ed. Michael Frede. Indianapolis: Hackett.

Galen. 1992. *On Semen*. Ed. and trans. Phillip de Lacy. Berlin: Akademie Verlag.

Gassendi, Pierre. 1649 [1993]. *Syntagma Philosophiae Epicuri*. Trans. Paul Westholm Johnson. In *A Syntagma of the Philosophy of Epicurus of Pierre Gassendi*. Ph.D. diss., University of Cincinnati.

Gassendi, Pierre. 1651. *Viri illustris Nicolai Claudii Fabricii de Peiresc vita*. Hagae Comitis: Sumptibus A. Vlacq.

Gassendi, Pierre. 1658 [1964]. *Opera omnia*. 6 vols. Lyon. Repr., Stuttgart-Bad Canstatt: Friedrich Frommann. (Cited as O).

Gassendi, Pierre. 1675. *Animadversiones in Decimum Librum Diogenis Laertii, qvi est de vita, moribus, placitisque Epicuri*. 3rd ed. Lyon: Barbier.

Gassendi, Pierre. 1966. *De generatione animalium*. Trans. Howard B. Adelmann. In Adelmann, *Marcello Malpighi and the Evolution of Embryology*. Ithaca: Cornell University Press.

Gassendi, Pierre. 1972. *Selected Works*. Trans. and Ed. C. B. Brush. New York: Johnson Reprints.

Gassendi, Pierre. 2003. *Pierre Gassendi (1592–1655): Lettres latines*. Trans. Sylvie Taussig. Turnhout: Brepols.

Gesner, Conrad. 1552. *Conradi Gesneri historia animalium*. Tigure: Apud Christ. Froschoverum.

Glanvill, Joseph. 1665 [1978]. *Scepsis Scientifica: or, Confest Ignorance, the Way to Science: in an Essay on the Vanity of Dogmatizing and Confident Opinion*. London. Repr., New York: Garland.

Glisson, Francis. 1641 [1993]. *English Manuscripts of Francis Glisson*. Ed. Andrew Cunningham. Cambridge: Cambridge Wellcome Unit.

Goetz, Johan Christoph. 1729. *Scripta D. Georgii Ernesti Stahlii*. 2nd ed. Norimbergae: Impensis B. W. M. Endteri filiae Mayerin.

Haller, Albrecht von. 1739–44. Editor. *Praelectiones academicae in proprias institutiones rei medicae*. Authored by Herman Boerhaave. Notes added by Albrecht von Haller. 6 vols. Göttingen: A. Vandenhoeck.

Haller, Albrecht von. 1747. *Primae lineae physiologiae in usum praelectionum academicarum*. Göttingen: A. Vandenhoeck.

Haller, Albrecht von. 1750 [1981]. Preface to *Allgemeine Historie der Natur*, vol. 1. Hamburg and Leipzig: G. C. Grund and A. H. Holle. Reprinted in Shirley A. Roe, *The Natural History of Albrecht von Haller*. New York: Arno Press.

Haller, Albrecht von. 1751 [1981]. *Réflexions sur le système de la génération de M. de Buffon*. Geneva: Barrillot Reprinted in Shirley A. Roe, *The Natural History of Albrecht von Haller*. New York: Arno Press.

Bibliography

Haller, Albrecht von. 1752 [1936]. *A Dissertation on the Sensible and Irritable Parts of Animals*. Introduced by Owsei Temkin. *Bulletin of the History of Medicine* 4: 651–99.

Haller, Albrecht von. 1752 [1981]. Preface to *Allgemeine Historie der Natur*, vol. 2. Louis. *Bibliothèque Raisonnée* 45: 68–88. Reprinted in Shirley A. Roe, *The Natural History of Albrecht von Haller*. New York: Arno Press.

Haller, Albrecht von. 1757–66. *Elementa physiologiae corporis humani*. 8 vols. Lausanne: M. M. Bousquet, S. d'Arnay, F. Grasset. (Cited as EP, followed by volume and page number.)

Haller, Albrecht von. 1757–65. *Elementa physiologiae corporis humani*. 8 vols. Lausannae: Sumptibus M. M. Bousquet & Sociorum; Bern: Sumptibus Societatis Typographiae.

Haller, Albrecht von. 1758. *Sur la formation du coeur dans le poulet; sur l'oeil, sur la structure du jaune*. 2 vols. Lausanne: M. M. Bousquet.

Haller, Albrecht von. 1760. "Review of *Theoria generationis*, by Caspar Friedrich Wolff." *Göttingische Anzeigen von gelehrten Sachen*, 1226–31.

Haller, Albrecht von. 1777. "Oeconomie animale." In *Supplément à l'Encyclopédie*, vol. 4. Amsterdam: M. M. Rey.

Haller, Albrecht von. 2000. *Commentarius de formatione cordis in ovo incubato*. Ed. M. T. Monti. Basel: Schwabe & Co.

Hartmann, Georg Volcmar. 1733. *Epistola de bruto ex homine, ad virum illustrem, excellentissimum, consultissimum, amplissimum et experietissimum dominum D. Georgium Ernestum Stahlium.* . . . Erfordiae: Sumptibus C. F. Jungnicolii.

Hartmann, Georg Volcmar. 1735. *Schediasma apologeticum quo sententia illustris Stahlii de natura humana in epistola de bruto ex nomine defensa . . . ita confirmatur.* . . . Erfordiae: Sumptibus C. F. Jugnicolii.

Harvey, William. 1628. *Exercitatio anatomica de motu cordis et sanguinis in animalibus*. Francofurti: Sumptibus Guilelmi, Fitzer.

Harvey, William. 1651 [1981]. *Disputations Touching the Generation of Animals*. Translated and introduced by Gweneth Whitteridge. Oxford: Blackwell Scientific Publications.

Harvey, William. 1651. *Exercitationes de generatione animalium. Quibus accedunt quaedam de partu: de membranis ac humoribus uteri: & de conceptione*. London: Octavianus Pulleyn.

Harvey, William. 1847. *On the Generation of Animals*. In *The Works of William Harvey*. London: Sydenham Society.

Hegel, Georg Wilhelm Friedrich. 1817 [1970]. *Philosophy of Nature*. Oxford: Oxford University Press.

Heister, Lorenz. 1738 [1743]. *De medicina mechanica praestantia prae stahliana*. In *Compendium medicinae practicae*. Amstelaedami: Apud Janssonio-Waesbergios.

Heliodorus. 1605. *An Æthiopian historie: fyrst written in Greeke by Heliodorus, and translated into English, by T.V. No lesse witty then pleasant: being newly corrected and augmented, with diuers new additions by the same author*. London: William Cotton.

Herder, Johann Gottfried von. 1910. *Gesammelte Schriften*. Ed. Akademie der Wissenschaften. Berlin: Reimer. Later Berlin: de Gruyter.

Herder, Johann Gottfried von. 1989. *Ideen zur Philosophie der Geschichte der Menschheit* Ed. Martin Bollacher. Frankfurt: Verlag Deutscher Klassiker.

Bibliography

Hertwig, Oscar. 1894 [1977]. *The Biological Problem of To-day: Preformation or Epigenesis? The Basis of a Theory of Organic Development.* Translated by P. Chalmers Mitchell and introduced by Joseph Anthony Mazzeo. Oceanside, NJ: Dabor Science Publications.

Highmore, Nathaniel. 1651. *History of Generation.* London: John Martin.

Hoffmann, Friedrich. 1739 [1749]. *Commentarius de differentia inter ejus doctrinam medico-mechanicum, et Georgii Ernesti Stahlii medico-organicam.* In Friedrich Hoffmann, *Operum omnium physico-medicorum supplementum,* vol. 1. Genevae: Apud Fratres de Tournes. Italian translation by F. P. de Ceglia in *Stahl e Hoffmann* (forthcoming).

Holbach, Paul-Henri Dietrich. d'. 1868. *System of Nature, or Laws of the Physical and Moral World.* Trans. H. D. Robinson. New York: Burt Franklin.

Hooke, Robert. 1665. *Micrographia; or, some Physiological Descriptions of Minute Bodies made by Magnifying Glasses.* London: Martyn & Allestry.

Houel, Jean. 1782. *Voyage pittoresque des isles de la Sicile, de Malte, et de Lipari.* Paris.

Jazzar, Ibn al-. 1997. *Ibn al-Jazzar on Sexual Diseases and Their Treatment: A Critical Edition of Zad al-musafir wa-qut al hadir.* Ed. and Trans. Gerrit Bos. London: Kegan Paul International.

Kant, Immanuel. 1790 [1928]. *The Critique of Teleological Judgement.* Trans. J. C. Meredith. Oxford: Oxford University Press.

Kant, Immanuel. 1790 [1928]. *Von den verschiedenen Racen der Menschen* [AA 2:427–44]. Translated as *The Critique of Teleological Judgement* by J. C. Meredith. Oxford: Oxford University Press.

Kant, Immanuel. 1965. *Kritik der reinen Vernunft* [A version (1781), AA 4:1–252; B version (1787), AA 3:1–552]. Translated as *Critique of Pure Reason* by Norman Kemp Smith. New York: St. Martin's.

Kant, Immanuel. 1968. *Gesammelte Schriften, Herausgegeben von der Königlich Preußischen Akademie der Wissenschaften.* Berlin: de Gruyter. (= Akademie Ausgabe; cited as AA, followed by volume and page number.)

Kant, Immanuel. 1987. *Critique of Judgement.* Trans. Werner S. Pluhar. Indianapolis: Hackett.

Kant, Immanuel. 1987. *Kritik der Urteilskraft* [AA 5:165–486]. Translated as *Critique of Judgment* by Werner Pluhar. Indianapolis: Hackett.

Kant, Immanuel. 1992. *Der einzig mögliche Beweisgrund zu einer Demonstration des Daseins Gottes.* Translated as *The Only Possible Argument in Support of a Demonstration of the Existence of God* in Kant, *Theoretical Philosophy 1755–1770,* trans. and ed. David Walford and Ralf Meerbote. Cambridge: Cambridge University Press.

Kant, Immanuel 1992. *Träume eines Geistersehers* [AA 2:315–84]. Translated as *Dreams of a Spirit-Seer* in Kant, *Theoretical Philosophy 1755–1770,* trans. and ed. David Walford and Ralf Meevebote. Cambridge: Cambridge University Press.

Kant, Immanuel. 1993. *Opus Postumum.* Ed. Eckart Förster. Cambridge: Cambridge University Press.

Kant, Immanuel. 1996. *Critique of Pure Reason* Trans. Werner S. Pluhar. Indianapolis: Hackett.

La Forge, Louis de. 1664. *Remarques sur le Traité de l'Homme.* Paris: Michel Bobin, Nicolas Le Gras.

Bibliography

Laurens, André du. 1600. *Historia anatomica.* Frankfurt. *L'histoire anatomique.* French translation by F. Sizé. Paris, J. Bertault, 1610.

Le Grand, Antoine. 1694. *An Entire Body of Philosophy according to the Principles of Renate Des Cartes.* London: Roycroft.

Leibniz, Gottfried Wilhelm. 1765. *Opera omnia.* 6 vols. Ed. Ludovici Dutens. Geneva: Apud Fratres De Tournes.

Leibniz, Gottfried Wilhelm. 1880. *Die philosophischen Schriften von G. W. Leibniz.* Ed. C. I. Gerhardt. Berlin: Weidmann.

Leibniz, Gottfried Wilhelm. 1923–96. *Sämtliche Schriften und Briefe.* Berlin: Akademie-Verlag. (Citations give series, volume, and page, e.g., A VI ii 229.)

Leibniz, Gottfried Wilhelm. 1954. *Principes de la nature et de la grâce fondés en raison. Principes de la philosophie ou Monadologie.* Ed. André Robinet. Paris: Presses Universitaires de France.

Leibniz, Gottfried Wilhelm. 1969 [1976]. *Gottfried Wilhelm Leibniz: Philosophical Papers and Letters.* Ed. and trans. Leroy E. Loemker. Dordrecht: D. Reidel.

Leibniz, Gottfried Wilhelm. 1977. *Mathematische Schriften.* 7 vols. Ed. C. I. Gerhardt. Hildesheim: G. Olms. (Cited as GM.)

Leibniz, Gottfried Wilhelm. 1988. *Opuscules et fragments inédits de Leibniz.* Ed. Louis Couturat. Hildesheim: Olms. (Cited as C.)

Leibniz, Gottfried Wilhelm. 1992. *G. W. Leibniz. De Summa Rerum: Metaphysical Papers 1675–1676.* Ed. and trans. G. H. R. Parkinson. New Haven: Yale University Press.

Leibniz, Gottfried. 1996. *New Essays on Human Understanding,* Trans. and ed. Peter Remnant and Jonathan Bennett. Cambridge: Cambridge University Press.

Leibniz, Gottfried Wilhelm. 1998. *G. W. Leibniz: Philosophical Texts.* Ed. and trans. R. S. Woolhouse and Richard Francks. Oxford: Oxford University Press.

Leibniz, Gottfried Wilhelm, and Samuel Clarke. 2000. *Correspondence.* Ed. Roger Ariew. Indianapolis: Hackett.

Liceti, Fortunio. 1616. *De monstrorum causis, natura et differentiis.* Padua.

Lindeboom, G. A., ed. 1975. *The Letters of Jan Swammerdam to Melchisedek Thévenot.* Amsterdam: Swets Zeitlinger.

Linnaeus, Carl. 1735. *Systema Naturae.* Leyden.

Longolius, Johann Daniel. 1732. *Systema stahlianum de vita et morte corporis humani ab incongruis medicorum mechanizantium opprobriis vindicatum....* Budissae: Literis Richterianis.

Lucretius, Titus Carus. 1992. *De rerum natura.* Trans. W. H. D. Rouse and Martin Fergueson Smith. Cambridge, MA: Harvard University Press.

Lycosthenes, Conrad. 1557. *Prodigiorum ac ostentorum chronicon.* Basileae: Per Henricum Petri.

Maignan, Emmanuel. 1673. *Philosophia naturae.* In Maignan, *Cursus philosophicus.* 2nd ed. Lyon: Gregoire.

Maillet, Bernard de. 1968. *Telliamed.* Trans. Albert V. Carozzi. Urbana: University of Illinois Press.

Malebranche, Nicolas. 1958–66. *Oeuvres complètes de Malebranche.* 21 vols. Series director Henri Gouhier. Various editors. Paris: J. Vrin. (Cited as OCM, followed by volume and page number.)

Bibliography

Malebranche, Nicolas. 1997. *Dialogues on Metaphysics and on Religion*. Ed. Nicholas Jolley, trans. David Scott. Cambridge: Cambridge University Press. (Cited as DMR, followed by dialogue section and page number.)

Malpighi, Marcello. 1672 [1966]. *On the Formation of the Chick*. Trans. Howard B. Adelmann. In Adelmann, *Marcello Malpighi and the Evolution of Embryology*, vol. 2. Ithaca: Cornell University Press.

Malpighi, Marcello. 1683. *Dissertatio epistolica de formatione pulli in ovo*. Londinii: J. Martin. Published in Malpighi, Marcello, *Opere scelte*, ed. L. Belloni. Torino: Utet, 1967.

Maupertuis, Pierre-Louis Moreau de. 1751. *Essai de cosmologie*. Leyden.

Maupertuis, Pierre-Louis Moreau de. 1980. *Vénus physique suivi de la Lettre sur le progrès des sciences*. Ed. Patrick Tort. Paris: Aubier Montaigne.

Maupertuis, Pierre-Louis Moreau de. 1984. *Essai de cosmologie*. In Maupertuis, *Essai de cosmologie: Système de la Nature: Réponse aux Objections de M. Diderot*, ed. François Azouvi. Paris: Vrin.

Mauriceau, François. 1668. *Des maladies des femmes grosses et accouchées*. Paris.

Monro, Alexander. 1785. *The Structure and Physiology of Fishes Explained and Compared with Those of Man and Other Animals*. Edinburgh.

Montaigne, Michel de. 1992. *Les Essais*. Paris: Arléa.

More, Henry. 1662 [1978]. *A Collection of Several Philosophical Writings*. London: James Flesher. Repr. New York: Garland Publishing.

Münster, Sebastian. 1552. *Cosmographia universalis*.

Nagy Borosnyai, Martin. 1729. *De potentia et impotentia animae humanae in corpus organicum....* Halae: Typis I. C. Hilligeri.

Nenter, Georg Philipp. 1714. *Theoria hominis sani sive physiologia medica*. Argentorati: Typis et sumt. J. Beckii.

Newton, Isaac. 1704. *Opticks*. London: Sam. Smith and Benj. Walford.

Newton, Isaac. 1999. *Mathematical Principles of Natural Philosophy*. Trans. and ed. I. Bernard Cohen and Anne Whitman. Berkeley: University of California Press.

Oldenburg, Henry. 1967. *The Correspondence of Henry Oldenburg*. Ed. A. R. Hall and M. B. Hall. London.

Paley, William. 1802 [1819]. *Natural Theology*. Vol. 4 of the *Collected Works*. London: Rivington.

Paracelsus (Theophrastus Bombastus von Hohenheim). 1537 [1975]. *De natura rerum*. Bk. 9. In *Sämtliche Werke*. 4 vols. Leipzig: Zentralantiquariat der Deutschen Demokratischen Republik.

Paracelsus. 1951. *Selected Writings*. Ed. Jolande Jacobi. New York: Bollingen Foundation.

Pardies, Ignace-Gaston. 1672 [1972]. *Discours de la connoissance des bestes*. Paris: Sebastien Mabre-Cramoisy. Repr. with introduction by Lenora Cohen Rosenfield, New York: Johnson Reprint Corporation.

Paré, Ambroise. 1840/1 [1970]. *Oeuvres complètes*. Paris. Repr., Geneva: Slatkine.

Paré, Ambroise. 1982. *On Monsters and Marvels*. Trans. Janis L. Pallister Chicago: University of Chicago Press.

Pascal, Blaise. 1966. *Pensées*. Ed. A. J. Kraisheimer. Harmondsworth: Penguin.

Plempius, V. F. 1638. *Fundamentes medicinae*. Louvain: Jacobi Zegresii.

Pliny the Elder (Caius Plinius). 1956–63. *Historia naturalis*. Cambridge, MA: Harvard University Press.

Régis, Pierre-Sylvain. 1654. *Philosophia naturalis*. 2nd ed. Amsterdam: Apud Ludovicum Elzevirium.

Régis, Pierre-Sylvain. 1691 [1970]. *Cours entier de philosophie ou Système général selon les principes de Descartes*. Lyon. Repr., New York: Johnson Reprint Corporation.

Reimarus, H. S. 1766. *The Principle Truths of Natural Religion*. Trans. R. Wynne. London.

Richter, Christian Friedrich. 1710. *Die Höchst-Nöthige Erkenntnis des Menschen: Zum Drittenmal vermehret und verbessert heraus gegeben*. Leipzig: J. F. Gleditsch und Sohn.

Riolan, Jean. 1625. *Schola anatomica*. Geneva: Joannis Celerii.

Riolan, Jean. 1629. *Anthropographie*. Paris: Denys Moreau.

Riolan, Jean. 1649. *Opuscula anatomica*. London: Milonis Flesher.

Rohault, Jacques. 1723 [1969]. *A System of Natural Philosophy*. Illustrated with Dr Samuel Clarke's Notes. London: Knapton. Repr., New York: Johnson Reprint Corporation.

Rorario, Girolamo. 1648. Ed. Naudé Gabriel, *Quod animalia bruta ratione utantur melius homine*.

Rüff, Jakob. 1637. *De conceptu et generatione hominis*. London: E. Griffin.

Sacco, N. 1628. *De calore naturali*. Papiae: J. Rubeum Baptistam.

Scaliger, Julius Caesar. 1557. *Exotericarum exercitationum liber*. . . . Lutetiae: Ex officina M. Vascosani.

Scaliger, Julius Caesar. 1619. *Aristotelis historia de animalibus. Iulio Caesare Scaligero interprete*. . . . Tolosae: Typis R. Colomerij.

Schacher, Polycarp Gottlieb. 1715. *De anima rationali, an sit corporis vitali principio*. Lipsiae: Fleischer.

Schenk, Johann Georg. 1609. *Monstrorum historia memorabilis*.

Sennert, Daniel. 1619. *De chymicorum cum Aristotelicis et Galenicis consensu ac dissensu*. Witterberg: Zachariah Schurer.

Sennert, Daniel. 1651. *Hypomnemata physica*. Venice: Francis Baba.

Sennert, Daniel. 1659. *Thirteen Books of Natural Philosophy*. Trans. Nicholas Culpeper. London: Abdiah Cole. Online photo reproduction at http://dewey.library.upenn. edu/sceti/printedbooksNew/index.cfm.

Sennert, Daniel. 1676. *Opera omnia*. Lyon.

Sennert, Daniel. 1683. *Hypomnemata physica*. Francofurti: Schleichus.

Sigorgne, Pierre. 1768. *Institutions leibnitiennes ou Précis de la monadologie*. Lyon: Frères Périsse.

Solinus. 1538. *De rerum toto orbe*. Basilae: Michaelem Insignum, Henricum Petri.

Spallanzani, Lazzaro. 1769. *Nouvelles recherches sur les découvertes microscopiques et la génération des corps organisés*. . . . *Avec des notes . . . par M. de Needham*. London and Paris: Lacombe.

Sperling, Johann. 1641. *Tractatus physicus de formatione foetus in utero*. Vitembergae: Apud haeredes T. Meni.

Stahl, Georg Ernst. 1701. *De differentia rationis et ratiocinationis*. Halae: Litteris C. Henckelii.

Bibliography

Stahl, Georg Ernst. 1705. *De frequentia morborum in homine prae brutis.* Halae: Litteris C. Henckelii.

Stahl, Georg Ernst. 1706. *De haereditaria dispositione ad varios affectus.* Halae: Litteris C. Henckelii.

Stahl, Georg Ernst. 1708. *Theoria medica vera, physiologiam et pathologiam, tanquam doctrinae medicae partes vere contemplativas, e naturae et artis veris fundamentis intaminata ratione et inconcussa experientia sistens.* Halae Magdeburgicae: Impensis Orphanotrophei. Quotations are from the second edition (1737), ed. Johann Juncker. Halae Magdeburgicae: Impensis Orphanotrophei. Third and last Latin edition (1831–35), ed. Ludwig Choulant, 2 vols. Lipsiae: Sumptibus Vossii. Partial German translation by Wendelin Ruf (1802), *Stahl's Theorie der Heilkunde*, mit einer Vorrede v. Kurt Sprengel. Halle: Gebauer. Another partial German translation by Karl Wilhelm Ideler (1831), *Georg Ernst Stahls Theorie der Heilkunde.* Berlin: Enslin. French translation in Georg Ernst Stahl, 1859–64, vols. 1–7.

Stahl, Georg Ernst. 1708. *De abortu et foetu mortuo.* . . . Halae Magdeburgicae: Literis C. Henckelii.

Stahl, Georg Ernst. 1859–64. *Oeuvres médico-philosophiques de G. E. Stahl.* 6 vols. French translation by Théodose Blondin. Paris: Baillière.

Stahl, Georg Ernst – Gottfried Wilhelm Leibniz. 1720. *Negotium otiosum seu Skiamachia adversus positiones fundamentales Theoriae medicae verae a viro quodam celeberrimo intentata, sed armis conversis enervata.* Halae: Litteris Orphanotrophei. French translation in Georg Ernst Stahl, 1859–64, 6: 2–362. Partial English translation in L. J. Rather and J. B. Frerichs, "The Leibniz-Stahl controversy," pts. 1 and 2, *Clio Medica* 3: 21–40; 5: 53–67.

Stenzel, Christian Gottfried. 1729. *Bini tractatus.* . . . Vitembergae: Stanno B. Gaeberdti.

Storch, Johann. 1752. *Leitung zur Historie des Höchsten Gottes.* . . . Eisenach.

Suárez, Francisco. 1965. *Disputationes metaphysicae.* Hildesheim: Olms.

Themistius. 1899. *In libros De anima paraphrasis.* Ed. R. Heinze. Berolini: G. Reimerum.

Tissot, Joseph. 1861. *La vie dans l'homme.* 2 vols. Paris: Victor Masson et Fils.

Trembley, Abraham. 1744. *Mémoires pour servir à l'histoire d'un genre de polypes d'eau douce, en bras en forme de cornes.* Leiden: Jean & Herman Verbeek.

Tyson, Edward. 1699. *Orang-outang, sive Homo sylvestris; Or the Anatomy of a Pygmie compared with That of a Monkey, an Ape, and a Man.* London.

Vasse, L. 1544. *Tabulae anatomicae.* Venetiae: Vincentium Vaugris.

Vesalius, Andreas. 1543. *De humani corpris fabrica.* Basel.

Weinrich, Martin. 1595. *De ortu monstrorum commentarius.* Sumptibus M. Osthesii.

Weisbach, Christian. 1712. *Wahrhaffte und gründliche Cur aller dem menschlichen Leibe zustossenden Kranckheiten.* . . . Strassburg: Dulsecker.

Willis, Thomas. 1672. *De anima brutorum, quae hominis vitalis ac sensitiva est, exercitationes duae.* Oxford: Richard Davis.

Wolff, Caspar Friedrich. 1759 [1966]. *Theoria generationis.* Halle. Reprinted in *Theorie von der Generation in zwei Abhandlungen erklärt und bewiesen.* Introduced by Robert Herrlinger. Hildesheim: Olms.

Wolff, Caspar Friedrich. 1759–77. Letters to Haller. Translated by Shirley Roe in *Matter, Life, and Generation: Eighteenth-Century Embryology and the Haller-Wolff Debate.* Cambridge: Cambridge University Press.

Bibliography

Wolff, Caspar Friedrich. 1759 [1896]. *Theoria generationis*. 2 vols. German translation by Paul Samassa. Leipzig: Verlag Wilhelm Engelmann.

Wolff, Caspar Friedrich. 1764. *Theorie von der Generation*. Berlin: Friedrich Wilhelm Birnstiel.

Wolff, Caspar Friedrich. 1789.*Von der eigenthümlichen und wesentlichen Kraft der vegetabilischen sowohl als auch der animalischen Substanz*. St. Petersburg: Kayserliche Academie der Wissenschaften.

Wolff, Caspar Friedrich. 1973. *Objecta meditationum pro theoria monstrorum*. Translated as *Predmety razmyshlenii v sviazi s teorii urodov* by Iu. Kh. Kopelevich and T. A. Lukina. Leningrad: Izdatel'stvo Nauka.

SECONDARY SOURCES

Aarsleff, Hans. 1989. The Berlin Academy and Frederick the Great. *History of the Human Sciences* 2: 193–206.

Adelmann, Howard B. 1942. *The Embryological Treatises of Hieronymus Fabricius of Aquapendente*. Ithaca: Cornell University Press.

Adelmann, Howard B. 1966. *Marcello Malpighi and the Evolution of Embryology*. 5 vols. Ithaca: Cornell University Press.

Albrecht, Michael. 1994. *Eklektik: Eine Begriffsgeschichte mit Hinweisen auf die Philosophie- und Wissenschaftsgeschichte*. Stuttgart: Frommann-Holzboog.

Allison, Henry. 1991. Kant's Antinomy of Teleological Judgment. *Southern Journal of Philosophy* 30: 25–42.

Anderson, Lorin. 1982. *Charles Bonnet and the Order of the Known*. Dordrecht: D. Reidel.

Anstey, Peter. 2000. *The Philosophy of Robert Boyle*. London: Routledge.

Appel, T A. 1987. *The Cuvier-Geoffroy Debate: French Biology in the Decades before Darwin*. New York: Oxford University Press.

Ariew, Roger. 1999. *Descartes and the Last Scholastics*. Ithaca: Cornell University Press.

Ariew, Roger, and Marjorie Grene. 1997. The Cartesian Destiny of Form and Matter. *Early Science and Medicine* 2: 300–25.

Arthur, Richard T. W. 1989. Russell's Conundrum: On the Relation of Leibniz's Monads to the Continuum: An Intimate Relation. In *Studies in the History and Philosophy of Science*, ed. James Robert Brown and Jürgen Mittelstrass. Dordrecht: Kluwer, 171–201.

Arthur, Richard T. W. 1998. Infinite Aggregates and Phenomenal Wholes. *Leibniz Society Review* 8 (12): 25–45.

Arthur, Richard T. W., ed. and trans. 2001. *G. W. Leibniz: The Labyrinth of the Continuum*. New Haven: Yale University Press.

Arthur, Richard T. W. 2003. The Enigma of Leibniz's Atomism. *Oxford Studies in Early Modern Philosophy* 1: 243–302.

Asma, Stephen T. 1996. *Following Form and Function: A Philosophical Archaeology of Life Science*. Evanston, IL: Northwestern University Press.

Aucante, Vincent. 1999. Les médecins et la médecine. In *La biografia intelletuale di René Descartes attraverso la Correspondance*. Naples: Vivarium, 607–25.

427

Bibliography

Aucante, Vincent. 2000. Le rôle du coeur de Fludd à Descartes et Harvey. In *Pour une civilisation du coeur*. Paris, 51–61.

Aucante, Vincent. 2002. La théorie de l'âme de Jean Fernel. *Corpus* 41: 9–42.

Aucante, Vincent. 2004. La vision chez Descartes et Plempius. In *Il Seicento e Descartes: Dibattiti cartesiani*. Florence: Le Monnier, 233–54.

Aulie, Richard. 1961. Caspar Friedrich Wolff and His 'Theoria Generationis', 1759. *Journal of the History of Medicine* 16: 124–44.

Balan, Bernard. 1988. Génération, organisation, développement: L'enjeu de l'épigénèse. In *Entre forme et histoire: La formation de la notion de développement à l'âge classique*. Paris: Meridiens Klincksieck.

Ballantyne, J. W. 1897. *Teratogenesis: An Inquiry into the Causes of Monstrosities: History of the Theories of the Past*. Edinburgh: Oliver and Boyd.

Balme, D. M. 1962. Development of Biology in Aristotle and Theophrastus: Theory of Spontaneous Generation. *Phronesis* 7: 91–104.

Balme, David. 1987. Aristotle's Biology Was Not Essentialist. In Gotthelf and Lennox.

Balme, David. 1987. The Place of Biology in Aristotle's Philosophy. In Gotthelf and Lennox.

Balme, D. M. 1990. Ανθρωπος ανθρωπον γεννα: Human Is Generated by Human. In *The Human Embryo: Aristotle and the Arabic and European Traditions*, ed. G. R. Dunstan, Exeter: University of Exeter Press, 20–31.

Balme, D. M. 1991. *Aristotle: History of Animals, Books VII–X*. Cambridge, MA: Harvard University Press.

Balme, D. M. 1992. *Aristotle: De Partibus Animalium I and De Generatione Animalium I* (with passages from II. 1–3). Oxford: Clarendon Press.

Balme, D. M. 2002. *Aristotle: Historia Animalium, Volume I: Books I–X* (prepared for publication by Allan Gotthelf). Cambridge: Cambridge University Press.

Baroncini, Gabriele. 1992. *Forme di esperienza scientifica*. Florence: Olschki.

Bartuschat, Wolfgang. 1972. *Zum systematischen Ort von Kants Kritik der Urteilskraft*. Frankfurt: Klostermann.

Bates, Don. 2000. Machina Ex Deo: William Harvey and the Meaning of Instrument. *Journal of the History of Ideas* 61: 577–93.

Baxter, Alice Levine. 1976. Edmund B. Wilson as a Preformationist: Some Reasons for His Acceptance of the Chromosome Theory. *Journal of the History of Biology* 9(1): 29–57.

Beeson, David. 1992. *Maupertuis: An Intellectual Biography*. Oxford: Voltaire Foundation.

Bernardi, Walter. 1986. *Le metafisiche dell' embrione: Scienze della vita e filosofia da Malpighi a Spallanzani, 1672–1793*. Florence: Leo S. Olschki Editore.

Bitbol-Hespériès, Annie. 1990. *Le principe de vie chez Descartes*. Paris: Vrin.

Bitbol-Hespériès, Annie. 1991. Le dualisme dans la correspondance entre Henry More et Descartes. In *Autour de Descartes, Le problème de l'âme et du dualisme*, ed. J.-L. Vieillard-Baron. Paris: Vrin, 141–58.

Bitbol-Hespériès, Annie. 1996. Connaissance de l'homme, connaissance de Dieu. *Les Etudes Philosophiques* 4: 507–33.

Bitbol-Hespériès, Annie. 1999. Descartes, Harvey et la Renaissance. In *Descartes et la Renaissance*, ed. E. Faye. Paris: H. Champion, 323–47.

Bibliography

Bitbol-Hespériès, Annie. 2000. Cartesian Physiology. In *Descartes' Natural Philosophy*, ed. Stephen Gaukroger, John Schuster, and John Sutton. London: Routledge, 349–82.

Bloch, Olivier R. 1971. *La philosophie de Gassendi: Nominalisme, matérialisme et métaphysique*. The Hague: Martinus Nijhoff.

Bodemer, Charles W. 1964. Regeneration and the Decline of Preformationism in Eighteenth-Century Embryology. *Bulletin of the History of Medicine* 38 (1): 20–31.

Bodemer, Charles W. 1968. *Embryological Thought in Seventeenth-Century England*. Medical Investigations in Seventeenth-Century England. Los Angeles: William Andrews Clark Memorial Library.

Bolton, R. 1987. Definition and Scientific Method in Aristotle's *Posterior Analytics* and *Generation of Animals*. In Gotthelf and Lennox.

Bouillier, Francisque. 1862. *Du principe vital et de l'âme pensante: Examen des diverses doctrines médicales et physiologiques sur les rapports de l'âme et de la vie*. Paris: Baillière et Fils.

Bowler, Peter J. 1971. Preformation and Pre-existence in the Seventeenth Century: A Brief Analysis. *Journal of the History of Biology* 4: 221–44.

Bowler, Peter J. 1975. The Changing Meaning of "Evolution." *Journal of the History of Ideas* 36: 95–114.

Broad, Jacqueline. 2002. *Women Philosophers of the Seventeenth Century*. Cambridge: Cambridge University Press.

Brown, Harcourt. 1963. Maupertuis Philosophe: Enlightenment and the Berlin Academy. *Studies on Voltaire* 24: 255–69.

Brown, Theodore. 1974. From Mechanism to Vitalism in Eighteenth-Century Physiology. *Journal of the History of Biology* 7: 179–216.

Buchdahl, Gerd. 1969. *Metaphysics and the Philosophy of Science: The Classical Origins: Descartes to Kant*. Cambridge, MA: MIT Press.

Buscaglia, Marino, et al., eds. 1994. *Charles Bonnet savant et philosophe (1720–1793)*. Geneva: Éditions Passé Présent.

Büttner, Manfred. 1973. Zum Übergang von der teleologischen zur kausalmechanischen Betrachtung der geographisch-kosmologischen Fakten: Ein Beitrag zur Geschichte der Geographie von Wolff bis Kant. *Studia Leibnitiana* 5: 177–95.

Butts, Robert, ed. 1986. *Kant's Philosophy of Physical Science*. Dordrecht: D. Reidel.

Butts, Robert. 1990. Teleology and Scientific Method in Kant's *Critique of Judgment*. *Nous* 24: 1–16.

Bylebyl, J. 1973. The Growth of Harvey's *De Motu Cordis*. *Bulletin of the History of Medicine* 47: 427–70.

Bylebyl, J. 1977. *De Motu Cordis*: Written in Two Stages? Response. *Bulletin of the History of Medicine* 51: 140–50.

Bylebyl, J. 1979. The School of Padua: Humanistic Medicine in the Sixteenth Century. In *Health, Medicine and Mortality in the Sixteenth Century*, ed. C. Webster. Cambridge: Cambridge University Press, 335–70.

Calinger, Ronald. 1968. Frederick the Great and the Berlin Academy of Sciences (1740–1766). *Annals of Science* 24: 239–49.

Callot, Emile. 1965. *La philosophie de la vie au XVIIIe siecle*. Paris: Riviere.

Canguilhem, Georges. 2003. *Le normal et le pathologique*. Paris: Presses Universitaires de France.

Bibliography

Cassirer, Ernst. 1951. *Philosophy of the Enlightenment*. Princeton: Princeton University Press.

Cassirer, Ernst. 1981. *Kant's Life and Thought*. New Haven: Yale University Press.

Céard, Jean. 1977. *La nature et les prodiges: L'insolite au XVIe siècle*. Geneva: Droz.

Ceglia, Francesco Paolo de. 2000. *Introduzione alla fisiologia di Georg Ernst Stahl*. Lecce: Pensa.

Ceglia, Francesco Paolo de. 2002. Ipocondria ed isteria nel sistema medico di Georg Ernst Stahl. Medicinae & Storia 2 (4): 51–86.

Chang Ku-Ming, Kevin. 2002. The Matter of Life: Georg Ernst Stahl and the Reconceptualizations of Matter, Body and Life in Early Modern Europe. Ph.D. diss., University of Chicago.

Cherni, Amor. 1995. Haller et Buffon: À propos des réflexions. *Revue d'Histoire des Sciences* 48: 267–305.

Cherni, Amor. 1998. *L'épistémologie de la transparence: Sur l'embryologie de A. von Haller*. Paris: Vrin.

Churchill, Frederick B. 1970. The History of Embryology as Intellectual History. *Journal of the History of Biology* 3: 155–81.

Clericuzio, Antonio. 2000. *Elements, Principles, and Corpuscles: A Study of Atomism and Chemistry in the Seventeenth Century*. Archives Internationales d'Histoire des Idées. no. 171. Dordrecht: Kluwer.

Clucas, Stephen. 1997. "The Infinite Variety of Formes and Magnitude": 16th and 17th Century English Corpuscular Philosophy and Aristotelian Theories of Matter and Form. *Early Science and Medicine* 2: 251–71.

Cobb, Matthew. 2000. Reading and Writing the Book of Nature: Jan Swammerdam (1637–1680). *Endeavour* 24: 122–8.

Cobb, Matthew. 2002. Malpighi, Swammerdam and the Colourful Silkworm: Replication and Visual Representation in Early Modern Science. *Annals of Science* 59: 111–47.

Cole, Francis J. 1930. *Early Theories of Sexual Generation*. Oxford: Clarendon Press.

Coleman, W. 1964. *Georges Cuvier, Zoologist: A Study in the History of Evolution Theory*. Cambridge, MA: Harvard University Press.

Cornell, J. F. 1986. Newton of the Grassblade? Darwin and the Problem of Organic Teleology. *Isis* 77: 405–21.

Costabel, Pierre. 1979. L'affaire Maupertuis-Koenig et les "questions de fait": Arithmos-Arrythmos. In *Skizzen aus der Wissenschaftsgeschichte*, ed. K. Figala and E. Berninger. Munich: Minerva Publikation, 29–48.

Cottingham, John. 1992. Cartesian Dualism: Theology, Metaphysics, and Science. *The Cambridge Companion to Descartes*. New York: Cambridge University Press, 236–57.

Coudert, Alison. 1975. A Cambridge Platonist's Kabbalist Nightmare. *Journal of the History of Ideas* 36: 633–52.

Coudert, Alison P. 1999. *The Impact of the Kabbalah in the Seventeenth Century*. Leiden: Brill.

Crombie, A. C. 1957. P. L. Moreau de Maupertuis, F.R.S. (1698–1759): Précurseur du transformisme. *Revue de Synthèse* 58 (3): 35–56.

Cunningham, Andrew. 1985. Fabricius and the "Aristotle Project" in Anatomical Teaching and Research at Padua. In *The Medical Renaissance of the Sixteenth Century*, ed.

Bibliography

A., Wear, R. K. French, and I. M. Lonie. Cambridge: Cambridge University Press, 195–222.

Cunningham, Andrew. 1997. *The Anatomical Renaissance: The Resurrection of the Anatomical Projects of the Ancients*. Aldershot: Scholar Press.

Cunningham, Andrew. 2002–3. The Pen and the Sword: Recovering the Disciplinary Identity of Physiology and Anatomy before 1800. Pts. 1 and 2. *Studies in History and Philosophy of Biological and Biomedical Sciences* 33: 631–65; 34: 51–76.

De Angelis, Simone. 2003. *Von Haller zu Newton: Studien zum Naturbegriff zwischen Empirismus und deduktiver Methode in der Schweizer Frühaufklärung*. Tübingen: Max Niemeyer Verlag.

Delage, Yves. 1903. *L' hérédité et les grands problèmes de la biologie générale*. 3rd ed. Paris: Schliecher Frères.

De Ley, H. 1980. Pangenesis versus Panspermia: Democritean Notes on Aristotle's Generation of Animals. *Hermes: Zeitschrift für Klassische Philologie* 108: 129–53.

Des Chene, Dennis. 2001. *Spirits and Clocks: Machine and Organism in Descartes*. Ithaca: Cornell University Press.

Detel, Wolfgang. 1978. *Scientia rerum natura occultarum: Methodologische Studien zur Physik Pierre Gassendis*. Berlin: De Gruyter.

Detlefsen, Karen. 2003. Supernaturalism, Occasionalism, and Preformation in Malebranche. *Perspectives on Science* 11: 443–83.

Dobell, Clifford. 1932. *Antony van Leeuwenhoek and his "Little Animals."* London: Bale.

Driesch, Hans. 1914. *The History and Theory of Vitalism*. Trans. C. K. Ogden. London: MacMillan.

Duchesneau, François. 1979. Haller et les théories de Buffon et C. F. Wolff sur l'épigénèse. *History and Philosophy of the Life Sciences* 1 (1): 65–100.

Duchesneau, François. 1982. *La physiologie des lumières: Empirisme, modèles et théories*. The Hague: Martinus Nijhoff.

Duchesneau, François. 1985. Vitalism in Late Eighteenth-Century Physiology: The Cases of Barthez, Blumenbach and John Hunter. In *William Hunter and the Eighteenth-Century Medical World*, ed. W. F. Bynum and Roy Porter. Cambridge: Cambridge University Press.

Duchesneau, François. 1993. *Leibniz et la méthode de la science*. Paris: Presses Universitaires de France.

Duchesneau, François. 1994. *La dynamique de Leibniz*. Paris: Vrin.

Duchesneau, François. 1998. *Les modèles du vivant de Descartes à Leibniz*. Paris: Vrin.

Duchesneau, François. 2003. Louis Bourguet et le modèle des corps organiques: Antonio Vallisneri. In *L' edizione del testo scientifico d' età moderna*, ed. M. T. Monti. Florence: Olschki, 3–31.

Duchesneau, François. 2003. Charles Bonnet et le concept leibnizien d'organisme. *Medicina nei Secoli* 15: 349–67.

Duchesneau, François. 2003. Leibniz's Model for Analyzing Organic Phenomena. *Perspectives on Science* 11: 378–409.

Duran, Jane. 1989. Anne Viscountess Conway: A Seventeenth Century Rationalist. *Hypatia* 4 (1): 64–79.

Edelman, Bernard. 1997. *La maison de Kant / The House That Kant Built*. Trans. Graeme Hunter. Toronto: University of Toronto Press.

Bibliography

Engelhardt, Dietrich von. 1997. Vitalism between Science and Philosophy in Germany around 1800. In *Vitalisms from Haller to the Cell Theory*, ed. Guido Cimino and François Duchesneau. Florence: Olschki.

Fagot, Anne. 1975. Le "transformisme" de Maupertuis. In *Actes de la journée Maupertuis (Créteil, 1er décembre 1973)*. Paris: Vrin, 163–78.

Fantini, Bernardino. 1994. Le cristal comme métaphore de la vie. In *Charles Bonnet savant et philosophe (1720–1793)*, ed. M. Buscaglia et al. Geneva: Éditions Passé Présent, 105–19.

Farber, Paul Lawrence. 1982. Research Traditions in Eighteenth-Century Natural History. In *Spallanzani e la Biologia del Settecento*, ed. Giuseppe Montalenti and Paolo Rossi Lazzaro. Florence: Olschki.

Farley, John. 1977. *The Spontaneous Generation Controversy from Descartes to Oparin.* Baltimore: Johns Hopkins University Press.

Ferrari, Jean. 1999. Kant, Maupertuis et le principe de moindre action. In *Pierre-Louis Moreau de Maupertuis: Eine Bilanz nach 300 Jahren*, ed. Hartmut Hecht. Berlin: Arno Spitz, 225–34.

Feuer, Lewis. 1963. *The Scientific Intellectual.* New York: Basic Books.

Fisher, Saul. 2005. *Pierre Gassendi's Philosophy of Science.* Leiden: Brill.

Foote, Edward T. 1969. Harvey: Spontaneous Generation and the Egg. *Annals of Science* 25 (2): 139–63.

Foucault, Michel. 1970. *The Order of Things: An Archaeology of the Human Sciences.* New York: Pantheon.

Fouke, Daniel C. 1989. Mechanical and "Organical" Models in Seventeenth-Century Explanations of Biological Reproduction. *Science in Context* 3: 365–81.

Fournier, Marian. 1996. *The Fabric of Life: Microscopy in the Seventeenth Century.* Baltimore: Johns Hopkins University Press.

Frank, R., Jr. 1980. *Harvey and the Oxford Physiologists: Scientific Ideas and Social Interaction.* Berkeley: University of California Press.

French, Roger. 1994. *William Harvey's Natural Philosophy.* Cambridge: Cambridge University Press.

French, Roger. 1995. Two Natural Philosophies. In *William Harvey's Natural Philosophy*. Cambridge: Cambridge University Press.

Fricke, Christel. 1990. Explaining the Inexplicable: The Hypothesis of the Faculty of Reflective Judgement in Kant's Third Critique. *Noûs* 24.

Friedman, Michael. 1991. Regulative and Constitutive: System and Teleology in Kant's *Critique of Judgment. Southern Journal of Philosophy* 30 (suppl.): 73–102.

Friedman, Michael. 1992. *Kant and the Exact Sciences.* Cambridge, MA: Harvard University Press.

Friedman, Michael. 1992. Causal Laws and the Foundations of Natural Science. In *The Cambridge Companion to Kant*, ed. Paul Guyer. Cambridge: Cambridge University Press, 161–99.

Gaissinovitch, A. E. 1968. Le rôle du newtonianisme dans la renaissance des idées épigénétiques en embryologie du XVIIIe siècle. *Actes du XIe Congrès International d'Histoire des Sciences* 5: 105–10.

Gaissinovitch, A. E. 1982. Influence des travaux de Caspar Friedrich Wolff sur la biologie du XVIIIe siècle. In *Lazzaro Spallanzani e la Biologia del Settecento*, ed. Giuseppe Montalenti and Paolo Rossi. Florence: Olschki.

Gaissinovitch, A. E. 1990. C. F. Wolff on Variability and Heredity. *History and Philosophy of the Life Sciences* 12: 179–201.

Garber, Daniel. 1992. *Descartes' Metaphysical Physics*. Chicago: University of Chicago Press.

Garber, Daniel. 1995. Leibniz: Physics and Philosophy. In *The Cambridge Companion to Leibniz*. New York: Cambridge University Press, 270–352.

Gasking, Elizabeth. 1967. *Investigations into Generation: 1651–1828*. Baltimore: Johns Hopkins University Press.

Gaukroger, Stephen. 2000. The Resources of a Mechanist Physiology and the Problem of Goal-Directed Processes. In *Descartes' Natural Philosophy*, ed. Stephen Gaukroger, John Schuster, and John Sutton. London: Routledge.

Gaukroger, Stephen. 2002. *Descartes' System of Natural Philosophy*. Cambridge: Cambridge University Press.

Genova, A. 1974. Kant's Epigenesis of Pure Reason. *Kant-Studien* 65: 259–73.

Geyer-Kordesch, Johanna. 1985. Die *Theoria medica vera* und Georg Ernst Stahls Verhältnis zur Aufklärung. In *Georg Ernst Stahl (1659–1734)*, ed. Wolfram Kaiser and Arina Völker. Halle: Wissenschaftliche Beiträge der Martin-Luther-Universität Halle-Wittemberg, 89–98.

Geyer-Kordesch, Johanna. 2000. *Pietismus, Medizin und Aufklärung in Preußen im 18. Jahrhundert: Das Leben und Werk Georg Ernst Stahls*. Tübingen: M. Niemeyer Verlag.

Giglioni, Guido. 1999. Girolamo Cardano e Giulio Cesare Scaligero: Il dibattito sul ruolo dell'anima vegetativa. In *Girolamo Cardano: Le opere, le fonti, la vita*, ed. Marialuisa Baldi and Guido Canziani. Milan: Franco Angeli, 313–39.

Ginsborg, Hannah. 2001. Kant on Understanding Organisms as Natural Purposes. In *Kant and the Sciences*, ed. E. Watkins. Oxford: Oxford University Press, 231–58.

Glass, Bentley. 1959. Maupertuis, Pioneer of Genetics and Evolution. In *Forerunners of Darwin: 1745–1859*, ed. Bentley Glass et al. Baltimore: Johns Hopkins University Press, 51–83.

Gontier, Thierry. 1998. *De L'homme à l'animal*. Paris: Vrin.

Goodrum, Matthew R. 2002. Atomism, Atheism, and the Spontaneous Generation of Human Beings: The Debate over a Natural Origin of the First Humans in Seventeenth-Century Britain. *Journal of the History of Ideas* 63: 207–24.

Gossman, Lionel. 1960. Berkeley, Hume and Maupertuis. *French Studies* 14: 304–24.

Gotthelf, A. 1976–7. Aristotle's Conception of Final Causality. *Review of Metaphysics* 30: 226–54. Reprinted with postscript in Gotthelf and Lennox 1987.

Gotthelf, Allan. 1989. Teleology and Spontaneous Generation in Aristotle: A Discussion. *Apeiron* 22: 181–93.

Gotthelf, Allan, and James G. Lennox, eds. 1987. *Philosophical Issues in Aristotle's Biology*. Cambridge: Cambridge University Press.

Gottlieb, Bernhard Joseph. 1943. Bedeutung und Auswirkungen des hallischen Professors und kgl. preuß. Leibarztes Georg Ernst Stahl, auf den Vitalismus des XVIII. Jahrhunderts, insbesondere auf die Schule von Montpellier. *Nova Acta Leopoldina: Neue Folge* 89 (12): 423–502.

Gould, Stephen Jay. 1996. *Full House: The Spread of Excellence from Plato to Darwin*. New York: Three Rivers Press.

Gould, Stephen Jay. 1997. Foreword to Clara Pinto-Correia, *The Ovary of Eve: Egg, Sperm and Preformation*. Chicago: Chicago University Press.

Greene, John C. 1959. *The Death of Adam: Evolution and its Impact on Western Thought*. Iowa City: University of Iowa Press.

Grene, Marjorie, and David Depew. 2004. *The Philosophy of Biology: An Episodic History*. Cambridge: Cambridge University Press.

Gueroult, Martial. 1980. The Metaphysics and Physics of Force in Descartes. In *Descartes: Philosophy, Mathematics and Physics*, ed. Stephen Gaukroger. Sussex: Sussex Harvester Press.

Gunther, R. T., ed. 1923–67. *Early Science in Oxford*. 15 vols. Oxford: Oxford University Press.

Guthrie, W. K. C. 1969. *A History of Greek Philosophy*. Vol. 2. Cambridge: Cambridge University Press.

Guyer, Paul. 2000. Editor's introduction to Immanuel Kant, *Critique of the Power of Judgment*. Cambridge: Cambridge University Press.

Guyer, Paul. 2001. Organisms and the Unity of Science. In *Kant and the Sciences*, ed. E. Watkins. Oxford: Oxford University Press, 259–81.

Hall, T. S. 1968. On Biological Analogs of Newtonian Paradigms. *Philosophy of Science* 35: 6–27.

Hall, T. S. 1982. Spallanzani on Matter and Life, with Notes on the Influence of Descartes. In *Lazzaro Spallanzani e la Biologia del Settecento*, ed. Giuseppe Montalenti and Paolo Rossi. Florence: Olschki.

Hartz, Glenn. 1998. Why Corporeal Substances Keep Popping up in Leibniz's Later Philosophy. *British Journal for the History of Philosophy* 6: 193–207.

Hatfield, Gary. 1979. Force (God) in Descartes' Physics. *Studies in History and Philosophy of Science* 10: 113–40.

Hatfield, Gary. 1992. Descartes' Physiology and Its Relation to Psychology. In *The Cambridge Companion to Descartes*. New York: Cambridge University Press, 335–70.

Hecht, Hartmut. 1994. Pierre-Louis Moreau de Maupertuis oder die Schwierigkeit der Unterscheidung von Newtonianern und Leibnitianern. In *Leibniz und Europa: Vorträge des VI. Internationalen Leibniz-Kongresses. Hannover: 18. bis 23 Juli 1994*. Hannover: Leibniz Gesellschaft, 331–8.

Hecht, Hartmut, ed. 1999. *Pierre-Louis Moreau de Maupertuis: Eine Bilanz nach 300 Jahren*. Berlin: Arno Spitz.

Hein, Hilde. 1972. The Endurance of the Mechanism-Vitalism Controversy. *Journal of the History of Biology* 5: 159–88.

Helm, Jürgen. 2000. Das Medizinkonzept Georg Ernst Stahls und seine Rezeption im Halleschen Pietismus und in der Zeit der Romantik. *Berichte zur Wissenschaftsgeschichte* 23: 167–90.

Henry, John. 1986. Occult Qualities and the Experimental Philosophy: Active Principles in Pre-Newtonian Matter Theory. *History of Science* 24: 335–81. Reprinted in Vere Chappell, ed., *Essays on Early Modern Philosophers from Descartes and Hobbes to Newton and Leibniz: Seventeenth-Century Natural Scientists*. New York: Garland Publishing, 1992; 1–47.

Hewson, M. A. 1975. *Giles of Rome and the Medieval Theory of Conception*. London: Athlon Press.

Bibliography

Hirai, Hiroshi. 2004. *Le concept de semence dans les théories de la matière à la Renaissance: De Marsile Ficin à Pierre Gassendi.* Brepols: Turnhout.

Hochdoerfer, Margarete. 1932. *The Conflict between the Religious and the Scientific Views of Albrecht von Haller (1708–1777).* Reprinted in Shirley A. Roe, *The Natural History of Albrecht von Haller.* New York: Arno Press, 1981.

Hoffheimer, Michael. 1982. Maupertuis and the Eighteenth-Century Critique of Preexistence. *Journal of the History of Biology* 15: 119–44.

Hoffman, Paul. 1966. Atomisme et génétique: Étude de la pensée philosophique et physiologique de Gassendi appliquée à la définition de la féminité. *Revue de Synthèse* 87 (41–42): 21–44.

Holmes, S. J. 1948. Micromerism in Biological Theory. *Isis* 39: 145–58.

Hunemann, Philippe. 2002. *Métaphysique et biologie: Kant et la constitution du concept d'organisme.* Villeneuve: Presses Universitaires du Septentrion.

Hutchison, Keith. 1982. What Happened to Occult Qualities in the Scientific Revolution? *Isis* 73: 233–53.

Hutchison, Keith. 1983. Supernaturalism and the Mechanical Philosophy. *History of Science* 21: 297–333.

Hutton, Sarah. 1995. Anne Conway critique d'Henry More: L'esprit et la matière. *Archives de Philosophie* 58: 371–84.

Hutton, Sarah. 1996. Henry More and Anne Conway on Preexistence and Universal Salvation. In *"Mind senior to the world": Stoicismo e origenismo nella filosofia platonica del Seicento inglese*, ed. Marialuisa Baldi. Milan: Francoangeli, 113–25.

Hutton, Sarah. 1997. Anne Conway, Margaret Cavendish and Seventeenth-Century Scientific Thought. In *Women, Science and Medicine 1500–1700*, ed. Lynette Hunter and Sarah Hutton. Thrupp, England: Sutton Publishing, 218–34.

Ibrahim, Annie. 1992. Matière inerte et matière vivante: La théorie de la perception chez Maupertuis. *Dix-Huitième Siècle* 24: 95–103.

Iltis, Carolyn. 1973. The Leibnizian-Newtonian Debates: Natural Philosophy and Social Psychology. *British Journal for the History of Science* 6: 343–77.

Israel, Jonathan. 2001. *Radical Enlightenment: Philosophy and the Making of Modernity 1650–1750.* Oxford: Oxford University Press.

Jacob, François. 1973. *The Logic of Life: A History of Heredity.* New York: Pantheon.

Jacob, Margaret. 1976. *The Newtonians and the English Revolution, 1689–1720.* Ithaca: Cornell University Press.

Jedidi, Ali Bey. 1991. *Les fondements de la biologie cartésienne.* Paris: La Pensée Universelle.

Jouanna, Jacques. 1995. *Hippocrate.* Paris: Fayard.

Joy, Lynn S. 1987. *Gassendi the Atomist: Advocate of History in an Age of Science.* Cambridge: Cambridge University Press.

Kaiser, Wolfram. 1985. Der Lehrkörper der Medizinischen Fakultät in der halleschen Amtszeit von Georg Ernst Stahl. *Georg Ernst Stahl (1659–1734): Hallesches Symposium 1984*, ed. A. Völker and W. Kaiser. Halle: Martin-Luther-Universität Halle-Wittemberg, 59–66.

Kirk, G. S., J. E. Raven, and M. Schofield. 1991. *The Presocratic Philosophers: A Critical History with a Selection of Texts.* 2nd ed. Cambridge: Cambridge University Press (abbreviated as KRS).

Bibliography

Koch, Richard. 1926. War Georg Ernst Stahl ein selbständiger Denker? *Sudhoffs Archiv für Geschichte der Medizin und der Naturwissenschaften* 18: 20–50.

Koyré, Alexandre. 1957. *From the Closed World to the Infinite Universe*. Baltimore: Johns Hopkins University Press, 235–72.

Krolzik, Udo. 1980. Das physikotheologische Naturverständnis und sein Einfluß auf das naturwissenschaftliche Denken im 18. Jahrhundert. *Medizinhistorisches Journal* 15: 90–102.

Kruk, Remke. 1990. A Frothy Bubble: Spontaneous Generation in the Medieval Islamic Tradition. *Journal of Semitic Studies* 35 (2): 265–82.

Kubbinga, Henk. 2002. *L'histoire du concept de "molécule."* 3 vols. Paris: Springer.

Kuehn, Manfred. 2001. *Kant: A Biography*. Cambridge: Cambridge University Press.

Kulstad, Mark. 1991. *Leibniz on Apperception, Consciousness, and Reflection*. Munich: Philosophia Verlag.

Labrousse, Élisabeth. 1964. *Pierre Bayle*. Vol. 3. *Hétérodoxie et rigorisme*. The Hague: Martinus Nijhoff.

Labrousse, Élisabeth. 1982. Introduction historique. In Pierre Bayle, *Oeuvres diverses*. Hildesheim: Olms.

Labrousse, Élisabeth. 1985. *Pierre Bayle*. Vol. 1. *Du pays de foix à la cité d'Érasme*. 2nd ed. Dordrecht: Martinus Nijhoff.

Larson, James L. 1994. *Interpreting Nature: The Science of Living Form from Linnaeus to Kant*. Baltimore: Johns Hopkins University Press.

Laudan, Rachel. 1987. *From Mineralogy to Geology: The Foundations of a Science, 1650–1830*. Chicago: University of Chicago Press.

Laywine, Alison. 1993. *Kant's Early Metaphysics and the Origins of the Critical Philosophy*. North American Kant Society Studies in Philosophy, no. 3. Atascadero, CA: Ridgeview.

Lemoine, Albert. 1864. *Le vitalisme et l'animisme de Stahl*. Paris: Germer Baillière.

Lennox, J. G. 1982. Teleology, Chance, and Aristotle's Theory of Spontaneous Generation. *Journal of the History of Philosophy* 20: 219–38.

Lennox, J. G. 1985. Are Aristotelian Species Eternal? In *Aristotle on Nature and Living Things*, ed. A. Gotthelf. Pittsburgh: Mathesis Publications, 67–94. Reprinted in J. G. Lennox, *Aristotle's Philosophy of Biology: Studies in the Origins of Life Science*. Cambridge: Cambridge University Press, 2001; 131–59.

Lennox, J. G. 1986. Aristotle, Galileo and the Mixed Sciences. In *Reinterpreting Galileo*, ed. W. Wallace. Washington, DC: Catholic University Press of America, 29–52.

Lennox, J. G. 1987. Divide and Explain: The Posterior Analytics in Practice. In *Philosophical Issues in Aristotle's Biology*, eds. A. Gotthelf and J. G. Lennox. Cambridge: Cambridge University Press: 90–119. Reprinted in J. G. Lennox. *Aristotle's Philosophy of Biology: Studies in the Origins of Life Science*. Cambridge: Cambridge University Press, 2001; 7–38.

Lennox, J. G. 1991. Between Data and Demonstration: The Analytics and the Historia Animalium. In *Science and Philosophy in Classical Greece*, ed. A. Bowen. New York: Garland Publishing, 261–94. Reprinted in J. G. Lennox. *Aristotle's Philosophy of Biology: Studies in the Origins of Life Science*. Cambridge: Cambridge University Press, 2001; 39–71.

Lennox, J. G. 1997. Nature does nothing in vain.... In *Beiträge zur antiken Philosophie: Festschrift für Wolfgang Kullmann*, ed. H.-C. Günther and A. Rengakos.

Stuttgart: Franz Steiner Verlag, 199–214. Reprinted in J. G. Lennox. *Aristotle's Philosophy of Biology: Studies in the Origins of Life Science*. Cambridge: Cambridge University Press, 2001; 205–24.

Lenoir, Timothy. 1980. Kant, Blumenbach, and Vital Materialism in German Biology. *Isis* 71: 77–108.

Lenoir, Timothy. 1981. The Göttingen School and the Development of Transcendental *Naturphilosophie* in the Romantic Era. *Studies in History of Biology* 5: 111–205.

Lenoir, Timothy. 1982. *The Strategy of Life: Teleology and Mechanics in Nineteenth Century German Biology*. Dordrecht: D. Reidel.

Lesky, Erna. 1950. *Die Zeugungs- und Vererbungslehren der Antike und ihr Nachwirken*. Mainz: Verlag der Akademie der Wissenschaft und der Literatur in Mainz. Abhandlungen der geistes- und sozialwissenschaftslichen Klasse. 19, 1225–1425.

Lewis, Eric. 2001. Walter Charleton and Early Modern Eclecticism. *Journal of the History of Ideas* 62: 651–64.

Lloyd, Geoffrey E. R. 1987. Empirical Research in Aristotle's Biology. In Gotthelf and Lennox.

Lloyd, G. E. R. 1996. *Aristotelian Explorations*. Cambridge: Cambridge University Press.

Loewisch, D. J. 1965. Kants *Kritik der reinen Vernunft* und Humes *Dialogues Concerning Natural Religion*. *Kant Studien* 56: 1170–1207.

Lohff, Brigitte. 1997. The Concept of Vital Forces as a Research Program: From Mid-XVIIIth Century to Johannes Müller. In *Vitalisms from Haller to the Cell Theory*, ed. Guido Cimino and François Duchesneau. Florence: Olschki.

Loptson, Peter. 1982. Introduction to *The Principles of the Most Ancient and Modern Philosophy*, by Anne Conway. The Hague: Martinus Nijhoff.

Lovejoy, A. O. 1911 [1959]. Kant and Evolution. In *Forerunners of Darwin*, ed. Bentley Glass, Owsei Temkin, and W. L. Strauss Jr. Baltimore: Johns Hopkins University Press.

Lüthy, Christoph. 1996. Atoms, Microscopes, and Causality: "Visual Reductionism" in the 17th Century. Ph.D. diss. Harvard University.

Magner, Lois N. 1994. *A History of the Life Sciences*. New York: Marcel Dekker.

Maienschein, Jane. 1985. Preformation or New Formation – Or Neither or Both? In *The Eighth Symposium of the British Society for Developmental Biology: A History of Embryology*, ed. T. J. Horder, J. A. Witkowski, and C. C. Wylie. Cambridge: Cambridge University Press.

Marc-Wogau, Konrad. 1938. *Vier Studien zu Kants Kritik der Urteilskraft*. Leipzig: Otto Harrassowitz, 73–108.

McLaughlin, Peter. 1982. Blumenbach und der Bildungstrieb: Zum Verhältnis von epigenetischer Embryologie und typologischem Artbegriff. *Medizinhistorisches Journal* 17: 357–72.

McLaughlin, Peter. 1990. *Kant's Critique of Teleology in Biological Explanation*. Lewiston: Edwin Mellen.

McRobert, Jennifer. 2000. Anne Conway's Vitalism and her Critique of Descartes. *International Philosophical Quarterly* 40 (1): 21–35.

Meinel, Christoph. 1988. Early Seventeenth-Century Atomism: Theory, Epistemology, and the Insufficiency of Experiment. *Isis* 79: 68–103.

Bibliography

Ménard, Pierre. 1937. L'esprit de la physiologie cartésienne. *Archives de Philosophie* 13: 191.

Mendelsohn, Everett I. 1976. Philosophical Biology versus Experimental Biology: Spontaneous Generation in the Seventeenth Century. In *Topics in the Philosophy of Biology*, ed. Marjorie Grene and Everett Mendelsohn. Dordrecht: D. Reidel, 37–65.

Mercer, Christia. 2001. *Leibniz's Metaphysics: Its Origins and Development*. Cambridge: Cambridge University Press.

Merchant, Carolyn. 1979. The Vitalism of Anne Conway: Its Impact on Leibniz's Concept of the Monad. *Journal of the History of Philosophy* 17: 255–69.

Meschini, F. 1998. *Neurofisiologia cartesiana*. Florence: Leo S. Olschki Editore.

Michael, Emily. 1997. Daniel Sennert on Matter and Form: At the Juncture of the Old and the New. *Early Science and Medicine* 2: 272–300.

Michael, Emily. 2001. Sennert's Sea Change: Atoms and Causes. In *Late Medieval and Early Modern Corpuscular Matter Theories*, ed. Christoph Lüthy, John E. Murdoch, and William R. Newman. Leiden: Brill, 331–62.

Michael, Emily, and Fred S. Michael. 1988. Gassendi on Sensation and Reflection: A Non-Cartesian Dualism. *History of European Ideas* 9: 583–95.

Michael, Emily, and Fred S. Michael. 1989. Two Early Modern Concepts of Mind: Reflecting Substance vs. Thinking Substance. *Journal of the History of Philosophy* 27: 29–48.

Mocek, Reinhard. 1995. Caspar Friedrich Wolffs Epigenesis-Konzept: Ein Problem im Wandel der Zeit. *Biologisches Zentralblatt* 114: 179–90.

Monti, Maria Teresa. 1988. Difficultés et arguments de l'embryologie d'Albrecht von Haller: La reconversion des catégories de l'anatome animata. *Revue des Sciences, Philosophiques et Théologiques* 72: 301–12.

Monti, Maria Teresa. 1989. Théologie physique et mécanisme dans la physiologie de Haller. In *Science and Religion / Wissenschaft und Religion: Proceedings of the Symposium of the XVIIIth International Congress of History of Science*. Bochum: Universitätsverlag Dr. N. Brockmeyer.

Monti, Maria Teresa. 1990. *Congettura ed esperienza nella fisiologia di Haller*. Florence: Olschki.

Monti, Maria Teresa. 1997. Les dynamismes du corps et les forces du vivant dans la physiologie de Haller. In *Vitalisms from Haller to the Cell Theory*, ed. Guido Cimino and François Duchesneau. Florence: Olschki.

Müller, Gerhard H. 1984. La conception de l'épigénèse chez Caspar Friedrich Wolff (1734–1794). *Revista di Biologia* 77: 343–62.

Müller-Sievers, Helmut. 1997. *Self-Generation: Biology, Philosophy, and Literature around 1800*. Stanford: Stanford University Press.

Musallam, Basim. 1990. The Human Embryo in Arabic Scientific and Religious Thought. In *The Human Embryo: Aristotle and the Arabic and European Traditions*, ed. G. R. Dunstan. Exeter: University of Exeter Press, 32–46.

Needham, Joseph. 1959. *A History of Embryology*. Second edition, revised with the assistance of Arthur Hughes. New York: Abelard-Schuman.

Neubauer, John. 1983. Albrecht von Haller's Philosophy of Physiology. *Studies on Voltaire and the Eighteenth Century* 215: 320–2.

Neubauer, John. 1984. La philosophie de la physiologie d'Albrecht von Haller. *Revue de Synthèse* 113–14: 135–42.

Bibliography

Nicolson, Marjorie Hope, ed. 1992. *The Conway Letters*. Revised edition by Sarah Hutton. Oxford: Clarendon Press.

Niebyl, Peter H. 1971. Sennert, van Helmont, and the Medical Ontology. *Bulletin of the History of Medicine* 45: 115–37.

Northwood, Heidi. 2003. The Passive Female Role in Aristotle's Embryology. Unpublished manuscript.

Oake, Roger. 1940. Did Maupertuis Read Hume's *Treatise of Human Nature? Revue de Littérature Comparée* 20: 81–7.

Oppenheimer, Jane M. 1967. *Essays in the History of Embryology and Biology.* Cambridge, MA: MIT Press.

Osler, Margaret. 1979. Descartes and Charleton on Nature and God. *Journal of the History of Ideas* 40: 445–56.

Osler, Margaret. 1994. *Divine Will and the Mechanical Philosophy: Gassendi and Descartes on Contingency and Necessity in the Created World.* Cambridge: Cambridge University Press.

Osler, Margaret. 2001. How Mechanical Was the Mechanical Philosophy? Non-Epicurean Themes in Gassendi's Atoms. In *Late Medieval and Early Modern Corpuscular Matter Theory*, ed. Christoph Lüthy, John Murdoch, and William R. Newman. Leiden: Brill.

Ostoya, Paul. 1954. Maupertuis et la biologie. *Revue d'Histoire des Sciences et de Leurs Applications* 7: 60–78.

Ottosson, P.-G. 1984. *Scholastic Medicine and Philosophy.* Naples: Bibliopolis.

Outram, D. 1984. *Georges Cuvier: Vocation, Science and Authority in Post Revolutionary France.* Manchester: Manchester University Press.

Pagel, Walter 1967. *William Harvey's Biological Ideas.* Basel: S. Karger.

Pagel, Walter. 1986. Aristotle and Seventeenth-Century Biological Thought. In *From Paracelsus to Van Helmont: Studies in Renaissance Medicine and Science.* London: Variorum Reprints, 489–509.

Palmerino, Carla-Rita. 1998. Atomi, meccanica, cosmologia: Le lettere Galileiane di Pierre Gassendi. Ph.D. diss., University of Florence.

Park, Katharine, and Lorraine Daston. 1981. Unnatural Conceptions: The Study of Monsters in Sixteenth- and Seventeenth-Century France and England. *Past and Present* 92: 20–54.

Park, Katharine, and Lorraine Daston. 1998. *Wonders and the Order of Nature.* New York: Zone Books.

Parkinson, G. H. R. 1969. Science and Metaphysics in the Leibniz-Newton Controversy. *Studia Leibnitiana Supplementa* 2: 79–112.

Partington, J. R. 1961–2. *A History of Chemistry.* 4 vols. New York: St. Martin's Press.

Perfetti, S. 1999. Three Different Ways of Interpreting Aristotle's *De Partibus Animalium*: Pietro Pomponazzi, Niccolò Leonico Tomeo and Agostino Nifo. In *Aristotle's Animals in the Middle Ages and Renaissance*, ed. C. Steel, G. Guldentops, and P. Beullens. Leuven: Leuven University Press, 297–316.

Perl, Margula. 1969. Physics and Metaphysics in Newton, Leibniz, and Clarke. *Journal of the History of Ideas* 30: 507–26.

Perler, Dominik. 1996. *Repräsentation bei Descartes.* Klostermann: Frankfurt am Main.

Petersen, Peter. 1921. *Geschichte der aristotelischen philosophie im protestantischen Deutschland.* Leipzig: F. Meiner.

Bibliography

Phemister, Pauline. 1999. Leibniz and the Elements of Compound Bodies. *British Journal for the History of Philosophy* 7(1–2): 57–78.

Philipp, Wolfgang. 1967. Physicotheology in the Age of Enlightenment: Appearance and History. *Studies on Voltaire and the Eighteenth Century* 57: 1233–67.

Pichot, André. 1993. *Histoire de la notion de vie.* Paris: Gallimard.

Pinto-Correia, Clara. 1997. *The Ovary of Eve: Egg and Sperm and Preformation.* Chicago: University of Chicago Press.

Plaass, Peter. 1994. *Kant's Theory of Natural Science.* Dordrecht: Kluwer.

Plamondon, Ann. 1975. The Contemporary Reconciliation of Mechanicism and Organicism. *Dialectica* 29(4): 213–21.

Plochmann, G. 1963. William Harvey and His Methods. *Studies in the Renaissance* 10: 192–210.

Porter, Roy. 1977. *The Making of Geology: Earth Science in Britain, 1660–1815.* Cambridge: Cambridge University Press.

Poser, Hans. 2000. Are Leibniz's Monads Corporeal Substances? Old Answers to a New Problem. Paper presented at the workshop on "Corporeal Substances and the Labyrinth of the Continuum in Leibniz" at the Florence Center for the History and Philosophy of Science, Florence, Italy, November 23–25.

Preus, Anthony. 1970. Science and Philosophy in Aristotle's *Generation of Animals. Journal of the History of Biology* 3: 1–52.

Priestley, F. E. L. 1970. The Clarke-Leibniz Controversy. In *The Methodological Heritage of Newton*, ed. Robert Butts and John Davis. Toronto: University of Toronto Press, 34–56.

Purcell, Rosamond. 1997. *Special Cases: Natural Anomalies and Historical Monsters.* San Francisco: Chronicle Books.

Pyle, Andrew. 1987. Animal Generation and Mechanical Philosophy: Some Light on the Role of Biology in the Scientific Revolution. *Journal for the History and Philosophy of the Life Sciences* 9(2): 225–54.

Pyle, Andrew. 1997. *Atomism and Its Critics.* Bristol, England: Thoemmes Press.

Pyle, Andrew. 2003. *Malebranche.* London: Routledge.

Rather, Lelland J. 1961. G. E. Stahl's Psychological Physiology. *Bulletin of the History of Medicine* 35: 37–49.

Rather, L. J. 1967. Thomas Fienus' (1567–1631) Dialectical Investigation of the Imagination as Cause and Cure of Bodily Disease. *Bulletin of the History of Medicine* 41: 349–67.

Reeve, C. D. C. 1981. Anaxagoras' Panspermism. *Ancient Philosophy* 1: 89–108.

Reill, Peter Hanns. 1986. Science and the Science of History in the Spätaufklärung. In *Aufklärung und Geschichte*, ed. H. E. Boedeker et al., Göttingen: Vandenhoeck & Ruprecht.

Reill, Peter Hanns. 1989. Anti-Mechanism, Vitalism and Their Political Implications in Late Enlightened Scientific Thought. *Francia* 16(2): 195–212.

Reill, Peter Hanns. 1992. Between Mechanism and Hermeticism: Nature and Science in the late Enlightenment. In *Frühe Neuzeit – Frühe Moderne?* ed. Rudolf Vierhaus. Göttingen: Vandenhoeck & Ruprecht, 393–421.

Reill, Peter Hanns. 1994. Science and the Construction of the Cultural Sciences in Late Enlightenment Germany: The Case of Wilhelm von Humboldt. *History and Theory* 33: 345–66.

440

Reill, Peter Hanns. 1995. Anthropology, Nature and History in the Late Enlightenment: The Case of Friedrich Schiller. In *Schiller als Historiker*, ed. Otto Dann, Norbert Oellers, and Ernst Osterkamp. Stuttgart: Metzler, 243–265.

Reill, Peter Hanns. 1998. Analogy, Comparison, and Active Living Forces: Late Enlightenment Responses to the Skeptical Critique of Causal Analysis. In *The Skeptical Tradition around 1800*, ed. Johann van der Zande and Richard Popkin. Dordrecht: Kluwer, 203–11.

Reill, Peter Hanns. Forthcoming. *Vitalizing Nature in the Enlightenment*. Berlekey: University of California Press.

Rey, Roselyne. 1997. Gassendi et les sciences de la vie au XVIIIᵉ siècle. In *Gassendi et l'Europe*, ed. Sylvia Murr. Paris: Vrin.

Richards, Robert. 2000. Kant and Blumenbach on the *Bildungstrieb*: A Historical Misunderstanding. *Studies in the History and Philosophy of Biology and the Biomedical Sciences* 31: 11–32.

Richards, Robert. 2003. *The Romantic Conception of Life: Science and Philosophy in the Age of Goethe*. Chicago: University of Chicago Press.

Rieppel, Olivier 1987. Organization in the Lettres Philosophiques of Louis Bourguet Compared to the Writings of Charles Bonnet. *Gesnerus* 44: 125–32.

Rieppel, Olivier. 1988. The Reception of Leibniz's Philosophy in the Writings of Charles Bonnet (1720–1793). *Journal of the History of Biology* 21: 119–45.

Robinet, André. 1970. *Malebranche de l'Académie des Sciences*. Paris: Vrin.

Robinet, André. 1975. Place de la polémique Maupertuis-Diderot dans l'oeuvre de Dom Deschamps. In *Actes de la journée Maupertuis (Créteil, 1er décembre 1973)*. Paris: Vrin, 33–46.

Rocci, Giovanni. 1975. *Charles Bonnet: Filosofia e scienza*. Florence: G. C. Sansoni.

Rochot, Bernard. 1944. *Les travaux de Gassendi sur Épicure et sur l'atomisme, 1619–1648*. Paris: J. Vrin.

Roe, Shirley. 1975. The Development of Albrecht von Haller's Views on Embryology. *Journal of the History of Biology* 8: 167–90.

Roe, Shirley. 1979. Rationalism and Embryology: Caspar Friedrich Wolff's Theory of Epigenesis. *Journal of the History of Biology* 12: 1–43.

Roe, Shirley. 1981. *Matter, Life, and Generation: Eighteenth-Century Embryology and the Haller-Wolff Debate*. Cambridge: Cambridge University Press.

Roe, Shirley. 1984. Anatomia Animata: The Newtonian Physiology of Albrecht von Haller. In *Transformation and Tradition in the Sciences*, ed. Everett Mendelsohn. Cambridge: Cambridge University Press.

Roe, Shirley. 1985. Voltaire versus Needham: Atheism, Materialism, and the Generation of Life. *Journal of the History of Ideas* 46: 65–87.

Roger, Jacques. 1963. *Les sciences de la vie dans la pensée française au XVIIIe siècle*. Paris: A. Colin. Second edition 1971.

Roger, Jacques. 1988. La notion de développement chez les naturalistes du XVIIIe siècle. In *Entre forme et histoire: La formation de la notion de développement à l'âge classique*. Paris: Meridiens Klincksieck.

Roger, Jacques. 1997. *The Life Sciences in Eighteenth-Century French Thought*. Trans. Robert Ellrich, ed. Keith R. Benson. Stanford: Stanford University Press.

Ross, G. MacDonald. 1983. *Occultism and Philosophy in the Seventeenth Century*. Paper presented at a conference of the Royal Institute of Philosophy.

Rudolph, Gerhard. 1991. La méthode hallérienne en physiologie. *Dix-Huitième Siècle* 23: 75–84.

Ruestow, Edward. 1985. Piety and the Defense of Natural Order: Swammerdam on Generation. In *Religion, Science and Worldview*, ed. Margaret Osler and Paul Farber. Cambridge: Cambridge University Press.

Ruse, Michael. 1979. *The Darwinian Revolution: Science Red in Tooth and Claw.* Chicago: University of Chicago Press.

Ruse, Michael. 1996. *Monad to Man: The Concept of Progress in Evolutionary Biology.* Cambridge, MA: Harvard University Press.

Ruse, Michael. 2003. *Darwin and Design: Does Evolution Have a Purpose?* Cambridge, MA: Harvard University Press.

Ruse, Michael. 2005. *On a Darkling Plain: The Evolution-Creation Struggle.* Cambridge, MA: Harvard University Press.

Russell, E. S. 1930. *The Interpretation of Development and Heredity.* Oxford: Clarendon Press.

Russo, François. 1978. Théologie naturelle et sécularisation de la science au XVIIIe siècle. *Recherches de Science Religieuse* 66: 27–62.

Salomon-Bayet, Claire. 1975. Maupertuis et l'institution. In *Actes de la journée Maupertuis (Créteil, 1er décembre 1973).* Paris: Vrin, 183–202.

Sandler, Iris. 1983. Pierre Louis Moreau de Maupertuis: A Precursor of Mendel? *Journal of the History of Biology* 16: 101–36.

Sanhueza, Gabriel. 1997. La pensée biologique de Descartes dans ses rapports avec la philosophie scolastique: Le cas Gomez Péreira. *La philosophie en commun.* Paris and Montréal: l'Harmattan.

Savioz, Raymond. 1948. *Mémoires autographes de Charles Bonnet de Genève.* Paris: Vrin.

Savioz, Raymond. 1948. *La philosophie de Charles Bonnet de Genève.* Paris: Vrin.

Scala, Aidée. 2004. *Girolamo Rorario: Un umanista diplomatico del Cinquecento e i suoi "Dialoghi."* Florence: Olschki.

Schechner-Genuth, Sara. 1997. *Comets, Popular Culture, and the Birth of Cosmology.* Princeton: Princeton University Press.

Schierbeek, Abraham. 1967. *Jan Swammerdam, 1637–1680: His Life and Works.* Amsterdam: Swets Zeitlinger.

Schmitt, Charles. 1983. *Aristotle and the Renaissance.* Cambridge, MA: Harvard University Press.

Schmitt, Charles. 1984. Science in the Italian Universities in the Sixteenth and Early Seventeenth Centuries. In *The Aristotelian Tradition and Renaissance Universities*, ed. Charles Schmitt. London: Warburg Institute, 35–56.

Schmitt, Charles. 1985. Aristotle among the Physicians. In *The Medical Renaissance of the Sixteenth Century*, ed. A. Wear, R. French, and I. Lonie. Cambridge: Cambridge University Press, 1–16.

Schmitt, C. 1989. *Reappraisals in Renaissance Thought.* London: Warburg Publications.

Scholem, Gershom. 1974. *Kabbalah.* New York: Dorset Press.

Schönfeld, Martin. 2000. *The Philosophy of the Young Kant: The Precritical Project.* Oxford: Oxford University Press.

Schott, Robin May. 1983. *Cognition and Eros.* Boston: Beacon.

Schrecker, Paul. 1938. Malebranche et le préformisme biologique. *Revue Internationale de Philosophique* 1(1): 77–97.

Sepper, Dennis L. 1996. *Descartes' Imagination: Proportion, Images, and the Activity of Thinking.* Berkeley: University of California Press.

Shapin, Steven. 1981. Of Gods and Kings: Natural Philosophy and Politics in the Leibniz-Clarke Disputes. *Isis* 72: 187–215.

Shell, Susan. 1996. *The Embodiment of Reason: Kant on Spirit, Generation, and Community.* Chicago: University of Chicago Press.

Simmons, Alison. 2001. Sensible Ends: Latent Teleology in Descartes' Account of Sensation. *Journal of the History of Philosophy* 39: 49–75.

Sloan, Phillip R. 1977. Descartes, the Sceptics, and the Rejection of Vitalism in Seventeenth-Century Physiology. *Studies in the History and Philosophy of Science* 8: 1–28.

Sloan, Philip R. 1979. The Historical Interpretation of Biological Species. *British Journal for the History of Science* 12: 109–53.

Sloan, Phillip R. 1985. The Question of Natural Purpose: Historical Review. In *Evolution and Creation*, ed. Ernan McMullin. Notre Dame: University of Notre Dame Press, 121–44.

Sloan, Phillip R. 2002. Preforming the Categories: Eighteenth Century Generation Theory and the Biological Roots of Kant's A Priori. *Journal of the History of Philosophy* 40: 229–53.

Smith, C. U. M. 1976. *The Problem of Life: An Essay in the Origins of Biological Thought.* New York: Wiley.

Smith, Justin E. H. 2004. Christian Platonism and the Metaphysics of Body in Leibniz. *British Journal for the History of Philosophy* 12(1): 43–59.

Sonntag, Otto. 1974. Albrecht von Haller on the Future of Science. *Journal of the History of Ideas* 35: 313–22.

Sonntag, Otto. 1977. The Mental and Temperamental Qualities of Haller's Scientist. *Physis* 19: 173–84.

Stefani, Marta. 2002. *Corruzione e generazione: John T. Needham e l' origine del vivente.* Florence: Olschki.

Steigerwald, Joan. 2002. Instruments of Judgement: Inscribing Organic Processes in Late Eighteenth-Century Germany. *Studies in History and Philosophy of Biological and Biomedical Sciences* 33: 79–131.

Terrall, Mary. 1996. Salon, Academy, and Boudoir: Generation and Desire in Maupertuis's Science of Life. *Isis* 87: 217–29.

Terrall, Mary. 2002. *The Man Who Flattened the Earth: Maupertuis and the Sciences in the Enlightenment.* Chicago: University of Chicago Press.

Thiele, Rüdiger. 1982. *Leonhard Euler.* Leipzig: Teubner.

Thorndike, Lynn. 1958. *History of Magic and Experimental Science.* New York: Columbia University Press.

Toellner, Richard. 1971. *Albrecht von Haller: Über die Einheit im Denken des letzten Universalgelehrten.* Wiesbaden: Franz Steiner Verlag.

Toellner, Richard. 1982. Die Bedeutung des physico-theologischen Gottesbeweises für die nachcartesianische Physiologie im 18. Jahrhundert. *Berichte zur Wissenschaftsgeschichte* 5: 75–82.

Toellner, Richard. 1997. Principles and Forces of Life in Haller. In *Vitalisms from Haller to the Cell Theory*, ed. Guido Cimino and François Duchesneau. Florence: Olschki.

Tonelli, Giorgio. 1959. Der Streit über die mathematische Methode in der Philosophie in der ersten Hälfte des 18. Jahrhunderts und die Entstehung von Kants Schrift über die "Deutlichkeit." *Archiv für Geschichte der Philosophie* 9: 57–66.

Tonelli, Giorgio. 1975. Conditions in Königsberg and the Making of Kant's Philosophy. In *Bewußtsein: Gerhard Funke zu eigen*, ed. Alexius Bucher, Hermann Drüe, and Thomas Seebohm. Bonn: Bouvier, 126–44.

Tort, Patrick, ed. 1980. Bio-bibliographie. In *Vénus physique*. Paris: Aubier-Montaigne.

Tymieniecka, Anna Teresa. 1965. *Leibniz's Cosmological Synthesis*. Netherlands, The Hague: Van Gorcum & Co.

Vartanian, Aram. 1953. *Diderot and Descartes: A Study of Scientific Naturalism in the Enlightenment*. Princeton: Princeton University Press.

Vartanian, Aram. 1984. Díderot and Maupertuis. *Revue Internationale de Philosophie* 38: 46–66.

Vernière, Paul. 1954. *Spinoza et la pensée Française avant la Révolution*. Paris: PUF.

Vlastos, Gregory. 1950. The Physical Theory of Anaxagoras. *Philosophical Review* 59: 31–57.

Waschkies, Hans-Joachim. 1987. *Physik und Physikotheologie des jungen Kant*. Amsterdam: B. R. Grüner.

Watkins, Eric, ed. 2001. *Kant and the Sciences*. Oxford: Oxford University Press.

Wear, A. 1983. William Harvey and the "Way of the Anatomists." *History of Science* 21: 225–49.

Weissmann, August. 1893. *Germplasm*. Walter Scott's Contemporary Science Series no. 14.

Westfall, Richard S. 1971. *Force in Newton's Physics*. New York: American Elsevier.

Wheeler, W. M. 1899. *Caspar Friedrich Wolff and the "Theoria Generationis."* Wood's Hole Biological Lectures for 1898. Boston.

Whitteridge, G. 1971. *William Harvey and the Circulation of the Blood*. London: MacDonald Publishers.

Whitteridge, G. 1981. *William Harvey: Disputations Touching the Generation of Animals*. London: Blackwell Scientific Publications.

Wilkie, J. S. 1967. Preformation and Epigenesis: New Historical Treatment. *History of Science* 6: 138–50.

Williams, Elizabeth. 1994. *The Physical and the Moral: Anthropology, Physiology, and Philosophical Medicine in France, 1750–1850*. Cambridge: Cambridge University Press.

Williams, Raymond. 1983. *Keywords: A Vocabulary of Culture and Society*. London: Flamingo.

Wilson, Catherine. 1993. Interaction with the Reader in Kant's Transcendental Theory of Method. *History of Philosophy Quarterly* 10: 87–100.

Wilson, Catherine. 1995. *The Invisible World*. Princeton: Princeton University Press.

Wilson, Philip K. 1992. Out of Sight, Out of Mind? The Daniel Turner-James Blondel Dispute over the Power of the Maternal Imagination. *Annals of Science* 49: 63–85.

Wright, L. 1973. Functions. *Philosophical Review* 82: 139–68.

Zammito, John H. 1992. *The Genesis of Kant's* Critique of Judgment. Chicago: University of Chicago.

Zammito, John. 1992. Kant versus Eighteenth Century Hylozoism. In *The Genesis of Kant's "Critique of Judgment."* Chicago: University of Chicago Press, 189–99.

Zammito, John. 1998. "Method" vs "Manner"? Kant's Critique of Herder's "Ideen" in the Light of the Epoch of Science, 1790–1820. In *Herder Jahrbuch / Herder Yearbook 1998*, ed. Hans Adler and Wölf Koepke. Stuttgart: Metzler, 1–26.

Zammito, John H. 2002. *Kant, Herder and the Birth of Anthropology.* Chicago: University of Chicago Press.

Zammito, John. 2003. This inscrutable "Principle" of an Original "Organization": Epigenesis and "Looseness of Fit" in Kant's Philosophy of Science. *Studies in the History and Philosophy of Science* 34: 73–109.

Zöller, Günter. 1988. Kant on the Generation of Metaphysical Knowledge. In *Kant: Analysen – Probleme – Kritik*, ed. H. Oberer and G. Seel. Würzburg: Königshausen & Neumann, 71–90.

Zumbach, Clark. 1984. *The Transcendent Science: Kant's Conception of Biological Methodology.* The Hague: Martinus Nijhoff.

Index

Agliata, Ferdinando Francesco Gravina, 393
Alberti, Michael, 277
  On the activation of *ratio*, 279
  Defense of Stahlism, 277, 278
Albert the Great (Albertus Magnus), 47
  On spontaneous generation, 183
Albinism, 51
Albrecht, Michael, 125
Aldrovandus, Ulysses, 54, 72
Ambrosini, 54
Animal, concept of, 216–217
Animal husbandry, 105
Aquinas, *see* Thomas Aquinas
Arantius, 71, 78
Aristotelianism
  At Cambridge, 29
  In Padua, 8, 29, 30, 45–46
  In Renaissance teratology, 58
Aristotle, 47, 130, 253
  actual, 22–23
  Against pangenesis, 3, 104
  Analogy of art and nature in, 42, 44
  De Anima, 41
  Automata in, 230–231
  On birds, 25
  Conditional necessity in, 23
  On eternity in kind, 83
  On generated and perishable beings, 21
  De Generatione animalium, 22, 41, 46, 49
  Generative emergence in, 28
  On hematogenesis, 5
  Historia animalium, 22, 24, 46
  Historia vs. causal investigation in, 24, 25–26
  On hybrids, 53
  On menstrual blood, 82

Metaphysics, 21
  Methodology of, 25
  On male and female contributions in gen., 26, 40, 41
  On mixture, 128, 129, 130
  'Nature does nothing in vain', 24, 41
  On nutrition, 27
  On observation of eggs, 72
  On order of development of organs, 202
  De Partibus animalium, 22, 46
  Parva Naturalia, 40
  Politics, 83
  On *pneuma*, 27
  Pre-existent cognition in, 127
  On the priority of actuality, 22–23
  Place of zoology in, 21–22
  On spontaneous generation, 182–183
  On stages of soul, 28
  Teleology in, 45
  On uniform and non-uniform parts, 28
  On women, 82
  Zoological program of, 24
Aromatari, Guiseppe, 50
Arnauld, Antoine
Arthur, Richard T. W.
Atomism, 129
Aucante, Vincent
Augustine of Hippo, 47, 49, 54, 224
  On spontaneous generation, 183
Averroes, *see* Ibn Rushd
Averroism
  Active intellect in, 141

Bacon, Francis, 65, 150
  On final causes, 403
  On spontaneous generation, 183

Baer, Karl Ernst von
Basson, Sébastien, 147, 150
 Apparent Democriteanism of, 167
 Atomism of, 166
 Platonism of, 166
 Theory of compounds of, 166, 168
Bauhin, Caspar, 55
Bayesianism, 210
Bayle, Pierre, 215
 On animal souls, 217, 220, 223
 On limits of mechanical explanation, 218,
  226
 On reflexivity of sensation, 222, 223
 Rejection of pre-established harmony, 43,
  218
 On Rorarius, 218
 'Third way' of, 216, 218
Beeckman, Isaac, 150
Beeson, David, 333, 336
Bernard, Cardinal of Cles, 215
Bernoulli, Jean, 322
Blumenbach, Johann Friedrich, 314, 317, 395
 Animism in, 358
 Bildungstrieb of, 249, 348, 356, 357
  As causal force, 361
  Contrasted with Needham's, 357
  Vis plastica
  Contrasted with Wolff's, 358
  vis essentialis
  As emergent property, 361
  Excitation by fertilization, 359
  As primitive feature of matter, 359
  Question of independence from matter,
   360–361
 Contributions of, 356
 Critique of ovist preformationism, 359
 Critique of spermist preformationism, 359
 Epigenesis in, 348, 357
 On hylozoism, 348
 Inconclusiveness of, 371
 Influence on Herder of, 346
 Influence on Kant of, 317, 349
 Newtonianism of, 361–362
 On regeneration, 357, 358
 On teratomas, 393
 Vital materialism of, 362, 371–372
Boerhaave, Herman, 237, 376
Boiastuau, Pierre, 47, 54
Bonnet, Charles, 80, 194, 257, 285
 Chain of being in, 287, 297, 298
 On combinations of elements, 294–295

Empiricism of, 288, 295
On envelopment, 309
On 'evolution', 303
On the finite complexity of organic bodies,
 310–311
Germs of restitution in, 296
Hypothesis of dissemination in, 292, 302
Hypothesis of *emboitement* in, 292, 294, 302
On immortality of organisms, 306, 308–309
Influence of Haller on, 294, 300, 303
Leibnizianism of, 286, 287, 288, 296, 300,
 302, 305
Machines of nature in, 295, 300–301
Newtonianism of, 295
On organic preformation, 304–305
On plants, 298–299
Principle of continuity in, 288, 289, 290,
 292, 293, 298, 311
Principle of sufficient reason in, 292, 295
On repairer germs, 303
On reproduction, 301
On the science of vital phenomena, 300
On the unity of species, 297
On zoophytes, 299, 302
Bordeu, Theophile, 376
Borel, Pierre, 163, 165
Borelli, Giovanni, 14
Bourguet, Louis, 289, 293, 301, 302
Boyle, Deborah, 16
Boyle, Robert, 148, 163, 175
 On crystals, 201
Buchdal, Gerd, 351
Buffon, Georges-Louis Leclerc, Comte de,
 240, 253, 285, 299, 320, 376, 378
 Cosmogony of, 380
 On the emergence of species, 379, 390, 391,
  396
 Epicureanism in, 383, 386
 Internal forms of, 343
 Relations with Maupertuis of, 323, 333
 System of nature of, 385
Burgmann, Peter Christoph, 276
 Critique of Stahl, 276, 277–278
Burke, Edmund, 385

Camper, Peter, 54, 376, 390, 391, 399
Cannibalism, 157, 163
Charleton, Walter, 124
 On animal consciousness, 137–138
 Appeal to ordinary language in, 129
 On blood, 135–136

Christianity of, 138
On Commistion, 128–129
Common notions in, 126–127, 128
Contrasted with Gassendi, 126
Eclecticism of, 125
Epicureanism of, 138
On immortality of the soul, 138
Innate ideas in, 126, 139
Materialism of, 136
Methodology of, 124, 126
On milk, 134
On nutrition, 131–132, 133–134
Similarity to Gassendi, 127, 129, 130, 139
On sympathy between breasts and uterus, 135
On venom, 136–137
On vital heat, 132–133, 136
Voluntarism of, 124, 142
Chrysippus, 129
Churchill, Frederick B., 240
Cicero, Marcus Tullius, 56, 127
Clerselier, Claude, 199
Coimbran Aristotle commentaries, 86, 223
Coiter, Volcher, 72
Commerson, Philibert, 391
Condillac, Étienne de, 376, 379
Congenital traits, 81
Conjoined twins, 48
Conway, Anne, 175
    Critique of dualism, 184–185
    Critique of Hobbes, 180
    Critique of More, 179, 181, 186–187
    Galenism of, 189–190
    And Kabbalah, 176, 191
    On male and female principles, 189–190
    Matter-spirit monism of, 16, 175, 179–182
    On modes, 178
    On mutability of creatures, 178–179, 191–192
    On personal identity, 193
    And Quakerism, 176
    On ruling spirits, 178, 193
    On similiarity of sexual and spontaneous generation, 188–189
    On spontaneous generation, 182–183
    On transformation of matter into spirit, 177–178
    'Trialism' of, 177, 180
    On vital motion, 192
Conway, Edward, 176
Crusius, Christian August, 320

Cudworth, Ralph, 200
    Plastic natures of, 228
Cuvier, Georges, 378, 403
    Against evolution, 414, 415
    On the 'conditions of existence', 411
    On the 'correlation of parts', 412
    On the 'subordination of characters', 412–413
    Empiricism of, 413–414
    Influence of, 403

D'Alembert, Jean, 322, 328
Daniel, Gabriel, 226–227
Darmanson, Jean, 227
Darwin, Charles, 3, 18, 377, 392, 403, 415
Darwin, Erasmus, 406, 408
de Ceglia, Francesco Paolo, 14
Deism, 225
Democritus, 150, 168
Depew, David, 240
Des Bosses, Bartholomaeus, 155
Descartes, René, 65, 118, 138, 139, 175, 242, 387
    Against final causes, 61
    Against occult forces, 78
    Against the vacuum, 66
    Automata in, 231
    On blood, 76
    Christianized Epicureanism in, 377
    Cosmogony of, 197, 380
    On death, 230
    Denial of animal souls by, 184–185, 216, 217, 224–225
    On the difficulty of mechanist physiology, 66, 198
    Dualism of, 175, 334
    Early Hippocratism of, 70
    Epigenesis in, 105–106, 199
    On the female contribution to generation, 68
    On fermentation, 61, 69, 77
    On formation of images on the retina, 66
    On the geometrical character of embryology, 69–70, 106
    On the importance of dissection, 253–254
    On mechanist physiology, 88
    On monsters, 60
    On nature, 59
    On the nature of heat, 61, 69
    On the nature of seed, 67–68
    Optics of, 65

Descartes, René (*cont.*)
  Pangenesis in, 106, 107
  On the physiology of imagination, 92–93
  On physiology as part of physics, 59–60, 197–198
  Preformationism in, 106
  Rejection of Aristotelian immaterial principles, 67
  Rivalry with Gassendi, 200–201
  On the role of the pineal gland, 66
  On sensation, 220, 222
  On sexual dimorphism, 89
  On spontaneous generation, 183, 198
  On sympathy, 91
Des Chene, Dennis, 88, 94
Description vs. explanation in generation theory, 242, 248–249, 253
De Volder, Burchard, 12
Diderot, Denis, 324, 333, 336, 376, 386
  System of nature of, 385
Digby, Kenelm, 150
Diogenes Laertius, 5
Driesch, Hans, 248
Duchesneau, François, 111, 241, 259
Dutens, Louis, 287

Eggs
  Observation of development in, 72
  Comparison of observations of, 72–75
  Descartes, Coiter, and Fabricius on
Elisabeth, Princess of Bohemia, 77
Êmboitement, 194, 235
Empiricism, 127
Endeavor (*conatus*), 155, 156
Epicureanism, 150
  Prolepsis in, 127
  Epigenesis, 42, 194, 235, 253
Epicurus, 338, 384
epigenesis, 42
Esteve, Pierre, 380
Euler, Leonhard, 322
  Rivalry with Wolff of, 323

Faber, Johann, 10
Fabricius d'Acquapendente, Hieronymus, 57, 70, 72, 78
  Against authority of the ancients, 58
  Aristotle Project, 29–30
  De Formatione ovi e pulli, 31, 35
  De Formato foetu, 31
  Relation to Harvey, 31

Fernel, Jean, 70, 78
  Similarity to Descartes of, 70
Ferrari, Jean, 321
Feyens, Thomas, 120, 121, 275
Fisher, Saul, 158
Fontenelle, Bernard de, 60
Formey, Samuel, 322
Forster, Georg, 346, 391
Foucault, Michel, 403
Freytag, Johann, 154
Friedman, Michael, 319, 353
Froidmont, Libert (Fromondus), 76
Functions (of organs), 203–204

Galen, 49, 51, 78, 83, 130, 268
  Dissections performed by, 52
  On incomprehensibility of generation, 59
  On male and female contributions in generation, 189
  Minima naturalia in, 150
  On mixture, 128
  Teleology in, 53
  On the three stages of generation, 71
Galilei, Galileo, 59, 65
Gassendi, Pierre, 95, 150, 253
  Against Cartesian *res extensa*, 160
  On *animulae*, 104, 108, 111, 112, 114–119, 159
  Christian faith of, 121
  On common notions, 127–128
  On dominant traits, 115–116
  Epicureanism of, 123
  On the flower of matter, 104, 122, 161
  On generation of plants, 112
  On influence of both parents' imagination, 95, 116–118
  Influence on Leibniz, 148, 157
  Materialism of, 103, 108, 120, 123, 159
  Matter theory of, 109, 123
  Mechanism of, 119
  Micromerism of, 111
  On molecular structure of semen, 108, 112
  Preformationism of, 148
  Pre-organized generation in, 109, 110, 113
  On seminal force, 111, 119
  On spontaneous generation, 109, 110, 112–113, 183, 188
  On traduction, 158
  On ubiquity of seeds, 159–160
  Vitalism of, 119
  Weak preformationism of, 119–120

Index

'Generation', etymology of, 50
Geoffroy St. Hilaire, Étienne, 331
Gesner, Conrad, 47, 56
Glanvill, Joseph, 200
Goethe, Johann Wolfgang von, 409
Goorle, David van, 150
Gould, Stephen Jay
Grene, Marjorie, 240
Guyer, Paul, 318

Hall, Thomas, 245
Haller, Albrecht von, 194, 237, 287, 300, 376
    Baconian method of, 256
    Defense of pre-existence in, 238, 285, 343
    Disavowal of vitalism in, 247
    Empiricism of, 259, 260
    Importance of experiment in, 257
    Influence on Kant of, 345
    On irritability, 245, 246, 251, 258
    Newtonianism of, 285
    Poetry of, 380
    Reception of Descartes in, 253, 254
    On role of God in generation, 259, 260, 261
    On sensibility, 245, 251, 258
    Vivisections performed by, 245–246
Harriot, Thomas, 150
Hartmann, Georg Volcmar, 282
    Teratology of, 282
Hartsoeker, Nicolaas, 229
Harvey, William, 8–10, 72, 134, 242, 269
    Analogy of art and nature in, 42
    anatomical similarities, 43
    On blood, 40, 200
    On consistency of nature, 34
    Criticism of Fabricius, 33, 35
    De Conceptione, 43
    De Motu cordis, 31, 40, 59, 61
    On dissection, 34
    On epigenesis, 42, 236, 258, 330
    Epistemology of, 32, 35
    Exercitationes de generatione animalium, 31, 32, 35, 58
    Hylomorphism in, 40
    On innate heat, 40
    On internal formative agents, 38
    On male and female contributions in gen., 38–39, 40, 41, 105
    Methodology of, 32
    On natural body, 39

    Nominalism of, 33
    On nutrition of the fetus, 135, 140
    On oviparous and viviparous generation, 42
    On primigenial moisture, 40
    On role of eggs in generation, 36, 38, 39, 42
    On scientia, 32
    On similarity between brain and uterus, 43–44
    On spontaneous generation, 183
    Teleology in, 45
Hegel, Georg Wilhelm Friedrich, 409
Heliodorus, 47, 51
Helmont, Francis Mercury van, 176
    On airborne spirits, 193
Helmont, Jean-Baptiste van, 148, 268
Herder, Johann Gottfried, 346, 383, 391, 392
    On human origins, 391
    Influence on Kant of, 347
Heredity, 80–81
Hermann, Jakob (?), 286
Hertwig, Oscar, 241
Highmore, Nathaniel, 58
    On spontaneous generation, 184
Hill, Nicholas, 150
Hippocrates, 47, 51
    On nutrition of the fetus
Hippocratism, 3
History of nature in the eighteenth century, 327, 377
Hobbes, Thomas, 175
    Materialism of, 175, 186
Hoffheimer, Michael, 330
Hoffmann, Friedrich, 263, 272
d'Holbach, Paul Henry Thiry, 383, 385, 386
Hooke, Robert, 175, 201, 209, 210
    On spontaneous generation, 184
Hume, David, 378, 379, 383, 387
    Critique of natural theology of, 384
Hunemann, Philippe, 344
Hutton, Sarah, 179

Ibn al-Jazzar, 83
Ibn Rushd (Averroes)
    Doctrine of refraction of forms in, 168
Ibn Sina (Avicenna), 8
    Imagination, power of, 87

Johann Friedrich, Duke of Brunswick, 157, 162
Jung, Joachim (Jungius), 148

Kant, Immanuel, 13, 314
  Antinomy of teleological judgment in, 368,
    369, 370, 371, 404–405
  On apes, 394
  On the *Bildungstrieb*, 351
  On bipedalism, 390
  On caloric, 399
  On contingency of living forms, 341
  Continuities in, 318
  Developmental shifts in, 318
  Distinction between organic and inorganic
    in, 340–341
  On educt and product, 347–348, 350
  Epicureanism in, 397
  On epigenesis, 347, 353
  On the fixity of species, 348–349, 392
  On the formative power of organisms,
    367–368
  On *generatio aequivoca*, 349, 406
  On *generatio heteronyma*, 349, 350, 394,
    407
  On *generatio homonyma*, 349, 350, 407
  On the grotesqueness of living beings,
    392–393
  On hylozoism, 345, 349, 350, 365, 395
  On the judgment of natural beauty, 397
  On the law of inertia, 365–366
  On life, 366
  On materialism, 383, 384, 386, 388–389,
    398
  On matter, 388
  On necessity and contingency, 338–339
  Notion of mechanism in, 369–370
  On noumena, 388
  On occasionalism, 350
  On order in nature, 340
  On the origin of species, 375–376, 391, 392,
    395, 399, 402–403, 406
  On the phases of the universe, 380–381
  Physicotheology of, 337, 344, 381–382
  On preformation, 341, 347, 350, 371
  On prestabilism, 350
  On procreation, 384–385
  On purposiveness in nature, 364, 395
  Range of interests of, 375
  Reading of Maupertuis of, 343–344, 346
  Reception of Blumenbach of, 362–363, 371,
    372
  Reception of Newtonianismin, 339, 354,
    380, 383, 387
  Reform of metaphysics of, 328
  Rejection of pantheism by, 338
  On speculative cosmology, 382
  On the sublimity of destruction, 381–382
  On supernatural generation, 343
  Supersensible ground in, 370, 372, 396
  On the systematic unity of nature, 364–365
  On teleology, 355, 366, 367–368, 371, 404,
    406, 407
  On vital forces, 400–401
Kircher, Athanasius, 10
Knutzen, Martin, 320
König, Samuel, 287, 289, 313, 322, 323
Kuehn, Manfred, 318, 320

La Condamine, Charles-Marie, 336
Lambert, Johann Heinrich, 384, 386
La Mettrie, Julien Offray de, 323, 333, 336
  Cartesianism of, 379
  Epicureanism in, 386
  System of nature of, 385
Laurens, André du (Laurentius), 49, 54, 84
  On incomprehensibility of generation, 59
  On laws of nature, 60
Lavater, Johann Kaspar, 287
Laywine, Alison, 319
Leeuwenhoek, Antoni van, 164, 171, 201, 207,
    229
  Discovery of microorganisms by, 11, 15,
    209
Le Grand, Antoine, 97
  On lactation, 97–98
Leibniz, Gottfried Wilhelm, 147
  Against Cartesian *res extensa*, 160
  Autonomy of substances in, 231
  Availability of texts in 18th century, 286
  On *bullae*, 155
  Conception of physiology of, 312
  Concept of species in, 291
  Controversy with Clarke of, 326, 328,
    387
  Corporeal substance in, 149, 170
  Defense of animal souls, 216
  Doctrine of domination in, 16–17, 171
  Doctrine of transformation in, 148, 164
  Early Hobbesian account of atoms in, 155
  On the flower of substance, 161, 162, 170
  On force, 228
  'Hypermechanism' of, 228
  On immortality of animal souls, 170–171,
    228, 230, 305–306
  Influence on life sciences of, 286, 314

Index

Leibniz, Gottfried Wilhelm (*cont.*)
'Machines of nature' and 'machines of art'
in, 291, 300
Opposition to materialism, 160
Mechanism of, 293
On the perpetual embodiment of souls, 172,
173–174, 308
Preformationism of, 225, 307
On preservation of substantial forms, 149
On the *principium multitudinis*, 173
On the *principium passionis*, 173
Principle of continuity in, 289, 290, 293
Reconciliation of ancient and modern
philosophy in, 147
Reconciliation of Catholicism and
Protestantism in, 157
Reintroduction of forms by, 165
Rejection of atoms of, 165
On spontaneous generation, 170
Theory of cohesion of, 163
Theory of monads of, 149–150, 387
Theory of pre-existence in, 293
On traduction, 156, 157
On transformation, 229
On the unity of substances, 228
Lennox, James G.
Lenoir, Timothy, 360, 361
Lewis, Eric, 125
Liceti, Fortunio, 56–57
Atomism of, 151
Influence on Sennert, 148, 159
Linnaeus, Carl, 389, 390
Locke, John, 379, 387
Nominal essences in, 298
Nominalism of, 87
Longolius, Johann Daniel, 280
Critic of Burgmann, 280–281
Defense of Stahl, 281
'Germanic' style of, 280
Views on preformation of, 281
Look, Brandon, 14
Lovejoy, Arthur O., 390–391, 394
Lucretius, Titus Carus, 53, 338, 377, 386, 398
On the extinction of species, 377–378
On hybrids, 53
On the *primordia rerum*, 377
Lycosthenes, Conrad, 47, 56

Maienschein, Jane, 241
Maignan, Emmanuel, 222
Maillet, Bernard de, 376, 382, 394

On the age of the earth, 378–379
On evolution, 393
On orangutans, 390
Malebranche, Nicolas, 96, 118, 166, 229
Advocacy of pre-existence, 196, 205–206,
207, 209
Against non-mechanical agencies, 207
Defense of 'genetic' approach, 202
Disagreement with Descartes, 202–203
Doctrine of *êmboitement* in, 207
Expositor of Descartes, 202
On function and structure, 205
Interest in findings of microscopists, 197,
209, 210–214
Occasionalism of, 342
On power of maternal imagination, 208–209
Pre-existence in, 108, 236
On propagation of species, 96
On softness of the fetus, 97
On supernatural creation of organisms, 204
Malpighi, Marcello, 197, 201, 229, 269, 270
Dissections performed by,
Observations on eggs, 209, 271
Marques d'envie, 76, 90
Maupertuis, Pierre-Louis Moreau de, 285,
314, 318
On 'aptness' in organisms, 335
Cosmogony of, 380
Critique of Newtonian natural theology of,
329
Epicureanism in, 383, 386
On epigenesis, 331
On geological catastrophes, 335
On heredity, 332
On hybridism, 332
Hylozoism of, 331, 333, 336
Influence on Kant of, 321, 328–329, 337,
345
On instinct, 334
Leibnizianism of, 323
Physicotheology of, 325–326, 330
On polydactyly, 324
On preformationism, 330
Principle of least action of, 322
Purported Spinozism of, 336
Reception of Newton by, 320, 322
Rivalry with Wolff of, 323
System of nature of, 385
On universal dynamism of nature, 334–336
Mendelssohn, Moses, 287, 313
Mercer, Christia, 164

453

Mérian, Jean-Baptiste, 325
Mersenne, Marin, 90, 198, 222, 254
Meyssonier, Lazare, 90
Michael, Emily, 168, 169
Monro, Alexander, 376, 379, 391
'Monster', etymology of, 49, 56
Montaigne, Michel de, 48
Monti, Maria Teresa, 260
More, Henry, 175
   On airborne spirits, 193
   Archaeus of, 228
   Contact with Anne Conway of, 176
   On Conway's book, 176
   On ensouledness of semen, 225
   On perpetual embodiment of souls, 179
   On spontaneous generation, 188
Moscati, 390
Münster, Sebastian, 47

Natural Selection, 411
Naudé, Gabriel, 215
Needham, John Turberville, 285, 299, 314
   Epigenesis in, 299
Neoplatonism, emanationism in, 173
Neubauer, John, 260
Newton, Isaac
   'General Scholium' of, 327
   On gravity, 247
   Influence on eighteenth-century, 245
   Life science of
   Influence on Maupertuis, 326
   Law of inertia of, 353

Oldenburg, Henry, 157
Orangutan, 379
Osler, Margaret, 124
Ovid, 47

Paley, William, 415
Paracelsus (Theophrastus Bombastus von
   Hohenheim)
   Naturalism in
Pardies, Ignace, 221
   On animal reason, 221
   On sensation, 222
Paré, Ambroise, 47, 49, 50, 51, 56, 84
   On chambers in the uterus, 53
   On importance of experiments, 53
Parmenides, 1
Parthenogenesis, 80

Pascal, Blaise, 327
Peiresc, Nicolas-Claude Fabri de, 72
Pereira, Gómez, 225
Perrault, Claude, 292
Philo of Alexandria, 378
Philosophia vulgaris refutata (anonymous),
   229
Philosophical Transactions, 87
Plato, 1, 150
Plempius, Vopiscus Fortunatus, 76
Pliny the Elder, 47, 52
Poiret, Pierre, 229
Pre-existence, 194, 235, 253
   Distinguished from preformation, 195, 208
Preformationism, 107–114, 119
Pyle, Andrew J.
   On supernaturalism in pre-existence theory,
   243

Raspe, Rudolph Eric, 287
Réaumur, René Antoine Ferchault de, 194
Redi, Francesco, 50
   On spontaneous generation, 184
Régis, Pierre-Sylvain (Regius), 76, 229
   On malleability of the fetus, 93–94
   Support of pre-existence of, 201
   On universality of maternal imagination's
   influence, 94
Reimarus, H. S., 386
Roberts, Richard, 317, 360, 361
Riolan the Younger, 55
Roe, Shirley, 239–240, 247, 251, 252, 254,
   259
Roger, Jacques, 152, 154, 198, 243
Rohault, Jacques, 201
Rorario, Girolamo (Rorarius), 215
   On animal reason, 219
Rousseau, Jean-Jacques, 386
Rudolph, Gerhard, 256
Rüff, Jakob, 47, 49, 50, 51, 53, 55
Ruse, Michael, 17

Scaliger, Julius Caesar, 148, 154, 266, 275
   Animism of, 267
   Anticipation of Sennert, 167
   Possible influence on Stahl, 267
   Theory of mixtion of, 149, 155
Schedel, Hartmann, 47
Schegk, Jacob, 169
Schelling, Friedrich Wilhelm Joseph, 400, 409

Schenck, Johann Georg, 56
Scholasticism
  Theory of *inanimatio* in, 140
  Theory of tripartite soul in, 217, 219
Schönfeld, Martin, 319, 320, 337
Schoock, Martin, 75
Sennert, Daniel, 141, 225, 275
  Anticipation of Leibniz, 148, 161, 171
  On chemical operations, 151
  On compound forms, 169
  Conciliatory stance of, 147
  On ensouledness of semen, 225
  On human generation, 152–153
  Importance of, 150
  On innate heat, 161, 168
  Minima naturalia in, 150
  On origin of human soul, 152
  On plant generation, 152
  Rejection of mechanistic account of forms
    of, 167, 168
  On spontaneous generation, 184
  On traduction, 153
Shell, Susan, 344
Sigorgne, Pierre, 287, 313
Simplicius, 4
Smith, C. U. M., 317
Smith, Justin E. H., 150
Spallanzani, Lazzaro, 194, 299, 302, 303, 313
Spener, Philipp Jakob, 282
Sperling, Johann, 275
Spinoza, Baruch, 218, 386, 398
  Conatus in, 228
'Spinozism', 328, 338
Stahl, Georg Ernst, 14, 262
  On *adsuetudines*, 274
  Analogy of fetal development and nutrition
    in, 268
  Aristotelianism of, 268
  Concept of soul in, 283–284
  Epigeneticism of, 270
  'Hydraulicism' of, 270, 272
  On inconclusiveness of observation, 269
  On influence of maternal desires in
    pregnancy, 272–273
  Misreading of Malpighi, 271
  On monstrosities, 273
  Ovist preformationism in, 271
  'Ovi-spermatism' of, 271
  On passivity of organic matter, 264
  Purported Pietism of, 282

On purpose of medicine, 262
  On *ratio*, 275, 276, 279
  On *ratiocinatio*, 275, 279
  Significance of religious beliefs for, 266
  On soul as architect of body, 265
  On soul as the source of life, 265
Stoicism, 54
  Common notions in, 127, 133
  On mixture, 128, 129
  Pneuma in, 141
Suárez, Francisco
  On animal instinct, 219
  On reflection and sensation, 222
Sulzer, Johann Georg, 325
Superfetation, 50
Swammerdam, Jan, 108, 164, 197, 201, 210,
    229
  Dissections of insects by, 208, 209

Telliamed, *see* Maillet, Bernard de
Terrall, Mary, 323, 332, 336
Thomas Aquinas, 65
  On the species of angels, 231
  On spontaneous generation, 188
  On the unity of animal souls, 229
Toellner, Richard, 260
Tonelli, Giorgio, 318, 319–320
Traducianism, 148, 149
Trembley, Abraham, 80, 289, 299, 302, 313
Tyson, Edward, 390, 391

Velthuysen, Lambert van, 156
Vesalius, Andreas, 52, 53
  Teleology in, 53
Voltaire, 60, 321, 322

Waschkies, Hans-Joachim, 319, 320
Watkins, Eric, 319
Weinrich, Martin, 54
Weissmann, August, 241
Wilkie, J. S., 242
Willis, Thomas, 137, 268
Wilson, Catherine, 162, 408, 409
Wolff, Caspar Friedrich, 237, 314
  Defense of epigenesis in, 238, 285, 345
  Disagreement with Blumenbach, 249–250
  On importance of observation, 255
  Influence on Herder of, 346
  Newtonianism of, 249
  Reception of Descartes in, 254–255, 256

Wolff, Caspar Friedrich (*cont.*)
  Principle of sufficient reason in, 249
  Qualified vegetable matter of, 250–251
  Rationalism of, 259
  Rejection of mechanical medicine, 247
  On role of God in generation, 259, 261
  On secretion and solidification, 243, 244
  Theory of pre-existence in, 252
  On *vis essentialis*, 243, 244, 247, 257–258
Wolff, Christian, 243
  On historical knowledge, 243
  On philosophical knowledge, 243
Wolffianism, 383, 389

Womb-soil analogy
  In Aristotle, 83
  In Medieval Islamic philosophy, 84
Wright, Larry, 410
Wright, Thomas, 320, 380
Zabarella, Jacopo, 167, 169
  Influence on Sennert, 167, 170
  Doctrine of refraction of forms in, 168
  On relationship of natural phil. and
    medicine, 30
  On self-multiplication of form, 153
Zammito, John, 355
Zumbach, Clark, 352

Printed in the United States
By Bookmasters